Glutathione in the Nervous System

Glutathione in the Nervous System

Edited by

Christopher A. Shaw

*Departments of Ophthalmology
and Physiology and Neuroscience Program
University of British Columbia, Vancouver
British Columbia, Canada*

CRC Press
Taylor & Francis Group
Boca Raton London New York

CRC Press is an imprint of the
Taylor & Francis Group, an **informa** business

GLUTATHIONE IN THE NERVOUS SYSTEM

First published 1998 by Taylor & Francis

Published 2019 by CRC Press
Taylor & Francis Group
6000 Broken Sound Parkway NW, Suite 300
Boca Raton, FL 33487-2742

© 1998 by Taylor & Francis Group, LLC
CRC Press is an imprint of Taylor & Francis Group, an Informa business

First issued in paperback 2019

No claim to original U.S. Government works

ISBN 13: 978-0-367-44788-5 (pbk)
ISBN 13: 978-1-56032-643-4 (hbk)

Visit the Taylor & Francis Web site at
http://www.taylorandfrancis.com

and the CRC Press Web site at
http://www.crcpress.com

This book was set in Times Roman. Composition and editorial services by TechBooks. Cover design by Norm Myers.

A CIP catalog record for this book is available from the British Library.

Library of Congress Cataloging-in-Publication Data

Glutathione in the nervous system / edited by Christopher A. Shaw
    p. cm.
      1. Glutathione – Physiological effect. 2. Glutathione – Pathophysiology.
    3. Neurochemistry. I. Shaw, Christopher A.
    (Christopher Ariel)
      [DNLM: 1. Glutathione – metabolism. 2. Nervous System – metabolism.
    QU 68 G5672 1998]
    QP552.G58G57   1998
    612.8'042 – dc21
    DNLM/DLC
    for Library of Congress   97-22575
                                               CIP

For JB and AS

Contents

Contributing Authors

Jaswinder S. Bains, Ph.D. *Departments of Ophthalmology and Physiology and Neuroscience Program, The University of British Columbia, Vancouver, British Columbia, Canada V6T 1Z3*

Gianni Benzi, *Institute of Pharmacology, Faculty of Science, University of Pavia, Italy*

Robert H. Brown, *Day Neuromuscular Research Center and Neurology Service, Massachusetts General Hospital, Charleston, Massachusetts 02109*

Arthur J. L. Cooper, *Departments of Biochemistry and of Neurology and Neuroscience, Cornell University Medical College, New York, New York and Burke Medical Research Institute, Cornell University Medical College, White Plains, New York*

Merit E. Cudkowicz, *Day Neuromuscular Research Center and Neurology Service, Massachusetts General Hospital, Charleston, Massachusetts 02109*

Michel Cuénod, *Brain Research Institute, University of Zürich, Zürich, Switzerland*

Kenneth Curry, *Precision Biochemicals Inc., Gerald McGavin Building, 206-2386 East Mall, Vancouver, British Columbia, Canada V6T 1Z3*

Kim Quang Do, *Brain Research Institute, University of Zürich, Zürich, Switzerland*

Glenn Dryhurst, *Department of Chemistry and Biochemistry, University of Oklahoma, Norman, Oklahoma 73019*

M. Gerlach, *Division of Clinical Neurochemistry, Department of Psychiatry, University of Würzburg, 97080 Würzburg, Germany and Division of Clinical Neurochemistry, Department of Neurology, St. Josef's Hospital, University of Bochum, Bochum, Germany*

Ole P. Hjelle, *Department of Anatomy, Institute of Basic Medical Sciences, University of Oslo, P.O. Box 1005, Blindern, N-0317 Oslo, Norway*

D. Huster *Department of Neurophysiology, Paul-Flechsig Institute for Brain Research, University of Leipzig, Jahnallèe 59, D-04109 Leipzig, Germany*

Réka Janáky, *Tampere Brain Research Center, University of Tampere Medical School, Box 607, FIN-33101 Tampere, Finland*

Zsolt Jenei, *Tampere Brain Research Center, University of Tampere Medical School, Box 607, FIN-33101 Tampere, Finland and Department of Animal Physiology, Kossuth Lajos University of Science, Box 18, H-4010 Debrecen, Hungary*

Ram Kannan, Ph.D. *Division of GI & Liver Diseases, and Department of Neurosurgery, USC School of Medicine, Los Angeles, California 90033*

Neil Kaplowitz, *Division of GI & Liver Diseases, and Department of Neurosurgery, USC School of Medicine, Los Angeles, California 90033*

Howard M. Lenhoff, *Department of Developmental and Cell Biology, University of California, Irvine, California 92697*

Hiroko Maeda, *Department of Pharmacology, Setsunan University, Hirakata, Osaka 573-01, Japan*

Wolfgang Maret, Ph.D. *Center for Biochemical and Biophysical Sciences and Medicine, Harvard Medical School, Boston, Massachusetts 02115*

Takao Minami, *Department of Pharmacology, Setsunan University, Hirakata, Osaka 573-01, Japan*

Antonio Moretti, *Institute of Pharmacology, Faculty of Science, University of Pavia, Italy*

Kiyokazu Ogita, Ph.D. *Department of Pharmacology, Setsunan University, Hirakata, Osaka 573-01, Japan*

Simo S. Oja, *Tampere Brain Research Center, University of Tampere Medical School, Box 607, FIN-33101 Tampere, Finland and Department of Clinical Physiology, Tampere University Hospital, Box 2000, FIN-33521 Tampere, Finland*

Ole P. Ottersen, *Department of Anatomy, Institute of Basic Medical Sciences, University of Oslo, P.O. Box 1005, Blindern, N-0317 Oslo, Norway*

Bryce A. Pasqualotto, *Department of Physiology, University of British Columbia, Vancouver, British Columbia, Canada V6T 1Z3*

Rajiv R. Ratan, M.D., Ph.D. *Neurology Laboratories, Harvard Institutes of Medicine, Rm. 857, 77 Avenue Louis Pasteur, Boston, Massachusetts 02115*

W. Reichelt, *Department of Neurophysiology, Paul-Flechsig Institute for Brain Research, University of Leipzig, Jahnallèe 59, D-04109 Leipzig, Germany*

P. Riederer, *Division of Clinical Neurochemistry, Department of Psychiatry, University of Würzburg, 97080 Würzburg, Germany*

E. Rinvik, *Department of Anatomy, Institute of Basic Medical Sciences, University of Oslo, P.O. Box 1005, Blindern, N-0317 Oslo, Norway*

Pirjo Saransaari, *Tampere Brain Research Center, University of Tampere Medical School, Box 607, FIN-33101 Tampere, Finland*

Christopher A. Shaw, *Departments of Ophthalmology and Physiology and Neuroscience Program, University of British Columbia, Vancouver, British Columbia, Canada V6T 1Z3*

Makoto Shuto, *Department of Pharmacology, Setsunan University, Hirakata, Osaka 573-01, Japan*

J. Sian, *Division of Clinical Neurochemistry, Department of Psychiatry, University of Würzburg, 97080 Würzburg, Germany*

Richard A. Smith, *The Center for Neurologic Study, La Jolla, California 92037*

Vince Varga *Tampere Brain Research Center, University of Tampere Medical School, Box 607, FIN-33101 Tampere, Finland and Department of Animal Physiology, Kossuth Lajos University of Science, Box 18, H-4010 Debrecen, Hungary*

Jian-R. Yi, *Division of GI & Liver Diseases, and Department of Neurosurgery, USC School of Medicine, Los Angeles, California 90033*

Yukio Yoneda, *Department of Pharmacology, Setsunan University, Hirakata, Osaka 573-01, Japan*

Khalequz Zaman, Ph.D. *Neurology Laboratories, Harvard Institutes of Medicine, Rm. 857, 77 Avenue Louis Pasteur, Boston, Massachusetts 02115*

Berislav V. Zlokovic, *Division of GI & Liver Diseases, and Department of Neurosurgery, USC School of Medicine, Los Angeles, California 90033*

Preface

When the great innovation appears, it will almost certainly be in a muddled, incomplete form. To the discoverer himself it will be only half-understood; to everyone else it will be a mystery. For any speculation which does not at first glance look crazy, there is no hope.

—Freeman J. Dyson[*]

We all agree that your theory is crazy. The question which divides us is whether it is crazy enough to have a chance of being correct.

—Niels Bohr[†]

The history of the study of glutathione in the nervous system owes something to serendipity. In his classic study of the effects of GSH on feeding behavior in *Hydra*, W. F. Loomis (1955) noted that his work originated following a chance remark by a colleague.

My own interest in glutathione arose from an apparently anomalous experimental result. We had been studying the developmental profiles of various neurotransmitter receptor populations in the sensory cortex. Quite early on, we had observed that the adult receptor distribution was radically different from that in early postnatal development. Seemingly, stimulation by a trophic factor or neurotransmitter was the causal agent in this redistribution of receptor location. It then followed that raising the cortical slices in culture, thus isolating them from all such influences, should prevent the receptor redistribution from occurring. To our consternation, however, the receptor redistribution proceeded normally in the supposedly inadequate environment of the culture medium. The surprises were not to end there, as further work on the dozens of media ingredients revealed that the necessary stimulation was likely being provided by the presumably innocuous molecule glutathione, usually added to the media simply as an antioxidant. These results suggested that in addition to its well-known antioxidant role, GSH was acting as a neurotransmitter or neuromodulator in some neural pathways.

This early working hypothesis was distinctly unpopular with some peer review committees, which voiced various criticisms. Chief amongst these was the claim that GSH's ubiquity precluded a selective neurotransmitter role; a second was that its known action as antioxidant was evidence against other more neurally specific actions. In the absence of any hard data for rebuttal at the time, we had to console ourselves with quotations such as those at the beginning of this preface.

A part of both criticisms is correct: GSH is ubiquitous and is an important antioxidant everywhere in the body, including the nervous system. The antioxidant–free-

---

[*]Dyson, F. J. 1958. Innovation in physics, *Sci. Am.* 199:74–82, p. 80.
[†]Quoted in the preceding reference.

radical scavenger action of GSH is one of its essential roles, the failure of which may play a large part in the events leading to various neurological disorders. One role, however, does not necessarily preclude another. Much of what the reader will find in the following chapters touches upon these additional roles of GSH, including possible neurotransmitter action, redox modulation of ionotropic receptor function, and neuroprotection against the excitotoxic actions of glutamate.

I believe that the multiple roles of GSH, potential and demonstrated, make it one of the most fascinating molecules yet described in the literature of the nervous system. My sense is that ongoing and future studies of GSH in the nervous system will soon put us on the threshold of novel insights into normal and abnormal synaptic function. This statement makes me guilty of hyperbole, but one characteristic that marks those who study GSH in the nervous system is a fierce enthusiasm for this molecule.

The goal of *Glutathione in the Nervous System* is to capture some of this enthusiasm by focusing the reader's attention on three major areas: (i) the origin and localization of GSH in the nervous system, (ii) the multiple effects of GSH on neural health and activity, and (iii) the potential for alterations in GSH status to lead to neurological damage of the type observed in amyotrophic lateral sclerosis, Parkinson's disease, and other neurological disorders. The reader will note significant points of disagreement between authors in several of the sections. This is to be expected at an early stage of a still embryonic field where almost any direction leads to unknown territory.

This book could not have been completed without the efforts of many individuals. To all the contributors, I offer my gratitude for your outstanding chapters. Dr. Ole P. Hjelle, Dr. Ole P. Ottersen, and colleagues kindly allowed me to use part of one of their beautiful electron micrographs for the book cover. I am grateful also to Dr. Fred Hollinger and Dr. Howard Lenhoff, who offered timely and critical advice at various stages of this project, and to Dr. Richard O'Grady, Ms. Laura Haefner and colleagues at Taylor and Francis and Mr. Andrew Wilson and colleagues at TechBooks for their support and help. I thank my colleagues Ms. Jill McEachern, Dr. Bryce Pasqualotto, Dr. Jaswinder Bains, Dr. Ken Curry, and Dr. Marge Langmuir for their insightful comments on the present book and on glutathione-related topics. Dr. Ken Curry, Mr. Craig Martin, and Mr. Raman Mahabir helped to prepare the original cover design. Finally, I thank the members of the Department of Anatomy, UBC: I could not have asked for better hosts or colleagues.

C.A.S.
Deep Cove, British Columbia
May 1997

PART I

History, Evolution, and Localization of
Glutathione in the Nervous System

Glutathione in the Nervous System
Edited by Christopher A. Shaw
Copyright © 1998 Taylor & Francis

1

Multiple Roles of Glutathione in the Nervous System

Christopher A. Shaw

*Departments of Ophthalmology and Physiology and Neuroscience Program,
University of British Columbia, Vancouver, British Columbia, Canada, V6T 1Z3*

ABBREVIATIONS

ALS amyotrophic lateral sclerosis
AD Alzheimer's disease
Cys cysteine
Cys-Gly cysteinylglycine
Glu-Cys glutamylcysteine
GSH reduced glutathione
GSSG oxidized glutathione
GSH-Px glutathione peroxidase
H_2O_2 hydrogen peroxide
•OH hydroxyl radical

PD Parkinson's disease
ROS reactive oxygen species
SOD superoxide dismutase
$O_2^{\bullet-}$ superoxide radical.

1. INTRODUCTION

Glutathione (γ-L-glutamyl, L-cysteinylglycine), a tripeptide of the amino acids glutamate, cysteine, and glycine, has been well known to biochemists for generations. Glutathione was first described over one hundred years ago (De Rey-Pailhade 1888a,b) and called "philothion." Hopkins (1921) further described the reducing properties of the molecule, renaming it glutathione.

Both the reduced form (GSH) and its oxidized dimer (GSSG) have been implicated in a great variety of molecular reactions (for reviews see Kosower and Kosower 1978; Meister 1989, 1991; Deneke and Fanburg 1989; Lomaestro and Malone 1995; Anderson 1997). Many of these reactions are crucial to cell survival (see Table 1), so much so that glutathione has been termed "the most important non-protein thiol" (Kosower and Kosower 1978). One hypothesis has even suggested that glutathione is responsible for the origin of life (Holt 1993). While this latter view seems likely to reflect scientific hyperbole, it may be difficult to overestimate the central importance of this molecule in the biochemistry of living cells. Some examples of the actions of GSH are shown in Tables 1 and 2.

With the last point in mind, it is somewhat surprising that glutathione is so little known by neuroscientists and that its roles in the nervous system are still only vaguely understood. A book celebrating the centennial of the discovery of glutathione (Taneguchi et al. 1989) made only passing reference to actions in the nervous system. This is all the more astonishing in that decades ago leading cell biologists, biochemists, and neuroscientists described potential roles for glutathione as a neurotransmitter (Loomis 1955; Lenhoff 1961; see also Lenhoff, chapter 2, this volume) and

TABLE 1. *Metabolic functions*
of glutathione

DNA synthesis and repair
Protein synthesis
Prostaglandin synthesis
Amino acid transport
Metabolism of toxins and carcinogens
Enhancement of immune-system function
Prevention of oxidative cell damage
Enzyme activation
Source of cysteine
Cofactor of redox reactions
Transport of metals across membranes
Delivery of metals between ligands

TABLE 2. *Glutathione actions in the nervous system*

Free-radical scavenging action:

Controls oxidative stress due to free-radical formation (e.g., in presence of transition metals such as Cu, Fe; diminished GSH status linked to ALS (Al-Chalabi, Powell, and Leigh 1995), PD (Jenner and Olanow 1996), and AD (Freidlich and Butcher 1994)

Neurotransmitter breakdown products, e.g., dopamine (Wei et al. 1996) and serotonin metabolites (Hirata et al. 1995)

Calcium or electrical activity of neurons (Orrenius et al. 1992)

β-Amyloid in AD (Harris et al. 1995; Schubert et al. 1995)

Redox action:

Neuroprotection against EAA excitotoxicity [e.g., stroke (Ortolani et al. 1995), ischemia (Sirsjo et al. 1996), epilepsy (Niketic et al. 1992)]

Modulation of ionotropic receptor action (Ogita et al. 1995)

Neurotransmitter action:

Excitatory neurotransmitter (Shaw et al. 1996), neural signaling (Brown 1994), and control of tropic factors (Spina et al. 1992)

Neural activity for behavior/cognition (Grosvener, Rhoads, and Kass-Simon 1996; Perez-Pinzon and Rice 1995)

Other actions:

Control of free radicals produced by radiation or chemotherapy (Bump and Brown 1990).

---

in the processes underlying memory (Werman, Carlen, and Kosower 1971; Kosower and Werman 1971; Kosower 1972). Much of whatever interest attended these original data remained dormant until relatively recently. Many of the original ideas were quite prescient: Kosower's 1972 model of memory, based on the bistable nature of glutathione, was well considered and plausible; perhaps it is time to revisit this model from an experimental perspective.

In this chapter, I will attempt to describe some of the actions of glutathione in the nervous system. Some of the more conventional aspects of GSH action, e.g., as a free-radical scavenger, will be dealt with in a somewhat cursory manner, inasmuch as they are discussed at length in some of the following chapters. I have, however, attempted to provide adequate references for those wishing to find more information.

A final caveat: Much of the following discussion in this chapter and those that come after is highly speculative—in my view, appropriately so, as suits a new, wide open, and exciting field. Many of these ideas will not survive the rigors of experimental testing; those that do may change our thinking about fundamental processes in the nervous system.

2. BASIC GSH CHEMISTRY AND REACTIONS

Much has been written on all aspects of the synthesis and degradation of GSH and on its many reactions in biological systems. For a comprehensive perspective on

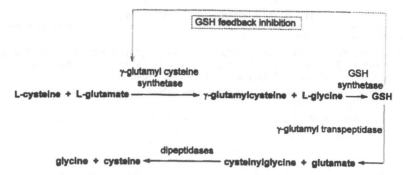

FIG. 1. Pathway for GSH synthesis and degradation. (For more details, see Meister 1989, 1991.)

this molecule the reader is referred to various reviews on GSH chemistry (Kosower and Kosower 1978; Meister 1989; 1991; Deneke and Fanburg 1989; Lomaestro and Malone 1995; Anderson 1997; see also several of the following chapters in this volume, notably those by Cooper, chapter 5, and Curry, chapter 10).

Glutathione is synthesized and degraded in many cell types by a series of enzymatic reactions illustrated in Fig. 1 (Meister 1989). As noted above, glutathione exists in both the reduced form (GSH) and the oxidized disulfide form (GSSG), the former by far the larger fraction. GSH and GSSG are interconvertible, as shown in Fig. 2. The ratio of GSH to GSSG is termed the *GSH status* and is a major determinant of cell health and viability (Kosower and Kosower 1978). As noted by these latter authors, the GSH status is not static, but rather reflects a dynamic system that responds to alterations in the cellular environment. Changes in GSH status have important biological consequences. For example, in most cells, including those of the nervous system, GSH constitutes approximately 98% of total glutathione (Slivka, Spina, and Cohen 1987); changes in this ratio can affect any or all of the cellular functions of GSH shown in Tables 1 and 2. In regard to this last point, GSH is crucial for the control of reactive oxygen species (ROS) in mitochondria, which arise as a by-product of oxidative metabolism. Mitochrondria import GSH from the cytoplasm (Meister 1989); alterations in mitochrondrial GSH are implicated in neuronal cell death in various neurological disorders (for references, see Benzi and Moretti, chapter 11, and Bains and Shaw, chapter 17, this volume).

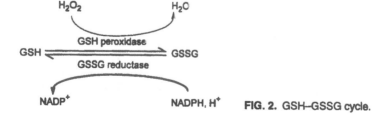

FIG. 2. GSH–GSSG cycle.

2.1 Antioxidant and Free-Radical Scavenger

Free radicals are atoms or molecules capable of independent existence that contain one or more unpaired electrons. Reactive oxygen species (ROS) include radicals and nonradicals centered on oxygen. Examples of the former are the superoxide radical ($O_2^{\bullet-}$) and the highly damaging hydroxyl radical ($\bullet OH$); the latter include oxygen derivatives such as singlet oxygen and hydrogen peroxide (H_2O_2). Free radicals and ROS are produced by biological processes such as electron transfer reactions during oxidative metabolism in mitochondria and by external events such as radiation. Biological systems have evolved a number of antioxidant defenses to protect themselves from the effects of free radicals. These include those defenses designed to prevent the generation of free radicals as well as those whose role is to scavenge free radicals that may be generated. In the first category are various special proteins as well as enzymes such as catalase and glutathione peroxidase. Free-radical scavengers include the enzyme superoxide dismutase (SOD), reduced glutathione, ascorbic acid (vitamin C), and α-tocopherol (a member of the vitamin E family). ROS and antioxidant defenses are normally in balance. Increases in the former or decreases in the latter change the balance to favor ROS action, and *oxidative stress* results (Halliwell 1992a,b, 1996; Cheeseman and Slater 1993; Evans 1993). Consequences of oxidative stress for neurons, in particular in various neuronal disorders, will be dealt with at length in various of the following chapters (e.g., Cooper, chapter 5; Zaman and Ratan, chapter 6; see also Part III).

GSH can participate as an antioxidant either indirectly or directly. In the first case, glutathione serves as a substrate for GSH peroxidase to reduce hydrogen peroxide to produce GSSG and water:

$$2GSH + H_2O_2 \xrightarrow{\text{GSH-Px}} GSSG + 2H_2O$$

GSH may also participate directly in three types of oxidation–reduction reactions; for details, see (Kosower and Kosower 1978). In the first, GSH donates a proton to a free radical in a reaction of the form:

$$GSH + R\bullet \rightarrow RH + GS\bullet$$

$$2GS\bullet \rightarrow GSSG$$

A typical case of this reaction involves the hydroxyl radical:

$$2GSH + 2\bullet OH \rightarrow GSSG + 2H_2O$$

This may be one of the most important reactions involving GSH in the nervous system, as the hydroxyl radical is one of the most damaging (Cheeseman and Slater 1993), producing its effects via a cascade of self-perpetuating free radicals. Free radicals can attack all major groups of biological molecules, but lipids appear to be the most vulnerable to direct and indirect damage as a consequence of lipid peroxidation. For a variety of reasons, neurons may be particularly vulnerable to free-radical damage

(Evans 1993; Olanow 1992; Halliwell 1992a,b; 1996). To complete the cycle, the enzyme glutathione reductase reconverts GSSG to GSH (Fig. 2).

The second type of reaction mediated by GSH is a thiol–disulfide interchange; a third type involves two-electron oxidation. The former may be involved in GSH actions on ionotropic receptor function (see Part 3). Readers interested in more details of these reactions are referred to the classic review by Kosower and Kosower (1978).

2.2 Interactions with Other Antioxidants

GSH works in synergy with the antioxidants ascorbic acid and α-tocopherol and plays a crucial role in reactions mediated by SOD. GSH regulates, at least in part, ascorbic acid levels (Meister 1992):

$$\text{dehydroascorbate} + 2\text{GSH} \rightarrow \text{ascorbic acid} + \text{GSSG}$$

A decrease in GSH leads to decreased ascorbic acid levels. However, ascorbate can partially compensate for diminution of GSH (Martensson and Meister 1991; see also references in Meister 1991, 1992).

The interrelationship with SOD may be of particular interest for understanding the origin of several neurological disorders: SOD is involved in the reduction (by a dismutation reaction) of the superoxide radical, the latter produced by various reactions involving oxygen. Of particular importance for the nervous system, the superoxide radical can be produced by reactions involving oxygen and neurotransmitters such as adrenaline and dopamine (Halliwell 1996) as well as those involving cytosolic calcium (Olanow 1992). SOD converts the superoxide radical to hydrogen peroxide, which usually breaks down harmlessly to produce water and molecular oxygen. A noncatalyzed reaction of hydrogen peroxide and superoxide can generate hydroxyl radicals. Similarly, a reaction with H_2O_2 catalyzed by a transition metal (e.g., Fe^{2+}, Cu^{2+}) can also generate the hydroxyl radical. An example is the Fenton reaction:

$$Fe^{2+} + H_2O_2 \rightarrow Fe^{3+} + \bullet OH + OH^-$$

Fe^{3+} can also react with superoxide in the reaction:

$$Fe^{3+} + O_2^{\bullet} \rightleftarrows (Fe^{3+} - O_2^{\bullet} \leftrightarrow Fe^{2+} - O_2) \rightleftarrows Fe^{2+} + O_2$$

leading to the iron-catalyzed Haber–Weiss reaction:

$$O_2^{\bullet-} + H_2O_2 \xrightarrow{\text{Fe catalyst}} \bullet OH + OH^- + O_2$$

These reactions (see Halliwell 1992a) may be critical components of oxidative stress and neurological damage in cases where iron is released (e.g., stroke, ischemia, head trauma) (Willmore and Triggs 1991; see also Halliwell 1992a). It may also be noteworthy that aluminum, sometimes associated with Alzheimer's disease, can form complexes with the superoxide radical or can lead to iron release (see Bains and Shaw, chapter 17, this volume).

As described in the previous section, GSH can neutralize the •OH radical, thus providing an essential level of defense against oxidative stress produced by such reactions.

2.3 GSH's Role in Cancer

GSH appears to have biphasic effects on cancer. First, GSH may prevent the development of some cancers (Flagg et al. 1993). On the other hand, just as the GSH ratio is crucial for the well-being of normal cells, so too it protects cancer cells from the effects of oxidative stress. In the latter case, oxidative stress may arise from the generation of free radicals as products of chemo- or radiation therapy (Meister 1989, 1991,1992). Normal GSH levels thus have the potential to compromise anticancer treatments. One strategy to prevent this has been to lower GSH levels by blocking synthesis. Buthionine sulfoximine (BSO), an inhibitor of γ-glutamylcysteine synthetase (Meister 1991; Anderson 1997) lowers GSH levels and has been reported to enhance the efficacy of cancer treatments. Such treatment appears to have been successfully used in various clinical settings (Meister 1989, 1991).

Given the multiple roles of GSH in normal cells (Table 1), it is difficult to imagine that BSO treatment, effective as it may be for enhancing cancer treatment, would not have deleterious effects on other cell populations [e.g., the lens, where it induces cataract in development (Meister 1991)], and perhaps especially for neural cells (Meister 1991; Jain et al. 1991) (see Table 2). As will be detailed in section 5 of this chapter and in the chapters on GSH in neurological disease (Part III), changes in GSH status may be a factor in the genesis of various disorders. If correct, prolonged alterations in GSH levels by BSO may turn out to have decidedly contrary effects, raising the prospect that successful treatment of cancer may come at the expense of later neural function and survival. To the best of my knowledge, no epidemiological studies have examined this issue, although one article published in a clinical journal (Holt 1993) has reported that an increased GSSG–total glutathione ratio following GSSG injections during cancer treatment led to a temporary form of epilepsy. As cited below, GSH can serve as neuroprotectant (see Cooper, chapter 5; Zaman and Ratan, chapter 6); GSH may further serve to diminish some neurotoxic side effects of anticancer drugs such as cisplatin (Cascinu et al. 1995). While clearly preliminary, such data suggest that concerns for long-term effects of GSH depletion on the central nervous system need to be seriously considered (see also Kosower and Kosower 1978 and Meister and Larsson 1989 for a description of developmental disorders associated with alterations in GSH synthesis).

3. GSH MODULATION OF RECEPTOR FUNCTION

A number of factors are able to affect the function of the various brain receptor populations. Included in these are well-known phenomena such as homologous and heterologous receptor regulation, both of which may depend on phosphorylation–

dephosphorylation reactions mediated by a great variety of protein kinases and phosphatases (for a review see Pasqualotto and Shaw 1996). Other, less well documented forms of covalent modification may also control aspects of receptor function, for example, palmitoylation and methylation. In addition, various of the phospholipases (PLC, PLA) are also reported to play a role in receptor function (Massicotte et al. 1991), as do various of the long-chain fatty acids (Nisikawa, Kimura, and Akaike 1994).

3.1 Redox Reactions at Ionotropic Receptors

In recent years, a number of studies have demonstrated that reduction–oxidation (redox) reactions also control aspects of receptor function. Reducing agents such as dithiothreitol (DTT), ascorbic acid and GSH have all been shown to alter membrane currents in various receptor populations, including the NMDA (Levy, Sucher, and Lipton 1990; Gilbert, Aizenmann, and Reynolds 1991; Leslie, Brown, and Trent 1992; Jánaky et al. 1993; Gozlan and Ben Ari 1995) as well as $GABA_A$ and glycine receptors (Pan et al. 1995). The effects of these reducing agents are not uniform across receptor populations, so that reduction of the receptor may enhance or diminish receptor-mediated currents (e.g., Pan et al. 1995). GSH actions may often, but not always, be opposite from those of GSSG. Of the reducing agents, GSH is among the weakest in its ability to alter receptor function, with effects not observed until a millimolar concentration range is reached. Millimolar concentrations may occur in the nervous system overall, although it is unclear if such high levels would be achieved in the extracellular space where GSH might act directly on receptors. In view of this last point, it is uncertain if GSH would normally exert profound actions on the various ionotropic receptors, except perhaps in circumstances where serious tissue damage had occurred. In such cases, GSH may serve a neuroprotective role (see Cooper, Pulsinelli, and Duffy 1980; also Cooper, chapter 5, and Zaman and Ratan, chapter 6, this volume). Such neuroprotective actions may be related to GSH–GSSG modulation of NMDA receptor function, which has been proposed as one of the controlling factors on a continuum of NMDA receptor activity between normal, enhanced, and pathological function (Gozlan and Ben Ari 1995). GSH may indeed have multiple actions as a neuroprotective agent, acting both to control ionic currents through NMDA and other ionotropic receptors, and also in its antioxidant capacity to control calcium-mediated free-radical production (Orrenius et al. 1992).

A more comprehensive review of the above literature, as well as experimental evidence for GSH as a redox modulator, is presented by Ogita et al. (chapter 7) and Jánaky et al. (chapter 8, this volume).

3.2 Second-Messenger Activation by GSH

Almost nothing is known about the potential for GSH to stimulate second-messenger activity in neural cells. Our own work on GSH receptors in cultured cortical astrocytes has shown a large increase in IP$_3$ activity following exposure of cortical astrocytes to GSH (Guo, McIntosh, and Shaw 1992). These data suggest that GSH can activate a population of metabotropic receptors, leading to alterations in second-messenger levels. An earlier study in starfish showed GSH-stimulated activation of adenylate cyclase (Gentleman and Mansour 1974). Whether adenylate cyclase, cAMP, cGMP, or other second messengers are activated by GSH in the nervous system is not known. This area is certain to be a fruitful avenue for future research.

3.3 GSH as a Cascading Neuromodulator

As summarized above, GSH may act to alter the redox state of various neurotransmitter receptors to affect receptor-mediated membrane currents. GSH activation may also stimulate second-messenger activity, potentially with consequences for activation of other receptor populations which share common second messengers. I will discuss in the section that follows the evidence that GSH may itself be an excitatory neurotransmitter acting on its own class of ionotropic receptors (Shaw, Pasqualotto and Curry 1996; see also Pasqualotto, Curry, and Shaw, chapter 9, this volume). In this section, however, I want to consider briefly the possibility that the sequential degradation of the tripeptide into its component amino acids (see Fig. 1) can yield a succession of molecules that may individually, and perhaps synchonously, affect neural activity.

The glutathione tripeptide consists of L-glutamate, L-cysteine, and glycine. Of these, the effects of glutamate are well characterized and have been shown to activate the ionotropic AMPA, NMDA, and kainate receptor populations (for a complete discussion of these points with corresponding references, see Ogita et al., chapter 7, this volume) as well as the metabotropic glutamate receptors (mGluRs) (Prezeau et al. 1994). Cysteine appears to act at the NMDA receptor (Olney et al. 1990; Pace et al. 1992), since its depolarizing actions can be mostly blocked by the NMDA competitive and noncompetitive antagonists, AP5 and MK 801, respectively (see also Pasqualotto, Curry, and Shaw, chapter 9, this volume). Glycine has multiple roles in the nervous sytem: in spinal cord and other regions, glycine acts at strychnine-sensitive ionotropic glycine receptors to activate an inhibitory chloride current (see Aprison and Werman 1965 for references; Kuhse, Betz, and Kirsch 1995). In cortex and hippocampus and elsewhere in the CNS, glycine binds to a modulatory site of the NMDA receptor complex (Johnson and Ascher, 1987).

In addition, two of the three dipeptide combinations of GSH are themselves neurally active (glutamylcysteine, cysteinylglycine), apparently by activation of the GSH receptor (see following section and Pasqualotto, Curry, and Shaw, chapter 9, this volume).

One possible effect of the degradation of GSH into its component amino acids would be the sequential activation of GSH receptors by GSH and the dipeptides of GSH, glutamate receptor subtypes by glutamate and cysteine, and finally NMDA receptors by all of the three amino acids. The outcome might be a prolonged depolarization acting to enhance neural activity in some circumstances. Such circumstances may favor the conditions underlying some forms of activity-dependent neural modifications.

It must be stressed, however, that such a mechanism has not been demonstrated at a synaptic level by electrophysiological methods. Partially, this may be due to the fact that most studies tend to focus on the early temporal events after synaptic stimulation. Further, some stimulation paradigms may employ agonists of glutamate to substitute for glutamate's excitatory neurotransmitter role.

This last point notwithstanding, several important issues would have to be resolved before the above scheme might be accepted as plausible. The first of these is that it is not clear if GSH is broken down in the extracellular space by the action of the degradative enzymes (e.g., the action of acetylcholine esterase on acetylcholine) or is taken up intact into target cells (e.g., as for noradrenaline). The former would clearly be required for the multistaged mechanism proposed above. The second point is that GSH and glutamatergic receptors would have to be colocalized at the same, or at least closely adjacent synapses. Hjelle et al. present evidence for such an arrangement of adjacent GSH-positive and GSH-negative synapses in spinal motor neurons in chapter 4 of this volume. Notably, the GSH-positive terminal was also glutamate-positive.

The above remains speculation, but perhaps also the subject of potentially far-reaching and novel discoveries about synaptic function in the brain.

4. GSH AS A NEUROTRANSMITTER IN THE CNS

Historically, the criteria for the identification of a neurotransmitter have varied in number and have reflected emerging knowledge of the processes underlying synaptic transmission. An early formulation was made by Werman and colleagues (Werman 1966) in their characterization of glycine as the inhibitory neurotransmitter of spinal interneurons released onto spinal motor neurons (Aprison and Werman 1965). Their criteria included eight items. More recently, McGeer, Eccles, and McGeer (1987) defined 11 criteria. There is insufficient space here to debate the relative merits of the different criteria. Therefore, for the purpose of the following I will adopt and paraphrase the criteria of Kandel, Schwartz, and Jessell (1991) from their well-known textbook. These are the following:

1. The putative neurotransmitter should be synthesized in the releasing neuron (see below for caveats to this).
2. It must be present in the presynaptic terminal and be released by stimulation in amounts that can have the required effect on the postsynaptic site.
3. Experimentally applied, the putative neurotransmitter should mimic exactly the action of the endogenous neurotransmitter, activating the same receptor populations, the same ion channels, second messengers (if applicable), etc. (Note that this requirement is similar to Werman's original formulation of the "criterion of identity of action.")

4. A specific mechanism will exist for removing the putative neurotransmitter from the site of action.

Although many substances have been proposed as neurotransmitters in the central nervous system, it has proven difficult to satisfy these criteria for most of them. This doesn't mean that they are not neurotransmitters, merely that the minimum conditions have not been satisfied. After all these years, glycine remains the most rigorously characterized neurotransmitter in any neural pathway (Aprison and Werman 1965).

Where does the work on GSH place it in regard to possible neurotransmitter status? Taking the above criteria in order:

1. GSH is found in neurons and glia (see Hjelle et al., chapter 4, for citations; see also Yudkoff et al. 1990; Philbert et al. 1991), Pow and Crook (1995); GSH is synthesized in neural cells, at least in astrocytes (Yudkoff et al. 1990). This does not, of course, imply that GSH is synthesized by the same cells that might use it as neurotransmitter, but perhaps this is a false criterion. Much GSH synthesis in the body is in hepatocytes; transport into CNS is via several blood–brain-barrier transporters (see Kannan et al. 1992, 1995; also chapter 3, this volume). Similarly, inside the nervous system, GSH might be made in glia and transported to neurons for use. Neither mode of acquisition would necessarily preclude a role in synaptic transmission if GSH were shown to be present in synaptic terminals. The latter, as cited above, appears to be the case with the demonstration of GSH in terminals of various neurons (see Hjelle, Chaudhry, and Ottersen 1994; Hjelle et al., chapter 4, this volume).

2. As cited above, GSH can be found in synaptic terminals. Zangerle et al. (1992) have shown that GSH and cysteine (Keller et al. 1989) are released from rat cortical and hippocampal slices in a calcium-dependent manner, although from which cells is not known.

3. GSH application to cortical cells in vitro elicits a depolarizing field potential which appears to be sodium-dependent (Shaw et al. 1996; see also Pasqualotto, Curry, and Shaw, chapter 9, this volume). Figure 3 illustrates GSH responses in this preparation. The response is independent of other known excitatory ionotropic receptor populations, as neither AMPA nor NMDA antagonists block the GSH response. The GSH response is not the same as that of cysteine, which acts through the NMDA receptor (Olney et al. 1990; Pace et al. 1992; Pasqualotto et al., chapter 9, this volume). At a pharmacological level, radiolabeled GSH binds to a population of binding sites distinct from those of the glutamatergic excitatory amino acid receptors: Other than cysteine, neither glutamate nor NMDA has any significant competitive effect on [^{3}H]GSH binding; AMPA antagonists have limited effects, but only at high concentrations (see Pasqualotto et al., chapter 9, this volume). In contrast, GSH is able to compete with relatively high affinity for agonists and antagonists of NMDA receptors (Ogita et al., chapter 7, this volume), suggesting that GSH might act on its own receptors as well as some glutamatergic receptor subtypes (see also Jánaky et al., chapter 8, this volume). These data are intriguing, but hardly conclusive for GSH neurotransmitter status for two main reasons. First, single-cell recording is required

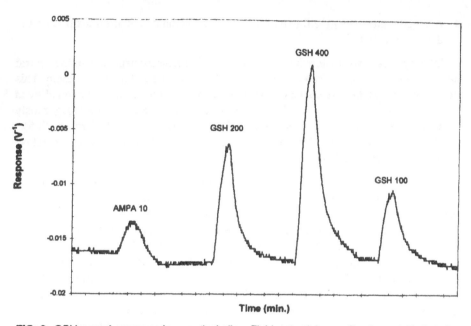

FIG. 3. GSH neural response in a cortical slice. Field potential recording in a cortical wedge preparation following application of GSH or AMPA. For details see Pasqualotto et al., chapter 9, this volume, or Shaw et al. (1996). GSH application gives rise to a depolarizing potential that cannot be blocked by AMPA or NMDA antagonists (not shown). Different concentrations of GSH are shown, as is a response to AMPA for comparison. Numbers above each trace indicate micromolar concentrations. Each tick mark on the x-axis equals 1 min.

to demonstrate that GSH can elicit membrane currents with a time course consistent with neurotransmission. Such recording has not been achieved for adult neurons, although such experiments are in progress in our laboratory. Second, it is not clear in what pathway GSH might be a neurotransmitter. Receptor autoradiography, however, offers several potential neural circuits, including regions of neocortex, hippocampus, and retina (Fig. 4). In particular, GSH receptors appear to be densely packed into layer 4 in the visual cortex, perhaps indicating a role in geniculostriate neurotransmission (Fig. 5 A). Complementary data for this suggestion come from transport experiments in which GSH can be shown to be retrogradely transported to the lateral geniculate nucleus following injection into the visual cortex (Fig. 5 C). A crucial demonstration of GSH neurotransmission in these circuits, however, would require that both GSH-induced action on potential target cells and responses to electrical stimulation of afferent pathways be blockable by GSH receptor antagonists. No such compounds yet exist, so such experiments cannot be performed at present.

4. Mechanisms for GSH removal from synaptic regions have not been demonstrated; it is not even clear if the GSH degradative dipeptidases are present in postsynaptic membranes or if uptake mechanisms exist.

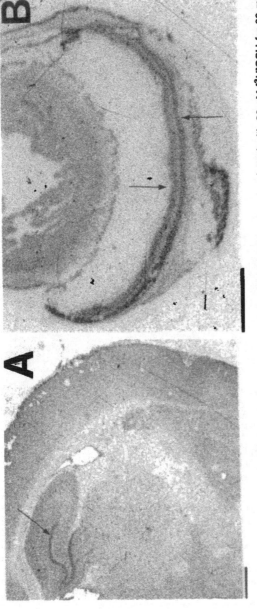

FIG. 4. Localization of [³H]GSH binding sites. Cryostat-cut sections (20 μM) were incubated with 20 nM [³H]GSH for 90 min in 4°C Tris acetate buffer. Labeled sections were used to expose Amersham Hyperfilm. Nonspecific binding (not shown) was determined by coincubation with 10⁻³ M unlabeled GSH and was less than 10% of total binding. A. Section containing the rat hippocampus. Note the dense label over the dentate gyrus (arrow). B. Section of the whole rat eye. Note dense labelling across the retinal layers (arrow). (Ogita and colleagues performed homogenate binding of GSH in various brain regions; see Ogita et al., chapter 7 for references. Those areas described as having high levels of GSH binding sites in these studies have been confirmed in our autoradiographic studies). Calibration, both panels: 1 mm.

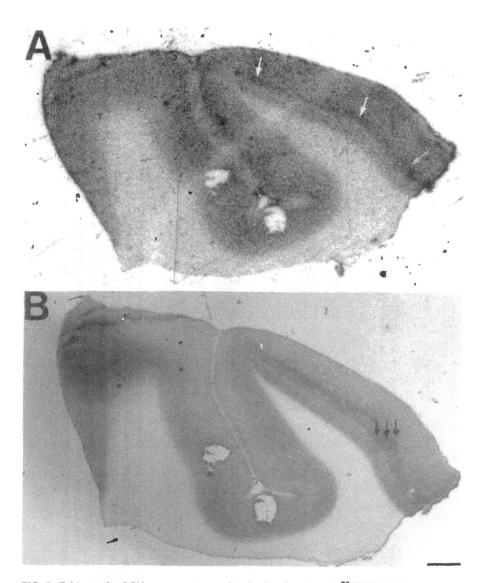

FIG. 5. Evidence for GSH as a neurotransmitter in visual cortex. A. [^{35}S]GSH binding sites in a 20-μM-thick section of macaque monkey visual cortex. Incubation conditions: 10 nM for 30 min at room temperature. White arrows indicate layer 4°C, the primary input layer from the lateral geniculate nucleus. B. Adjacent section stained for cytochrome C to reveal input layer 4 (arrows demarcate the laminar boundaries with 4A at the top, 4Cα in the middle, and 4Cβ at the bottom). C. Uptake of [^{3}H]GSH injected into rat visual cortex. Rats were injected with 20 μCi in 4 μL injected at several depths at the center of area 17 (OC1) (arrow) and allowed a 24-h survival before sacrifice. In panels A and C, sections were exposed to tritium-sensitive film to generate autoradiograms. Note that the dorsal lateral geniculate nucleus (DLG) is densely labeled. Similar experiments using [^{35}S]cysteine gave comparable results, and injections of [^{3}H]glutamate or [^{3}H]glycine showed showed considerably less uptake to DLG (not shown). Calibration, all panels: 1 mm.

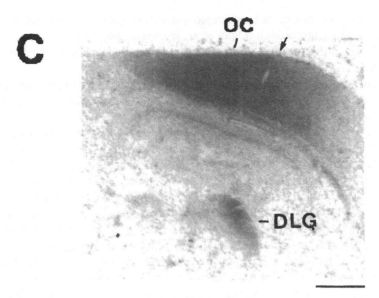

FIG. 5. *Continued.*

For some, the issue of GSH's ubiquity may still militate against a potential role as neurotransmitter. This is one of the hoariest of criticisms and one dealt with years ago by Aprison and Werman (1965). Their response to similar comments made concerning glycine still has validity: The mere fact that a molecule is present inside a cell does not imply that the same distribution will be found externally; the latter only occurs when it is released following neural activation in doses sufficient to activate receptors on target neurons.

We are at an early stage in the work designed to determine if GSH can function as a neurotransmitter in the CNS. To date, all the data have supported this hypothesis. However, much remains to be done before this hypothesis will gain widespread acceptance.

5. IMPLICATIONS FOR MULTIPLE ROLES OF GSH IN NORMAL SYNAPTIC FUNCTION

It will be clear from the foregoing that GSH has the potential to play many roles in neural function. Of those described above, it is clear that the traditional role of GSH as antioxidant and free-radical scavenger occurs in the nervous system. Added to this are all the other roles in cellular biochemistry (see Table 1). It is also reasonably well established that GSH may act as a neuroprotective agent against the effects of glutamatergic stimulation, although the mechanism is still not completely resolved (see Cooper, chapter 5; Zaman and Ratan, chapter 6). In addition, the likelihood that

GSH redox modulation of various receptor populations plays a certain role in shaping of neural activity seems high (Ogita et al., chapter 7; Jánaky et al., chapter 8, this volume), even if disagreements persist about the magnitude and even the direction of such effects.

Most speculative is the hypothesis that GSH may serve as a neurotransmitter at its own population of receptors. There is still much to be done before this hypothesis will be accepted, but the data generated to date support such a role, at least for some neural populations at specific ages. Equally obviously, this hypothesis may be completely wrong, the observed neurotransmitterlike actions being attributable to one of GSH's other actions or to some other nonspecific mechanism. Even were that true, the previously established roles of GSH would make it one of the most interesting and potentially important molecules yet described in the nervous system.

If, however, GSH could be demonstrated to have a neurotransmitter role, the implications would be fairly shocking to at least some part of the neuroscience community. The discovery of a major new neurotransmitter this late in the history of the neurosciences would be a large surprise to those who hold the view that most of the pieces that make up the milieu of the brain have been described. If the new neurotransmitter were sufficiently novel in its actions, the discovery might suggest the further, likely more disturbing, possibility that other unique molecules awaited discovery. Keeping in mind an estimate that the human genome may encode $10^{30,000}$ gene products (Kauffman 1991) and that a huge number of these are likely to be found in the nervous system, the latter is a very distinct, if to some unwelcome, possibility.

Not only would the discovery of a novel excitatory neurotransmitter be in itself a surprise; it would also force a reevaluation of the types, importance, and proportions of known or suspected neurotransmitters and of their interactions in normal synaptic activity, during development, in neuroplasticity, and, not least, in neuropathological conditions.

The above would be the possibilities and questions raised by the simple fact of the existence of a major new neurotransmitter. What if the story were more complicated and the putative neurotransmitter had a variety of guises and roles that could both vary with and affect the outcome of synaptic transmission? While still unproven, if GSH is a neurotransmitter in the brain, it is doing exactly this. Imagine the following: Synaptic GSH release activates its own receptors, modulates the action of local ionotropic receptors, and then possibly breaks down to activate the glutamatergic subtypes it has just modified by redox reactions; additional GSH release by neurons or glia then serves a neuroprotective role against the effects of hyperstimulation that it itself has created. It also cleans up the free radicals generated by the neural activity. Under normal conditions, such a novel form of neurotransmission might provide for many levels of sensitivity as well as complexity. If, in addition, GSH were to effect some control over the synthesis or action of other molecules, e.g., the cytokines, then the levels of complexity would be further expanded. A scheme of the possible roles of GSH in neural function, and their interactions, is shown in Fig. 6.

Finally, how aspects of neural function might be affected by changing GSH levels during development (Chen, Richie, and Lang 1989; Kudo et al. 1990; Shivakumar,

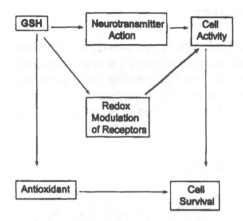

FIG. 6. Multiple roles of GSH in the nervous system. Some of GSH's known and possible roles in the nervous system and their possible relations are shown.

Anandatheerthavarada, and Ravindranath 1991; Kannan et al. 1992) remains an unresolved issue.

6. GSH IN NEUROLOGICAL DISEASE

It is obvious that any molecule with multiple roles, each essential for a different aspect of cell function, may provide multiple paths for cellular malfunction as normal actions are compromised. Given the multiple roles proposed above for GSH in the nervous system, the failure of any of these could be a direct cause of some forms of neural pathology. As we will discuss in chapter 17 (Bains and Shaw), a diminution of GSH for any reason (e.g., decreased synthesis, decreased activity of GSH reductase, decreased levels of selenium controlling the actions of GSH peroxidase) might increase oxidative stress, alter ionotropic receptor modulation, and diminish GSH transmitter action. The latter not only would alter neural activity, but would also inhibit the release of essential trophic factors.

Oxidative stress alone has been demonstrated to induce necrotic, apoptotic, and excitotoxic neuron death in ALS, AD, PD, and stroke and ischemia (see the chapters in Part III); failure to correctly control ionotropic receptor currents could lead to excitotoxic damage (Gozlan and Ben Ari 1995); loss of neurotransmitter action could cause disuse degeneration as well as deprive other neural cells of trophic substances, in turn leading to the demise of target neurons. Alterations in GSH status of whatever origin could thus be fatal to neural cells in a variety of different ways. The type of cell destroyed might depend on the site of the GSH deficit, the nature of the GSH local action(s), and the relative vulnerability of particular cells to different noxious stimuli (e.g., sensitivity to calcium as a consequence of intracellular buffering capacity and the presence of NMDA receptors). These and related topics will be discussed in greater detail in Part III.

7. SUMMARY

GSH serves multiple roles in the nervous system. It is an antioxidant–free-radical scavenger, a redox modulator of ionotropic receptor currents, a neuroprotector, and possibly a novel neurotransmitter. These roles and their possible interactions may provide the basis for a clearer understanding of synaptic transmission in the normal nervous system, with additional implications for alterations in neural function and survival in neurological disease states.

ACKNOWLEDGMENTS

This work was supported by grants from the ALS Association and the Natural Science and Engineering Research Council (Canada). I thank Ms. J. C. McEachern, Dr. B. A. Pasqualotto, Dr. J. S. Bains, and Dr. K. Curry for valuable comments on a draft of this chapter. Thanks are due also to Dr. A. Hendrickson for providing the primate tissue shown in Fig. 5, and to Mr. S. A. Bowlsby, whose M.Sc. thesis provided the data for this same figure.

REFERENCES

Al-Chalabi, A., Powell, J. F., and Leigh, P. N. 1995. Neurofilament, free radicals, excitotoxins and amyotrophic lateral sclerosis. *Muscle and Nerve*.18:540–5.

Anderson, M. E. 1997. Glutathione and glutathione delivery compounds. *Advances in Pharmacology* 38:65–78.

Aprison, M. H., and Werman, R. 1965. The distribution of glycine in cat spinal cord and roots. *Life Sciences* 4:2075–83.

Ballatori, N. 1994. Glutathione mercaptides as forms of metals. *Advances in Pharmacology* 27:271–98.

Brown, L. A. 1994. Glutathione protects signal transduction in type II cells under oxidant stress. *American Journal of Physiology* 266:172–7.

Bump, E. A., and Brown, J. M. 1990. Role of glutathione in the radiation response of mammalian cells in vitro and in vivo. *Pharmacology and Therapeutics* 47:117–36.

Cascinu, S., Cordella, L., Del Ferro, E., Fronzoni, M., and Catalano, G. 1995. Neuroprotective effect of reduced glutathione on cisplatin-based chemotherapy in advanced gastic cancer: a randomized double-blind placebo-controlled trial. *Journal of Clinical Oncology* 13:26–32.

Cheeseman, K. H., and Slater, T. F. 1993. An introduction to free radical biochemistry. *British Medical Bulletin* 49:481–93.

Chen, T. S., Richie, J. P., and Lang, C. A. 1989. The effect of aging on glutathione and cysteine levels in different regions of the mouse brain. *Proceedings of the Society for Experimental Biology and Medicine* 190:399–402.

Cooper, A. J. L., Pulsinelli, W.ˈA., and Duffy, T. E. 1980. Glutathione and ascorbate during ischemia and postischemic reperfusion in rat brain. *Journal of Neurochemistry* 35:1242–1245.

Deneke, S. M., and Fanburg, B.L. 1989. Regulation of cellular glutathione. *American Journal of Physiology* 257:L163–73.

De Rey-Pailhade, J. 1988a. Sur un corps d'origine organique hydrogénant le soufre à froid. *Comptes Rendus Hebdomadaire Séances de l'Académie de Sciences* 106:1683–4.

De Rey-Pailhade, J. 1988b. Nouvelle recherches physiologique sur la substance organique hydrogénant le soufre à froid. *Comptes Rendusde l'Académie de Sciences* 107:43–4.

Evans, P. H. 1993. Free radicals in brain metabolism and pathology. *British Medical Bulletin* 49:577–87.

Flagg, E. W., Coates, R. J., Jones, D. P., Eley, J. W., Gunter, E. W., Jackson, B., and Greenberg, R. 1993. Plasma total glutathione in humans and its association with demographic and health-related factors. *British Journal of Nutrition* 70:797–808

Freidlich, A. L., and Butcher, L. L. 1994. Involvement of free radicals in β-amyloidosis: an hypothesis. *Neurobiology of Aging* 15:443–55.

Gentleman, S., and Mansour, T. E. 1974. Adenylate cyclase in a sea anemone: implications for chemoreception. *Biochimica Biophysica Acta* 343:469–79.

Gilbert, K. R., Aizenmann, E., and Reynolds, I. J. 1991. Oxidized glutathione modulates N-methyl-D-aspartate- and depolarization-induced increases in intracelluar Ca^{2+} in cultured rat forebrain neurons. *Neuroscience Letters* 133:11–14.

Gozlan, H., and Ben Ari, Y. 1995. NMDA receptor redox sites: are they targets for selective neuronal protection? *Trends Pharmacol Sci* 16:368–74.

Grosvenor, W., Rhoads, D. E., and Kass-Simon, G. 1996. Chemoreceptor control of feeding processes in hydra. *Chemical Senses* 21:313–21.

Guo, N., McIntosh, C., and Shaw, C. 1992. Glutathione: new candidate neuropeptide in the central nervous system. *Neuroscience* 51:835–42.

Halliwell, B. 1992a. Reactive oxygen species and the central nervous system. *Journal of Neurochemistry* 59:1609–23.

Halliwell, B. 1992b. Oxygen radicals as key mediators in neurological disease: fact or fiction? *Annals of Neurology* 32:S10–S15.

Halliwell, B. 1996. Cellular stress and protection mechanisms. *Biochemical Society Transactions* 24:1023–27.

Harris, M. E., Hensley, K., Butterfield, D. A., Leedle, R.A., and Carney, J.M. 1995. Direct evidence of oxidative injury by the Alzheimer's β-amyloid peptide (1–40) in cultured hippocampal neurons. *Experimental Neurology* 131:193–202.

Hirata, H., Ladenheim, B., Rothman, R. B., Epstein, C., and Cadet, J. L. 1995. Methamphetamine-induced serotonin neurotoxicity is mediated by superoxide dismutase. *Brain Research* 677:345–7.

Hjelle, O. P., Chaudhry, F. A., and Ottersen, O.P. 1994. Antisera to glutathione: characterization and immunocytochemical application to the rat cerebellum. *European Journal of Neuroscience* 6:793–804.

Holt, J. A. G. 1993. The glutathione cycle is the creative reaction of life and cancer. Cancer causes oncogenes and not vice versa. *Medical Hypotheses* 40:262–6.

Hopkins, F. G. 1921. On an autooxidisable constituent of the cell. *Biochemical Journal* 15:286–305.

Jain, A., Martensson, J., Stole, E., Auld, P. A. M., and Meister, A. 1991. Glutathione deficiency leads to mitochondrial damage in brain. *Proceedings of the National Academy of Sciences U.S.A.* 88:1913–17.

Jánaky, R., Varga, V., Saransaari, P., and Oja, S. S. 1993. Glutathione modulates the N-methyl-D-aspartate receptor-activated calcium influx into cultured rat cerebellar granule cells. *Neuroscience Letters* 156:153–7.

Jenner, P., and Olanow, C. W. 1996. Oxidative stress and the pathenogenesis of Parkinson's disease. *Neurology* 47:S161–71.

Johnson, J. W., and Ascher, P. 1987. Glycine potentiates the NMDA response in cultured mouse brain neurones. *Nature* 325:529–31.

Kandel, E. R., Schwartz, J. H., and Jessell, T. M., eds. 1991. *Principles of neural science* 3rd ed., 214–215. New York: Elsevier.

Kannan, R., Kuhlenkamp, J. F., Ookhtens, M., and Kaplowitz, N. 1992. Transport of glutathione at blood–brain barrier of the rat: inhibition by glutathione analogs and age-dependence. *Journal of Pharmacology and Experimental Therapeutics* 263:964–68.

Kannan, R., Jian, R. Y., Zlokovic, B. V., and Kaplowitz, N. 1995. Molecular characterization of a reduced glutathione transporter in the lens. *Investigative Ophthalmology and Visual Science* 36:1785–92.

Kauffman, S. A. 1991. Antichaos and adaptation. *Scientific American* 265:78–84.

Keller, H. J., Do, K. Q., Zollinger, M., Winterhalter, K. H., and Cuénod, M. 1989. Cysteine: depolarization-induced release from rat brain in vitro. *Journal of Neurochemistry* 52:1801–6.

Kosower, E. M. 1972. A molecular basis for learning and memory. *Proceedings National Academy of Sciences of the U.S.A.* 69:3292–6.

Kosower, N. S., and Kosower, E. M. 1978. The glutathione status of cells. *International Review of Cytology* 54:109–60.

Kosower, E. M., and Werman, R. 1971. New step in transmitter release at the myoneural junction. *Nature New Biology* 233:121–3.

Kudo, H., Kokunai, T., Kondoh, T., Tamaki, N., and Matsumoto, S. 1990. Quantitative analysis of glutathione in rat central nervous system: comparison of GSH in infant brain with that in adult brain. *Brain
Research* 511:326–8.

Kuhse, J., Betz, H., and Kirsch, J. 1995. The inhibitory glycine receptor: architecture, synaptic localization
and molecular biology of a postsynaptic ion-channel complex. *Current Opinion in Neurobiology* 5:318–
23.

Lenhoff, H. M. 1961. Activation of the feeding reflex in *Hydra littoralis*. I. Role played by reduced
glutathione, and quantitative assay of the feeding reflex. *Journal of General Physiology* 45:331–44.

Leslie, S. W., Brown, L. M., and Trent, R. D. 1992. Stimulation of N-methyl-D-aspartate receptor-mediated
calcium entry into dissociated neurons by reduced and oxidized glutathione. *Molecular Pharmacology*
41:308–14.

Levy, D. I., Sucher, N. J., and Lipton, S. A. 1990. Redox modulation of NMDA receptor-mediated toxicity
in mammalian central neurons. *Neuroscience Letters* 110:291–6.

Lomaestro, B. M., and Malone, M. 1995. Glutathione in health and disease: pharmacological issues. *Annals
of Pharmacotherapy* 29:1263–73.

Loomis, W. F. 1955. Glutathione control of the specific feeding reactions of *Hydra*. *Journal of General
Physiology* 62:209–28.

Martensson, J., and Meister, A. 1991. Glutathione deficiency decreases tissue ascorbate levels in newborn
rats: ascorbate spares glutathione and protects. *Proceedings of the National Academy of Sciences of the
U.S.A.* 88:4656–60.

Massicotte, G., Vanderklish, P., Lynch, G., and Baudry, M. 1991. Modulation of D,L-α-amino-3-hydroxy-
5-methyl-4-isoxasole-propionic acid/quisqualate receptors by phospholipase A$_2$: a necessary step in
long-term potentiation? *Proceedings of the National Academy of Sciences of the U.S.A* 88:1893–7.

McGeer, P. L., Eccles, J. C., and McGeer, E. G., eds. 1987. *Molecular neurobiology of the mammalian
brain.* 2nd ed. Plenum Press, New York.

Meister A. 1989. On the biochemistry of glutathione. In *Glutathione centennial: molecular perspectives
and clinical implications*, eds. N. Taneguchi, T. Higashi, Y. Sakamoto, and A. Meister, 3–21. New York:
Academic Press.

Meister, A. 1991. Glutathione deficiency produced by inhibition of its synthesis, and its reversal: applications in research and therapy. *Pharmacology and Therapeutics* 51:155–94.

Meister, A. 1992. On the antioxidant effects of ascorbic acid and glutathione. *Biochemical Pharmacology*
44:1905–15.

Meister, A., and Larsson, A. 1989. Glutathione synthetase deficiency and other disorders of the gamma-
glutamyl cycle. In *The metabolic basis of inherited disease*, ed. C. R. Scriver, A. L. Beaudet, W. S. Sly,
and D. Valle, 855–68. New York: McGraw-Hill.

Niketic, V., Besla, D., Raicevic, S., Sredic, S., and Stojkovic, M. 1992. Glutathione adduct of hemoglobin
(Hb ASSG) in hemolysates of patients on long-term antileptic therapy. *International Journal of Biochemistry* 24:503–7.

Nisikawa, M., Kimura, S., and Akaike, N. 1994. Facilitatory effect of docosahexaenoic acid on N-methyl-
D-aspartate response in pyramidal neurones of rat cerebral cortex. *Journal of Physiology (London)*
475:83–93.

Ogita, K., Enomoto, R., Nakahara, F., Ishitsubo, N., and Yoneda, Y. 1995. A possible role of glutathione
as an endogenous agonist at the N-methyl-D-aspartate recognition domain in rat brain.*Journal of Neurochemistry* 64:1088–96.

Olanow, C. W. 1992. A radical hypothesis for neurodegeneration. *Trends in Neuroscience* 16:439–44.

Olney, J. W., Zorumski, C., Price, M. T., and Labruyere, J. 1990. L-Cysteine, a bicarbonate-sensitive
endogenous excitotoxin. *Science* 251:596–99.

Orrenius, S., Burkitt, M. J., Kass, G. E. N., Dypbukt, J. M., and Nicotera, P. 1992. Calcium ions and
oxidative cell injury. *Annals of Neurology* 32:S33–S42.

Ortolani, O., Caggiano, M., Mannelli, R., Gogliettino, A., and Tufano, R. 1995.Protection from ischemia-
reperfusion damage in patients with stroke: the role of rutin and GSH, *Transplantation Proceedings*
27:2877–8.

Pace, J. R., Martin, B. M., Paul, S. P., and Rogawski, M. A. 1992. High concentrations of neutral amino
acids activate NMDA receptor currents in rat hippocampal slices. *Neuroscience Letters* 141:97–100.

Pan, Z.-H., Bahring, R., Grantyn, R., and Lipton, S. A. 1995. Differential modulation by sulfhydryl redox
agents and glutathione of GABA- and glycine-mediated currents in rat retinal ganglion cells. *Journal of
Neuroscience* 15:1384–91.

Pasqualotto, B. A., and Shaw, C. A. 1996. Regulation of ionotropic receptors by protein phosphorylation.
Biochemistry and Pharmacology 51:1417–25.

Perez-Pinzon, M. A. and Rice, M. E. 1995. Seasonal- and temperature-dependent variation in CNS ascorbate and glutathione levels in anoxia-tolerant turtles. *Brain Research* 705:45–52.

Philbert, M. A., Beiswanger, C. M., Waters, D. K., Reuhl, K. R., and Lowndes, H. E. 1991. Cellular and regional distribution of reduced glutathione in the nervous system of the rat: histochemical localization by mercury orange and *o*-phthaldialdehyde-induced histofluorescence. *Toxicology and Applied Pharmacology* 107:215–27.

Pow, D. V., and Crook, D. K. 1995. Immunocytochemical evidence for the presence of high levels of reduced glutathione in radial glial cells and horizontal cells in the rabbit retina. *Neuroscience Letters* 193:25–8.

Prezeau, L., Carrette, J. Helpap, B., Curry, K., Pin, J. P., and Bockaert, J. 1994. Pharmacological characterization of metabotropic glutamate receptors in several types of brain cells in primary cultures. *Molecular Pharmacology* 45:570–7.

Schubert, D., Behl, C., Lesley, R., Brack, A., Dargusch, R., Sagara, Y., and Kimura, H. 1995. Amyloid peptides are toxic via a common oxidative mechanism. *Proceedings National Academy of Sciences of the U.S.A.* 92:1989–93.

Shaw, C. A., Pasqualotto, B. A., and Curry, K. 1996. Glutathione-induced sodium currents in neocortex. *Neuroreport* 7:1149–52.

Shivakumar, B. R., Anandatheerthavarada, H. K., and Ravindranath, V. 1991. Free radical scavenging systems in developing rat brain. *International Journal of Developmental Neuroscience* 9:181–5.

Sirsjo, A., Kadedal, B., Arstrand, K., Lewis, D. H., Nylander, G., and Gidlof, A. 1996. Altered glutathione levels in ischemic and postishemic skeletal muscle: difference between severe and moderate ischemic insult. *Journal of Trauma* 41:123–8.

Slivka, A., Spina, M. B., and Cohen, G. 1987. Reduced and oxidized glutathione in human and monkey brain. *Neuroscience Letters* 74:112–8.

Spina, M. B., Squito, S.P., Miller, J., Lindsay, R. M., and Hyman, C. 1992. Brain-derived neurotrophic factor protects dopamine neurons against 6-hydroxydopamine and *N*-methyl-4-phenylpyridinium iontotoxicity:involvement of the glutathione system. *Journal of Neurochemistry* 59:99–106.

Taneguchi, N., Higashi, T., Sakamoto, Y., and Meister, A., eds. 1989. *Glutathione centennial: molecular properties and clinical applications.* New York: Academic Press.

Wei, Q., Yeung, M., Jurma, O. P., and Anderson, J.K. 1996. Genetic elevation of monoamine oxidase levels in dopaminergic PC12 cells results in increased free radical damage and sensitivity to MPTP. *Journal of Neuroscience Research* 46:666–73.

Werman, R. 1966. Criteria for the identification of a central nervous system transmitter. *Comparative Biochemistry and Physiology* 18:745–66.

Werman, R., Carlen, P. L., and Kosower, E.M. 1971. Effect of the thiol-oxidizing agent, diamide, on acetylcholine release at the frog endplate. *Nature New Biology* 233:120–1.

Willmore, L. J. and Triggs, W. J. 1991. Iron-induced lipid peroxidation and brain injury responses. *International Journal of Developmental Neuroscience* 9:175–80.

Yudkoff, M., Pleasure, D., Cregar, I., Lin, Z. P., Nissim, I., Stern, J., and Nissim, I. 1990. Glutathione turnover in cultured astrocytes: studies with [^{15}N]-glutamate. *Journal of Neurochemistry* 55:137–45.

Zangerle, L., Cuénod, M., Winterhalter, K. H., and Do, K. Q. 1992. Screening of thiol compounds: depolarization-induced release of glutathione and cysteine from rat brain slices. *Journal of Neurochemistry* 59:181–9.

Glutathione in the Nervous System
Edited by Christopher A. Shaw
Copyright © 1998 Taylor & Francis

2

The Discovery of the GSH Receptor in Hydra and Its Evolutionary Significance

Howard M. Lenhoff

Department of Developmental and Cell Biology, University of California, Irvine, California 92697

1. INTRODUCTION

It seems fitting that this volume on reduced glutathione as a possible neurotransmitter have a chapter on the glutathione receptor of hydra. First, it was in hydra that the discovery of a glutathione receptor was made (Loomis 1955). Second, cellular receptors to peptides among the eucaryotes probably first evolved in diploblastic organisms at the tissue level of construction, such as hydra, rather than in organisms at the triploblastic organ level. Third, the structural simplicity of hydra may make it

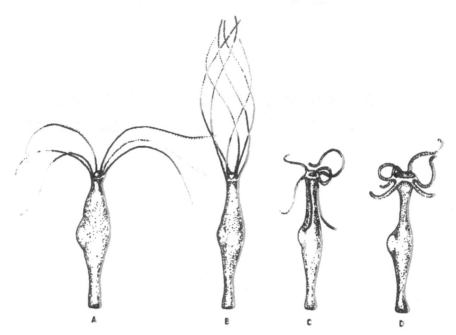

FIG. 1. Stages of the feeding behavior of hydra in response to reduced glutathione. A, a hydra before the addition of GSH; B, immediately after the addition of GSH; C, after being exposed to GSH for 30 s; D, during the feeding response (30 s to about 30 min), which ends with the animal returning to position A (Lenhoff 1961b).

possible to unravel some of the secrets of peptide receptors that are not amenable to experimentation in more complex organisms.

To understand how GSH acts in the hydra, one needs to know a little of the biology and natural history of the animal. A hydra is shaped like a two-ply hollow tube, about 8 by 1 mm when extended, made up of both outer (ectodermal) and inner (endodermal) epithelial layers. At the posterior end of the tube is a basal disk with which the hydra usually attaches to a surface, and at the anterior end is a mouth surrounded by a ring of tentacles (Fig. 1 A). The tentacles are armed with many nematocytes, one of the seven basic cell types of the hydra. The nematocytes contain nematocysts (stinging capsules), which start the feeding process by everting a coiled tube, which pierces and wounds the prey. The reduced glutathione (GSH) oozing out of the wounded prey initiates a specific feeding behavior in hydra.

Feeding in hydra, i.e., the capture and ingestion of food, consists of a number of separate steps: (i) A prey organism that accidentally blunders into an outstretched tentacle is captured, wounded, and poisoned through the action of the deadly ne-matocysts that line the tentacle; consequently, the wounded prey leaks body fluids containing reduced glutathione into the aquatic environment; (ii) following capture of the prey, the tentacles contract toward the mouth, and the mouth opens; (iii) on contact with the mouth, the food is ingested. The second step (hereafter called the feeding response), that is, the contraction of the tentacles toward the mouth and the opening of the mouth, has been shown to be controlled specifically by the presence of trace amounts of reduced glutathione in the fluids emanating from the wounded prey (Loomis 1955). That crude extracts of hydra prey elicit a feeding response in hydra had been known (Ewer 1947). Once GSH was identified as an activator of the feeding response, our early work on glutathione receptors focused on the mechanism by which reduced glutathione triggered the glutathione receptor of the hydra to activate the coordinated movements involved in the hydra's feeding response.

A casual analysis of the feeding behavior of hydra might suggest that the mechanisms by which they detect chemicals in potential food are analogous to those observed in taste and olfaction in higher terrestrial organisms. Yet on closer examination, these chemically evoked behavioral and physiological responses may show more similarity to hormonal and neurotransmitter systems of higher forms. Questions arising from categorizing chemosensory phenomena cannot be put off as simply a matter of semantics. The mere labeling of a phenomenon often determines how the problem will be investigated and funded, and by whom. The experimental approach taken usually depends upon whether the investigator is a psychologist or a biochemist, an ethologist or a physiologist. In the case of the lower invertebrates, the demarcation between behavior and physiology becomes gossamer-thin, because many lower invertebrates lack endocrine systems, well-defined nervous systems, and structures usually associated with more highly organized animals.

That the feeding behavior of hydra and of some other cnidarians is controlled by specific exogenous chemicals points out some of the hazy areas of categorizing chemosensory phenomena. On the other hand, such research also gives insight into the evolution of receptors and sensory integration and into the mechanism of activation of receptors to specific chemicals (see section 8 of this chapter). Although chemical activation of cellular receptors most likely evolved in single-cell organisms, it is in the lower metazoans that these chemical signals were first transduced from cell to cell to effect a coordinated organismal response.

In our research, we have focused primarily on the common feature of all chemosensory systems—the activation of a receptor by a specific molecule. In this chapter I describe (i) the discovery of the activation by GSH of the feeding response in hydra, (ii) the quantification of the GSH-activated feeding response in vivo, (iii) the properties of the feeding response as determined from those measurements, (iv) the kinetics of the GSH-activated feeding response, (v) the conformation of GSH at the receptor site, (vi) other behavioral responses in cnidarians activated by GSH or by specific amino acids, and (vii) a view on the evolution of receptor sites to peptides and amino acids in higher organisms.

2. DISCOVERY OF THE ACTIVATION BY GSH OF THE FEEDING RESPONSE IN HYDRA

My postdoctoral mentor, the late W. F. Loomis, discovered that hydra's feeding behavior was activated by reduced glutathione while he was "taming" the hydra, as he would say, to make it a laboratory animal subject to systematic investigation. Loomis, trained as a physician, and as a biochemist with Fritz Lipmann and David E. Green, first brought a hydra into his laboratory in 1953. In his first major publication on hydra (Loomis 1954), he described a method for growing hydra in the laboratory. For the hydra's food, he gave them live nauplii of the brine shrimp, *Artemia salina*. In an attempt to simplify the growing of hydra in the laboratory, he exposed a hydra to dilute freshly prepared extracts of brine shrimp. To his surprise, he observed the hydra's tentacles to writhe and contract and its mouth to open and remain open (Fig. 1 C, D). Similar behaviors had been described earlier by Ewer (1947).

Loomis noticed these behaviors in hydras exposed to very dilute solutions of the extract and concluded that some substances present in low concentrations were responsible for this induced behavior, which resembled the process of feeding. He found further that the "feeding activator" was heat-stable, dialyzable, and oxidizable (Loomis 1955). Hence he concluded that a small autooxidizable substance was responsible for the observed behaviors, and commenced to make a list of possible candidates. On his list was reduced glutathione.

As chance would have it, during this period Loomis was visited by one of the few other researchers on the hydra at that time, Dr. Helen Park of NIH. Park, who was studying the hydra's response to radiation, once treated hydra with solutions of glutathione and also noticed the strange writhings of the tentacles and mouth opening. When Loomis began to hone in on the small oxidizable molecule, Park's comments got him to focus on reduced glutathione.

Having found that commercially available reduced glutathione activated what appeared to be a feeding behavior in hydra did not satisfy Loomis, because commercial GSH is isolated from yeast extract. Because the GSH seemed to be active in such small amounts, e.g., concentrations of 10^{-5} M, Loomis presumed that perhaps another small molecule from yeast extract might have contaminated the GSH and thereby caused the observed feeding behavior. He eliminated that possibility when he obtained a trace of synthetic GSH from the Nobel Laureate Vincent DuVigneaud, and found that the synthetic tripeptide also activated the feeding response (Loomis 1955).

A question could be asked, however: although pure synthetic GSH stimulated the feeding behavior of hydra, in nature was it GSH that was responsible for activating the response, or another substance? To answer this question, I demonstrated that by adding glutamic acid, shown later to be a competitive inhibitor of the GSH-activated response (Lenhoff and Bovaird 1961), to extracts of the hydra's prey, the feeding response could be inhibited. Added GSH could reverse the inhibition caused by glutamate (Lenhoff 1961a).

3. QUANTIFICATION OF THE GSH-ACTIVATED FEEDING RESPONSE IN VIVO

Despite that dramatic discovery that GSH activated the feeding response of hydra, research into the mechanism of activation required that a means be determined to quantify the degree of the response to the concentration of GSH and to various other factors influencing cells. To develop such a quantitative assay of the feeding response to GSH became my primary objective. It would have been nice to have had a response that could be measured objectively and simply, either electrophysiologically or with some instrument. To the contrary, however, the assay we developed was based upon our knowledge of the biology of the hydra. We were gratified that the in vivo assay we developed allowed us to explore even some subtle physicochemical parameters of the interaction of the tripeptide GSH at its receptor site on the hydra.

3.1 Description and Measurement of the Feeding Response

The following are the main features of the assay we developed: A hydra in the absence of glutathione is shown in Fig. 1 A: the mouth is closed, and the tentacles are outstretched and nearly motionless. After glutathione is added, the tentacles begin to writhe and sweep inwards toward the central vertical axis of the animal (Fig. 1 B). Next, the tentacles bend toward the mouth, and the mouth opens (Fig. 1 C). Shown in this composite drawing (Fig. 1 C) are the various positions that a tentacle takes before contracting. These movements, culminating in mouth opening, usually all take place within one-half minute. Figure 1 D shows how a hydra looks during the greater portion of the feeding response, its mouth open wide and the tentacles in various phases of contraction. Frequently, the tips of the tentacles are observed within the hydra's mouth, as shown in Fig. 1 C and D.

To assure objectivity and to obtain quantitative data on the feeding response, we measured the times that elapsed between the moment at which the hydra were placed in the glutathione solution and the moments at which the mouths initially opened (initial time t_i) and finally closed (final time t_f). The length of time that a mouth was open in response to glutathione was therefore, $t_f - t_i$ (Lenhoff 1961a). Each value for $t_f - t_i$ that we reported is the mean for a group of five animals.

We have found that, using carefully controlled conditions for growing, maintaining, and selecting the hydra and for carrying out the experiments, each set of animals used in an experiment usually responded to glutathione in near-synchrony and gave $t_f - t_i$ values having small standard deviation. The initial time t_i probably represents the time it takes for glutathione to combine with the receptor and for all of the physiological events to occur in the interim between activation of the receptor and the opening of the mouth. Thus, $1/t_i$ would represent the rate at which these events take place. The time that the mouth remains open, $t_f - t_i$, is interpreted to represent the time that it takes for some limiting substrate to be consumed. Alternatively, it could indicate the time required for an inhibitor to be released and become effective. The time $t_f - t_i$ can be shortened or lengthened by varying the temperature. Hence, some event(s) affecting the duration of the feeding response may be thermochemical (Lenhoff 1961b).

The study of the activation of the feeding behavior of hydra by GSH provides the investigator with important advantages. First, the molecule GSH itself is sufficiently complex, though not too large, to allow one to make a significant number of analogs for determining the conformation(s) and structure–activity relationships of GSH necessary to activate the receptor. Second the very nature of the hydra itself simplifies matters for the experimenter: (i) the animal has a simple tissue-level structure; (ii) the receptor is on the surface of the outer epithelium; (iii) the biological response can be readily quantified; (iv) pure clones of the animals can be easily grown in the laboratory in kilogram quantities (wet weight); and (v) the fluid environment (medium) surrounding the receptor can be accurately controlled within a pH range of about 4–8 and over wide ranges of ion concentrations.

Thus, by quantifying the extent of the hydra's feeding response to a wide range of environmental factors, such as pH and the concentration of various ions, it is possible to deduce many physicochemical parameters of both the receptor and the activator. For these reasons, special attention was given to development of this behavioral assay and to having animals that could respond to GSH reliably and in virtual synchrony.

4. SOME PROPERTIES OF THE FEEDING RESPONSE

Using such an assay, we have found, for example, that both environmental calcium ions (Lenhoff and Bovaird 1959) and sodium ions (Asbill, Danner, and Lenhoff, personal communication) are absolute requirements for activation by glutathione to take place. Environmental potassium ions, on the other hand, are inhibitory (Lenhoff 1965). All of these effects can be easily reversed by either adding or removing the respective ions. Perhaps these ions act by affecting the cellular membrane potential of the hydra.

An indication of the relative speed at which the equilibrium between glutathione and the receptor is attained was determined by means of a simple set of experiments. Hydra placed in a glutathione solution would open their mouths within a minute, and would close their mouths within a minute after the glutathione was removed (Lenhoff 1961b). These same animals could repeat this opening and closing sequence many times during an hour (Lenhoff 1961b). Hence we can conclude (i) that glutathione has to be present constantly in the solution, and thus at the receptor site, in order for a response to take place, and (ii) that the equilibrium between glutathione and the receptor is rapidly attained.

Other in vivo experiments indicate that glutathione is not consumed during activation of the feeding response, that the response eventually stops for some intrinsic reason, and that following this cessation there is a recovery period for the animal to respond fully to glutathione again (Lenhoff 1961b).

5. KINETICS OF THE GSH-ACTIVATED FEEDING RESPONSE

Early experiments suggested that, on the surface of the hydra's outer epithelium, there were receptor sites that become activated only in the presence of GSH. A saturable receptor was indicated by analysis of plots of the duration of the feeding response against the concentration of GSH to which the hydra was exposed (Lenhoff 1961a). Analysis of the resultant curve with concepts borrowed from enzymology suggests that there is a receptor that is saturated and gives a maximal response at

GSH concentrations of 5×10^{-6} M and greater. A maximal response is considered analogous to the maximal velocity of an enzyme-catalyzed reaction; both occur during saturation of an active site.

5.1 Determination of the Apparent Dissociation Constant

The assumptions made in determining the apparent dissociation constant K_A between the activator (A) and the receptor (R) have been reported elsewhere (Lenhoff 1965, 1968). The effect of the activation is signified by e, and the maximum effect by e_M. The equation derived,

$$\frac{[A]}{e} = \frac{1}{e_M}[A] + \frac{K_A}{e_M},$$

is analogous to the second form of the Lineweaver–Burk plot, the equation developed by Beidler (1954) for mammalian taste chemoreception, which is a form of the Langmuir adsorption isotherm. This equation is useful in analyzing chemoreception phenomena because it minimizes deviations in individual animal responses that occur at very low levels of activator. Previous data (Lenhoff 1969) have shown that this equation can be used to interpret the plot of $[A]/e$ against $[A]$; we obtained straight lines at most glutathione concentrations (Fig. 2). From such plots, we can determine, for example, at pH 7, an apparent K_A of 10^{-6} M. Such a low K_A is meaningful

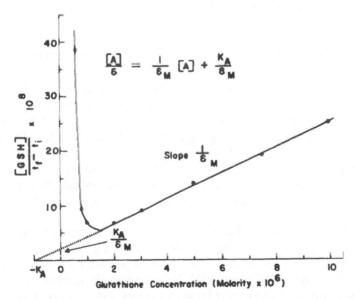

FIG. 2. Plot for determining constants of the combination of GSH with its receptor (Lenhoff 1965, 1969).

from at least three viewpoints: (i) it indicates a high affinity of the receptor for glutathione, (ii) concentrations around 10^{-6} M are well within the physiological range expected under natural conditions of feeding, and (iii) this constant provides a means of characterizing the receptor, that is, the glutathione receptor of *H. littoralis* may be said to have an apparent dissociation constant of 10^{-6} M under the given conditions. The constant is characteristic of the receptor and remains nearly the same whatever the nutritional state of the hydra (Lenhoff 1961a,b). Similarly, experiments in which the buffer anion is varied alter the maximal response but not the dissociation constant (Lenhoff 1969).

5.2 In Vivo Determination of the pH Profile of the Glutathione Receptor

Changes in K_A with pH can be used to determine the pK's of the ionizable groups on glutathione or at the receptor site that are involved in the combination with glutathione. The pK measurements of ionizable groups on the glutathione receptor were made by a means analogous to those used in enzymology in determining the pK's of ionizable groups at the active sites of enzymes. For our purposes, we needed an equilibrium equation similar to Dixon's (Dixon 1953; Dixon and Webb 1964) for enzymes, which would take into account the influence of pH on the dissociation constant. This modified equation (Lenhoff 1969) assumed that if the activator (A), the receptor site (R), or the activator–receptor complex (AR) ionizes, then each component in the expression for equilibrium (A, R, AR) equals its concentration multiplied by a coefficient that is a function of pH. For example, if the activator ionized, then the total concentration of the activator, A_t, would be A times the pH function of A, or f_a (pH). The logarithmic form of the equation is

$$pK_A = pK_A^\bullet + \log f_{ar}(\text{pH}) - \log f_r(\text{pH}) - \log f_a(\text{pH}).$$

Here pK_A refers to the negative logarithm of the dissociation constant of AR, whereas p$K_A^\bullet$ is the same constant if none of the components has ionic groups; if no component ionizes, then pK_A and p$K_A^\bullet$ are equal. The derivation of this equation is explained elsewhere (Lenhoff 1968). The foregoing equation indicates that a plot of pK_A against pH will consist of a series of straight lines joined by short curved parts. The data indicate that this relationship holds true for the glutathione–hydra system (Fig. 3). The results (Lenhoff 1969) followed almost exactly the predictions from the modified Dixon equations. The following interpretations were made (Lenhoff 1969):

1. The concave downward inflections at pH 4.6, 4.8, 6.5, and 7.6 represent pK's of ionizable groups at the receptor site; these pK's probably do not represent ionizable groups of glutathione, which have pK's either below 4 (2.1 and 3.5) or above8 (8.7 and 9.6) (Wieland 1954). If the receptor site is protein, then the pK's may represent two β carboxyls of peptide aspartic acid (or γ carboxyls of peptide glutamic acid), an imidazole group, and a terminal α amino group, respectively.
2. The horizontal lines indicate pH values that do not affect the combination of glutathione with the receptor site.

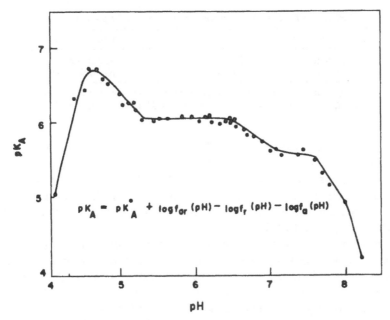

FIG. 3. Effect of pH on the dissociation constant K_A of GSH and its receptor (Lenhoff 1965, 1969).

3. The quenching of the charges (Dixon and Webb 1964) at about pH 4 and 8 indicate that receptor-site groups having pK's of 4.6 and 7.6 may be associated with complementary charged groups of glutathione.
4. From these three findings, we conclude that ionizable groups of the receptor site participate in binding glutathione.

6. CONFORMATION OF GSH AT THE RECEPTOR SITE

6.1 Earlier Work on the Active Structure of Glutathione

To determine the structure of GSH present at the receptor site, it was necessary to quantify the ability of different structural analogs of GSH either to activate or to inhibit a response. The aforementioned bioassay experiments were used to determine the effectiveness (K_A and e_M) of synthetic agonists (Lenhoff 1961b). The relative ability of analogs that bind but do not activate, i.e., antagonists, was determined by measuring their ability to inhibit competitively the activity of GSH in eliciting the feeding response (Lenhoff and Bovaird 1961).

Data from these earlier investigations established that:

1. The thiol is not required for activation, because ophthalmic acid (γ-Glu-Abu-Gly), norophthalmic acid (γ-Glu-Ala-Gly), and S-methylglutathione (S-Me-GSH) also activated the response (Cliffe and Waley 1958; Lenhoff and Bovaird 1961).

TABLE 1. *Effect of GSH analogs with substitutes for glycine moiety[a]*

Good activators	Moderate activators	Poor activators
Tyrosine	Proline	Glycinamide
Leucine	Glycylalanine	Aminoisobutyric acid

[a]To avoid complications resulting from the presence of the thiol, these analogs have α-aminobutyric acid substituted for cysteine in position 2.

2. Activation of the response requires the intact tripeptide backbone of glutathione, because the just-mentioned tripeptide analogs activated feeding, whereas single or combinations of amino acids and dipeptides did not. Furthermore, a number of tripeptide analogs with large and charged substituents at the sulfhydryl group did not activate, but bound sufficiently to be competitive inhibitors (Lenhoff 1961b; Lenhoff and Bovaird 1961).
3. The receptor has a high affinity for the γ-glutamyl part of the tripeptide because glutamic acid and glutamine were the only amino acids to show competitive inhibition (Lenhoff and Bovaird 1961), and the tripeptide asparthione (β-Asp-Cys-Gly) did not initiate the response (Loomis 1955; Lenhoff and Bovaird, 1961).
4. The α amino of glutathione is probably required for association of glutathione with the receptor (Lenhoff 1961a; Lenhoff and Bovaird 1961).

A summary of research on the specificity of the glutathione receptor of *H. attenuata* is presented in Tables 1–3 (Lenhoff 1981). This work involves the use of over 70 analogs, many of which were synthesized specifically for this project (Cobb et al. 1982).

6.2 Position-3 (Glycine) Analogs

Analogs of glutathione were synthesized having substituents for glycine in the 3 position. The results (Table 1) show that when glycine was replaced by either tyrosine or leucine, the resultant tripeptide activated a perfectly good feeding response. Tripeptides with the other substituents listed also initiated feeding behaviors, but of

TABLE 2. *Effects of analogs of GSH with substitutes for the cysteine moiety*

Strong activators	Moderate activators	Inhibitors
S-Me-cysteine	Alanine	Cysteine-S-S-G
Aminobutyric acid	Serine	Cysteine-S-(N-ethylsuccinimido)
Valine	Aminoisobutyric acid	Tyrosine
Norvaline		Phenylalanine
Leucine		S-acetylcysteine
		Cysteine sulfinate
		Cysteine sulfonate
		Glycine[a]

[a]γ-Glutamylglycylglycine was only a weak inhibitor.

TABLE 3. *Effect of glutamic acid and its analogs on response activated by GSH*

Analog class	Strong inhibitor	Moderate inhibitor	Noninhibitory
A	L-Glutamic acid	—	D-Glutamic acid
B	γ-Me-glutamic acid	β-Me-glutamic acid	α-Me-glutamic acid
C	—	Isoglutamine	γ-Aminobutyric acid
D	—	Glutamine	Norvaline
E	α-Aminoadipic acid	—	Aspartic acid
	—	—	Proline
	—	—	Pyroglutamic acid
F	—	—	Glutamic acid
	—	—	α-Ketoglutaric acid
	—	—	β-Aminoglutamic acid
	—	—	N-methyl glutamate
G	Trans-(Cyclopropyl)glycine	—	cis-(Cyclopropyl)glycine

lesser to barely detectable degrees. We tentatively concluded from these experiments that the nature of the side chain of the 3-position amino acid is not critical for the tripeptide to activate, but that the presence of a charge at that end of the peptide is important. It is of interest that although the hydra GSH receptor responded to tripeptides having either tyrosine or leucine in the 3 position, the γ-glutamyltranspeptidase of hydra did not bind or react with those analogs (Danner et al. 1976).

6.3 Position-2 (Cysteine) Analogs

Investigations of these analogs showed that many substituents in the 2 position gave peptides with about the same activity as GSH (Table 2). An even greater number of analogs, however, bound sufficiently well to behave as antagonists (competitive inhibitors), but were not able to activate a feeding response.

From these findings, it is apparent that neither the reducing nor the hydrogen-bonding properties of the thiol group of cysteine are needed to activate the receptors. The size and hydrophobicity of the 2-position side chain, however, may play a role in affecting the activating properties of tripeptide analogs. For example, quantitative examination of the data summarized in Table 2 shows a corresponding loss in activation and binding potency with norvaline > α-amino-n-butyric acid > alanine > glycine. Within this latter series of tripeptides, the potency probably does not depend on size alone, but on the degree of lipophilicity and conformation flexibility inherent in the substituted groups. It is interesting that γ-glutamylglycylglycine did not activate, but instead was a weak antagonist. Hence, it is not simply the tripeptide structure that is required for activation. Possibly the substitution of glycine in 2 position affects the conformation of the tripeptide so it will not activate. Addition of a small side chain, such as an ethyl or a methyl, to that glycine might lead to a conformation of the tripeptide more compatible with the receptor.

6.4 Conformation of the 1 Position (γ-Glutamyl) Needed for Activation

Because glutamic acid is the only amino acid component of the tripeptide that acted as an antagonist (Lenhoff and Bovaird 1961), thereby inhibiting feeding response in the presence of GSH, we were able to develop a reliable bioassay for illuminating the conformation of the γ-glutamyl moiety at the receptor site. We determined the concentration of glutamic acid or its analogs that inhibited the response activated by concentrations of GSH close to that of the K_A. With this assay, approximately 2×10^{-5} M glutamic acid inhibited the response by 50%. We interpreted this inhibition as a reflection of the glutamic acid's ability to bind at the site normally occupied by the γ-glutamyl moiety of GSH.

Examination of the data in Table 3 shows that: (i) The asymmetric γ carbon was important for binding (inhibition became weaker as the extra methyl was moved from the γ to the β to the α carbon); (ii) the α amino group had to be present and in the α carbon (inactive as inhibitors were glutamic acid, α-ketoglutaric acid, N-methylglutamic acid, and β-aminoglutamic acid); (iii) the α carboxyl must be present (GABA was noninhibitory, and isoglutamine was only partially inhibitory); (iv) the three charged groups must attach to the receptor (L-glutamate inhibited, whereas D-glutamate did not); (v) the three charged groups must be a specified distance from each other (aspartic, proline, and pyroglutamic acid did not inhibit, whereas α-aminoadipic acid was a good inhibitor).

Our investigation of two glutamic acid analogs having cyclopropane rings to restrict rotation around the β-carbon–γ-carbon bond made it possible to determine the approximate conformation of the glutamyl moiety of GSH at the receptor site. The analog trans-α-(carboxycyclopropyl)glycine (Tcg) proved to be as effective an inhibitor as is glutamic acid, whereas the cis form (Ccg) did not inhibit.

As shown in Table 3, four compounds were selected, from which we calculated the distances between their atoms. The strong inhibitors, glutamic acid and Tcg, were selected to demonstrate the widest possible ranges of interatomic distances possible. The noninhibitors, aspartic acid and Ccg, which also had the three active ionic groups, were selected to represent interatomic distances that should be excluded from the calculations. From this information, it was possible to approximate the active distances between the atoms while the analogs fitted effectively into the receptor site. The results of these calculations are shown in Table 4.

Further calculations were carried out by constructing plots similar to the kind developed by Ramachandran. These calculations show the potential energies of glutamic

TABLE 4. *Approximate distances between atoms in glutamic acid while it binds to the receptor*

Stand	Distances (Å) of probable active range
α C → δ C [a]	2.89 → 3.87
α C = 0 → δ C	4.38 → 4.97
α N → δ C	4.38 → 4.97

[a]The δ C refers to the γ carboxyl of glutamic acid.

acid and of Tcg as functions of the rotations between α-carbon–β-carbon bonds and the β-carbon–γ-carbon bonds. The area of overlap suggests that most likely torsional angles are $(120 \pm 30)°$ for the α–β bond and $180°$ for the β–γ one. From these calculations and experiments, we are able to approximate the conformation for glutamic acid acting as an antagonist, as well as the conformation of the γ-glutamyl moiety of GSH while activating the glutathione receptor site.

7. OTHER BEHAVIORAL RESPONSES IN CNIDARIANS ACTIVATED BY GSH OR BY SPECIFIC AMINO ACIDS

Because only GSH activates the feeding behavior of all species of hydra that have been tested thus far, as well as behavioral responses in a tick (Galun and Kindler 1965), a hagfish (Døving and Holmberg 1974), and a snail (Kater and Rowell 1973), there is no reason to believe *a priori* that something unique about this molecule makes it the only suitable activator of feeding among the cnidarians. A survey of over 25 cnidarians showed that (i) other compounds could serve as specific activators of the feeding behavior, and (ii) in some cases, animals responded to two or more feeding activators in diverse, but specific, ways.

Animals tested from every class and most families of the Cnidaria exhibit a feeding response to either one or a few small molecules (Table 5). Those found most commonly to initiate a feeding response are the tripeptide GSH and the imino acid proline. In the Hydrozoa, the feeding response of each organism investigated was

TABLE 5. *Chemical activators of feeding in the cnidaria*

Type of organism	Activator
I. Hydrozoans	
A. Hydroids	
1. Five species of *Hydra*	GSH
2. Four species of colonial marine hydroids (without theca)	Proline
B. Siphonophores	
Two species (including Portuguese man-of-war)	GSH
II. Anthozoans	
A. Sea anemones	
1. *Anthopleura*	GSH
2. *Boloceroides*	Valine
3. *Actinia*	Glutamate
4. *Haliplanella*	Leucine
5. *Calliactis*	GSH, proline
B. Colonial anemones	
1. *Palythoa*	Proline and/or GSH
2. *Zoanthus*	GSH
C. Corals	
1. Six species	Proline or GSH
2. One species	4 amino acids
III. Scyphozoa	
One species (large jellyfish)	20 amino acids, GSH, glyciglycine

induced by a single specific compound (Lenhoff 1974; Lenhoff and Heagy 1977). Proline is especially prevalent as an activator among the athecate colonial marine hydroids. For example, all members of these groups thus far tested, i.e., *Cordylophora* (Fulton 1963), *Pennaria* (Pardy and Lenhoff 1968), *Tubularia* (Rushforth 1976), and *Proboscidactyla* (Spencer 1974), responded only to proline. All hydras tested responded only to GSH (Loomis 1955; Lenhoff 1974), and the only other hydrozoans tested, two siphonophores, also responded to GSH (Lenhoff and Schneiderman 1959; Mackie and Boag 1963).

In the Anthozoa, we see three trends. In general, although more than one compound may elicit feeding behaviors, the animals exhibit varying ranges of specificity. For example, the most specificity is seen among the sea anemones: *Boloceroides* responds primarily to valine (Lindstedt, Muscatine, and Lenhoff 1968), *Anthopleura* to GSH (Lindstedt 1971a), *Haliplanella* to leucine (Lindstedt 1971b), *Actinia* to glutamic acid (Steiner 1957), and *Calliactis* to GSH (Reimer 1973). The specificity broadens with the colonial anemones: whereas *Zoanthus* responds primarily to GSH (Reimer 1971a), *Palythoa* responds to high concentrations of either GSH or proline or to low concentrations of these two activators acting synergistically (Reimer 1971b). Lastly, corals seem to respond best to proline alone or GSH alone, as well as to numerous other amino acids at higher concentrations (Mariscal and Lenhoff 1968; Mariscal 1971; Lehman and Porter 1973).

Chrysaora, the only large jellyfish tested, seems to respond to GSH and to a large number of amino acids (Loeb and Blanquet 1973). More species need to be tested before making any generalizations about this group.

I am not necessarily convinced, however, that all the compounds reported in Table 5 are true activators of feeding. Likewise, I also feel that other feeding activators for those organisms listed may yet be detected. Lenhoff and Heagy (1977) posed three criteria for a compound to be accepted as a natural activator of a feeding behavior: (i) the compound should be active in low concentrations, i.e., 10^{-4} M or less; (ii) sufficient analogs of the presumed activator should be tested to show that the receptor has some degree of specificity; and (iii) an analog that is a nontoxic competitive inhibitor should reversibly inhibit the response activated by natural tissue extracts. Thus far, in only one instance, that of hydra, have all these criteria been met. The next most thoroughly studied case is Fulton's (1963) work showing that the proline activation of feeding in *Cordylophora lacustris* could take place in less than 10^{-4} M and that the specificity resided in the imino region of a heterocyclic α-imino acid, which is neither substituted nor unsaturated in such a way as to affect the imino acid group.

Nonetheless, in addition to the cases where some anthozoans respond to either GSH or proline, there are two other unusual cases in which cnidarians respond to two substances. In both cases, each substance initiates a separate response. Yet in the presence of both substances, a very specific behavioral pattern is initiated.

Lindstedt (1971a) describes an unusual case in which two phases of the feeding response of the sea anemone *Anthopleura elegantissima* are controlled by different chemical activators. Asparagine controls the contraction and bending of tentacles that brings food to the mouth; reduced glutathione controls the ingestion of food once it contacts the mouth. A complete feeding response occurs only when both chemical activators are present.

An even more complex case occurs in hydra. In addition to the GSH receptor, there also exists a receptor to tyrosine on the surface of cells lining the hydra's gut. When hydra are activated by GSH present in the external environment and by tyrosine present in the gut, the animal exhibits a "neck response," that is, a constriction of the upper one-third of the body tube. These neck constrictions apparently allow the hydra to retain previously ingested food in the gut while swallowing newly captured prey. No other natural amino acid, including phenylalanine, could substitute for tyrosine. Analogs of tyrosine having either the α amino or the α carboxyl blocked were inactive (Blanquet and Lenhoff 1968).

From these experimental results we conclude that in addition to its external glutathione receptor, hydra has an enteroreceptor specific for tyrosine. The hydroxyl, the α amino, and the α carboxyl groups must all be present in order for the amino acid to be active (Blanquet and Lenhoff 1968).

The existence in hydra of two chemoreceptor systems that must act in harmony represents, to our knowledge, the first report of two integrated, chemically mediated responses in the lower invertebrates. This system differs from the asparagine–glutathione system in the sea anemone *Anthopleura* (Lindstedt 1971a) in which the molecules act in two sequential steps to activate feeding.

8. ON THE EVOLUTION OF RECEPTOR SITES TO PEPTIDES AND AMINO ACIDS IN MORE COMPLEX ORGANISMS

The view that an organism evolves receptor sites in response to some ubiquitous molecules that are adapted to special tasks seems especially applicable to the feeding activators of cnidarians. So long as a molecule is widely present in prey organisms and has properties distinguishing it from closely related substances, it might serve as a feeding activator. One might therefore, expect some coelenterates to have evolved receptor sites for compounds other than glutathione emitted by the captured prey. Such is the case. As described in the preceding section, a number of different amino acids have been shown to effect feeding in a wide variety of cnidarians.

After a receptor site for a specific compound had been acquired during evolution, further modification of the receptor site itself might have occurred. For example, Fulton (1963) has suggested that the evolution of a receptor site for glutathione into one for the α-imino acid proline may have proceeded by means of slight structural changes in the receptor site. He postulated this because one of the possible cyclized forms of glutathione in solution is close in structure to an α-imino acid. And, since proline is also present in the fluids released from prey organisms, the change in structure of the receptor site was not disadvantageous to *Cordylophora* but, under some circumstances, advantageous, and so persisted. Perhaps the zoanthid *Palythoa psarnophilia*, which responds synergistically to a mixture of glutathione and proline (Reimer 1971b), and the coral *Cyphastrea*, which responds to either glutathione or proline (Mariscal and Lenhoff 1968), represent forms retaining intermediate forms of the receptors during the evolution of a purely proline receptor from a glutathione one.

I find it reasonable to suppose that receptor sites for amino acids or glutathione in hydra, and for transmitter substances and peptide hormones in higher organisms, may have evolved in a similar fashion. But the intriguing question is, "Did chemical receptors to environmental compounds, such as those found in lower invertebrates, give rise to the chemical receptors (olfactory, gustatory, hormonal, or transmitter) in higher forms?" For example, because dopamine and norepinephrine are formed from tyrosine, would it not be efficient for organisms to retain and utilize modifications of a primitive tyrosine receptor similar to the one found in hydra? Would not the same argument apply to the evolution of receptors for the neurotransmitters glutathione, glutamic acid, and glycine? Such an argument could be taken back one more step into a consideration of chemical receptors for compounds stimulating pinocytosis in single cells. These receptors may truly be the evolutionary precursor of many receptors in higher forms.

Obviously, such evolutionary questions are difficult to answer; one possible approach would be to purify the receptors in question and to see if there is a pattern of conservation of the primary structure of those proteins. Nevertheless, perhaps there may be an argument in support of this evolutionary view, at least in the case of peptide receptor sites. Both lower and higher organisms possess receptor sites for specific peptides, which, when activated, lead to contraction responses. Would it not seem simpler for organisms, during evolution, to have modified existing receptor–effector systems to perform new tasks than to have developed a completely new receptor–effector system?

The cnidarians appear to offer a pivotal point in the evolution of chemical receptor systems into the more specialized ones of higher forms. The chemoreceptors of cnidarians always face the environment, i.e., the external fluids or the fluids in the gut, because these diploblastic animals are composed of basically two epithelial layers of cells separated by a thin acellular mesolamella. Thus, it is possible that as triploblastic animals with organ systems evolved, existing surface chemoreceptors of ancestral diploblastic metazoans became internalized and took on new functions, responding to circulating substances rather than to substances released from captured prey. Perhaps it is at this point in evolution that chemoreceptors and their corresponding effector systems started on the pathway toward specialization.

9. CONCLUSIONS

In this chapter I have tried to: (i) present a historical perspective on how research on glutathione receptors in hydra began; (ii) provide an example of how quantitative data describing the physical–chemical properties of a receptor surface and the conformation of a peptide at an active receptor site can be determined from carefully controlled behavioral observations; (iii) suggest a means whereby receptors to small molecules and peptides may have evolved.

A reader may rightfully ask, "If the hydra system is so amenable to experimentation, why don't current researchers use hydra for their investigations on the mechanism of activation of receptors?" There could be a number of valid reasons, such as difficulty of growing hydra in the laboratory, or the small amount of tissue they offer for biochemical studies. Actually, those objections are not valid; methods for growing and harvesting large numbers of hydra have been described (Loomis and Lenhoff 1956; Lenhoff and Brown 1970; Lenhoff 1983).

In my opinion, the reason animals like hydra have not been used for modern research in receptor studies stems to some degree from the anthropocentricity of researchers and granting agencies. To their way of thinking, the receptors of hydra to glutathione may explain some interesting natural history, but have little to do with understanding receptors in humans, or with advancing the progress of medicine. Such appears to be the attitude in much of the current era of biological research, and most likely will remain so as more and more biologists study fewer and fewer phenomena.

Research on the whole animal, and especially on animals presently not considered accessible for laboratory research, may hold many secrets that remain to be unlocked. If we seek only to explain further the processes that we already know of, we may be missing many exciting opportunities in that vast unknown. At least that is how I was challenged when I left classical enzymology in 1954 to join Loomis when he first "tamed" the hydra and observed that its feeding behavior was controlled by GSH released from its captured prey.

ACKNOWLEDGMENTS

I dedicate this paper to the genius and memory of my mentor, W. F. Loomis, M.D., who first discovered the GSH-activated response in hydra, and who, in 1955, helped break down the barriers that existed between behavioral zoology and biochemistry. This research is a collaborative effort of some of my outstanding colleagues—Drs. Wyrta Heagy, Garland R. Marshall, Melanie H. Cobb, and Jeanne Danner. We thank the National Institutes of Health, the National Science Foundation, the Howard Hughes Medical Institute, and the Campbell Soup Company for research support.

REFERENCES

Beidler, L. M. 1954. A theory of taste stimulation. *Journal of General Physiology* 38:133–9.

Blanquet, R. S., and Lenhoff, H. M. 1968. Tyrosine enteroreceptor of hydra: its function in eliciting a behavior modification. *Science* 159:633–4.

Cliffe, E. E., and Waley, S. G. 1958. Effect of analogues of glutathione on the feeding reaction of hydra. *Nature* 183:804–5.

Cobb, M. H., Heagy, W., Danner, J., Lenhoff, H. M., and Marshall, G. 1982. Structural and conformational properties of peptides interacting with the glutathione receptor of hydra. *Molecular Pharmacology* 21:629–36.

Danner, J., Lenhoff, H. M., Cobb, M. H., Heagy, W., and Marshall, G. R. 1976. *Biochemical and Biophysical Research Communications* 73:180–6.

Dixon, M. 1953. The effect of pH on the affinities of enzymes for substrates and inhibitors. *Biochemical Journal* 55:161–70.

Dixon, M., and Webb, E. C. 1964. *Enzymes*, 2nd ed. Academic Press, New York, 154–65.

Døving, K. B., and Holmberg, K. 1974. A note on the function of the olfactory organ of the hagfish. *Myxine glutinosa. Acta Physiologica Scandinavica* 91:430–2.

Ewer, R. F. 1947. On the functions and mode of action of the nematocyst of *Hydra. Proceedings of the Zoological Society of London* 117:365–76.

Fulton, C. 1963. Proline control of the feeding reaction of *Cordylophora. Journal of General Physiology* 46:823–37.

Galun, R., and Kindler, S. N. 1965. Glutathione as an inducer of feeding in ticks. *Science* 147:166–76.

Kater, S., and Rowell, C. 1973. Integration of sensory and centrally programmed components in generation of cyclical feeding activity of *Helisoma trivolvis. Journal of Neurophysiology* 36:142–55.

Lehman, J. T., and Porter, J. W. 1973. Chemical activation of feeding in the Caribbean reef building coral *Montastrea cavernosa. Biological Bulletin (Woods Hole, Mass)* 145:140–9.

Lenhoff, H. M. 1961a. Activation of the feeding reflex in *Hydra littoralis*. I. Role played by reduced glutathione and quantitative assay of the feeding reflex. *Journal of General Physiology* 45:331–44.

Lenhoff, H. M. 1961b. Activation of the feeding reflex in *Hydra littoralis*. In *The biology of hydra and of some other coelenterates*, ed. H. M. Lenhoff and W. F. Loomis, 203–32. Coral Gables, FL: University of Miami Press.

Lenhoff, H. M. 1965. Some physicochemical aspects of the microenvironments surrounding hydra during activation of their feeding behavior. *American Zoologist* 5:515–24.

Lenhoff, H. M. 1968. Chemical perspectives on the feeding response, digestion and nutrition of selected coelenterates. In *Chemical zoology*, ed. M. Florkin and B. Scheer, 2:157–221. New York: Academic Press.

Lenhoff, H. M. 1969. pH profile of a peptide receptor. *Comparative Biochemistry and Physiology* 28:571–86.

Lenhoff, H. M. 1974. On the mechanism of action and evolution of receptors associated with feeding and digestion. In *Coelenterate biology: reviews and new perspectives*, ed. L. Muscatine and H. M. Lenhoff, 359–89. New York: Academic Press.

Lenhoff, H. M. 1981. Biology and physical chemistry of feeding response of hydra. In *Biochemistry of Taste and Olfaction*, ed. R. Cagan, 475–497. New York: Academic Press.

Lenhoff, H. M. 1983. Culturing large numbers of hydra. In *Hydra: Research Methods*, ed. H. M. Lenhoff, 53–62. New York: Plenum Press.

Lenhoff, H. M., and Bovaird, J. 1959. Requirement of bound calcium for the action of surface chemoreceptors. *Science* 130:1474–1476.

Lenhoff, H. M., and Bovaird, J. 1961. Action of glutamic acid and glutathione analogue on the hydra glutathione receptor. *Nature* 189:486–7.

Lenhoff, H. M., and Brown, R. D. 1970. Mass culture of hydra: An improved method and its application to other invertebrates. *Laboratory Animals.* 4:139–154.

Lenhoff, H. M., and Heagy, W. 1977. Aquatic invertebrates: Model systems for study of receptor activation and evolution of receptor proteins. *Annual Review of Pharmacology and Toxicology* 17:243–58.

Lenhoff, H. M., and Schneiderman, H. A. 1959. The chemical control of feeding in the Portuguese man-of-war, *Physalia physalia* L., and its bearing on the evolution of the cnidaria. *Biological Bulletin (Woods Hole, Mass.)* 116:452–60.

Lindstedt, K. J. 1971a. Biphasic feeding response in a sea anemone: control by asparagine and glutathione. *Science* 173:333–4.

Lindstedt, K. J. 1971b. Chemical control of feeding behavior. *Comparative Biochemistry and Physiology* 39:553–81.

Lindstedt, K. J. Muscatine, L., and Lenhoff, H. M. 1968. Valine activation of feeding in the sea anemone *Boloceroides. Comparative Biochemistry and Physiology* 26:567–72.

Loeb, M., and Blanquet, S. 1973. Feeding behavior in polyps of the Chesapeake Bay sea nettle, *Chrysaora quinquecirrha* (Desor, 1848). *Biological Bulletin (Woods Hole, Mass.)* 145:150–8.

Loomis, W. F. 1954. Environmental factors controlling growth in hydra. *Journal of Experimental Zoology.* 126:223–34.

Loomis, W. F. 1955. Glutathione control of the specific feeding reactions of hydra. *Annals of the New York Academy of Sciences* 62:209–28.

Loomis, W. F., and Lenhoff, H. M. 1956. Growth and sexual differentiation of *Hydra* in mass culture. *Journal of Experimental Zoology* 132:555–74.

Mackie, G. O., and Boag, D. A. 1963. Fishing, feeding and digestion in siphonophores. *Publication Station Zoological Napoli* 33:178–96.

Mariscal, R. N. 1971. The chemical control of the feeding behavior in some Hawaiian corals. In *Experimental coelenterate biology*, ed. H. M. Lenhoff, L. Muscatine, and L. Davis, 100–18. Honolulu: University of Hawaii Press.

Mariscal, R. N., and Lenhoff, H. M. 1968. The chemical control of feeding behavior in *Cyphastrea ocella* and some other Hawaiian corals. *Journal of Experimental Biology* 49:689–99.

Pardy, R. L., and Lenhoff, H. M. 1968. The feeding biology of the gymnoblastic hydroid, *Pennaria tiarella*. *Journal of Experimental Zoology* 168:197–202.

Reimer, A. A. 1971a. Feeding behavior in the Hawaiian zoanthids *Palythoa* and *Zoanthus*. *Pacific Science* 25:512–20.

Reimer, A. A. 1971b. Chemical control of feeding behavior in *Palythoa* (Zoanthidea, Coelenterata). *Comparative General Pharmacology* 2:383–96.

Reimer, A. A. 1973. Feeding behavior in the sea anemone *Calliactis polypus* (Forskal, 1775). *Comparative Biochemistry and Physiology* 44:1289–301.

Rushforth, N. B. 1976. Electrophysiological correlates of feeding behavior in the *Tubularia*. In *Coelenterate ecology and behavior* (G. O. Mackie, ed.). Plenum Press, New York, 729–38.

Spencer, A. N. 1974. Behavior and electrical activity in the hydrozoan *Proboscidactyla flavicirrata* (Brandt). *Biological Bulletin (Woods Hole, Mass.)* 146:100–15.

Steiner, G. 1957. Uber die chemische nahrungswahl von *Actinia equina* (L.). *Naturwissenschaften* 44:70–1.

Wieland, T. 1954. Chemistry and properties of glutathione. In *Glutathione*. ed. S. Colowick, A. Lazarow, E. Racker, and D. R. Schwartz, 45–59. New York: Academic Press.

Glutathione in the Nervous System
Edited by Christopher A. Shaw
Copyright © 1998 Taylor & Francis

3

Carrier-Mediated GSH Transport at the Blood–Brain Barrier and Molecular Characterization of Novel Brain GSH Transporters

Ram Kannan, Jian-R. Yi, Berislav V. Zlokovic, and Neil Kaplowitz

*Division of GI & Liver Diseases, and Department of Neurosurgery,
USC School of Medicine, Los Angeles, California 90033*

ABBREVIATIONS

BBB   blood–brain barrier
BSO   buthionine sulfoximine
BSP-GSH   bromosulfophthalein-GSH
BUI   brain uptake index
DTT   dithiothreitol
GGT   gamma glutamyl transpeptidase
GSSG   oxidized glutathione
GSH   reduced glutathione

GSH-MEE glutathione monoethyl ester
HPLC high-performance liquid chromatography
SAAM simulation analysis and modeling

1. INTRODUCTION

In most mammalian tissues, including the brain, glutathione is by far the most prevalent acid-soluble thiol and a key component of intracellular and extracellular antioxidant defense (Meister and Anderson 1983). GSH deficiency is associated with a
number of neurological disorders, and the role of glutathione in neuroprotection is
well known (Rehncrona et al. 1980; Meister 1978; Berl et al. 1959; Perry, Godin,
and Hansen 1982; Larsson et al. 1983; Droge et al. 1994). Recently, an additional
role for GSH as a neuromodulator has been proposed (Guo, McIntosh, and Shaw
1992; Nedergaard 1994; Leslie et al. 1992; Levy, Sucher, and Lipton 1991). It has
been known for some time that GSH may regulate the feeding response of *Hydra*
through a receptor on its primitive nervous system (Miner 1955; Venturini; 1987;
Lenhoff, chapter 2, this volume). GSH binding and competitive antagonism of glutamate binding in rat brain synaptasomes has been described (Ogita and Yoneda 1987,
1988). Understanding GSH homeostasis in the brain, therefore, is of importance. Although the mammalian brain possesses the enzymatic machinery to synthesize GSH
from its precursors, the turnover rate of GSH in the whole brain is considered very
slow except for tiny pools with rapid turnover such as the choroid plexus (Anderson
et al. 1989; Griffith and Meister 1979). In this review, we have summarized studies
from our laboratory on transport of GSH as the intact tripeptide across the blood–
brain barrier (BBB) and describe recent findings on characterization of novel GSH
transporters in brain capillaries. For convenience, the discussion is divided into three
categories: (i) carrier-mediated GSH transport in animal models in vivo, (ii) transport in cultured brain endothelial cells and neonatal astrocytes, and (iii) molecular
characterization of GSH transporters in the *Xenopus laevis* oocyte expression cloning
system.

2. GSH TRANSPORT ACROSS THE BBB

Plasma levels of GSH in rodents and humans is in the 10–20 μM range (Flagg et al.
1993; Lash and Jones 1985). Plasma GSH is cleared by several organs (including the
brain), essentially by two mechanisms: (i) direct uptake of GSH by carrier-mediated
transport and (ii) breakdown of GSH by γ-glutamyl transpeptidase (GGT) and dipeptidases followed by transport of constituent amino acids (Meister and Anderson 1983).
Although there exists a transport system for cysteine, the major sulfur amino acid precursor of GSH, it has to compete with the abundance of other plasma amino acids
for the L-system transport across the BBB (Wade and Brady 1981; Hargreaves and
Pardridge 1988). We hypothesized that GSH may be taken up intact by the brain,
as has been suggested for other organs such as the kidney, intestine, lungs, and lens

(Lash and Jones 1984; Visarius et al. 1996; Hagen and Jones 1987; Hagen, Brown, and Jones 1986; Bai, Brown, and Jones 1994; Zlokovic et al. 1994a; Kannan et al. 1995; 1996a).

2.1 Carrier-Mediated Transport in Vivo and Inhibitor Specificity

Transport of GSH in vivo was examined in two animal models, viz. rat and guinea pig, using rapid bolus intracarotid artery administration (Oldendorf 1970) and in situ vascular brain perfusion (Zlokovic et al. 1986). Our early studies employed multiple approaches (see below) to exclude the GGT-mediated hydrolysis and resynthesis and to prove unequivocally that intact GSH transport does occur (Kannan et al. 1990; Kannan et al. 1992a,b). Uptake of GSH, measured as the brain uptake index (BUI) in male Sprague–Dawley rats, was significantly higher than that of impermeant marker sucrose and somewhat lower than that of phenylalanine–cysteine (Fig. 1 A). The oxidized form of GSH (GSSG) had a BUI similar to that of impermeant sucrose. Uptake of intact GSH was also found in vascularly perfused guinea-pig brain (Fig. 1 B). The molecular form of uptake by HPLC was predominantly (>95%) as GSH in both models under conditions of GGT inhibition either with acivicin or serine borate. Uptake determined in dual labeled experiments ([^{35}S]cysteine-labeled GSH and [^{3}H]glycine-labeled GSH) gave the same ^{35}S/^{3}H ratio as the administered dose, confirming that uptake was as intact GSH (Zlokovic et al. 1994b). Glutathione uptake determined by the BUI technique in the rat was saturable with an apparent K_m of 5.84 mM. Uptake was by a specific mechanism, since various amino acids, amino acid analogs, GSSG, and γ-glutamyl compounds did not affect GSH uptake (Kannan et al. 1992a). On the other hand, a variety of GSH analogs and organic anions inhibited GSH transport (Fig. 2). The inhibition of GSH uptake by GSH monoethyl ester (GSH-MEE) was concentration-dependent (Kannan et al. 1992a). The decrease in uptake cannot be explained by inhibition of GGT, because the ester is neither a substrate (Anderson et al. 1985) nor an inhibitor of this enzyme (Kannan and Kaplowitz, unpublished observations).

In addition to strong inhibition by GSH conjugates and S-alkyl derivatives, GSH uptake by the brain was inhibited by unconjugated bilirubin (UCB), a physiologically relevant organic anion (Tiribelli, Lunazzi, and Sottocasa 1990). Unconjugated bilirubin complexed to bovine serum albumin to give a 2:1 UCB:BSA ratio caused a 60% decrease in GSH uptake (Fig. 3). The inhibition of GSH uptake by UCB was specific for GSH, since the same concentration of UCB (and BSA) did not alter cysteine uptake across the BBB (Fig. 3).

2.2 Age Dependence of Transport

One interesting aspect of GSH uptake at the BBB was that the uptake was developmentally regulated (Kannan et al. 1992a, 1995). GSH uptake declined from a very

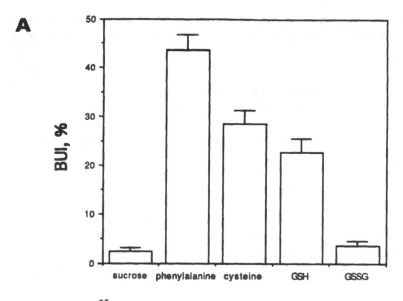

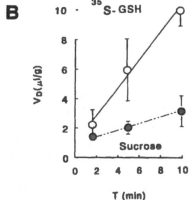

FIG. 1. (A) BUI of radiolabeled substrates in rats pretreated with acivicin. BUIs were determined 15 s after intracarotid artery injection of a mixture of 4 μCi (tracer only) of either [^{14}C]sucrose, [^{14}C]phenylalanine, [^{35}S]cysteine, [^{35}S]GSH, or [^{35}S]GSSG and 15 μCi ^{3}H$_2$O in Ringer–Hepes buffer with 1 mM DTT, except for GSSG, for which DTT was omitted. Values are means ± SEM for five to seven rats per group. (B) Uptake of [^{35}S]GSH (4 nM) and [^{14}C]sucrose into perfused guinea-pig brain. Perfusions were carried out in the presence of serine borate to inhibit GGT. Each point represents mean ±SEM ($n = 3$–6). T is the perfusion time in minutes.

high BUI (≈45%) in 2-week-old rats to ≈5% in mature adult (6-month-old) rats (Fig. 4 A). The BUI of the impermeable marker sucrose remained unaffected during this period, while the GSH BUI of mature rats fell almost to that of sucrose. GGT enzyme activity measurements revealed unaltered GGT activity during this period. A decline in GSH transport with maturation was also observed in the guinea-pig perfused-brain model (Fig. 4 B). These findings suggest that decreased transport of GSH during the developmental and maturational stages of growth may result in observed low GSH in brain compartments of aged rats (Ravindranath, Shivakumar, and Anandatheerthavarada 1989; Benzi et al. 1988; Vali Pasha and Vijayan 1989).

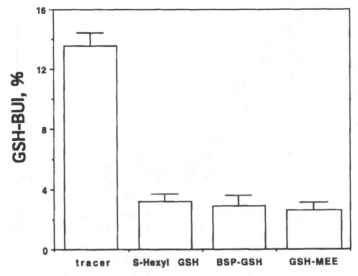

FIG. 2. Effect of GSH analogs on BUI of tracer GSH in acivicin-pretreated rats. The inhibitor (5 mM) was mixed with tracer [^{35}S]GSH (0.025 pmol) before the intracarotid administration and determination of BUI. Data are mean ± SEM for four to six rats per group. GSH–MEE, GSH moneoethyl ester. Taken from Kannan et al. (1992a).

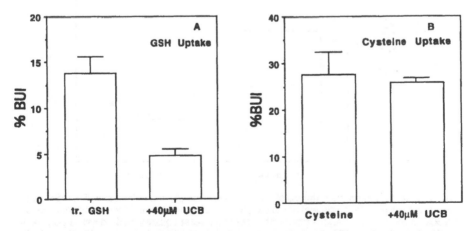

FIG. 3. Effect of unconjugated bilirubin (UCB) on GSH (A) and cysteine (B) uptake. BUI of tracer doses of [^{35}S]GSH and [^{35}S]cysteine were determined with or without 40 μM UCB complexed to 20 μM BSA. Data are mean ± SEM for four or five rats per group.

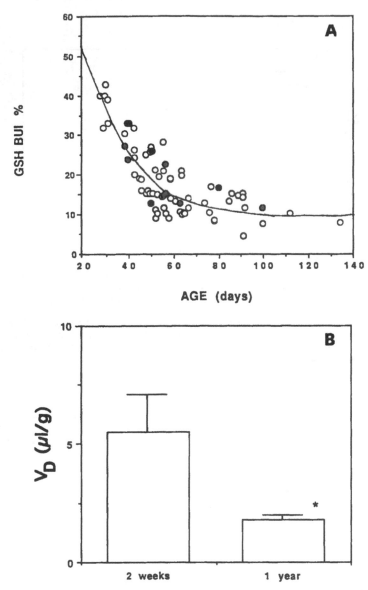

FIG. 4. Effect of age on brain uptake of tracer GSH in rats (A) and guinea pigs (B). GSH BUI in acivicin-pretreated (closed circles) untreated (open circles) and brains 15 s after arterial injection of tracer is shown. The trend is fitted with a nonlinear least-squares method with a polynomial function by using the SAAM program. Guinea-pig brain uptake data are from 10-min perfusion with tracer doses of [^{35}S]GSH in 2-week- and 1-year-old guinea pigs. Asterisk indicates statistical difference at $p < 0.05$.

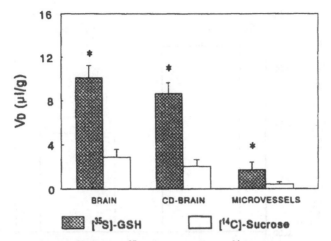

FIG. 5. Compartmental distribution of [^{35}S]GSH (4 nM) and [^{14}C]sucrose in guinea-pig neocortical homogenate (brain), capillary-depleted (CD) brain (parenchyma), and microvessels after 10 min of brain perfusion. Values are mean ± SEM ($n = 3$–6). Asterisk indicates $p < 0.05$ versus sucrose. Taken from Zlokovic et al. (1994b).

2.3 Transendothelial Transport

The question arose whether the GSH taken up from plasma remained in the endothelial compartment or underwent transcytosis (Triguero et al. 1990). Our results showed that the bulk of [^{35}S]GSH taken up by the guinea-pig brain capillaries was found in brain parenchyma and not in capillaries after 10-min vascular perfusion. Thus, the major portion of the radioactivity of [^{35}S]GSH in the brain homogenate was recovered in the capillary-depleted supernatant fraction (Fig. 5). This finding supports transcytosis and suggests that GSH is simply not trapped in the endothelial cells. Indeed, the results support the existence of a luminal transporter for uptake and abluminal transporter for GSH in brain endothelium (see section 3).

3. TRANSPORT IN CULTURED BRAIN ENDOTHELIAL CELLS AND NEONATAL ASTROCYTES

Glutathione is distributed abundantly in brain endothelial cells and in astrocytes, but its level in neurons is reported to be very low (Yudkoff et al. 1990; Raps et al. 1989). The physiological implication of this observation is that neurons have a relatively low antioxidant defense, which may account for their vulnerability to oxidative stress, and that glial cells may serve to protect neurons through release of GSH (Makar et al. 1994; Guo et al. 1992). Using mouse brain endothelial cells in culture, we have been able to show that GSH is transported intact into the cell. The uptake was significantly inhibited by removal of sodium as studied at two GSH concentrations

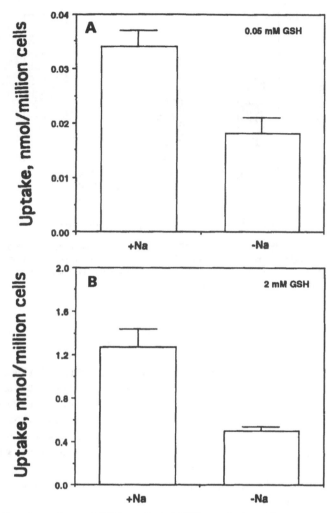

FIG. 6. Sodium dependence of GSH uptake in MBEC-4 cells. Net GSH uptake at two concentrations (0.05 and 2 mM GSH) is shown. Uptake was performed for 1 h in either NaCl or choline chloride medium with 2–3 million cells/well that were pretreated with acivicin and BSO. Data are mean ± SEM from three individual experiments each done in triplicate.

(Fig. 6). Since these cells exhibit polarity (Tatsuta et al. 1994), it would be of interest to study whether the sodium-dependent and sodium-independent GSH transporters are localized on the basolateral or apical membranes or both. In a recent study, luminal and abluminal membrane vesicles were isolated from bovine brain endothelial cells, and it was shown that GGT was located on the luminal membrane, whereas amino acid transport system A was reported to be on the abluminal side (Sanchez del Pino, Hawkins, and Peterson 1995).

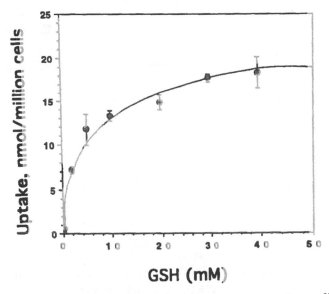

FIG. 7. Effect of unlabeled GSH on GSH uptake in neonatal astrocytes. Tracer [^{35}S]GSH in 10 mM DTT, alone or with varying concentrations of GSH (1–40 mM), was incubated for 1 h in NaCl medium, pH 7.4, at 37°C. The ordinate represents GSH uptake in nanomoles per million cells, and the abscissa GSH concentration in millimolar. Mean (±SEM, $n = 3$) values for each concentration after subtracting the 4°C values are shown. The astrocytes were pretreated with 1 mM acivicin and 10 mM BSO for 30 min before uptake studies were performed.

On the other hand, unlike that in brain endothelial cells, GSH uptake in neonatal astrocytes was not affected by sodium removal and exhibited saturation kinetics with a K_m of 3.3 mM and a V_{max} of 0.62 (nmol/10^6 cells)/min (Fig. 7). The K_m of efflux in rat astrocytes was very high ($\approx$30 mM) according to a recent report (Sagara, Makino, and Bannai 1996). GSH transport in astrocytes not being sodium-dependent, a high-K_m bidirectional GSH transporter, as in other cell types (Fernandez-Checa et al. 1992; Garcia-Ruiz, Fernandez-Checa, and Kaplowitz 1992; Lu et al. 1996), may be present in astrocytes. The presence of additional GSH transporters in astrocytes cannot be excluded.

4. MOLECULAR EVIDENCE FOR THE PRESENCE OF MULTIPLE GSH TRANSPORTERS IN BRAIN CAPILLARIES

Supportive evidence for in vivo findings on GSH transport was obtained from molecular studies in the *Xenopus laevis* oocyte expression system (Fernandez-Checa et al. 1993). Initial studies revealed that *X. laevis* oocytes injected with mRNA from bovine brain capillaries expressed GSH transport. GSH uptake did not differ in oocytes pretreated or untreated with acivicin to inhibit GGT. As shown in Fig. 8, oocytes injected with bovine and guinea-pig capillary mRNA expressed GSH transport, and

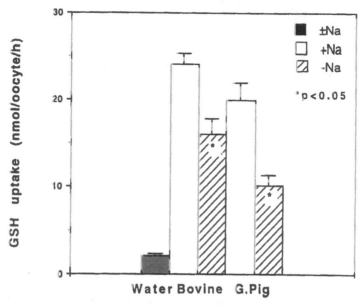

FIG. 8. Uptake of intact GSH into oocytes injected with bovine brain capillary mRNA injected and guinea-pig capillary mRNA. Oocytes were injected with either water or mRNA (30 ng) and maintained at room temperature for 3 days. Injected oocytes were washed and incubated in either NaCl or Na$^+$-free (sucrose) medium in the presence of [^{35}S]GSH (5 μCi in 10 mM DTT plus 10 mM GSH) for 1 h. Uptake is expressed in nanomoles per oocyte per hour. Oocytes injected with both bovine and guinea-pig mRNA showed a significant ($p < 0.05$, indicated by asterisk) decrease in uptake in sucrose medium (hatched bars) as compared to NaCl medium (open bars). Water-injected oocytes showed similar, low uptake in both media (solid bars). Taken from Kannan et al. (1996b).

the uptake was partially inhibited by replacement of NaCl with choline chloride ($\approx$49- and $\approx$36-percent inhibition in uptake in oocytes injected with bovine capillary mRNA and guinea-pig capillary mRNA, respectively). This finding suggested that GSH transport may occur by sodium-dependent and -independent mechanisms. When the bovine capillary mRNA was size-fractionated and each size fraction checked for GSH transport, transport activity was found in three distinct fractions expressing intact GSH transport (Fig. 9). Considering the difference in amount of cRNA versus poly(A)$^+$ RNA injected, fractions 5, 7–8, and 11–12 showed 10–25-fold enrichment in expressed activity versus total RNA. Both fractions 5 and 11 showed sodium-independent GSH transport, and a transcript for GGT was present in fraction 11. Fraction 7, on the other hand, exhibited sodium-dependent GSH transport and did not contain GGT transcript. We had initially identified a clone from rat liver that appeared to be a GSH transporter and appeared to be expressed in brain (Yi et al. 1994). We have subsequently found that its identity was an artifact due to *E. coli* genes in our library. Therefore, the molecular identity of liver and brain facilitative GSH transporters is currently unknown.

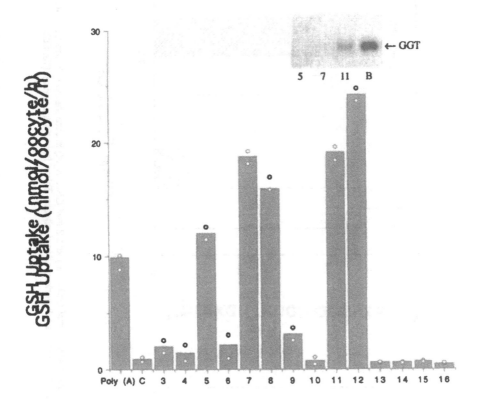

Size Fraction

FIG. 9. GSH transport activity in mRNA size fractions isolated by size fractionation of cDNA. The complementary RNAs from each fraction (3 ng/oocyte) were injected into two different oocyte preparations. After 3 days, oocytes were pretreated with acivicin and were incubated in NaCl medium containing 10 mM GSH. GSH uptake in oocytes after injection of bovine capillary poly(A) RNA (30 ng) and in water-injected oocytes (marked C) are shown on the left side. Among the active fractions, fraction 11 contained a transcript for GGT (inset). Adapted from Kannan et al. (1996b).

5. IDENTIFICATION OF NA⁺-DEPENDENT GSH TRANSPORT AND ITS INHIBITOR SPECIFICITY AND KINETICS

In acivicin-pretreated oocytes injected with cRNA from fraction 7 or 11, uptake of GSH studied at two concentrations, viz. 0.05 and 2 mM, was Na^+-dependent in fraction 7 and not in fraction 11 (Fig. 10). Oocytes injected with fraction-7 cRNA also exhibited striking inhibition of GSH uptake by 2 mM BSP–GSH. The inhibition by BSP–GSH was ≈90 percent at 0.05 mM GSH and ≈39 percent at 2 mM GSH. As in

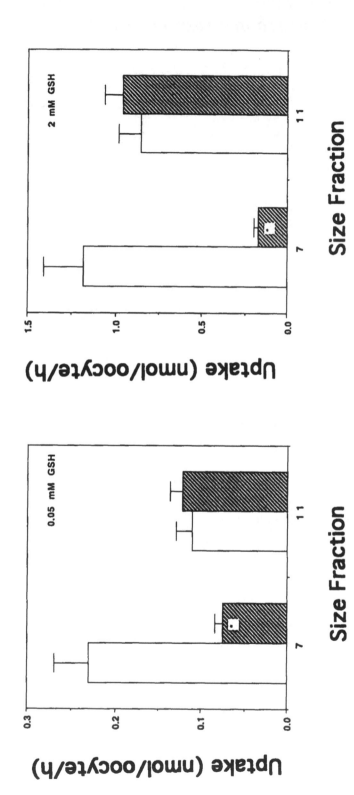

FIG. 10. Effect of replacement of Na^+ with choline chloride on GSH uptake at 0.05 mM GSH (left) and 2 mM GSH (right) in size fractions 7 and 11. Data are mean ± SEM from three separate experiments on different oocyte preprations, each performed in duplicate. Taken from Kannan et al. (1996b).

in vivo studies, several lines of evidence were gathered to dissociate transport activity in fraction 7 from GGT-mediated degradation and resynthesis: (i) the net uptake of GSH was similar under GGT-inhibited and -uninhibited conditions, (ii) the molecular form of uptake of ^{35}S- and ^{3}H-labeled GSH in oocytes injected with fraction-7 cRNA was predominantly GSH with and without acivicin, and the ^{35}S/^{3}H ratio in GSH peak in oocytes was similar to that in the uptake medium, and (iii) uptake of [^{35}S]cysteine (plus 50 μM unlabeled cysteine in 10 mM DTT) showed that nearly all the radioactivity was associated with cysteine and no conversion to GSH had taken place in the 1-h incubation period.

Kinetics of GSH uptake in the presence of sodium in oocytes injected with cRNA from fraction 7 showed concentration dependence with two components, a high-affinity Michaelis–Menten component with a K_m of 0.40 mM, and a sigmoid low-affinity component with a K_m of 10.8 mM (Kannan et al. 1996b).

6. A HYPOTHETICAL MODEL FOR GSH TRANSPORT IN THE BRAIN

The exact role of the GSH transporters in brain physiology is yet to be fully understood. Some speculations can, however, be made, based on above findings. A hypothetical model depicting BBB and astrocyte GSH transport is shown in Fig. 11. According to this model, GSH transport from circulation into the BBB can be envisioned to proceed by luminal influx that is Na$^+$-dependent, GGT-independent (this is not to exclude separate hydrolysis of GSH), and inhibited by BSP–GSH and other organic anions. This system has the possibility of being concentrative, so that low plasma GSH (10–20 μM) can lead to net accumulation in endothelial cells, due to coupling to the entry of sodium. An efflux mechanism that proceeds by the facilitative Na$^+$-independent GSII transporter from the abluminal surface of the endothelium and from the astrocyte compartment is also shown. A second Na$^+$-independent transporter may also exist at these locations. It is also conceivable that one or both the sodium-independent facilitative transporters are present on the luminal pole of endothelium (not shown). These facilitative Na$^+$-independent transporters would have the potential for bidirectional operation as observed in cell culture studies (Lu et al. 1996), but under physiological conditions would serve as efflux transporters, because net transport would be determined by the concentration gradient (high GSH in cells and low extracellular GSH). The model also includes the possibility of cysteine transport into the endothelial cells and astrocytes for GSH synthesis followed by GSH efflux into the brain. The oxidation of interstitial cysteine according to Bannai (Sagara, Miura, and Bannai 1993), or, as we propose, more likely incorporation into GSH and release as GSH, is also indicated.

A second model shows the possible role of GGT. GGT on the luminal or abluminal pole of endothelial cells or on astrocytes might operate in parallel, hydrolyzing GSH (along with dipeptidase) to cysteine. The efflux of GSH at the same location as GGT would insure high availability of GSH to the active site of GGT (ectoenzyme) to

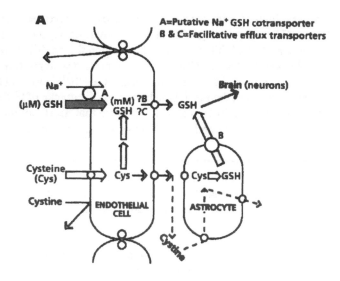

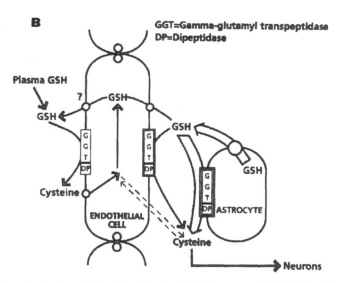

FIG. 11. A hypothetical model depicting the role of GSH transport in the brain. Model A stresses the possibility of rapid GSH uptake from the capillary lumen driven by Na^+ coupling and slower abluminal efflux. Cysteine is utilized for GSH synthesis in endothelial cells and astrocytes. The Bannai hypothesis of extracellular cysteine autooxidation and astrocyte uptake, reduction, and release is shown as dashed lines. Model B depicts the potential role of GSH hydrolysis, stressing the possibility that abluminal GGT and DP maintain extracellular cysteine that is utilized for GSH and protein synthesis, especially in neurons. Alternatively, extracellular cysteine autooxidizes and is handled as proposed in model A.

produce transpeptidation products and cysteine. Since cysteine autooxidizes rapidly, a mechanism for continually generating cysteine in a paracrine–autocrine fashion through the combined action of GSH transport (efflux) and GGT–dipeptidase would be analogous to the proposals for interorgan and intraorgan GSH homeostasis of Meister (Griffith and Meister 1979). Since cystine is not taken up by neurons (Sagara et al. 1993), such a mechanism for its supply may be critical. A sodium-driven uptake of GSH at the luminal pole, however, is extremely novel and would have to be distinct from this GGT–GSH efflux model, although they are by no means mutually exclusive.

The models shown are speculative in view of our current state of knowledge. However, the evidence of multiple GSH transporters in the brain (capillary endothelial cells and astrocytes) is convincing. In view of the importance of GSH, it stands to reason that understanding the action of GSH transporters will be of importance in elucidating overall brain GSH homeostasis. Our laboratory is actively engaged in research leading to development of reagents for the transporters (viz., cDNA and polyclonal antisera) which would permit elucidation of localization and assessment of regulation (development, inducers). Once the transporters are localized and the transport physiology of expressed clones in oocytes fully characterized with respect to kinetics, specificity, ion requirements, driving forces, and so forth, it will be possible to develop more refined and accurate models that can be further tested in cell culture and in in vivo systems and that weigh the relative contributions of the schemes depicted in Fig. 11 A and B or some synthesis of the two.

One intriguing corollary of model A in Fig. 11 relates to organic anion transport, since some organic anions inhibit GSH transport in various cells and brain in vivo. It is conceivable that the GSH transporters also transport organic anions, among which are endogenous degradation products (e.g. bilirubin) and drugs (Levine, Frederick, and Rapoport 1982; Ives and Gardiner 1990). The capacity of newly cloned GSH transporters to transport organic anions will be important to ascertain; its implications would be enormous and, in view of the developmental pattern of expression of transport, might include kernicterus, the hyperbilirubinemia of the newborn (Odell and Schuetta 1985).

ACKNOWLEDGMENTS

This work was supported by National Institutes of Health grants DK 30312 and DK48522 and Veterans Administration Medical Research Funds.

REFERENCES

Anderson, M. E., Powrie, F., Puri, R. N., and Meister, A. 1985. Glutathione monoethyl ester. Preparation, uptake by tissues and conversion to glutathione. *Archives of Biochemistry and Biophysics* 239:538–48.

Anderson, M. E., Underwood, M., and Bridges, R. J. 1989. Glutathione metabolism at the blood–cerebrospinal fluid barrier. *FASEB Journal* 3:2527–31.

Bai, C., Brown, L. S., and Jones, D. P. 1994. Glutathione transport by type II cells in perfused rat lung. *American Journal of Physiology* 267(Lung Cell Molecular Biology 11):L447–55.

Benzi, G., Pastoris, O., Marzatico, F., and Villa, R. F. 1988. Influence of aging and drug treatment on the cerebral glutathione system. *Neurobiology of Aging* 9:371–5.

Berl, S., Purpura, D. P., Girardo, M., and Waelsch, H. 1959. Amino acid metabolism in epileptogenic and non-epileptogenic lesions of the neocortex (cat). *Journal of Neurochemistry* 4:311–7.

Droge, W., Schulze-Osthoff, K., Mihm, S., Galter, D., Schenk, H., Eck, H. -P., Roth, S., and Gmundr, H. 1994. Functions of glutathione and glutathione disulfide in immunology and immunopathology. *FASEB Journal* 8:1131–8.

Fernandez-Checa, J. C., Takikawa, H., Horie, M., Ookhtens, M., and Kaplowitz, N. 1992. Canalicular transport of reduced glutathione in normal and mutant Eisai hyperbilirubinemic rats. *Journal of Biological Chemistry* 267:1667–73.

Fernandez-Checa, J. C., Yi, J.-R., Garcia-Ruiz, C., Knezic, Z., Tahara, S. M., and Kaplowitz, N. 1993. Expression of rat liver glutathione transport in *Xenopus laevis* oocytes. *Journal of Biological Chemistry* 268:2324–8.

Flagg, E. W., Coates, R. J., Jones, D. P., Eley, J. W., Gunter, E. W., Jackson, B., and Greenberg, R. S. 1993. Plasma total glutathione in humans and its association with demographic and health-related factors. *British Journal of Nutrition* 70:797–808.

Garcia-Ruiz, C., Fernandez-Checa, J., and Kaplowitz, N. 1992. Bidirectional mechanism of plasma membrane transport of reduced GSH in hepatocytes and membrane vesicles. *Journal of Biological Chemistry* 267:22256–64.

Griffith, O. W., and Meister, A. 1979. Glutathione: interorgan translocation, turnover and metabolism. *Proceedings of the National Academy of Sciences of the U.S.A.* 76:5606–10.

Guo, N., McIntosh, C., and Shaw, C. 1992. Glutathione: new candidate neuropeptide in the central nervous system. *Neuroscience* 51:835–42.

Hagen, T. M., Brown, L. A., and Jones, D. P. 1986. Protection against paraquat-induced injury by exogenous GSH in pulmonary alveolar Type II cells. *Biochemical Pharmacology* 75:4537–42.

Hagen, T. M., and Jones, D. P. 1987. Transepithelial transport of glutathione in vascularly perfused small intestine of rat. *American Journal of Physiology* 252:G687–93.

Hargreaves, K. M., and Pardridge, W. M. 1988. Neutral amino acid transport at the blood brain barrier. *Journal of Biological Chemistry* 263:19392–7.

Ives, N. K., and Gardiner, R. M. 1990. Blood–brain permeability to bilirubin in the rat studied using intracarotid bolus injection and in situ perfusion techniques. *Pediatric Research* 27:436–41.

Kannan, R., Kuhlenkamp, J. F., Jeandidier, E., Trinh, H., Ookhtens, M., and Kaplowitz, N. 1990. Evidence for carrier-mediated transport of glutathione across the blood–brain barrier in the rat. *Journal of Clinical Investigation* 85:2009–13.

Kannan, R., Kuhlenkamp, J. F., Ookhtens, M., and Kaplowitz, N. 1992a. Transport of GSH at blood–brain barrier of the rat: inhibition and age-dependence. *Journal of Pharmacology and Experimental Therapeutics* 263:964–70.

Kannan, R., Mackic, J. B., Kaplowitz, N., and Zlokovic, B. V. 1992b. Blood–brain barrier glutathione transfer in guinea-pigs of different ages. *Society for Neuroscience Abstracts* 18:1129A.

Kannan, R., Yi, J.-R., Zlokovic, B. V., and Kaplowitz, N. 1995. Molecular characterization of a reduced gluthione transporter in the lens. *Investigative Ophthalmology and Visual Science* 36:1785–92.

Kannan, R., Yi, Y.-R., Tang, D., Zlokovic, B. V., and Kaplowitz, N. 1996a. Identification of a novel, sodium-dependent reduced glutathione transporter in the rat lens epithelium. *Investigative Ophthalmology and Visual Science* 37:2269–75.

Kannan, R., Yi, J.-R., Tang, D., Li, Y., Zlokovic, B. V., and Kaplowitz, N. 1996b. Evidence for the existence of a sodium-dependent glutathione (GSH) transporter. Expression of bovine brain capillary mRNA and size fractions in *Xenopus laevis* oocytes and dissociaton from gamma-glutamyltranspeptidase and facilitative GSH transporters. *Journal of Biological Chemistry* 271:9754–8.

Larsson, A. S., Orrenius, S., Holmgren, A., and Mannervik, B. 1983. *Functions of glutathione: biochemical, physiological, toxicological, and clinical aspects.* New York: Raven Press.

Lash, L. H., and Jones, D. P. 1984. Renal glutathione transport: characteristics of the sodium-dependent system on the basal lateral membrane. *Journal of Biological Chemistry* 259:14508–14.

Lash, L. H., and Jones, D. P. 1985. Distribution of oxidized and reduced forms of glutathione and cysteine in rat plasma. *Archives of Biochemistry and Biophysics* 240:583–92.

Leslie, S. W., Brown, L., Trent, R. D., Kee, K.-H., Morris, J. L., Jones, T. W., Randall, P. K., Lau, S. S., and Monks, T. J. 1992. Stimulation of N-methyl-D-aspartate receptor-mediated calcium entry into dissociated neurons by reduced and oxidized glutathione. *Molecular Pharmacology* 41:308–14.

Levine, R. I., Frederick, W. R., and Rapoport, S. I. 1982. Entry of bilirubin into the brain due to opening of the blood-brain barrier. *Pediatrics* 69:255–9.

Levy, D. L., Sucher, N. J., and Lipton, S. A. 1991. Glutathione prevents N-methyl D-aspartate receptor-mediated neurotoxicity. *Neuropharmacology and Neurotoxicology* 2:345–7.

Lu, S. C., Sun, W.-M., Yi, J., Ookhtens, M., Sze, G., and Kaplowitz, N. 1996. Role of two recently cloned rat liver GSH transporters in the ubiquitous transport of GSH in mammalian cells. *Journal of Clinical Investigation* 97:1488–96.

Makar, T. K., Nedergaard, M., Preuss, A., Gelbard, M., Perumal, A. S., and Cooper, A. J. L. 1994. Vitamin E, ascorbate, glutathione disulfide, and enzymes of glutathione metabolism in cultures of chick astrocytes and neurons; evidence that astrocytes play an important role in antioxidant processes in the brain. *Journal of Neurochemistry* 62:45–53.

Meister, A. 1978. Relationship between ataxia and defects of the gamma glutamyl cycle. *Advances in Neurology* 21:289–302.

Meister, A., and Anderson, M. E. 1983. Glutathione. *Annual Review of Biochemistry* 52:711–60.

Miner, R. W. 1955. Glutathione control of the specific feeding reactions of *Hydra. Annals of the New York Academy of Sciences* 62:209–28.

Nedergaard, M. 1994. Direct signaling from astrocytes to neurons in cultures of mammalian brain cells. *Science* 263:1768–71.

Odell, G. B., and Schuetta, H. S. 1985. Bilirubin encephalopathy. In *Neural energy metabolism and metabolic encephalopathy.* (D. W. McCandless, ed.) Plenum, New York, 229–61.

Ogita, K., and Yoneda, Y. 1987. Possible presence of [^{3}H]glutathione (GSH) binding sites in synaptic membranes. *Neuroscience Research* 4:486–96.

Ogita, K., and Yoneda, Y. 1988. Temperature-dependent and independent apparent binding activities of [^{3}H]glutathione in brain synaptic membranes. *Brain Research* 463:37–46.

Oldendorf, W. H. 1970. Measurement of brain uptake of radiolabeled substrates using a tritiated water internal standard. *Brain Research* 24:372–6.

Perry, T. L., Godin, D. V., and Hansen, S. 1982. Parkinson's disease: a disorder due to nigral glutathione deficiency? *Neuroscience Letters* 33:305–10.

Raps, S. A., Lai, J. C. K., Hertz, L., and Cooper, A. J. L. 1989. Glutathione is present in high concentrations in cultured astrocytes but not in cultured neurons. *Brain Research* 493:398–401.

Ravindranath, V., Shivakumar, B. R., and Anandatheerthavarada, H. K. 1989. Low glutathione levels in brain regions of aged rats. *Neuroscience Letters* 101:187–90.

Rehncrona, S., Folbergova, J., Smith, D. S., and Siesjo, B. K. 1980. Influence of complete and pronounced incomplete cerebral ischemia and subsequent recirculation on cortical concentrations of oxidized and reduced glutathione. *Journal of Neurochemistry* 34:477–86.

Sagara J.-I., Miura, K., and Bannai, S. 1993. Cystine uptake and glutathione level in fetal brain cells in primary culture and suspension. *Journal of Neurochemistry* 61:1667–71.

Sagara, J., Makino, N., and Bannai, S. 1996. Glutathione efflux from cultured astrocytes. *Journal of Neurochemistry* 66:1876–81.

Sanchez del Pino, M. M., Hawkins, R. A., and Peterson, D. R. 1995. Biochemical discrimination between luminal and abluminal enzyme and transport activities of the blood–brain barrier. *Journal of Biological Chemistry* 270:14907–12.

Tatsuta, T., Naito, M., Oh-hara, T., Sugawara, I., and Tsuruo, T. 1992. Functional involvement of P-glycoprotein in blood–brain barrier. *Journal of Biological Chemistry* 267:20383–91.

Tiribelli, C., Lunazzi, G. C., and Sottocasa, G. L. 1990. Biochemical and molecular aspects of hepatic uptake of organic anions. *Biochimica Biophysica Acta* 1031:261–73.

Triguero, D., Buciak, J. B., Yang, J., and Pardridge, W. M. 1990. Capillary-depletion method for quanti-fying blood–brain transcytosis of circulating peptides and plasma proteins. *Journal of Neurochemistry* 54:1882–8.

Vali Pasha, K., and Vijayan, E. 1989. Glutathione distribution in rat brain at different ages and the effect of intraventricular glutathione on gonadotropin levels in ovariectomized steroid primed rats. *Brain Research Bulletin* 22:617–9.

Venturini, G. 1987. The hydra GSH receptor. Pharmacological and radioligand binding studies. *Comparative Biochemistry and Physiology* 87C:321–4.

Visarius, T. M., Putt, D. A., Schare, J. M., Pegouske, D. M., and Lash, L. H. 1996. Pathways of glu-tathione metabolism and transport in isolated proximal tubular cells from rat kidney. *Biochemistry and Pharmacology* 52:259–72.

Wade, L. A., and Brady, H. M. 1981. Cysteine and cystine transport at the blood–brain barrier. *Journal of Neurochemistry* 37:730–4.

Yi, J.-R., Lu, S., Fernandez-Checa, J., and Kaplowitz, N. 1994. Expression cloning of a rat hepatic reduced glutathione transporter with canalicular characteristics. *Journal of Clinical Investigation* 93:1841–5.

Yudkoff, M., Pleasure, D., Cregar, L., Lin, Z., Nissim, I., Stern, J., and Nissim, I. 1990. Glutathione turnover in cultured astrocytes: studies with ^{15}N-glutamate. *Journal of Neurochemistry* 55:137–45.

Zlokovic, B. V., Begley, D. J., Djuricic, B., and Mitrovic, D. M. 1986. Measurement of solute transport across the blood–brain barrier in the perfused guinea-pig brain: method and application to N-methyl-α-aminoisobutyric acid. *Journal of Neurochemistry* 46:1444–51.

Zlokovic, B. V., Mackic, J. M., McComb, J. G., Kaplowitz, N., and Kannan, R. 1994a. Blood to lens transport of reduced glutathione in an in situ perfused guinea-pig eye. *Experimental Eye Research* 59:487–96.

Zlokovic, B. V., Mackic, J. B., McComb, J. G., Weiss, M. H., Kaplowitz, N., and Kannan, R. 1994b. Evidence for transcapillary transport of reduced glutathione in vascular perfused guinea-pig brain. *Biochemical and Biophysical Research Communications.* 201:402–8.

Glutathione in the Nervous System
Edited by Christopher A. Shaw
Copyright © 1998 Taylor & Francis

4

Antibodies to Glutathione: Production, Characterization, and Immunocytochemical Application to the Central Nervous System

Ole P. Hjelle and E. Rinvik

Department of Anatomy, Institute of Basic Medical Sciences, University of Oslo, Oslo, Norway

D. Huster and W. Reichelt

Department of Neurophysiology, Paul-Flechsig Institute for Brain Research, University of Leipzig, Leipzig, Germany

Ole P. Ottersen

Department of Anatomy, Institute of Basic Medical Sciences, University of Oslo, Oslo, Norway

- · **Conclusions**
- · **References**

ABBREVIATIONS

GABA γ-aminobutyric acid
ILD incidental Lewy-body disease
PD Parkinson's disease
SNc pars compacta of the substantia nigra
SNr pars reticulata of the substantia nigra

1. INTRODUCTION

Many acute and chronic neurodegenerative diseases typically affect select neuronal populations while leaving others unharmed. Classical examples are the degeneration of hippocampal CA1 neurons following transient global ischemia, and the preferential vulnerability of substantia nigra neurons in Parkinson's disease and of motoneurons in amyotrophic lateral sclerosis. One of several factors that may contribute to the selective cell death seen in these and other conditions is the inability of the vulnerable neurons to cope with increased free-radical production and oxidative stress (Jenner et al. 1992a,b; Mizui, Kinouchi, and Chan 1992; Coyle and Puttfarcken 1993; Rosen et al. 1993; Owen et al. 1996).

The most abundant antioxidant in the mammalian brain is glutathione (Orlowski and Karkowsky 1976). We have developed antibodies to this tripeptide, in the belief that a precise and quantitative assessment of its cellular and subcellular localization would help elucidate the possible pathogenetic role of impaired or perturbed antioxidative mechanisms in neurodegenerative disease. It was also hoped that specific antibodies would serve as useful tools to explore other functions that have been attributed to glutathione, including its proposed signal function (Guo, McIntosh, and Shaw 1992; Zängerle et al. 1992; Shaw, Pasqualotto, and Curry 1996).

The aim of the present chapter is to review the immunocytochemical data that have been accumulated during the short time that has elapsed since the first glutathone antibodies were characterized (Hjelle, Chaudhry, and Ottersen 1994). An attempt will also be made to discuss some of the methodological problems that are associated with the production, characterization, and application of antibodies to such a small molecule as glutathione.

2. RAISING AND CHARACTERIZATION OF ANTIBODIES

The small size of the glutathione molecule had to be taken into account when designing the immunization procedure. We therefore adapted an immunization protocol

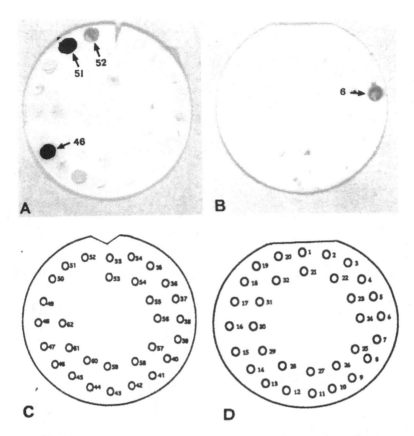

FIG. 1. Specificity testing against immobilized conjugates of amino acids and peptides. After the conjugates were spotted onto cellulose acetate filters, they were processed with a glutathione (GSH) antiserum (no: 744, diluted 1 : 300). All conjugates were prepared from a 9 mM concentration of amino acid or peptide [for detailed procedure see Ottersen and Storm-Mathisen (1984)], except GSH, which was used at two different concentrations (9 and 3 mM). Some of the conjugates were applied to more than one site. We tested the following molecules: 1: L-homocysteic acid; 2: D-homocysteic acid; 3: D-cysteine; 4: DL-homocysteine; 5: L-citrulline; 6: GSH, 9 mM; 7: β-alanine; 8: D-aspartate; 9: hypotaurine; 10: L-cysteine; 11: L-isoleucine; 12: D-arginine; 13: L-methionine; 14: γ-glutamylglutamate; 15: D-leucine; 16: L-phenylalanine; 17: L-serine; 19: L-leucine; 20: L-proline; 21: GABA; 22: L-glutamate; 23: taurine; 24: glycine; 25: L-aspartate; 26: L-glutamine; 27: adrenalin; 28: noradrenalin; 29: phosphoethanolamine; 30: 3-aminopropane sulphonic acid; 31: L-homocysteine sulphinic acid; 32: L-α-alanine; 33: L-valine; 34: L-tryptophan; 35: L-threonine; 36: L-cysteine; 37: L-tyrosine; 38: L-asparagine; 39: L-lysine; 40: L-arginine; 41: L-histidine; 42: L-cysteic acid; 43: γ-glutamyltaurine; 44: L-cysteine; 45: N-acetylaspartate; 46: oxidized glutathione (GSSG), 9 mM; 47: S-sulpho-L-cysteine; 48: leucine enkephalin; 49: methionine enkephalin; 51: GSH; 9 mM; 52: GSH, 3 mM; 53: GABA; 54: quinolinic acid; 56: cysteinylglycine; 57: L-cysteine; 61: tyrosylglycylglycine; 62: cysteinylglycine. Spots 18, 50, and 55 contained a crude rat brain protein extract, and spots 58, 59, and 60 contained a protein carrier (bovine serum albumin, ovalbumin, and keyhole limpet hemocyanin, respectively). Both the crude rat brain protein extract and the protein carriers were treated with formaldehyde and glutaraldehyde. Only the GSH and GSSG conjugates showed labeling significantly above background. γ-Glutamylcysteine was applied to a separate filter disk (not shown here) and did not display any affinity for the antiserum. This experiment was performed with a nonpurified antiserum. Parts C and D display the distribution of conjugates in A and B, respectively. Reproduced with permission from Hjelle et al. (1994).

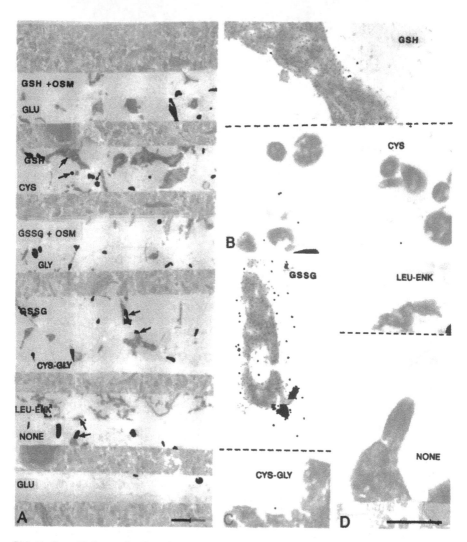

FIG. 2. Quantitative evaluation of selectivity. An ultrathin test sandwich, prepared as described by Ottersen (1987), was incubated in the same drops of immunoreagents as the tissue sections (GSH antiserum 744, dilution 1:500) for optimum reliability. The sandwich contains a series of different conjugates of amino acid, glutaraldehyde–formaldehyde, and brain protein (seen as electron-dense bodies surrounded by resin), separated by brain sections used as spacers. Abbreviations: GSH+OSM and GSSG+OSM, GSH or GSSG conjugates treated with OsO$_4$; LEU–ENK, leucine–enkephalin; NONE, crude extract of rat brain protein treated with formaldehyde–glutaraldehyde. The other abbreviations are standard for amino acids. The conjugates were prepared as for the spot test illustrated in Fig. 1. (A) Low-magnification view comprising all layers of the test section. Some of the electron-dense bodies (arrows) are shown at higher magnification in panels B–D. Note the selective distribution of gold particles over GSH and GSSG conjugates. Dashed lines indicate borders between sections. For quantitative data, see Fig. 3. Bars: 2 mm (A); 0.5 mm (D, also valid for B and C). Reproduced with permission from Hjelle et al. (1994).

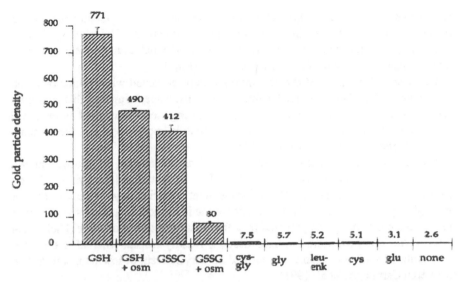

FIG. 3. Quantitative assessment of specificity based on analysis of the test sections shown in the preceding figure. Each column represents the gold particle density (particles per square micrometer) over an amino acid or peptide conjugate ($n = 4$; SEM and particle density are indicated over each column). Note that the values obtained for the GSH and GSSG conjugates that were not treated with OsO$_4$ are significantly different from each other and from conjugates not containing GSH or GSSG ($P < 0.05$; analysis of variance, Student–Newman–Keul procedure). The background labeling over empty resin was < 1 particle/μm^2 and was not subtracted. For abbreviations see Fig. 2. Reproduced with permission from Hjelle et al. (1994).

that was used previously for the generation of antibodies to amino acids (Storm-Mathisen et al. 1983). Reduced glutathione was coupled to bovine serum albumin by a mixture of glutaraldehyde and formaldehyde, and the complex was then injected intracutaneously or subcutaneously in rabbits (Hjelle et al. 1994). Following affinity purification or purification on a column containing glutaraldehyde-treated carrier protein, the antibodies proved to react selectively with reduced and oxidized glutathione (Figs. 1–3). The antibodies should thus be considered to reveal the total tissue contents of glutathione, oxidized as well as reduced.

The antibodies did not show any detectable binding to glutamate, cysteine, or glycine (the three amino acids of which glutathione is composed), nor was any labeling observed of γ-glutamylcysteine or cysteinylglycine (Fig. 1; Hjelle et al. 1994).

As the antibodies were produced with the aim of applying them to postembedding immunogold analyses, they were also characterized in an electron microscopic sandwich system (Fig. 2), designed as described previously (Ottersen 1987). The sandwich contains a series of test antigens, which were considered as potentially cross-reacting substances because of their structural resemblance to glutathione. To ensure that the test results were valid for the conditions of the present immunocytochemical study, the sandwich sections were always incubated together with the tissue sections. Quantitative evaluation of the sandwiches [by use of a computer program developed by Blackstad, Karagülle, and Ottersen (1990)] confirmed that the antibodies selectively recognized reduced and oxidized glutathione, although reduced glutathione produced almost twice as strong labeling as the oxidized form (Fig. 3). The labeling intensity of cysteinylglycine and the individual amino acids of which glutathione is composed was less than 1% of that of reduced glutathione (Fig. 3).

The possibility remained that the antibodies cross-reacted with an unknown substance that had not been included in the test systems. It speaks against this possibility that the labeling was strongly reduced by pretreating the experimental animals with buthionine sulfoximine (Hjelle et al. 1994), which is known to inhibit the synthesis of glutathione (Griffith and Meister 1979).

The sandwich test system revealed that the immunolabeling for glutathione was depressed by exposure to osmium tetroxide (Fig. 3), a standard fixative for electron microscopy. To avoid the use of osmium without suffering any undue loss of morphological preservation, the large majority of the immunogold analyses were carried out on tissue that had been prepared by a freeze substitution procedure (Hjelle et al. 1994). With this procedure the fixed tissue is rapidly frozen, whereafter the water is substituted with an organic solvent and an acrylic resin at low temperature (van Lookeren Campagne et al. 1991).

3. CELLULAR AND SUBCELLULAR DISTRIBUTION OF GLUTATHIONE

Our antibodies have so far been used to analyze the distribution of glutathione in the cerebellum (Hjelle et al. 1994), inner ear (Usami, Hjelle, and Ottersen 1996), retina (Huster et al. 1997), spinal cord (Ramirez-Leon et al. 1997) and substantia nigra (Rinvik unpublished observations). They have also been applied to cultured neurons and astrocytes (Langeveld et al. 1996). In the following we will summarize the results that have been obtained.

3.1 Cerebellum

Gold particles signaling glutathione immunoreactivity occured in glial well as in neuronal elements (Fig. 4). The highest labeling intensities were found in glial processes, particularly in perivascular glia, and in subpopulations of myelinated axons (Fig. 4 C). All neuronal populations exhibited significant immunolabeling. The labeling intensity of Purkinje-cell somata was relatively high (Fig. 4 D) and comparable to that of the adjacent glial cell bodies (Golgi epithelial cells). Mossy fiber terminals [putatively glutamatergic; (Ji et al. 1991)] also displayed relatively strong glutathione immunoreactivity (Fig. 4 A), whereas basket-cell axons (GABAergic) were weakly labeled.

3.2 Inner Ear

Glutathione is known to protect hair cells of the inner ear against the cytotoxic actions of gentamycin and similar compounds (Garetz, Altschuler, and Schacht 1994), and it

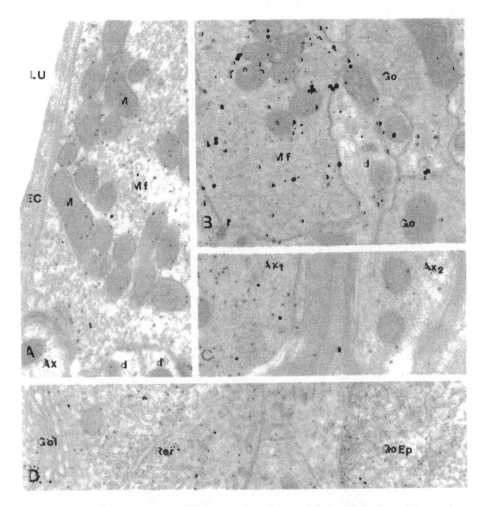

FIG. 4. A: Cerebellar cortex (rat). Note GSH immunoreactive mossy fiber terminal (Mf) and virtually unlabeled endothelial cell (EC) and axon (Ax). The mossy fiber terminals can be recognized
by their ultrastructure and their labeling properties for glutamate and glutamine shown in B. Other
abbreviations: d, granule cell dendrites; Lu, vessel lumen; M, mitochondria. B: Double-labeled
section from granule cell layer (method: see Ottersen, Zhang, and Walberg 1992). Small particles
(15 nm) signaling glutamate immunoreactivity are particularly concentrated over the mossy fiber
terminal (Mf), which also contains large particles (30 nm) representing glutamine (a precursor
of transmitter glutamate). Golgi-cell terminals (Go; putative GABAergic) are weakly labeled for
either amino acid. C: Two myelinated axons (Ax1 and Ax2) in the subcortical white matter show
very different labeling intensities for GSH. Asterisk indicates point of separation between the two
adjacent myelin sheaths. D: Gold particles signaling GSH occur at comparable densities over
Golgi epithelial cells (GoEp) and the adjacent Purkinje cell (left). Gol, Golgi apparatus; Rer, rough
endoplasmic reticulum. Modified from Hjelle et al. (1994).

was therefore of interest to identify the cells that could be responsible for glutathione synthesis in this organ. Very low levels of immunoreactivity were found in the hair cells and adjoining supporting cells of the organ of Corti (Usami et al. 1996). In contrast, strong immunoreactivity occurred in the stria vascularis and spiral ligament. The immunoreactivity at these sites was concentrated in basal and intermediate cells, and in fibrocytes, respectively (Fig. 5). Endothelial cells were also strongly labeled (Fig. 5 A).

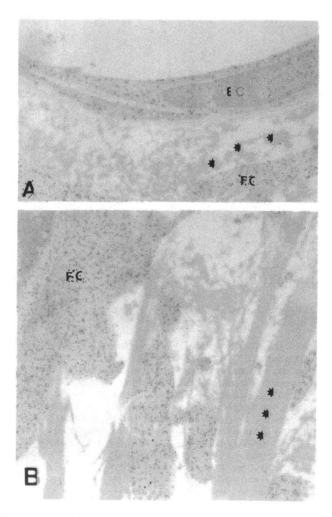

FIG. 5. GSH immunoreactivity in inner ear (guinea pig). A, B: Spiral ligament. The capillary endothelial cells (EC) and the fibrocytes (FC) are strongly labeled, in contrast to the immunonegative collagen fibers (asterisks) and ground substance. C: Stria vascularis. The intermediate cells (IC) are strongly labeled compared to the marginal cells (MC).

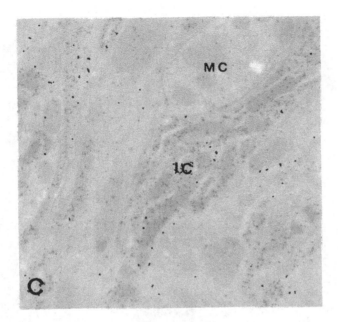

FIG. 5. *Continued.*

3.3 Retina

In agreement with the results of a previous study (Pow and Crook 1995), the highest concentration of glutathione immunoreactivity in the retina occurred in the pigment epithelial cells (Huster et al. 1997). The gold particles in these were distributed over mitochondria as well as the cytoplasmic matrix (Fig. 6), with a moderate enrichment in the former compartment (gold particle density ratio 1.6). In other cell types with significant labeling (Müller cells and receptor inner segments) most of the gold particles were confined to the mitochondria (mitochondrial/cytoplasmic gold particle ratios 16 and 25, respectively). The outer receptor segments were very weakly labeled (Fig. 6).

3.4 Spinal Cord

Motoneurons were generally weakly labeled for glutathione (Fig. 7), although some immunoreactivity occurred in the mitochondria. However, high densities of gold particles were found in a subpopulation of those terminals that were apposed to the dendrites or somata of the motoneurons (Fig. 7). Based on analyses of consecutive sections incubated with antisera to glutamate, GABA, and glycine (Ramirez-Leon et al. 1997), it was found that a large proportion of the glutathione immunolabeled terminals was enriched in glutamate and therefore probably glutamatergic.

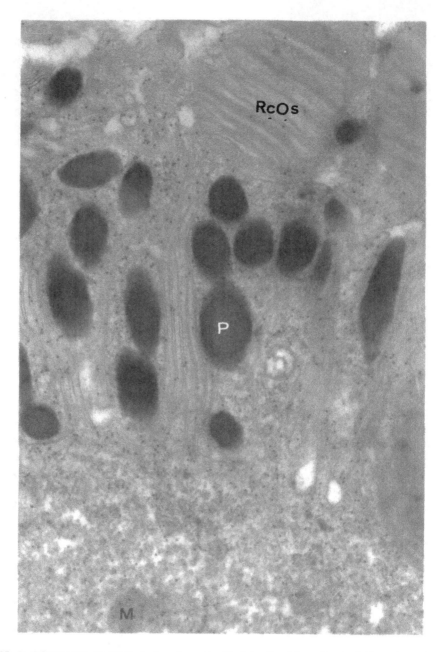

FIG. 6. GSH immunoreactivity in the pigment epithelial cells of the retina (rat). Note dense immunogold labeling of pigment epithelial-cell cytoplasm and very weak labeling of receptor outer segments (RcOs). P, pigment granula; M, mitochondrion.

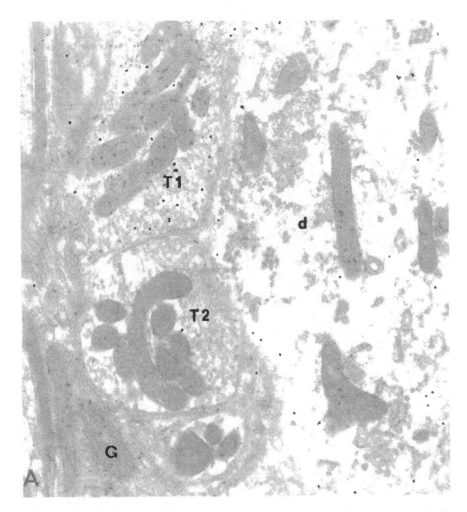

FIG. 7. GSH immunoreactivity in the ventral horn of the spinal cord (rat). A high concentration of gold particles occurs in subpopulations of nerve terminals (exemplified by T1), while other terminals are weakly labeled (T2). These terminals are apposed to a proximal dendrite (d), probably belonging to a motoneuron. A large proportion of the gold particles in this dendrite is confined to mitochondria (arrowheads in B). G indicates a glial profile; double arrows indicate the myelin sheath of a weakly labeled axon.

3.5 Substantia Nigra

We have seen a selective distribution of glutathione-like immunoreactivity in the pars reticulata of the substantia nigra (SNr), in the rat as well as in the cat. A considerable number of gold particles indicating glutathione are consistently seen over dendritic profiles of varying caliber (Figs. 8 A, 11 A). There is also a considerable glutathione immunolabeling over the axoplasm of selected myelinated axons (Fig. 8 A), and

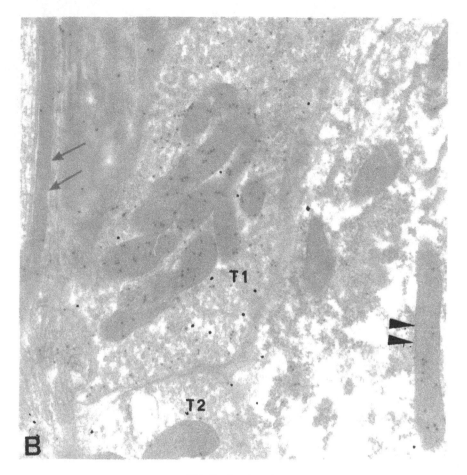

FIG. 7. *Continued.*

scattered gold particles over neuronal cell bodies (Fig. 10 B). On the other hand, only a few gold particles are seen over boutons. It is also apparent from our study that glial profiles show glutathione immunoreactivity (Fig. 8 B). The density of gold particles over glial profiles showed some variation from one experiment to another, but it never exceeded that seen in dendrites or in some myelinated axons. There is a slight overall tendency for the gold particles to cluster over mitochondria (Figs. 8 B, 9 A, 9 B, 10 B). This may be relevant in regard to the reports that suggest that the oxidative stress and damage observed in the substantia nigra of patients with Parkinson's disease could be caused by a mitochondrial defect (Schapira 1996).

The present study does not permit us to state with certainty wether the immunolabeled dendrites belong to dopaminergic neurons or to other cells in the substantia nigra. However, it is well known that dopaminergic cells in SNc send their dendrites into the SNr, where they are densely covered with GABAergic afferents from the striatum

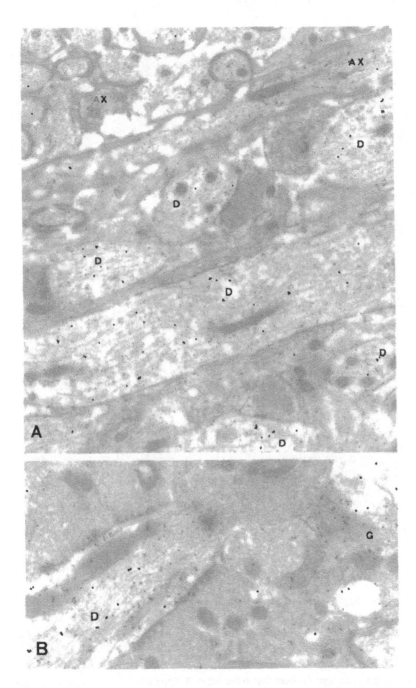

FIG. 8. Electron micrographs from SNr in the rat. A shows the selective distribution of 15-nm gold particles indicating glutathione-LI over dendritic profiles (D) and some myelinated axons (AX). B shows a dendrite (D) and glial profile (G) with an intensity of glutathione-LI considerably higher than that seen over the boutons synapsing upon D.

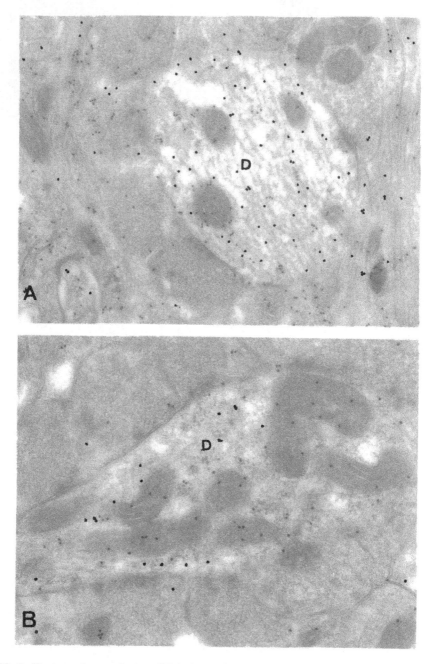

FIG. 9. Electron micrographs from SNr in the rat. A, B: The dendritic profiles (D) are labeled with 15-nm gold particles indicating glutathione-LI. Notice the tendency for the gold particles to cluster over mitochondria. B is from osmicated, Durcupan-embedded material and displays less intense immunolabeling than the other electron micrographs, which were all taken from freeze-substituted material.

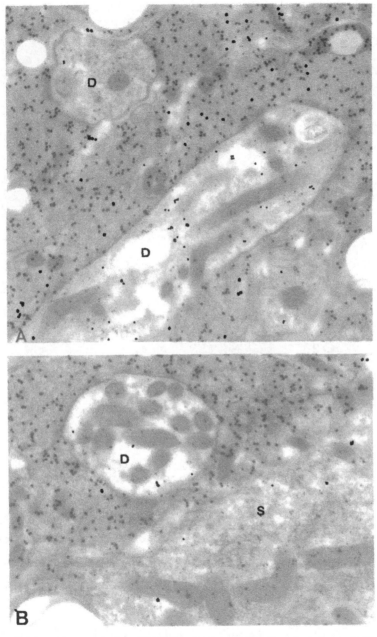

FIG. 10. Electron micrographs from SNr in the rat where the sections have been double-labeled with antibodies raised against glutathione (15-nm gold particles) and antibodies raised against GABA (30-nm gold particles). Notice the selective compartmentation of glutathione-LI in dendritic profiles (D) and neuronal cell somata (S), and of GABA-LI in the many boutons that characteristically cover the nigral dendrites. The two dendritic profiles in A could belong to different cell types, since the elongated profile displays some GABA-LI in addition to glutathione-LI, whereas the transversely sectioned dendrite only displays gluthathione-LI.

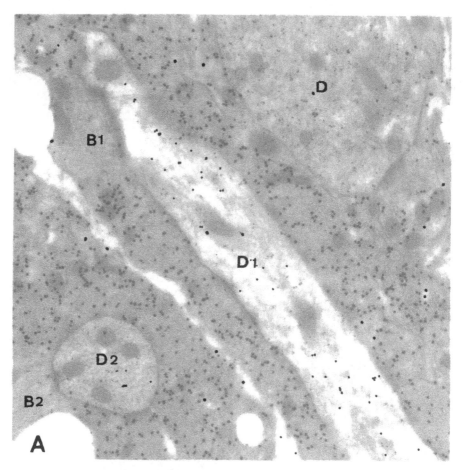

FIG. 11. Electron micrographs from SNr in the rat. In A the sections have been double-labeled with antibodies raised against glutathione (15-nm gold particles) and GABA (30-nm gold particles). In B this pattern was reversed. In A the typical SNr dendritic profiles (D1 and D2) are practically covered with GABA-containing terminals except for two boutons (B1 and B2) that establish asymmetrical synaptic contact with D1 and D2, respectively. B1 and B2 show some scattered 15-nm gold particles indicative of glutathione-LI. In B a similar pattern is seen. Bouton B2 is strongly GABA immunolabeled and establishes a characteristic symmetrical synapse with dendrite D. The neighboring bouton B1 is not GABAergic but displays a few 30-nm gold particles indicative of glutathione-LI over the mitochondria. Bouton B1 establishes an asymmetrical synaptic contact with D and is probably a glutamatergic terminal (Rinvik and Ottersen 1993). Bouton B3 is most probably a GABAergic terminal, but again, a few 30-nm gold particles are seen, related to the mitochondria.

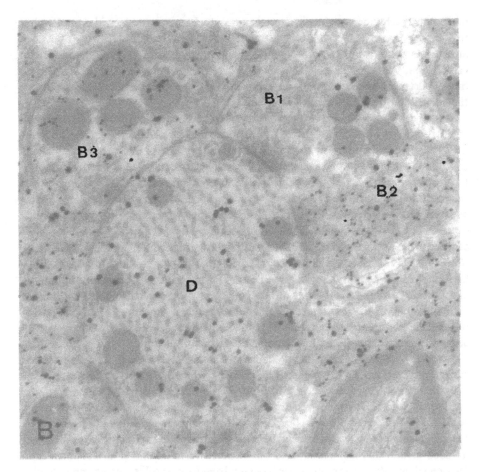

FIG. 11. *Continued.*

and the globus pallidus. In double-labeling experiments it is seen that glutathione-immunolabeled dendrites are synapsed upon by a large number of GABAlike immunoreactive boutons (Fig. 10 A, B). In the same experiments it is noted that the few boutons that establish asymmetric synaptic contacts with glutathione immunoreactive dendrites are not labeled with the antibodies raised against GABA (Figs. 11 A, B). These boutons are probably glutamatergic and could belong to afferents from the subthalamic nucleus (Rinvik and Ottersen 1993).

4. DISCUSSION

At the outset it needs to be emphasized that our knowledge of the glutathione compartmentation in the brain is still far from complete. However, this situation will undoubtedly change as glutathione immunocytochemistry finds wider application in the years to come. Thus the investigations that have been published so far have attested to the feasibility of producing selective antibodies to glutathione (Hjelle et al. 1994; Amara, Coussemacq, and Geffard 1994; Pow and Crook 1995), and have underscored the usefulness of such antibodies to assess the cellular as well as the subcellular distribution of this peptide.

It has been demonstrated in the present review and in previous studies (Hjelle et al. 1994; Usami et al. 1996) that fixed glutathione can be recognized by the antibodies even after embedding in an acrylic (Lowicryl) or epoxy (Durcupan) resin. This has important implications as it permits the application of postembedding immunogold procedures. Besides providing a high anatomical resolution, the latter procedures afford possibilities for quantitative analyses, based on particle counting (Usami et al. 1996; Huster et al. 1997). The particle counts can be regarded as reliable indicators of the glutathione concentration, because the postembedding labeling is restricted to the section surface and thus unaffected by penetration barriers in the tissue (Ottersen 1989a). The development of electron-microscopic calibration systems similar to those designed for other small molecules (Ottersen 1989b; Nagelhus, Lehmann, and Ottersen 1993) should allow the particle counts to be translated into millimolar concentrations of the peptide. Thus the postembedding procedure offers the potential of estimating the concentration of glutathione in subcellular compartments, at an anatomical resolution in the low nanometer range.

Another potential of immunocytochemistry is that of assessing the ratio between reduced and oxidized glutathione. If this could be done by means of the postembedding immunogold procedure, it would provide an indication of the redox status in individual cells and organelles (Viguie et al. 1993; DeMattia et al. 1994; Aukrust et al. 1995). This prospect was a major incentive to the development of our immunocytochemical approach. Unfortunately, it turned out that the antibodies raised against reduced glutathione also recognized the oxidized form (Hjelle et al. 1994; see also the preceding references). So far we have not been successful in our attempts to raise antibodies that react with oxidized glutathione but not with the reduced form (Hjelle et al., unpublished). This implies that the labeling patterns that have been reviewed here must be taken to represent the total pool of glutathione, reduced plus oxidized. It is likely, however, that the reduced form is responsible for the major share of the immunocytochemical signal, as the reduced form is known to predominate over the oxidized form by two orders of magnitude (Meister 1988).

The structural dissimilarity between oxidized and reduced glutathione should be sufficient to make them immunologically distinguishable. In support of this view it should be noted that the present immunization procedure was previously used to raise antibodies that distinguished between two stereoisomers of the same amino acid (Zhang, Storm-Mathisen, and Ottersen 1993; Gundersen et al. 1993) and between such closely related amino acids as glutamate and glutamine (Ottersen et al. 1992; see also Fig. 4 B) or homocysteic acid and glutamate (Zhang and Ottersen 1992). It is likely, therefore, that our failure to produce antibodies specific for the oxidized or reduced forms of glutathione reflects the instability of the conjugate used for immunization. Specifically, part of the reduced glutathione in the immunogen may spontaneously oxidize before or after inoculation, while oxidized glutathione may be subject to the converse process. According to this line of reasoning the addition of an antioxidant to the immunogen should favor the formation of antibodies with higher selectivities for the reduced form of glutathione. This appears to be borne out by the available experimental data (Pow and Crook 1995; Hjelle et al., unpublished).

Pending information on the selective distribution of the two forms of glutathione, the following discussion will focus on the cellular and subcellular localization of the total pool of this tripeptide. It will be seen that the compartmentation of glutathione does not adhere to a limited set of principles valid for all brain regions. Rather, one is struck by the pronounced regional and cellular heterogeneity of the labeling pattern— a heterogeneity that undoubtedly reflects the multifarious roles of glutathione in brain tissue.

4.1 Glial versus Neuronal Glutathione

A central issue is whether glutathione is predominantly neuronal or glial. The literature contains numerous statements to the effect that glutathione mainly resides in glial cells (Slivka, Mytilineou, and Cohen 1987; Raps et al. 1989; Yudkoff et al. 1990). Such statements are often accompanied by references to biochemical analyses of glial versus neuronal cultures.

The following factors must be taken into consideration when interpreting data obtained in cell cultures. First, the results may depend on the brain region of origin. Using well-defined media, Langeveld et al. (1996) demonstrated a predominant glial pool of glutathione when comparing neuronal and glial cultures from the cerebral cortex. However, cultures prepared from the mesencephalon or striatum revealed no significant difference in glutathione contents between the two cell types. Second, the sensitivity of the intracellular pool of glutathione to the extracellular levels of amino acids such as cysteine and glutamate (Kato et al. 1992; Kranich, Hamprecht, and Dringen 1996) underscores the importance of using standardized culture media. Third, it has been proposed that the glutathione homeostasis in the brain depends on a metabolic interaction between neurons and glia (Sagara, Miura, and Bannai 1993). This interaction will necessarily be perturbed under culture conditions.

The picture that has emerged from the immunocytochemical investigations that have been published to date is that glutathione occurs in neurons as well as in glia. This is true for the cerebellum and the substantia nigra as well as the spinal cord. In none of these brain regions does the glial labeling predominate. Figure 4 D serves to illustrate this point: the Purkinje cell body and the adjacent soma of a Golgi epithelial cell (a glial cell type specific for the cerebellum) show comparable labeling intensities. However, higher concentrations of gold particles were found in perivascular glial processes (Hjelle et al. 1994), pointing to possible transport functions of glutathione at these sites.

4.2 Glutathione in Neurons

Having concluded that a substantial fraction of the brain's glutathione pool occurs in neurons, the question arises whether some neuronal populations are preferentially enriched in this tripeptide. The available qualitative data suggest that neuronal elements may differ considerably in their glutathione contents. This is clearly shown in Fig. 4 C, which displays two contiguous axons, one strongly labeled for glutathione and one virtually devoid of immunolabeling. A similar heterogeneity is evident among nerve terminals in the spinal cord (Fig. 7; see also Ramirez-Leon et al. 1997). When confronted with heterogeneities of this kind one must ask whether they reflect differences among different neuronal populations or whether they reflect differences in the functional status of the respective cells. Correlative studies based on double labeling for glutathione and appropriate transmitters indicate that glutathione may be concentrated in terminals with specific transmitter signatures (Ramirez-Leon et al. 1997; see Section 3.4). The finding of glutathione in select nerve terminals is interesting in view of the proposed role of glutathione in signal transmission (Guo et al. 1992; Zängerle et al. 1992; Shaw et al. 1996).

Glutathione labeling also occurs in neuronal cell bodies and dendrites. Athough the dendrosomatic labeling appears to vary among brain regions (compare the spinal cord, Fig. 7, with the cerebellum, Fig. 4) the present data do not support the conclusion of Amara et al. (1994) that the concentration of glutathione generally is lower in neuronal somata than in axons. It is likely that the latter conclusion was based in part on a misinterpretation of the light-microscopic immunolabeling pattern, as their Fig. 2 indicates that many of the vertically oriented processes in the cerebral cortex are apical dendrites rather than axons.

4.3 Glutathione in Nonneural Cells

The inner ear was the first organ to be subjected to a quantitative analysis of glutathione immunoreactivity (Usami et al. 1996). As glutathione is thought to play an important role in the detoxification of ototoxic agents (Garetz et al. 1994), it was surprising to find that the hair cells were weakly labeled for glutathione compared to nonneural cells in the spiral ligament and stria vascularis. This labeling pattern invited speculations as to the existence of a perilymphatic transport of glutathione from the lateral wall tissue to the organ of Corti (Usami et al. 1996).

An analogous situation appears to exist in the retina in so far as the pigment epithelial cells are much more intensely labeled for glutathione than are the cells of the neural part of the retina (Pow and Crook 1995; Huster et al. 1997). While it is conceivable that the pigment epithelial cells serve to supply retinal neurons and glia with glutathione (Pow and Crook 1995) it is important to note that a correspondingly high level of glutathione also occurs in the pigment cells of the choroid (Huster et al. 1997). This suggests that the enrichment of glutathione is linked to the synthesis of melanin—a function that these two cell types have in common. It is well known that melanin synthesis is associated with the formation of toxic by-products (Riley 1988), and glutathione might offer protection against these.

As glutathione has been implicated in numerous transport functions, it was of interest to determine whether this tripeptide is enriched in endothelial cells. Such cells were found to be strongly labeled in the inner ear (Usami et al. 1996). In the cerebellum, however, the labeling of endothelial cells was weak (Fig. 4 A), and significant gold particle densities occurred in surrounding glial processes (see section 4.1). These differences point to possible regional heterogeneities in transendothelial amino acid transport.

4.4 Glutathione in the Substantia Nigra

The labeling pattern in the substantia nigra is of particular interest. There is a considerable body of evidence suggesting that oxidative stress may play an important role in Parkinson's disease (PD), although the causal relationship between the degeneration of the dopaminergic nerve cells in the pars compacta of the substantia nigra (SNc) and oxidative stress remains elusive (Jenner and Olanow 1996). Several years ago it was reported that reduced glutathione was significantly reduced in the substantia nigra of patients with PD (Perry, Godin, and Hansen 1982). This observation was confirmed in later studies, where it was shown that glutathione was diminished in the substantia nigra but not in other regions of the brain of patients with PD (Perry and Yong 1986; Perry, Hansen, and Jones 1988). These reports were largely ignored, as was another study that showed that the reduction of total glutathione in SNc in PD correlated with the severity of the disease (Riederer et al. 1989). Paying particular attention to the different forms of glutathione and to the methodological problems associated with biochemical investigations of post mortem material, Sofic et al. (1992) and Jenner (1993a,b) reported that reduced glutathione was diminished in SNc of patients with PD, whereas oxidized glutathione was unaffected or only marginally elevated (Sian et al. 1994a). It was also documented that glutathione was diminished in SNc of patients with incidental Lewy-body disease (ILD), a condition considered by many clinicians as a preclinical form of PD (Jenner et al. 1992a,b; Sofic et al. 1992; Jenner 1993a,b; Dexter et al. 1994). Importantly, there was no reduction of glutathione in SNc in autopsy brains of patients who had suffered from other neurodegenerative diseases than PD.

The majority of the published reports on glutathione in the substantia nigra have been based on pharmacological and biochemical investigations. To our knowledge very few studies have been undertaken with the purpose of assessing the distribution of glutathione within the cellular compartments of the substantia nigra. It has been known for a long time that the substantia nigra of the rat contains very high concentrations of glutathione peroxidase (Brannan et al. 1980; Mizuno and Ohta 1986).

In a light-microscopical immunohistochemical investigation, Damier et al. (1993) reported that glutathione peroxidase–like immunoreactivity was detected exclusively in glial cells in the substantia nigra of supposedly normal human brains. In brains of patients with PD there was an increased density of glutathione peroxidase immunostained cells surrounding the surviving dopaminergic neurons. Sian et al. (1994b), on the other hand, could not find any changes in glutathione peroxidase in SNc in PD. In a recent immunocytochemical investigation Knollema et al. (1996) studied the distribution of the enzyme glutathione reductase in various murine species. They found an intense immunoreactivity for this enzyme (which recycles oxidized glutathione back to the reduced form) in the cells in SNc, but few and scattered immunoreactive cells in SNr. Interestingly, the authors found a differential expression of the enzyme in the various murine species. Thus, some species showed a pronounced immunoreactivity for glutathione reductase in glial cells in regions that lacked neuronal immunoreactivity.

Previous studies have not permitted any definite conclusion on the cellular compartmentation of glutathione in the substantia nigra. The data reviewed here show that dendrites and somata of the substantia nigra generally display quite strong immunolabeling. Although the intensity of the labeling of glial profiles varied, in many instances it matched (though it never surpassed) the density of gold particles over dendrites and select myelinated axons. Some of the labeled neuronal profiles undoubtedly represent dendrites of dopaminergic cells, as judged by their characteristic covering of GABA immunopositive terminals. It was previously reported that cultured mesencephalic cells that were positive for tyrosine hydroxylase (i.e., putative dopaminergic neurons) were weakly stained for glutathione compared to other neurons in the culture (Langeveld et al. 1996). Although the present immunogold data are compatible with the idea that certain nondopaminergic cells may be somewhat more strongly labeled than the dopaminergic ones, the differences are not nearly as pronounced as those indicated in the material of Langeveld et al. (1996). Discrepancies of this kind may be explained by the culture conditions, as explained in section 4.1.

4.5 Subcellular Localization of Glutathione

The immunogold analyses indicate that glutathione occurs in the cytoplasm as well as in the nuclei. Its presence in the latter site is expected, because its small size should allow it to diffuse through the nuclear pores. In the cytoplasm one usually finds that glutathione is enriched in mitochondria compared to the other cell compartments. This enrichment, expressed as the particle density ratio between the mitochondria and the surrounding cytoplasmic matrix, varies considerably among different cell populations, as shown quantitatively in the retina (Huster et al. 1997; see also above). Mitochondrial labeling is often found even in cell profiles that are otherwise virtually devoid of immunoreactivity (see, e.g., Fig. 7). The presence of glutathione in mitochondria would be in line with biochemical analyses (Reichelt and Fonnum 1969) and is likely to reflect the need to neutralize reactive oxygen species formed during the process of oxidative phosphorylation (Huang and Philbert 1995).

5. CONCLUSIONS

The immunocytochemical data reviewed here have disclosed a pattern of glutathione distribution that is much more complex than hitherto assumed. To better interpret this pattern one needs to disentangle the different roles of glutathione; specifically it would seem important to distinguish the compartments where glutathione primarily serves as an antioxidant from the compartments where glutathione serves other roles. Correlative studies with antibodies to glutathione peroxidase, glutathione reductase, and related enzymes would be helpful in this regard. It is also essential that the immunolocalization studies be extended to experimental disease models or pathological material, to test the proposal that an absolute or relative glutathione deficiency might contribute to the development of neurodegenerative disease. Finally, as antibodies are now available to glutathione and γ-glutamylcysteine (Huster et al. 1997) as well as to the constituent amino acids (Storm-Mathisen et al. 1983; Dale et al. 1986; Ottersen 1989a; Huster et al. 1997), carefully designed in vitro experiments should help elucidate the factors that determine the intracellular level and turnover of glutathione.

ACKNOWLEDGMENTS

Dr. Shin-ichi Usami kindly provided the electron micrographs for Fig. 5. Thanks are also due to Dr. V. Ramirez-Leon and Dr. B. Ulfhake for helpful comments, and to B. Riber, K. M. Gujord, and G. Lothe for expert technical assistance. This work was supported by the Norwegian Research Council and the European Union Biomed Programme (contract BMH4-CT96-0851).

REFERENCES

Amara, A., Coussemacq, M., and Geffard, M. 1994. Antibodies to reduced glutathione. *Brain Research* 658(1–2):237–42.

Aukrust, P., Svardal, A. M., Muller, F., Lunden, B., Berge, R. K., Ueland, P. M., and Froland, S. S. 1995. Increased levels of oxidized glutathione in CD^{4+} lymphocytes associated with disturbed intracellular redox balance in human immunodeficiency virus type 1 infection. *Blood* 86(1):258–67.

Blackstad, T. W., Karagülle, T., and Ottersen, O. P. 1990. MORFOREL, a computer program for two-dimensional analysis of micrographs of biological specimens, with emphasis on immunogold preparations. *Computers in Biology and Medicine* 20:15–34.

Brannan, T. S., Maker, H. S., Weiss, C., and Cohen, G. 1980. Regional distribution of glutathione peroxidase in the adult rat brain. *Journal of Neurochemistry* 35(4):1013–4.

Coyle, J. T., and Puttfarcken, P. 1993. Oxidative stress, glutamate, and neurodegenerative disorders. *Science* 262:689–95.

Dale, N., Ottersen, O. P., Roberts, A., and Storm-Mathisen, J. 1986. Inhibitory neurones of a motor pattern generator in *Xenopus* revealed by antibodies to glycine. *Nature* 324:255–7.

Damier, P., Hirsch, E. C., Zhang, P., Agid, Y., and Javoy-Agid, F. 1993. Glutathione peroxidase, glial cells and Parkinson's disease. *Neuroscience* 52(1):1–6.

DeMattia, G., Laurenti, O., Bravi, C., Ghiselli, A., Iuliano, L., and Balsano, F. 1994. Effect of aldose reductase inhibition on glutathione redox status in erythrocytes of diabetic patients. *Metabolism: Clinical and Experimental* 43(8):965–8.

Dexter, D. T., Sian, J., Rose, S., Hindmarsh, J. G., Mann, V. M., Cooper, J. M., Wells, F. R., Daniel, S. E., Lees, A. J., Schapira, A. H., Jenner, P., and Marsden, C. D. 1994. Indices of oxidative stress and mitochondrial function in individuals with incidental Lewy body disease. *Annals of Neurology* 35(1):38–44.

Garetz, S. L., Altschuler, R. A., and Schacht, J. 1994. Attentuation of gentamicin ototoxicity by glutathione in the guinea pig in vivo. *Hearing research* 77(1–2):81–7.

Griffith, O. W., and Meister, A. 1979. Potent and specific inhibition of glutathione synthesis by buthionine sulphoximine (*S-n*-butyl homocysteine sulfoximine). *Journal of Biological Chemistry* 254:7558–60.

Gundersen, V., Danbolt, N. C., Ottersen, O. P., and Storm-Mathisen, J. 1993. Demonstration of glutamate/aspartate uptake activity in nerve endings by use of antibodies recognizing exogenous D-aspartate. *Neuroscience* 57:97–111.

Guo, N., McIntosh, C., and Shaw, C. 1992. Glutathione: new candidate neuropeptide in the central nervous system. *Neuroscience* 51:835–42.

Hjelle, O. P., Chaudhry, F. A., and Ottersen, O. P. 1994. Antisera to glutathione: characterization and immunocytochemical application to the rat cerebellum. *European Journal of Neuroscience* 6:793–804.

Huang, J., and Philbert, M. A. 1995. Distribution of glutathione and glutathione-related enzyme systems in mitochondria and cytosol of cultured cerebellar atrocytes and granule cells. *Brain Research* 680:16–22.

Huster, D., Haug, F.-M., Nagelhus, E., Reichelt, W., and Ottersen, O. P. 1997. Subcellular distribution of glutathione in retinal cells of rat and guinea pig: a high resolution, quantitative immunogold analysis. Submitted.

Jenner, P. 1993a. Presymptomatic detection of Parkinson's Disease. *Journal of Neural Transmission (suppl.)* 40:23–36.

Jenner, P. 1993b. Altered mitochondrial function, iron metabolism and glutathione levels in Parkinson's disease. *Acta Neurologica scandinavica (suppl.)* 146:6–13.

Jenner, P., and Olanow, C. W. 1996. Pathological evidence for oxidative stress in Parkinson's disease and related degenerative disorders. In *Neurodegeneration and neuroprotection in Parkinson's disease*, ed. C. W. Olanow, P. Jenner, and M. Youdim, 24–46. London; Academic Press.

Jenner, P., Dexter, D. T., Sian, J., and Schapira, A. H. 1992a. Oxidative stress as a cause of nigral cell death in Parkinson's disease and incidental Lewy body disease. *Annals of Neurology* 32:S82–7.

Jenner, P., Schapira, A. H., and Marsden, C. D. 1992b. New insights into the cause of Parkinson's disease. *Neurology* 42(12):2241–2250.

Ji, Z. Q., Aas, J. E., Laake, J., Walberg, F., and Ottersen, O. P. 1991. An electron-microscopic, immunogold analysis of glutamate and glutamine in terminals of rat spinocerebellar fibers. *Journal of Comparative Neurology* 307:296–310.

Kato, S., Negishi, K., Mawatari, K., and Kuo, C.-H. 1992. A mechanism for glutamate toxicity in the C6 glioma cells involving inhibition of cysteine uptake leading to glutathione depletion. *Neuroscience* 48(4)903–14.

Knollema, S., Hom, H. W., Schirmer, H., Korf, J., and Ter Horst, G. J. 1996. Immunolocalization of glutathione reductase in the murine brain. *Journal of Comparative Neurology* 373(2):157–72.

Kranich, O., Hamprecht, B., and Dringen, R. 1996. Different preferences in the utilization of amino acids for glutathione synthesis in cultured neurons and astroglial cells derived from rat brain. *Neuroscience Letters* 219:211–4.

Langeveld, C. H., Schepens, E., Jongelen, C. A. M., Stoof, J. C., Hjelle, O. P., Ottersen, O. P., and Drukarch, B. 1996. Presence of glutathione immunoreactivity in cultured neurones and astrocytes. *Neuroreport* 7:1833–6.

Meister, A. 1988. Glutathione metabolism and its selective modification. *Journal of Biological Chemistry* 263:17205–8.

Mizui, T., Kinouchi, H., and Chan, P. H. 1992. Depletion of brain glutathione by buthionine sulfoximine enhances cerebral ischemic injury in rats. *American Journal of Physiology* 262:H313–17.

Mizuno, Y., and Ohta, K. 1986. Regional distributions of thiobarbituric acid–reactive products, activities of enzymes regulating the metabolism of oxygen free radicals, and some of the related enzymes in adult and aged rat brains. *Journal of Neurochemistry* 46(5):1344–52.

Nagelhus, E. A., Lehmann, A., and Ottersen, O. P. 1993. Neuronal–glial exchange of taurine during hypo-osmotic stress: a combined immunocytochemical and biochemical analysis in rat cerebellar cortex. *Neuroscience* 54:615–31.

Orlowski, M., and Karkowsky, A. 1976. Glutathione metabolism and some possible functions of glutathione in the nervous system. *International Review in Neurobiology.* 19:75–121.

Ottersen, O. P. 1987. Postembedding light- and electronmicroscopic immunocytochemistry of amino acids. Description of a new model system allowing identical condtions for specificity testing and tissue processing. *Experimental Brain Research* 69:167–74.

Ottersen, O. P. 1989a. Quantitative electron microscopic immunocytochemistry of amino acids. *Anatomy and Embryology* 180:1–15.

Ottersen, O. P. 1989b. Postembedding immunogold labeling of fixed glutamate: an electron microscopic analysis of the relationship between gold particle density and antigen concentration. *Journal of Chemical Neuroanatomy* 2:57–66.

Ottersen, O. P., and Storm-Mathisen. 1984. Glutamate- and GABA-containing neurons in the mouse and rat brain, as demonstrated with a new immunocytochemical technique. *Journal of Comparative Neurology* 229:374–92.

Ottersen, O. P., Zhang, N., and Walberg. 1992. Metabolic compartmentation of glutamate and glutamine: Morphological evidence obtained by quantitative immunocytochemistry in rat cerebellum. *Neuroscience* 46:519–34.

Owen, A. D., Schapira, A. H., Jenner, P., and Marsden, C. D. 1996. Oxidative stress and Parkinson's disease. *Annals of the New York Academy of Sciences* 786:217–23.

Perry, T. L., and Yong, V. W. 1986. Idiopathic Parkinson's disease, progressive supranuclear palsy and glutathione metabolism in the substantia nigra of patients. *Neuroscience Letters* 67(3):269–74.

Perry, T. L., Godin, D. V., and Hansen, S. 1982. Parkinson's disease: a disorder due to nigral glutathione deficiency? *Neuroscience Letters* 33(3):305–10.

Perry, T. L., Hansen, S., and Jones, K. 1988. Brain amino acids and glutathione in progressive supranuclear palsy. *Neurology* 38(6):943–6.

Pow, D. V., and Crook, D. K. 1995. Immunocytochemical evidence for the presence of high levels of reduced glutathione in radial glial cells and horizontal cells in the rabbit retina. *Neuroscience Letters* 193:25–8.

Ramirez-Leon, V., Kullberg, S., Ottersen, O. P., Storm-Mathisen, J., and Ulfhake, B. 1997. Distribution of glutathione-, glutamate-, GABA- and glycine-immunoreactivities in the aged rat spinal cord motor nucleus. Submitted.

Raps, S., Lai, J. C. K., Hertz, L., and Cooper, A. J. L. 1989. Glutathione is present in high concentrations in cultured astrocytes but not in cultured neurons. *Brain Research* 493:398–401.

Reichelt, K. L., and Fonnum, F. 1969. Subcellular localization of N-acetyl-aspartyl-glutamate, N-acetyl-glutamate and glutathione in brain. *Journal of Neurochemistry* 16:1409–16.

Riederer, P., Sofic, E., Rausch, W. D., Schmidt, B., Reynolds, G. P., Jellinger, K., and Youdim, M. B. 1989. Transition metals, ferritin, glutathione, and ascorbic acid in Parkinsonian brains. *Journal of Neurochemistry* 52(2):515–20.

Riley, P. A. 1988. Radicals in melanin biochemistry. *Annals of the New York Academy of Sciences* 551:111–9.

Rinvik, E. and Ottersen, O. P. 1993. Terminals of subthalamonigral fibres are enriched with glutamate-like immunoreactivity: an electron microscopic immunogold analysis in the cat. *Journal of Chemical Neuroanatomy* 6:19–30.

Rosen, D. R., Siddique, T., Patterson, D., Figlewicz, D. A., Sapp, P., Hentati, A., Donaldson, D., Goto, J., O'Regan, J. P., Deng, H. X., Rahmani, Z., Krizus, A., McKenna-Yasek, D., Cayabyab, A., Gaston, S. M., Berger, R., Tanzi, R. E., Heplerin, J. J., Herzfeldt, B., Van den Bergh, R., Hung, W. Y., Bird, T., Deng, G., Mulder, D. W., Smyth, C., Laing, N. G., Soriano, E., Pericak-Vance, M. A., Haines, J., Rouleau, G. A. Gusella, J. S., Horvitz, H. R., and Brown, R. H. 1993. Mutations in Cu/Zn superoxide dismutase gene are associated with familial amyotrophic lateral sclerosis. *Nature* 362:59–62.

Sagara, J. I., Miura, K., and Bannai, S. 1993. Maintenance of neuronal glutathione by glial cells. *Journal of Neurochemistry* 61(5):1672–6

Schapira, A. H. 1996. Oxidative stress and mitochondrial dysfunction in neurodegeneration. *Current Opinion in Neurobiology* 9(4):260–4.

Shaw, C. A., Pasqualotto, B. A., and Curry, K. 1996. Glutathione-induced sodium currents in neocortex. *Neuroreport* 7(6):1149–52.

Sian, J., Dexter, D. T., Lees, A. J., Daniel, S., Jenner, P., and Marsden, C. D. 1994a. Glutathione-related enzymes in brain in Parkinson's disease. *Annals of Neurology* 36(3):356–61.

Sian, J., Dexter, D. T., Lees, A. J., Daniel, S., Agid, Y., Javoy-Agid, F., Jenner, P., and Marsden, C. D. 1994b. Alterations in glutathione levels in Parkinson's disease and other neurodegenerative disorders affecting basal ganglia. *Annals of Neurology* 36(3):348–55.

Slivka, A., Mytilineou, C., and Cohen, G. 1987. Histochemical evaluation of glutathione in brain. *Brain Research* 409:275–84.

Sofic, E., Lange, K. W., Jellinger, K., and Riederer, P. 1992. Reduced and oxidized glutathione in the substantia nigra of patients with Parkinson's disease. *Neuroscience Letters* 142(2):128–30.

Storm-Mathisen, J., Leknes, A. K., Bore, A. T., Vaaland, J. L., Edminson, P., Haug, F. M. S., and Ottersen, O. P. 1983. First visualization of glutamate and GABA in neurons by immunocytochemistry. *Nature* 301:517–20.

Usami, S.-I., Hjelle, O. P., and Ottersen, O. P. 1996. Differential cellular distribution of glutathione—an endogenous antioxidant—in the guinea pig inner ear. *Brain Research* 743:337–40.

van Lookeren Campagne, M., Oestreicher, B., van der Krift, T. P., Gispen, W. H., and Verkleij, A. J. 1991. Freeze-substitution and Lowicryl HM20 embedding of fixed rat brain: suitability for immunogold ultrastructural localization of neuronal antigens. *Journal of Histochemistry and Cytochemistry* 39:1267–1279.

Viguie, C. A., Frei, B., Shigenaga, M. K., Ames, B. N., Packer, L., and Brooks, G. A. 1993. Antioxidant status and indexes of oxidative stress during consecutive days of exercise. *Journal of Applied Physiology* 75(2):566–72.

Yudkoff, M. D., Pleasure, L., Cregar, L., Lin, Z.-P., Nissim, I., Stern, J., and Nissim, I. 1990. Glutathione turnover in cultured astrocytes: studies with [3H]glutamate. *Journal of Neurochemistry* 55:137–45.

Zängerle, L., Cuénod, M., Winterhalter, K. H., and Do, K. Q. 1992. Screening of thiol compounds: depolarization-induced release of glutathione and cysteine from rat brain slices. *Journal of Neurochemistry* 59:181–9.

Zhang, N., and Ottersen, O. P. 1992. Differential cellular distribution of two sulphur containing amino acids in rat cerebellum: an immunocytochemical investigation using antisera to taurine and homocystic acid. *Experimental Brain Research* 90:11–20.

Zhang, N., Storm-Mathisen, J., and Ottersen, O. P. 1993. A model system for specificity testing and antigen quantification in single and double labeling postembedding electron microscopic immunocytochemistry. *Elsevier neuroscience protocols* 93/050/13/01–20.

PART II

Multiple Roles of Glutathione in Neural Function

Glutathione in the Nervous System
Edited by Christopher A. Shaw
Copyright © 1998 Taylor & Francis

5

Role of Astrocytes in Maintaining Cerebral Glutathione Homeostasis and in Protecting the Brain Against Xenobiotics and Oxidative Stress

Arthur J. L. Cooper

Departments of Biochemistry and of Neurology and Neuroscience, Cornell University Medical College, New York, New York and Burke Medical Research Institute, Cornell University Medical College, White Plains, New York

1. INTRODUCTION

The weight of the average adult human brain is about 2% that of the whole body, yet the blood flow to the adult brain is about 20% of the cardiac output. The adult human brain relies almost entirely on the metabolism of glucose to meet its energy demands, and most (>90 percent) of this carbohydrate is oxidized to CO_2. Very little brain glucose (a few percent) is normally converted to lactate and released to the venous blood. The very high capacity of the brain to oxidize glucose suggests that this organ may be especially vulnerable to oxidative stress such as that induced by reflow following removal of an ischemic insult. Nevertheless, the brain does have the necessary machinery to combat oxidative stress provided it is not prolonged. Within the last few years it has become apparent that glutathione (GSH) plays a particularly important role in protecting the brain against oxidative stress and against potentially harmful xenobiotics, including some drugs and drug metabolites that can cross the blood–brain barrier. In this chapter the general biological properties of GSH will be discussed, especially in relation to normal brain functioning. Emphasis will be placed on the role of GSH as a neuroprotectant and on the recent findings that astrocytes may be particularly important in protecting the brain against noxious stimuli.

2. GENERAL BIOLOGICAL PROPERTIES OF GLUTATHIONE

This topic has recently been extensively reviewed by the author (Cooper 1997), so that only the salient features are presented here. GSH, along with ascorbate and α-tocopherol (vitamin E), is a major antioxidant in most tissues. Indeed, some evidence suggests that ascorbate and GSH act synergystically as antioxidants and that ascorbate can ameliorate the effects of GSH deficiency (see for example Jain et al. 1994 and references quoted therein).

In addition to its roles in protecting tissues against noxious compounds, GSH has other biological functions. The sulfhydryl of GSH is less reactive than that of free cysteine, and consequently GSH serves as a nontoxic storage form of cysteine. Cysteine can form relatively stable hemithioketals with ketones (e.g., α-keto acids). With aldehydes [e.g., glyoxylate and pyridoxal 5'-phosphate (PLP)], hemithioacetal formation is followed by cyclization to a thiazolidone derivative. High concentrations of cysteine inhibit some key PLP-containing enzymes, including glutamate decarboxylase, and this may contribute to the known neurotoxocity of cysteine. In contrast to cysteine, GSH does not form stable adducts with α-keto acids or with PLP, although it does form a stable adduct with glyoxylate. The concentration of GSH in mammalian tissues ranges from about 0.4 to 12 mM. The concentration of cyst(e)ine in tissues is generally <100 μM. Apparently, the high tissue concentration of GSH results in part from the unusual γ-glutamyl peptide linkage, which confers resistance to peptidases.

Orlowski and Meister (1970) pointed out that GSH may have a unique metabolic role in the transport of amino acids. The authors suggested that enzymes involved in GSH metabolism can be linked to form a process for the recycling of the three amino acid constituents of GSH. The authors termed this process the γ-glutamyl cycle (Fig. 1). It was suggested that the positioning of γ-glutamyltranspeptidase (Fig. 1, reaction 1) on the cell surface and of other enzymes of glutathione metabolism within the cell permits the γ-glutamyl cycle to play a role in the translocation of amino acids across cell membranes. While there is strong evidence to suggest that the

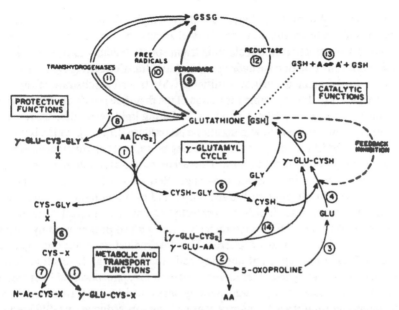

FIG. 1. Metabolism of glutathione (GSH). Reactions: (1) γ-glutamyltranspeptidase; (2) γ-glutamylcyclotransferase; (3) 5-oxoprolinase; (4) γ-glutamylcysteine synthetase; (5) glutathione synthetase; (6) dipeptidase; (7) L-cysteine-S-conjugate N-acetyltransferase; (8) glutathione S-transferase; (9) glutathione peroxidase; (10) free-radical quenching (probably nonenzymatic); (11) glutathione transhydrogenases; (12) glutathione disulfide (GSSG) reductase; (13) enzymatic reactions in which GSH is required as a cofactor but is not consumed; (14) transport of γ-glutamylcystine and reduction to γ-glutamylcysteine. AA = amino acids; Cys_2, cystine; X = compounds that form conjugates with GSH. From Meister (1986) (reproduced with permission of the publisher).

γ-glutamyl cycle does indeed operate in vivo, the cycle is now recognized not to play a key role in amino acid transport (except perhaps in the case of cystine). Rather, the cycle is thought to have a specialized role in metabolism (e.g., in transformations of leukotriene A, estrogens, and prostaglandins and in detoxification of xenobiotics). Nevertheless, the idea that enzymes associated with GSH metabolism can be linked in a catalytic cycle has stimulated a great deal of useful work and has led to a strong appreciation of the role of GSH in biological processes, including those in the brain.

Of major biological significance is the ready interconversion between GSH (a reduced form) and GSSG (glutathione disulfide, an oxidized form). Most tissues, including the brain, normally contain low concentrations of GSSG. As discussed in the next section, the normally low level of cerebral GSSG is consistent with the inherently high activity of the flavoprotein GSSG reductase in brain [Eq. (1)]. In most tissues the bulk of cellular GSH is in the cytosolic compartment. However, a small fraction (10–20 percent) is in the mitochondria, where it presumably serves as an important defense against endogenously generated reactive oxygen species. Small amounts of GSH are present in the cell in the form of mixed disulfides with low-molecular-weight thiols (e.g., cysteine, γ-glutamylcysteine) and with cysteinyl residues of proteins. The latter may be involved in the regulation of activity of some enzymes. GSH is also important in the production of deoxyribonucleotide triphosphate precursors of DNA. Intracellular GSH is a substrate of transhydrogenases (thiol transferases) (Fig. 1, reaction 11) that reduce low-molecular-weight disulfides. GSH is also a substrate of thiol transferases that catalyze the reversible formation of the disulfide bonds of certain proteins.

Hydrogen peroxide and other peroxides, including lipid peroxides, are reduced by the selenium-containing enzyme glutathione peroxidase [Eq. (2)], which is present in most tissues. In addition, many tissues, including brain, contain a distinct phospholipid hydroperoxide glutathione peroxidase. The two peroxidases function in the protection of cell membranes against oxidative damage. Reduction of lipid peroxides is also catalyzed by certain glutathione S-transferases and by a low molecular weight form of glutathione peroxidase. Lipids are protected against free radical damage in part by α-tocopherol (vitamin E), which quenches free-radical propagation by formation of the oxy radical form of α-tocopherol. This free radical does not propagate but is reduced by GSH in an apparent nonenzymatic reaction (Fig. 1, reaction 10). This reaction and those catalyzed by glutathione peroxidase, phospholipid hydroperoxide glutathione peroxidase, and the transhydrogenases lead to the formation of GSSG, which does not normally accumulate because it is rapidly reduced by GSSG reductase [Eq. (1)]:

$$GSSG + NADPH + H^+ \rightarrow 2GSH + NADP^+ \tag{1}$$

$$2GSH + H_2O_2 \text{ (or } RO_2H) \rightarrow GSSG + 2H_2O \text{ (or } ROH + H_2O) \tag{2}$$

As noted above, GSH forms S-conjugates (Fig. 1, reaction 8) with a number of compounds of endogenous origin such as leukotriene A, estrogens, and prostaglandins. GSH also forms S-conjugates with compounds of exogenous origin and is thus important in drug metabolism and in the detoxification of some other xenobiotics. In the mercapturate detoxification pathway, the GSH conjugate is typically hydrolyzed to the corresponding cysteinylglycine S-conjugate and thence to the cysteine S-conjugate (Fig. 1, reactions 8 $\rightarrow$ 1 $\rightarrow$ 6) by the consecutive action of γ-glutamyltranspeptidase and dipeptidase, respectively. The cysteine S-conjugate is then acetylated to the N-acetyl-S-cysteine conjugate (mercapturate), which is excreted in the urine. Cysteine S-conjugates can also undergo γ-glutamylation in a reaction catalyzed by γ-glutamyltranspeptidase (Fig. 1, reaction 1). Cysteine S-conjugates and glutathione S-conjugates often undergo a variety of chemical transformations, many of which are catalyzed by enzymes involved in other aspects of metabolism. For example, activities of cysteine S-conjugate β-lyases [which catalyze the reaction $RSCH_2CH(NH_3^+)CO_2^-$ + $H_2O \rightarrow$ pyruvate + NH_4^+ + RS^-] in rat liver and kidney have been shown to be associated with several PLP-containing enzymes, including kynureninase, kynurenine-pyruvate aminotransferase, and cytosolic glutamine transaminase K (Cooper and Tate 1997). In some cases, in which the eliminated fragment (RS^-) is very reactive, toxicity toward liver, brain, and especially kidney may result from the action of cysteine S-conjugate β-lyases. In this case, rather than serving as a mechanism for the detoxification of xenobiotics, the mercapturate pathway serves as a vehicle for a bioactivation (or toxification) process.

Finally, GSH is an essential and specific cofactor for a number of enzymes, including formaldehyde dehydrogenase, glyoxylase, maleylacetoacetate isomerase, dehydrochlorinases, and prostaglandin endoperoxide isomerase. In these reactions, GSH is not consumed, but presumably plays a catalytic role.

3. CONCENTRATION OF SULFUR-CONTAINING AMINO ACIDS IN VIVO

An important concept regarding GSH turnover in brain involves metabolic compartmentation. In order to appreciate this idea, a knowledge of the concentrations of key sulfur-containing amino acids in brain, blood, and cerebrospinal fluid is helpful. The concentrations of GSH and cyst(e)ine in whole brain are $\approx$1–3 mM and 30–70 μM, respectively, depending on the species. Within a given species the concentration of GSH in various brain regions tends to be in the following order: forebrain and cortex > cerebellum > brain stem $\geq$ spinal cord. The concentration in sciatic nerve is relatively low. In the cat, the concentration of GSH in the brain increases from 1.6 to about 2.2 mM during the first month of postnatal development. GSH levels are decreased in the aged rodent brain. [See Cooper (1997) for original references.] The concentration of total glutathione [i.e. GSH + (GSSG $\times$ 2)] in rat plasma is $\approx$15–20 μM, of which at least 80 percent is accounted for as GSH (Jain et al. 1991). The concentration of cyst(e)ine in rat blood is $\approx$70–100 μM (Semon et al. 1989), of which $\approx$90 percent is in the form of cystine (Saetre and Rabenstein 1978). The concentrations of total glutathione and cyst(e)ine in rat cerebrospinal fluid are $\approx$5 and 4 μM, respectively; in both cases the reduced form predominates (Anderson et al. 1989).

Glutathione is easily oxidized to its disulfide form, yet most of the glutathione in the brain is in the reduced form (see the next paragraph). Maintenance of a high GSH/GSSG ratio requires energy expenditure. As noted in the previous section, the adult brain relies almost exclusively on the oxidation of glucose to meet its energy demands. The pentose phosphate pathway is present in the brain, but only 3–5 percent of cerebral glucose carbon is converted to CO_2 via the oxidative branch of the pathway; the rest is oxidized via the tricarboxylic acid cycle (Hostetler and Landau 1967; Baquer et al. 1977, 1988). Nevertheless, the pentose phosphate pathway is important in brain as a means of providing NADPH for the reduction of GSSG to GSH (Hotta 1962; Hotta and Seventko 1968; Baquer et al. 1988). Interestingly, the pentose phosphate pathway can be upregulated in cultured astrocytes and neurons exposed to hydrogen peroxide (Ben-Yoseph, Boxer, and Ross 1994, 1996a,b). In addition to the pentose phosphate pathway, other possibly significant sources of NADPH for the maintenance of high GSH/GSSG ratios (at least in the retina) are the reactions catalyzed by the two malic enzymes and by isocitrate dehydrogenase (Winkler, De Santis, and Solomon 1986).

GSSG reductase [Eq. (1)] activity is high in normal brain. Therefore, Folbergrová, Rehncrona, and Siesjö (1979) assumed that this reaction is in equilibrium in normal brain. From this assumption and the known NADPH/NADP$^+$ ratio, these authors calculated that the concentration of GSSG in normal brain should be $\sim 10^{-10}$ M. It is technically difficult, however, to measure low levels of tissue GSSG in the presence of high levels of GSH. Some nonenzymatic oxidation of GSH to GSSG during tissue extraction cannot be avoided. Nevertheless, by taking several precautions, Cooper, Pulsinelli, and Duffy (1980) were able to show that the concentration of GSSG in rat brain is ≤ 2 μM—a value ≤ 0.2 percent of total glutathione. Other workers have shown that the concentration of GSSG in normal human and monkey brain is ≤ 1.2 percent of total glutathione (Slivka, Spina, and Cohen 1988). Many studies have been reported in which the GSSG/GSH ratio in the brain is used as an index of oxidative stress but in which the normal baseline levels of brain GSSG are reported as ≥ 50 μM. Because such baseline values are obviously artificially high, such studies should be interpreted with caution (Cooper 1997). Nevertheless, it is apparent that the actual in vivo brain GSSG content can indeed rise to values >1% of the total glutathione pool under conditions in which the production of hydrogen peroxide is stimulated (e.g., after reserpine administration) (Spina and Cohen 1989). As noted above, hydrogen peroxide is detoxified in part by the action of glutathione peroxidase with concomitant formation of GSSG [Eq. (2)]. Evidently, under the conditions of rapid hydrogen peroxide formation the pentose-phosphate-pathway–GSSG-reductase reaction in brain is unable to keep up with the oxidation of GSH to GSSG and GSSG accumulates.

4. SUBCELLULAR LOCALIZATION OF GLUTATHIONE IN THE BRAIN

Mitochondria contain a distinct pool of GSH but do not possess enzymes necessary for its biosynthesis (Griffith and Meister 1985). However, mitochondria are effective at importing GSH from the cytosol, and rat liver mitochondria contain at least two transport systems for GSH import (Mårtensson, Lai, and Meister 1990; Kurosawa et al. 1990). Some evidence suggests that GSSG can also be taken up by mitochondria to be reduced therein to GSH (Kurosawa et al. 1990). As in most tissues $\approx$10–20% of GSH in neural cells is in the mitochondrial compartment (Jain et al. 1991; Bonnefoi 1992; Werner and Cohen 1993; Huang and Philbert 1995, 1996). The importance of the mitochondrial pool of GSH is attested by the fact that depletion of this pool by administration of buthionine sulfoximine (BSO, an inhibitor of γ-glutamylcysteine synthetase and thus of GSH synthesis) to experimental animals leads to mitochondrial swelling and degeneration, and to myofiber necrosis in skeletal muscle, swelling of type-2-cell lamellar bodies and capillary endothelial cells in the lung, cataract formation, and gut epithelial degeneration (Mårtensson and Meister 1989). In preweanling rats treated with BSO, depletion of brain GSH leads to severe mitochondrial damage and death (Jain et al. 1991). BSO treatment of preweanling rats is associated with reduction of citrate synthase activity in isolated brain mitochondria (Jain et al. 1991), probably through disruption of membrane integrity and leakage of enzyme

(Heales et al. 1995). Huang and Philbert (1996) showed that treatment of astrocytes in culture with ethacrynic acid (a compound that penetrates mitochondria and depletes GSH by reacting with it to form a glutathione conjugate) leads to a marked reduction of GSSG reductase in mitochondria and to a lesser extent in the cytosol. A number of metabolic perturbations were noted, including decreases in glyceraldehyde 3-phosphate dehydrogenase activity, ATP levels, and cell membrane potential. Huang and Philbert (1996) suggested that loss of mitochondrial GSH renders astrocytes unable to combat oxidative stress, leading to perturbations of thiol-dependent enzymes, mitochondrial function, and energy metabolism. Bonnefoi (1992) showed that methyl iodide is toxic to cerebrocortical cells in culture. Cell death is associated with loss of mitochondrial GSH and mitochondrial function. Evidently, critical levels of GSH must be maintained in brain and other organs for normal mitochondrial function.

5. CELLULAR LOCALIZATION OF GLUTATHIONE IN THE BRAIN

In addition to its subcellular compartmentation, GSH is also compartmented in brain at the cellular level. Early tracer work with ^{15}N-labeled ammonia by Berl et al. (1962) led to the suggestion that the brain possesses two metabolically distinct compartments of glutamate. Thus, ammonia was shown to be converted to glutamine (amide) in a small compartment of rapidly turning-over glutamate that is kinetically distinct from a larger, more slowly turning-over glutamate compartment. At the time of that study the anatomical location of the small and large compartments was unknown. However, subsequent work from a number of laboratories showed that the small compartment is mostly associated with astrocytes and the large compartment with neurons [reviewed by Cooper and Plum (1987)]. Berl, Lajtha, and Waelsch (1961) noted a rapid labeling of cerebral GSH following a bolus administration of [^{14}C]glutamate into the cerebral ventricles of rats and suggested that GSH in brain may be preferentially synthesized in the small compartment. Later experiments with [^{13}N]ammonia also supported this idea (Cooper, Lai, and Gelbard 1988). The idea that GSH in the brain is preferentially synthesized in the small compartment is in agreement with computer simulation studies (Garfinkel 1966) and histochemical studies (Slivka, Mytilineou, and Cohen 1987). On the basis of results obtained with a histochemical staining procedure, Slivka et al. (1987) suggested that GSH in the brain may be primarily localized to nonneuronal elements, such as glia, and/or in axons or nerve terminals. Other workers have also used histochemical techniques to show that although GSH is well represented throughout the embryonic rat brain, it is largely absent from neurons (except dorsal root ganglia and cerebellar granule cells) in the adult rat brain (Philbert et al. 1991; Lowndes et al. 1994). GSH staining has been detected in the glomerular layer and granule-cell layer somata of the olfactory bulb (Kirstein et al. 1991). Based on the depolarization-induced release of glutathione from rat brain slices, it was suggested that glutathione may be present in neuronal elements (Zängerle et al. 1992). This suggestion is in accord with the previous finding that GSH is present in isolated nerve terminals (Reichelt and Fonnum 1969). Moreover, Amara, Coussemacq, and Geffard (1994) used an immunohistochemical technique to demonstrate the presence of GSH in nerve fiber tracts. Finally, Hjelle, Chaudhry, and Ottersen (1994) used antibodies to GSH to demonstrate that GSH in the brain is present in Golgi epithelial cells, and perivascular glia processes, and also in Purkinje cells, granule cell bodies, and mossy fiber tracts. The immunohistochemical studies confirmed previous studies that GSH in the brain is present in the cytosol and mitochondria. The studies also showed GSH to be in the nucleus (Hjelle et al. 1994). In conclusion, the available evidence suggests that in the intact adult central nervous system, GSH is strongly concentrated in astrocytes. The results of the above-mentioned studies also suggest that GSH is present in high concentrations in nerve endings and axons; with notable exceptions, however, the concentration is lower in the nerve bodies.

6. GSH IN NEURAL CELLS IN CULTURE

The concentration of GSH in cultured neuronal and astrocytic cells has been measured by several investigators. A consistent finding is that cultured astrocytes maintain a high level of GSH [20–200 nmol per milligram of protein ($\approx$2–20 mM)]—values at least equivalent to that found in whole brain (1–3 mM), and in many cases considerably greater (Raps et al. 1989; Cho and Bannai 1990; Yudkoff et al. 1990; Pileblad, Eriksson, and Hansson 1991; Devesa et al. 1993; Sagara, Miura, and Bannai 1993a,b; Sagara, Makino, and Bannai 1996; Makar et al. 1994; O'Connor et al. 1995; Huang and Philbert 1995, 1996; Juurlink, Schültke, and Hertz 1996). On the other hand, reported values for the concentration of GSH in cultured neurons are more variable [1–40 nmol per milligram of protein ($\approx$0.1–4 mM)] (Raps et al. 1989; Murphy et al. 1989; Bridges et al. 1991; Pileblad et al. 1991, Sagara et al. 1993b; Makar et al. 1994; Huang and Philbert 1995, 1996), but tend to be less than values reported for cultured astrocytes. The variability of reported GSH levels in cultured neurons may be due to the fact that neurons require cysteine for the maintenance of intracellular GSH (Sagara et al. 1993b; see also the following section) and that the culture media used by different investigators may have had variable levels of cyst(e)ine. Neurons cultured from different regions of brain may have different requirements for exogenous cyst(e)ine and exhibit different levels of GSH. In this regard, Langeveld et al. (1996) have recently shown that, depending on the region of origin within the brain, neurons either have smaller or similar concentrations of GSH relative to those in astrocytes of comparable origin. Neurons cultured from rat pup brains at different ages may also have different requirements for cyst(e)ine—as is known to be the case for oligodendrocytes (see the following paragraph). Finally, the absence or presence of growth factors may affect the GSH status of neurons. Thus, nerve growth factor stimulates the activity of GSH-synthesizing enzymes in pheochromocytoma PC12 cells, leading to increased levels of GSH (Pan and Perez-Polo 1993).

At least one study has shown that GSH is also present in oligodendrocytes in culture (Yonezawa et al. 1996). In this regard it is interesting to note that oligodendrocytes cultured from 1-day-old rat pups and grown for 3 days are markedly dependent on cystine for survival. However, in oligodendrocytes cultured from 6-day-old rats the dependence on cystine is not as great, and survival in cystine-depleted medium may be supported by a diffusible astrocytic factor (Yonezawa et al. 1996).

In summary, many studies have unanimously shown that cultured astrocytes contain high concentrations of GSH. This finding is in accord with tracer and histochemical studies of whole brain. The status of GSH within cultured neurons is less certain. Levels of GSH reported in the literature for cultured neurons vary widely. Probably, neurons in culture if provided with adequate cysteine will synthesize appreciable levels of GSH, although possibly not to the same extent as do astrocytes. The accumulated evidence, however, suggests that GSH in neurons in vivo (with some notable exceptions) is largely confined to nerve endings and axons.

7. INTERCELLULAR TRAFFICKING OF SULFUR-CONTAINING MOLECULES IN THE CNS

7.1 Glutathione

As mentioned above, GSH is found in micromolar quantities in the cerebrospinal fluid. In addition, microdialysis studies have shown GSH to be released from brain cells in vivo (Orwar et al. 1994; Yang et al. 1994). GSH is released from hippocampal slices, and the amount released is increased in the presence of the γ-glutamyltranspeptidase inhibitor acivicin (Li et al. 1996). Such studies raise the possibility that GSH (or its constituents) may be trafficked among different neural cells. It is well known that considerable movement of small molecules occurs between astrocytes and neurons. For example, much of neurotransmitter glutamate, GABA, and (possibly) aspartate released from nerve endings is taken up by astrocytes. Carbon and nitrogen are returned from astrocytes to the nerve terminals in the form of glutamine; carbon is also returned in the form of α-ketoglutarate and, possibly, of malate (Hertz 1979; Shank and Campbell 1984; Cooper and Plum 1987; Yudkoff et al. 1992). As noted above, Yudkoff et al. (1990) showed that astrocytes in culture contain high levels of GSH. These authors noted that the astrocytes in culture synthesized GSH at a high rate (comparable to that in the kidney), and that GSH was released to the medium. Other workers have also noted release of GSH from cultured astrocytes to the medium (Sagara et al. 1996) (but see below). It is tempting to speculate that the astrocytes provide nerve endings with GSH as a normal physiological response and that GSH exchange is yet another example of metabolite trafficking between astrocytes and neurons (Yudkoff et al. 1992). Such metabolic trafficking of GSH might be physiologically important. For example, some evidence has been presented that GSH may be a modulator of glutamatergic transmission by acting as a selective agonist for the N-methyl-D-aspartate (NMDA) recognition domain of the NMDA receptor ionophore complex (Ogita et al. 1995; see also Ogita et al., chapter 7 of this volume and Janáky et al., chapter 8 of this volume). Therefore, the trafficking of GSH from astrocytes to neurons may be involved not only in maintaining carbon, nitrogen, and sulfur balance in neurons but also in regulation of neurotransmitter response. However, Sagara et al. (1996) have recently presented evidence that neurons in culture do not take up intact GSH.

GSH may also be released from nerve endings as well as from astrocytes (Zängerle et al. 1992). In this regard, Guo and Shaw (1992) have presented strong evidence that astrocytes in culture possess high affinity binding sites for GSH. This finding suggests that release of GSH from nerves may regulate physiological responses in astrocytes (such as G-protein coupling, activation of second messengers, regulation of protein kinases, Ca^{2+} release, gene expression) (Guo and Shaw 1992). Although, as noted above, at least two studies suggest that viable astrocytes can release GSH, one study does not support this idea. Thus, Juurlink, Schültke, and Hertz (1996) reported that astrocytes in culture contain high levels of GSH (4.5–5 mM), but that the intact cells do not release this pool of GSH. In the absence of O_2 and glucose the cultured astrocytes begin to die, but GSH is released only after cell death. [In this regard it is interesting to note that GSH increases in the extracellular fluid after an ischemic insult to the brain (Orwar et al. 1994; Yang et al. 1994).] During prolonged ischemia, the concentration of intracellular GSH in the cultured astrocytes falls, presumably as a consequence of catabolism of GSH. Juurlink et al. (1996) speculated that loss of GSH due to turnover to the constituent amino acids—glutamate and cysteine—may be deleterious. This idea is of interest in light of previous findings that whole brain GSH is depleted during ischemia (Cooper et al. 1980; Rehncrona et al. 1980; Noguchi, Higuchi, and Matsui 1989; Lyrer et al. 1991; Mizui, Kinouchi, and Chan 1992), and that excess cysteine (Karlsen et al. 1981; Olney et al. 1990) and glutamate (Olney, Ho, and Rhee 1971) in the extracellular space are toxic to the central nervous system. Cysteine and its oxidized form (cystine) may be toxic to the central nervous system in part by interfering with glutamate uptake and neurotransmission. The astrocytic quisqualate-sensitive transporter has a high affinity for cystine (Murphy et al. 1989; Murphy, Schnarr, and Coyle 1990), and cysteine hyperactivates the NMDA receptor (Olney et al. 1990). Of special interest is the finding by Slivka and Cohen (1993) that whole-brain ischemia in the gerbil results in a 10- to 15-fold increase in the concentration of cysteine in striatum and hippocampus; the increase in cysteine is matched by an almost stoichiometric decrease in GSH. As discussed below, neurons are strongly dependent on exogenous cysteine to maintain cellular GSH levels, and some evidence suggests that this cysteine is derived from the astrocytes. Evidently, in this regard the astrocytes play an important role in maintaining sulfur homeostasis in the whole brain. However, because excess cysteine is neurotoxic, the astrocytes must possess a finely controlled mechanism for releasing the appropriate amount of cysteine that is necessary for neuronal survival (and function). Release of too much cysteine may result in neuronal injury; release of too little cysteine may result in lowered GSH and vulnerability to oxidative damage.

In conclusion, much evidence suggests that nerve endings release GSH. Some evidence suggests that astrocytes normally also release GSH, but due to a conflicting report this possibility needs to be further investigated. Release of GSH to the extracellular fluid may have important physiological roles in transmitter and messenger regulation. It is not clear what is the fate of the released GSH after the GSH-induced physiological response has been enacted. Some of the released GSH may be accumulated by the astrocytes (see below). Although of theoretical metabolic importance, a flow of GSH from the extracellular space to neurons has not been demonstrated. Metabolic trafficking of GSH/GSSG is well established for many other cell types (Griffith and Meister 1979). For example, GSH is released from liver cells to both blood and bile (Lauterburg, Adams, and Mitchell 1984). Evidence has been presented that GSH is transported across hepatocyte sinusoidal (Inoue et al. 1984) and canalicular membranes (Inoue et al. 1983). Evidently, metabolic trafficking of GSH between different cell types in the central nervous system (either intact or, more likely, as the constituent amino acids) is a fruitful area for further research.

7.2 Cysteine/Cystine

Depolarization of rat brain slices releases cysteine in addition to GSH (Keller et al. 1989; Zängerle et al. 1992). Zängerle et al. (1992) suggested that GSH and cysteine may play a role in synaptic transmission and may help control the redox state of synaptic elements. Indeed, some evidence has been presented that GSH and cysteine may be involved in the modulation of receptor responsiveness in accordance with their properties as reducing agents (Liu and Quirion 1992).

Astrocytes contain a high affinity Na^+-independent transport system (X_c^-) for the uptake of cystine (Cho and Bannai 1990; O'Connor et al. 1995). Cho and Bannai (1990) reported that inhibition of cystine uptake in cultured astrocytes by excess glutamate leads to GSH depletion and to eventual cell death. Murphy et al. (1990) reported that inhibition of cystine uptake in cultured neurons also leads to a decrease in cellular GSH and cell death. However, Sagara et al. (1993a,b) argued that this finding may be due to astroglial contamination of the neuronal cultures. Sagara et al. (1993a,b) have shown that cultured astrocytes and neurons are able to accumulate cysteine (probably by the Na^+-dependent ASC system), but that only astrocytes are able to take up cystine. Other workers have also shown that astroglial cells take up both cysteine and cystine and convert these amino acids to GSH (Dringen and Hamprecht 1996; Kranich, Hamprecht, and Dringen 1996a,b). Moreover, Kranich et al. (1996a,b) showed that astroglial cells in culture utilize cystine even more effectively than cysteine for the synthesis of GSH, but that neuron-rich primary cultures utilize cysteine more effectively than cystine for this purpose. Sagara et al. (1993a,b) believe that neurons depend on astrocytes for the maintenance of GSH levels and that neurons maintain their GSH level by taking up cysteine provided by glial elements.

8. MAINTENANCE OF SULFUR BALANCE IN THE BRAIN

As noted above, the concentration of GSH in brain is much greater than that in blood or cerebrospinal fluid. Therefore, the brain must possess an avid system for accumulating GSH or, more likely, of synthesizing it in situ. Some evidence has been presented that GSH (Kannan et al. 1990, 1992; Zlokovic et al. 1994; Kaplowitz et al. 1996) and GSH conjugates (Patel, Fullone, and Anders 1993) are transported intact across the blood–brain barrier, and a GSH transporter has been cloned and sequenced [Kaplowitz et al. (1996); see also Kannan et al., chapter 3, this volume]. A positive arterial–venous difference for GSH ($\approx$6 μM) has been measured across the cerebral cortex of 14-day-old rats and of adult rats (Jain et al. 1991). Combined with blood flow data, this difference yields a net utilization rate of 3.4 and 4.2 (nmol/min)/g for cerebral cortex of 14-day-old rats and adult rats, respectively (Jain et al. 1991). GSH is also exported from the choroid plexus to the cerebrospinal fluid (Anderson et al. 1989). Astrocytic processes surround capillaries and underlie the ependyma. Therefore, GSH (or its constituent cysteine) entering the brain from the blood and extracellular fluid may first be processed in astrocytes, thereby accounting in part for the high level of GSH in these cells. Such an arrangement would be biologically important because astrocytes utilize GSH as a neuroprotectant (see the next section). However, despite the ability of GSH to cross the blood–brain barrier, it is probable that the blood is not the major source of cerebral GSH. Generally, GSH is very poorly transported intact into mammalian cells (Anderson and Meister 1989), and brain levels are not raised appreciably upon administration of GSH to experimental animals (Jain et al. 1991; Anderson, Powrie, and Meister 1985). Thus, although the K_m for GSH transport into brain was reported to be 5.8 mM (Kannan et al. 1990) and the concentration of GSH in rat plasma is $\approx$15 μM, a 1000-fold rise in blood GSH did not result in increased brain GSH (Jain et al. 1991). Moreover, the rat brain contains at least three kinetically distinct compartments of GSH. The cerebral metabolic rate for GSH of $\approx$4 (nmol/min)/g is considerably less than the turnover rate of the most mobile pool of brain GSH [$\approx$15–25 (nmol/min)/g] (Jain et al. 1991). As discussed by Jain et al. (1991), at least two possible mechanisms exist for maintaining brain GSH sulfur: (i) cysteine moieties may be recovered during the turnover of GSH, and (ii) cysteine or cysteine-containing molecules (in addition to GSH) may be transported into brain.

Isolated brain capillaries contain an ASC-type transporter (i.e., a carrier that catalyzes the sodium- and energy-dependent translocation of alanine, serine, and cysteine) (Tayarani et al. 1987). However, this transporter is situated on the abluminal surface, suggesting that the carrier may be involved in the egress of amino acids from the brain. Nevertheless, rat brain is capable of taking up cysteine from the blood (Oldendorf 1971; Oldendorf and Szabo 1976; Wade and Brady 1981), but apparently not cystine per se (Wade and Brady 1981). Cysteine is transported into brain on the L-type carrier (Wade and Brady 1981). Some evidence suggests that human brain may also take up cyst(e)ine (Felig, Wahren, and Ahlborg 1973). Methionine is readily taken up across the blood–brain barrier in rats and humans (Oldendorf 1971; Felig et al. 1973; Oldendorf and Szabo 1976; Pardridge 1977). Pardridge (1977) has calculated a unidirectional flux of methionine into rat brain of 1.6 (nmol/min)/g. However, the extent to which methionine is a precursor of cysteine sulfur (transsulfuration pathway) in rat brain is not known, though it may be minor (O'Connor et al. 1995; Kranich

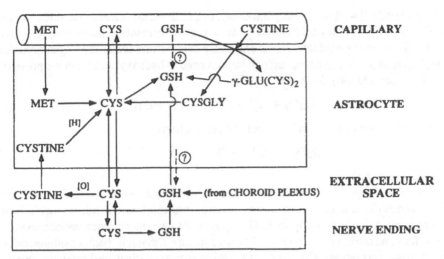

FIG. 2. Proposed routes for the compartmental metabolism of sulfur in the brain. Note the central importance of the astrocytes.

et al. 1996a). Although free cystine is not transported across the blood–brain barrier, it is possible that cystine moieties are transported into brain as (i) γ-glutamylcystine [after formation of this compound by γ-glutamyltranspeptidase (Fig. 1, reaction 1) (Thompson and Meister 1975; Anderson and Meister 1983) at the cell surface of capillaries/astrocytes], or as (ii) cystinyl bisglycine (Jain et al. 1991). In addition to its occurrence in brain capillaries, γ-glutamyltranspeptidase is present in astrocytes (Shine et al. 1981) and nerve endings (Minn and Besagni 1983). Enzymes of the γ-glutamyl cycle are especially well represented at the choroid plexus (Tate, Ross, and Meister 1973).

Based on the above discussion, a flow diagram representing possible origins of sulfur in the brain and the central role of the astrocytes in maintaining sulfur homeostasis in brain is presented in Fig. 2. The dynamics and regulation of sulfur homeostasis in the brain are interesting areas for future research.

9. ROLE OF ASTROCYTES AS NEUROPROTECTANTS

9.1 Protection against Oxidative Stress

In recent years it has become apparent that free radicals and/or reactive oxygen species are involved in the pathogenesis of several neurological diseases and that vitamin E and GSH are important protectants. In order to introduce the topic of the importance of GSH as an antioxidant in the brain, it is necessary to provide a brief description of free radicals and their relevance in the central nervous system. Free radicals are molecules or atoms that are often chemically reactive and that possess unpaired electrons in their outer orbitals. Two biologically important free radicals are superoxide anion radical ($O_2^{\bullet-}$) and hydroxyl radical (OH•). $O_2^{\bullet-}$ is not especially reactive, but its conjugate acid (HO_2•) is a strong oxidant. OH• is extremely reactive and toxic. Hydrogen peroxide (H_2O_2) is an oxidizing agent that can be converted to hydroxyl radical in the presence of iron [Eqs. (3), (4)] (Han et al. 1996):

$$H_2O_2 + O_2^{\bullet-} \rightarrow O_2 + OH\bullet + OH^- \tag{3}$$

(in the presence of Fe^{2+}/Fe^{3+}; Haber–Weiss reaction),

$$H_2O_2 + Fe^{2+} \rightarrow OH\bullet + OH^- + Fe^{3+} \tag{4}$$

(Fenton reaction).

Reactive oxygen species are generated during normal metabolism. For example, $O_2^{\bullet-}$ and H_2O_2 are generated during electron transport in the mitochondria from incomplete reduction of O_2 to H_2O. H_2O_2 is also produced from various oxidase reactions and during the synthesis of prostaglandins. During ischemia xanthine dehydrogenase is proteolytically clipped to yield an enzyme with altered catalytic activity. The modified enzyme is a xanthine oxidase which generates H_2O_2 and $O_2^{\bullet-}$. Free radicals are also generated by the autooxidation of various substances such as catecholamines and by the action of the microsomal cytochrome P-450 system (Han, Mytilineou, and Cohen 1996).

Most aerobic cells produce $O_2^{\bullet-}$, and many of these cells also produce short-lived nitric oxide (NO•) ($t_{1/2} \sim 1$ s). Tissues contain at least three isoforms of nitric oxide synthase (NOS), namely, a calcium-activated inducible form (iNOS) in macrophages and astrocytes, a vascular endothelial constitutive form (eNOS), and a calcium-dependent neuronal form (nNOS). [See Heales, Bolaños, and Clark (1996) for a discussion.] NO• is now regarded as a neurotransmitter (Lowenstein, Dinerman, and Snyder 1994). Its action is terminated by quenching with heme compounds. Alternatively, a species derived from NO• may react with cysteine and glutathione to form longer-lived nitrosothiols. NO• can also react extremely rapidly (almost at the diffusion limit) with superoxide to form peroxynitrite ($ONOO^-$) (Koppenol et al. 1992). $ONOO^-$, which has an estimated half-life of a few seconds (Benton and Moore 1970), is a powerful oxidant that can damage biological systems (Koppenol et al. 1992). For example, $ONOO^-$-mediated conversion of tyrosine to nitrotyrosine may prevent key phosphorylation reactions and disrupt cell signaling processes. Protonation of $ONOO^-$ yields peroxynitrous acid (ONOOH), which spontaneously decomposes to NO_2 and OH• (20–30 percent); the remainder isomerizes to nitrate (NO_3^-) + H^+ (Beckman et al. 1990).

A few recent studies implicate $ONOO^-$ as (i) a major reactive oxygen species in the brain and (ii) a potentially major neurotoxin. For example, Vatassery (1996) showed that $ONOO^-$ caused a very rapid oxidation of α-tocopherol and ascorbate in synaptosomes; sulfhydryl-containing compounds were more resistant. Deliconstantinos and Villiotou (1996) showed that synaptosomes possess both xanthine oxidase and NOS activities. In the presence of UV-B radiation, Ca^{2+} influx was stimulated. This influx triggered increased $O_2^{\bullet-}$ and NO• production, which in turn stimulated increased $ONOO^-$ production. Deliconstantinos and Villiotou (1996) suggested that $ONOO^-$ formation generates a neurotoxic response. Bolaños et al. (1994) showed that treatment of astrocytes in culture with lipopolysaccharide and interferon-γ resulted in induction of iNOS. Activities of both cytochrome c oxidase and succinate–cytochrome c reductase were reduced in the treated cells, suggesting that increased NO•/$ONOO^-$ production inhibited the mitochondrial respiratory chain. Nevertheless, the treated cells remained viable, presumably as a result of stimulated glycolysis.

Interestingly, the astrocytes in culture were completely resistant to acutely administered $ONOO^-$ even at extraordinary high levels (2 mM), whereas neurons in culture were selectively damaged at much lower concentrations (0.1 mM) (Bolaños et al. 1995). In the studies of Bolaños et al. (1995), the concentration of GSH in the neurons was about half that in the astrocytes. Acute treatment of the neurons with 0.1 mM $ONOO^-$ resulted in considerable loss of GSH, but 2 mM $ONOO^-$ had no affect on the pool of GSH in the astrocytes (Bolaños et al. 1995). The findings of Bolaños et al. (1995) suggest that the astrocytes are much more resistant to oxidative stress than are neurons. The extreme resistance of normal astrocytes to oxidative stress is presumably due in part to the ability of these cells to maintain high levels of GSH. If, however, cellular GSH is depleted, astrocytes become susceptible to the action of peroxynitrite and mitochondrial complexes I and II/III become severely damaged (Barker et al. 1996). In another study from the same laboratory, it was shown that BSO-induced reduction in GSH is accompanied by increased NOS activity both in whole brain in vivo and in neurons in culture (Heales et al. 1996). The loss of GSH and increased NOS activity resulted in cell death in the cultured neurons. The combined evidence from the experiments with $ONOO^-$ suggests that astrocytes under normal conditions provide a cellular line of defense against oxidative stress in the brain. Indeed, in culture, astrocytes have been shown to protect neurons (Langeveld et al. 1995; Desagher, Glowinski, and Prement 1996) and oligodendrocytes (Noble, Antel, and Yong 1994) against hydrogen peroxide–induced toxicity. In the study of Desagher et al. (1996) it was pointed out that high levels of catalase contribute to the protective effect of astrocytes. The findings do not negate the importance of GSH/glutathione peroxidase/reductase in preventing oxidative damage. Thus, as mentioned above, GSH protects astrocytes against peroxynitrite toxicity (Barker et al. 1996). Moreover, the strong upregulation of the pentose phosphate pathway in astrocytes exposed to hydrogen peroxide (Ben-Yoseph et al. 1994, 1996a,b) can be explained by a strong demand for NADPH to regenerate GSH through the action of GSSG reductase. Therefore, it is likely that both catalase and glutathione peroxidase contribute to the protective role of astrocytes against reactive oxygen species.

In addition to their resistance to $ONOO^-$, astrocytes are also resistant to other reactive oxygen species such as hydrogen peroxide. As with $ONOO^-$, neurons are more vulnerable to the toxic action of hydrogen peroxide than are astrocytes (Ben-Yoseph et al. 1994). Additional evidence for the resistance of astrocytes to reactive oxygen species has been documented. Thus, Murphy et al. (1989) showed that cystine depletion in mixed hippocampal neuronal–astrocytic cell cultures led to the death of neurons but not of astrocytes. Cystine depletion lowered GSH levels and hence increased oxidative stress. Han et al. (1996) showed that incubation of glia isolated from newborn rats with L-dopa or with other autooxidizable substances resulted in upregulation of GSH production. The upregulation provided significant protection against a further oxidative challenge. Similar results were also observed with fetal rat mesencephalon (which contains glia plus neurons), but not with pure neuronal cultures derived from the mesencephalon. Instead, the pure neuronal cultures were destroyed by application of L-dopa (Han et al. 1996). The melanization of dopa by autooxidation generates a remarkably reactive mixture of oxidizing agents and/or free radicals (including quinones, semiquinones, $O_2^{\bullet-}$, OH•, and H_2O_2) (Han et al. 1996). The data of Han et al. (1996) show that astrocytes are extraordinarily resistant to the deleterious effects of oxidative stress and that these cells may actually protect neurons from such stress. Protection, however, is limited. Thus, the oxidative stress caused by reperfusion following an ischemic episode results in selective neuronal damage in vivo. By contrast, astrocytes are resistant to the insult, although they undergo morphological changes accompanied by an increased number of mitochondria (Petito and Babiak 1982; Petito 1986). As noted above, the resistance of astrocytes to oxidative stress is due in part to the presence of catalase and to their ability to maintain high levels of GSH. In addition, however, chick astrocytes in culture have high levels of vitamin E (Makar et al. 1994) and high activity of γ-glutamylcysteine synthetase (the first of two steps involved in GSH biosynthesis). Finally, Huang and Philbert (1995) noted higher activities of glutathione peroxidase and GSSG reductase in cultured rat cerebellar astrocytes than in cultured cerebellar granule cells. Clearly, astrocytes have especially well developed machinery for combating oxidative stress.

Despite the protection to the brain against free radicals afforded by the antioxidative machinery of the astrocytes, the defense system may break down in many neurodegenerative diseases. For example, hydrogen peroxide is released from cells in the brain following global forebrain ischemia and reperfusion (Hyslop et al. 1995). This finding suggests that the defense systems of the brain cannot prevent the rise of potentially damaging hydrogen peroxide following reflow. It has been suggested that under certain pathological conditions iNOS may be induced in injured astrocytes and that instead of providing a protective role, these cells may generate $ONOO^-$, which damages nearby neurons (Barker et al. 1996). Evidence for a pathological role of free radicals is especially compelling in the case of Parkinson's disease (Dexter et al. 1989; Jenner et al. 1992; Di Monte, Chan, and Sandy 1992). GSH levels are decreased in affected brain regions (Jenner et al. 1992), and this loss is accompanied by mitochondrial damage and loss of respiratory enzyme (complex I) activities (Schapira et al. 1990; Di Monte et al. 1992). Loss of GSH may not of itself result in alterations of complex I activity, but the depletion may render the mitochondrial respiratory pathway more vulnerable to the action of toxins (Seaton, Jenner, and Marsden 1996). 1-Methyl-4-phenylpyridinium (MPP^+) induces Parkinsonism in higher primates. The toxicity is due in part to formation of free radicals, and GSH plays a protective role (Bhave et al. 1996; Wüllner et al. 1996). As noted above, administration of GSH to experimental animals does not result in increased brain GSH. However, GSH can be increased in rat brain by administration of L-2-oxothiazolidine-4-carboxylate, a prodrug of cysteine and hence of GSH (Anderson and Meister 1989;

Messina et al. 1989). Thus, prodrugs designed to increase brain GSH may be of therapeutic benefit in Parkinson's disease and possibly in other neurodegenerative diseases.

In summary, much evidence suggests that astrocytes are extremely resistant to oxidative stress, in part due to relatively high concentrations of GSH and vitamin E, and to relatively high activities of catalase, glutathione peroxidase, GSSG reductase, and GSH-synthesizing enzymes. Resistance may also be due in part to the ability of these cells to upregulate GSH synthesis. Moreover, their powerful antioxidant defense mechanisms enable astrocytes to play an important role in protecting neurons and oligodendrocytes against reactive oxygen species. This idea is in line with the fact that many reactive oxygen-containing species (e.g., $NO\bullet$, $ONOO^-$) have half-lives commensurate with a potential to diffuse over the distance of several cell diameters. Diseases leading to a reduction of the protective mechanisms in astrocytes and/or to the induction of iNOS activity in astrocytes would be expected to render neurons more susceptible to the action of reactive oxygen species. Therefore, regimens designed to boost the endogenous level of brain GSH (and especially that in the astrocytic compartment) may have clinical benefit in diseases exacerbated by direct (e.g., Parkinson's disease) or secondary (e.g., reflow following stroke) oxidative stress.

9.2 Protection against Glutamate Neurotoxicity

The possibility that astroglial cells can protect against glutamate toxicity to the brain has recently been suggested by Dringen and Hamprecht (1996). As noted above, astrocytes contain uptake systems for glutamate and cysteine. These cells also contain a transporter for glycine (Holopainen and Kontro 1989). Indeed, glutamate (or a suitable substitute that can be metabolized to glutamate), cysteine, and glycine must be present in the external medium for astroglial cells to efficiently synthesize GSH. Dringen and Hamprecht (1996) proposed that under physiological conditions the synthesis of GSH from glutamate (in astroglial cells) in the presence of glycine or cysteine would be a way of disposing of toxic glutamate. Evidently, the same system would also dispose of potentially neurotoxic cysteine.

9.3 Protection against Potentially Harmful Xenobiotics

As mentioned above, many potentially harmful foreign compounds can be detoxified through the mercapturate pathway. The first step of the mercapturate pathway is catalyzed by glutathione S-transferases (GSTs) (Fig. 1, reaction 8). These enzymes catalyze the conjugation of GSH with a variety of electrophilic substances of endogenous and exogenous origin (Cooper and Tate 1997). Much of the work on GSH conjugation has been carried out on liver cells, which have been shown to contain many related cytosolic GSTs, a single microsomal GST, and possibly a mitochondrial form. Recent work has shown that the brain also possesses many types of GSTs (reviewed by Johnson et al. 1993; Lowndes et al. 1994; Cooper 1997). Several immunohistochemical studies have shown that within the brain, cytosolic GSTs (depending on the type) are well represented in astrocytes (especially the end feet), ependymal cells, tany-cytes, the subventricular zone, the choroid plexus, and oligodendocytes. (See Cooper 1997 for original references.) Earlier studies did not reveal GSTs to be present in neurons. However, Lowndes et al. (1994) suggested that differences in fixation and staining procedures employed by various investigators may have led in some cases to an underestimation of the degree to which certain GSTs are represented in various cell types in the brain. Moreover, cultured chick neurons (Makar et al. 1994) and rat cerebellar granule cells (Huang and Philbert 1995) possess GST activity. Johnson et al. (1993) showed that microsomal GST is present in Purkinje cells throughout the cerebellar cortex and within neurons of the brain stem and hippocampus of the rat. Nuclei of Purkinje cells and nuclei of neurons in the brainstem, hippocampus, and cerebral cortex were also shown to be immunopositive for α-class GST 1-1 (Y_aY_a). Interestingly, α-class GST 2-2 (Y_cY_c) was detected in the nucleolus but not in the nucleus (Johnson et al. 1993). Johnson et al. (1993) used high-performance liquid chromatography coupled with sodium dodecylsulfate gel electrophoresis to show that the brain possesses regional differences in the pattern of α-, μ-, and π-class GST subunit expression. The authors pointed out that these differences may be of clinical significance. In the cerebellar cortex the concentration of the μ-class GST subunit 4 (Y_{b2}) was found to be greatest in the flocculus and lowest in the vermis, a pattern that coincides with the known greater susceptibility of the vermis than of the flocculus to toxic and metabolic insults (Johnson et al. 1993).

Several roles have been suggested for GSTs in brain. For example, it has been proposed that GSTs protect the myelin sheath against toxic substances (Cammer et al. 1989; Tansey and Cammer 1991). Of interest is the finding that human brain and testes contain a unique μ-GST (Campbell et al. 1990). The brain and testes are unusual among organs in possessing a tight blood–organ barrier. It was suggested that the unique μ-GST may be important in the normal functioning of the blood–brain barrier and of the blood–testes barrier (Campbell et al. 1990). GSTs may also remove epoxyeicosatrienoic acid derivatives from cells and participate in the biosynthesis of prostaglandins and leukotrienes in the brain (Cammer et al. 1989). Because Y_p and Y_b subunits are found in the ependymal cells and in astrocytic end feet, it is possible that GSTs within these cells form a first line of defense in the brain against potentially harmful xenobiotics and that these cells can participate in the transport of substances into and out of the brain (Cammer et al. 1989). Cammer et al. (1989) stated that such transport could affect hormonal control over processes such as myelination and neuronal growth and could facilitate removal of endogenous and exogenous toxins from the CNS.

In conclusion, the brain exhibits a complicated pattern of expression of GSTs some-what reminiscent of the distribution of GSH. The concentrated location of many GSTs within the blood–brain barrier, choroid plexus, astrocytic end feet, and ependymal cells presents a first line of defense against potentially harmful substances entering the brain. However, insofar as GSTs are also present in neurons and oligodendrocytes, these enzymes also constitute a second line of defense in brain. Because GSTs can act on several eicosonoates in the brain, it is also possible that these enzymes participate in local hormonal signaling within the brain. One could speculate that such signaling involves special astrocytic–neuronal interactions. Evidently, the role of GSTs in the normal functioning of the brain and in the detoxification of potentially harmful xenobiotics are additional fruitful areas for future research.

10. CONCLUSIONS

Within the last few years it has become apparent that GSH plays an important role in the central nervous system, both as a major antioxidant and as a protectant against potentially harmful foreign substances. It has become equally apparent that GSH plays specialized roles in transport, in short-term hormonal regulation, in protection against hyperexcitation, and in neurotransmitter modulation and cell signaling. Despite this central importance of GSH in the brain, very little is known about the mechanisms whereby cerebral GSH homeostasis is maintained. Especially intriguing is the finding that GSH metabolism is compartmented in the brain and that astrocytes contain the most concentrated pool of cerebral GSH. In the future, we can look forward to a further understanding of the role of astrocytic GSH in protecting the central nervous system against oxidative stress and against potentially harmful xenobiotics. We can also look forward to an increased understanding of the role of GSH in maintaining cerebral sulfur homeostasis, in the trafficking of sulfur between various cellular compartments, and in the regulation of normal physiological responses. Interventions designed to increase brain GSH, or to prevent its decline, may be of therapeutic benefit in some neurodegenerative diseases.

ACKNOWLEDGMENTS

I thank Drs. J. P. Blass, G. G. Gibson, and R. Dringen for their many helpful suggestions. I also thank Drs. S.-K. Han and G. Cohen for helpful discussions concerning reactive oxygen species.

REFERENCES

Amara, A., Coussemacq, M., and Geffard, M. 1994. Antibodies to reduced glutathione. *Brain Research* 659:237–42.

Anderson, M. E., and Meister, A. 1983. Transport and direct utilization of γ-glutamylcyst(e)ine for glutathione synthesis. *Proceedings of the National Academy of Sciences of the U.S.A.* 80:707–11.

Anderson, M. E., and Meister, A. 1989. Marked increase of cysteine levels in many regions of the brain after administration of 2-oxothiazolidine-4-carboxylate. *FASEB Journal* 3:1632–6.

Anderson, M. E., Powrie, F., and Meister, A. 1985. Glutathione monoethyl ester: preparation, uptake by tissues and conversion to glutathione. *Archives of Biochemistry and Biophysics* 239:538–48.

Anderson, M. E., Underwood, M., Bridges, R. J., and Meister, A. 1989. Glutathione metabolism at the blood–cerebrospinal fluid barrier. *FASEB Journal* 3:2527–31.

Baquer, N. Z., Hothersall, J. S., McLean, P., and Greenbaum, A. L. 1977. Aspects of carbohydrate metabolism in developing brain. *Developmental Medicine and Child Neurology* 19:81–104.

Baquer, N. Z., Hothersall, J. S., and McLean, P. 1988. Function and regulation of the pentose phosphate pathway in brain. *Current Topics in Cell Regulation* 29:265–89.

Barker, J. E., Bolaños, J. P., Land, J. M., Clark, J. B., and Heales, S. J. R. 1996. Glutathione protects astrocytes from peroxynitrite-mediated mitochondrial damage: implications for neuronal/astroglial trafficking and neurodegeneration. *Developmental Neuroscience* 18:391–6.

Beckman, J. S., Beckman, T. W., Chen, J., and Marshall, P. A. 1990. Apparent hydroxyl radical production by peroxynitrite: implications for endothelial injury from nitric oxide and superoxide. *Proceedings of the National Academy of Sciences of the U.S.A.* 87:1620–4.

Benton, D. J., and Moore, P. 1970. Kinetics and the mechanism of formation of and decay of peroxynitrous acid in perchloric acid solutions. *Journal of the Chemical Society* (A)3179–82.

Ben-Yoseph, O., Boxer, P. A., and Ross, B. D. 1994. Oxidative stress in the central nervous system: monitoring the metabolic response using the pentose phosphate pathway. *Developmental Neuroscience* 16: 328–36.

Ben-Yoseph, O., Boxer, P. A., and Ross, B. D. 1996a. Assessment of the role of glutathione and pentose phosphate pathways in the protection of primary cerebrocortical cultures from oxidative stress. *Journal of Neurochemistry* 66:2329–37.

Ben-Yoseph, O., Boxer, P. A., and Ross, B. D. 1996b. Nonivasive assessment of the relative roles of cerebral antioxidative enzymes by quantification of pentose phosphate pathway activity. *Neurochemical Research* 21:1005–12.

Berl, S., Lajtha, A., and Waelsch, H. 1961. Amino acid and protein metabolism. VI. Cerebral compartments of glutamic acid metabolism. *Journal of Neurochemistry* 7:186–97.

Berl, S., Takagaki, G., Clarke, D. D., and Waelsch, H. 1962. Metabolic compartments *in vivo*. Ammonia and glutamic acid acid metabolism in brain and liver. *Journal of Biological Chemistry* 237:2562–9.

Bhave, S. V., Johanessen, J. N., Lash, L. L., Wakade, T. D., and Wakade, A. R. 1996. Age-dependent sensitivity of cultured peripheral sympathetic neurons to 1-methyl-4-phenylpyridinium: role of glutathione. *Journal of Neurochemistry* 67:557–65.

Bolaños, J. P., Peuchen, S., Heales, S. J. R., Land, J. M., and Clark J. B. 1994. Nitric oxide–mediated inhibition of the mitochondrial respiratory chain in cultured astrocytes. *Journal of Neurochemistry* 63:910–16.

Bolaños, J. P., Heales, S. J. R., Land, J. M., and Clark, J. B. 1995. Effect of peroxynitrite on the mitochondrial respiratory chain: differential susceptibility of neurons and astrocytes in primary culture. *Journal of Neurochemistry* 64:1965–72.

Bonnefoi, M. S. 1992. Mitochondrial glutathione and methyl iodide–induced neurotoxicity in primary neural cell cultures. *Neurotoxicology* 13:401–12.

Bridges, R. J., Koh, J.-Y., Hitalski, C. G., and Cotman, C. W. 1991. Increased cytotoxic vulnerability of cortical cultures with reduced levels of glutathione. *European Journal of Pharmacology* 192:199–200.

Cammer, W., Tansey, F., Abramovitz, M., Ishigaki, S., and Listowsky, I. 1989. Differential localization of glutathione-*S*-transferase Y_p and Y_b subunits in oligodendrocytes and astrocytes of rat brain. *Journal of Neurochemistry* 52:876–83.

Campbell, E., Takahashi, Y., Abramovitz, M., Peretz, M., and Listowski, I. 1990. A distinct human testis and brain μ-class glutathione *S*-transferase. Molecular cloning and characterization of a form present even in individuals lacking hepatic type μ isozymes. *Journal of Biological Chemistry* 265:9188–93.

Cho, Y., and Bannai, S. 1990. Uptake of glutamate and cystine in C-6 glioma cells and in cultured astrocytes. *Journal of Neurochemistry* 55:2091–7.

Cooper, A. J. L. 1997. Glutathione in the brain: disorders of glutathione metabolism. In *The molecular and genetic basis of neurological disease*. 2nd ed., ed. R. N. Rosenberg, S. B. Prusiner, S. DiMauro, and R. L. Barchi, 1195–230. Boston:Butterworth-Heinemann.

Cooper, A. J. L., and Plum, F. 1987. Biochemistry and physiology of brain ammonia. *Physiological Reviews* 67:440–519.

Cooper, A. J. L., and Tate, S. S. 1997. Enzymes involved in processing of glutathione conjugates. In *Comprehensive toxicology: volume 3. Biotransformations*, ed., G. Sipes, C. A. McQueen, and A. J. Gandolfi. 329–363. Oxford, U.K.: Elsevier Press.

Cooper, A. J. L., Pulsinelli, W. A., and Duffy, T. E. 1980. Glutathione and ascorbate during ischemia and post ischemic reperfusion in rat brain. *Journal of Neurochemistry* 35:1242–5.

Cooper, A. J. L., Lai, J. C. K., and Gelbard, A. S. 1988. Ammonia and energy metabolism in normal and hyperammonemic rat brain. In *The biochemical pathology of the astrocytes*, ed. M. D. Norenberg, L. Hertz, and A. Schousboe, 419–34. New York: Alan R. Liss.

Deliconstantinos, G., and Villiotou, V. 1996. NO synthase and xanthine oxidase activities of rabbit brain synaptosomes: peroxynitrite formation as a causative factor of neurotoxicity. *Neurochemical Research* 21:51–61.

Desagher, S., Glowinski, J., and Prement, J. 1996. Astrocytes protect neurons against hydrogen peroxide toxicity. *Journal of Neuroscience* 16:2553–62.

Devesa, A., O'Connor, J. E., García, C., Puertas, I. R., and Viña, J. R. 1993. Glutathione metabolism in primary astrocyte cultures: flow cytometric evidence of heterogeneous distribution of GSH content. *Brain Research* 618:181–9.

Dexter, D. T., Crater, C. J., Wells, F. R., Javoy-Agid, F., Agid, Y., Lees, A., Jenner, P., and Marsden, C. D. 1989. Basal lipid peroxidation is increased in Parkinson's disease. *Journal of Neurochemistry* 52:381–9.

Di Monte, D. A., Chan, P., and Sandy, M. S. 1992. Glutathione in Parkinson's disease: a link between oxidative stress and mitochondrial damage? *Annals of Neurology* 32:S112–15.

Dringen, R., and Hamprecht, B. 1996. Glutathione content as an indicator for the presence of metabolic pathways of amino acids in astroglial cultures. *Journal of Neurochemistry* 67:1375–82.

Felig, P., Wahren, J., and Ahlborg, G. 1973. Uptake of individual amino acids by the human brain. *Proceedings of the Society for Experimental Biology and Medicine* 142:230–1.

Folbergrová, J., Rehncrona, S., and Siesjö, B. K. 1979. Oxidized and reduced glutathione in the rat brain under normoxic and hypoxic conditions. *Journal of Neurochemistry* 32:1621–7.

Garfinkel, D. 1966. A simulation study of the metabolism and compartmentation in brain of glutamate, aspartate, the Krebs cycle and related metabolites. *Journal of Biological Chemistry* 241:3918–29.

Griffith, O. W., and Meister, A. 1979. Glutathione: interorgan translocation, turnover and metabolism. *Proceedings of the National Academy of Sciences of the U.S.A.* 76:5606–10.

Griffith, O. W., and Meister, A. 1985. Origin and turnover of mitochondrial glutathione. *Proceedings of the National Academy of Sciences of the U.S.A.* 82:4668–72.

Guo, N., and Shaw, C. 1992. Characterization and localization of glutathione binding sites on cultured astrocytes. *Molecular Brain Research* 15:207–15.

Han, S.-K., Mytilineou, C., and Cohen, G. 1996. L-DOPA upregulates glutathione and protects mesencephalic cultures against oxidative stress. *Journal of Neurochemistry* 66:501–10.

Heales, S. J. R., Davies, S. E. C., Bates, T. E., and Clark, J. B. 1995. Depletion of brain glutathione is accompanied by impaired mitochondrial function and decreased N-acetyl aspartate concentration. *Neurochemical Research* 20:31–8.

Heales, S. J. R., Bolaños, J. P., and Clark, J. B. 1996. Glutathione depletion is accompanied by increased neuronal nitric oxide synthase activity. *Neurochemical Research* 21:35–9.

Hertz, L. 1979. Functional interactions between neurons and astrocytes. 1. Turnover and metabolism of putative neurotransmitters. *Progress in Neurobiology* 13:277–323.

Hjelle, O. P., Chaudhry, F. A., and Ottersen, O. P. 1994. Antisera to glutathione: characterization and immunohistochemical application to the rat cerebellum. *European Journal of Neuroscience* 6:793–804.

Holopainen, I., and Kontro, P. 1989. Uptake and release of glycine in cerebellar granule cells and astrocytes in primary culture: potassium stimulated release from granule cells is calcium dependent. *Journal of Neuroscience Research* 24:374 83.

Hostetler, K. Y., and Landau, R. 1967. Estimation of the pentose cycle contribution to glucose metabolism in tissue *in vivo*. *Biochemistry* 6:2961–4.

Hotta, S. 1962. Glucose metabolism in brain tissue: the hexosemonophosphate shunt and its role in glutathione reduction. *Journal of Neurochemistry* 9:43–51.

Hotta, S., and Seventko, J. M., Jr. 1968. The hexosemonophosphate shunt and glutathione reduction in guinea pig brain tissue. Changes caused by chlorpromazine, amytal, and malonate. *Archives of Biochemistry and Biophysics* 123:104–8.

Huang, J., and Philbert, M. A. 1995. Distribution of glutathione and glutathione-related enzyme systems in mitochondrial and cytosol of cultured cerebellar astrocytes and granule cells. *Brain Research* 680: 16–22.

Huang, J., and Philbert, M. A. 1996. Cellular responses of cultured cerebellar astrocytes to ethacrynic acid-induced perturbation of subcellular glutathione homeostasis. *Brain Research* 711:184–92.

Hyslop, P. A., Zhang, Z., Pearson, D. V., and Phebus, L. A. 1995. Measurement of striatal H_2O_2 by microdialysis following global forebrain ischemia and reperfusion in the rat: correlation with the cytotoxic potential of H_2O_2 *in vitro*. *Brain Research* 671:181–6.

Inoue, M., Kline, R., Tran, T., and Arias, I. M. 1983. The mechanism of biliary secretion of reduced glutathione. Analysis of transport process in isolated rat-liver canalicular membrane vesicles. *European Journal of Biochemistry* 134: 467–71.

Inoue, M., Kline, R., Tran, T., and Arias, I. M. 1984. Glutathione transport across hepatocyte plasma membranes. Analysis using isolated rat-liver sinusoidal-membrane vesicles. *European Journal of Biochemistry* 138:491–6.

Jain, A., Mårtensson, J., Stole, E., Auld, P. A. M., and Meister, A. 1991. Glutathione deficiency leads to mitochondrial damage in brain. *Proceedings of the National Academy of Sciences of the U.S.A.* 88:1913–7.

Jain, A., Buist, N. R. M., Kennaway, N. G., Powell, B. R., Auld, P. A. M., and Mårtensson, J. 1994. Effect of ascorbate or N-acetylcysteine treatment in a patient with hereditary glutathione synthetase deficiency. *Journal of Pediatrics* 124:229–33.

Jenner, P., Dexter, D. T., Sian, J., Schapira, A. H. V., and Marsden, C. D. 1992. Oxidative stress as a cause of nigral cell death in Parkinson's disease and incidental Lewy body disease. *Annals of Neurology* 32:582–7.

Johnson, J. A., Barbery, A. E., Kornguth, S. E., Brugge, J. F., and Siegel, F. L. 1993. Glutathione S-transferase isozymes in rat brain neurons and glia. *Journal of Neuroscience* 13:2013–23.

Juurlink, B. H. J., Schültke, E., and Hertz, L. 1996. Glutathione release and catabolism during energy substrate restriction in astrocytes. *Brain Research* 710:229–33.

Kannan, R., Kuhlenkamp, J. F., Jeandidier, E., Trinh, H., Ookhtens, M., and Kaplowitz, N. 1990. Evidence for carrier-mediated transport of glutathione across the blood–brain barrier in the rat. *Journal of Clinical Investigation.* 85:2009–13.

Kannan, R., Kuhlenkamp, J. F., Ookhtens, M., and Kaplowitz, N. 1992. Transport of glutathione at blood–brain barrier of the rat. *Pharmacology and Experimental Therapeutics* 263:964–72.

Kaplowitz, N., Fernández-Checa, J. C., Kannan, R., Garcia-Ruiz, C., Ookhtens, M., and Yu, J. R. 1996. GSH transporters: molecular characterization and role in GSH homeostasis. *Biological Chemistry Hoppe-Seyler* 377:267–73.

Karlsen, R. L., Grofova, I., Malthe-Sørenssen, D., and Fonnum, F. 1981. Morphological changes in rat brain induced by L-cysteine injection in newborn animals. *Brain Research* 208:167–80.

Keller, H. J., Do, K. Q., Zollinger, M., Winterhalter, K. H., and Cuénod, M. 1989. Cysteine: depolarization-induced release from rat brain *in vitro. Journal of Neurochemistry* 52:1801–6.

Kirstein, L. C., Coopersmith, R., Bridges, R. J., and Leon, M. 1991. Glutathione levels in olfactory and non-olfactory structures of rats. *Brain Research* 534:341–6.

Koppenol, W. H., Moreno, J. J., Pryor, W. A., Ischiropoulos, H., and Beckman, J. S. 1992. Peroxynitrite, a cloaked oxidant formed by nitric oxide and superoxide. *Chemical Research in Toxicology* 5:834–42.

Kranich, O., Hamprecht, B., and Dringen, R. 1996a. Utilization of cysteine and cysteine precursors for the synthesis of glutathione in astroglial cultures (abstract). *Journal of Neurochemistry* 66:S76.

Kranich, O., Hamprecht, B., and Dringen, R. 1996b. Different preferences in the utilization of amino acids for glutathione synthesis in cultured neurons and astroglial cells derived from rat brain. *Neuroscience Letters* 219:211–4.

Kurosawa, K., Hayashi, N., Sato, N., Kamada, T., and Tagawa, K. 1990. Transport of glutathione across the mitochondrial membrane. *Biochemical and Biophysical Research Communications* 167:367–72.

Langeveld, C. H., Jongenelen, C. A. M., Schepens, E., Stoof, J. C., Bast, A., and Drukarch, B. 1995. Cultured rat striatal and cortical astrocytes protect mesencephalic dopaminergic neurons against hydrogen peroxide toxicity independent of their effect on neuronal development. *Neuroscience Letters* 192:13–16.

Langeveld, C. H., Schepens, E., Jongenelen, C. A. M., Stoof, J. C., Hjelle, O. P., Ottersen, O. P., and Drukarch, B. 1996. Presence of glutathione immunoreactivity in cultured neurons and astrocytes. *Neuroreport* 7:1833–6.

Lauterberg, B. H., Adams, J. D., and Mitchell, J. R. 1984. Hepatic glutathione homeostasis in the rat: efflux accounts for glutathione turnover. *Hepatology* 4:586–90.

Li, X., Orwar, O., Revesjö, C., and Sandberg, M. 1996. γ-Glutamyl peptides and related amino acids in rat hippocampus *in vitro:* Effects of depolarization and γ-glutamyl transpeptidase inhibition. *Neurochemistry International* 29:121–8.

Liu, Y. F., and Quirion, R. 1992. Modulatory role of glutathione on μ-opiod, substance P/neurokinin-1, and kainic acid receptor binding sites. *Journal of Neurochemistry* 59:1024–32.

Lowenstein, C. J., Dinerman, J. L., and Snyder, S. H. 1994. Nitric oxide: a physiological messenger. *Annals of Internal Medicine* 120:227–37.

Lowndes, H. E., Beiswinger, C. M., Philbert, M. A., and Reuhl, K. R. 1994. Substrates for neural metabolism of xenobiotics in adult and developing brain. *Neurotoxicology* 15:61–74.

Lyrer, P., Landolt, H., Kabiersch, A., Langemann, H., and Kaesler, H. 1991. Levels of low molecular weight scavengers in the rat brain during focal ischemia. *Brain Research* 567:317–20.

Makar, T. K., Nedergaard, M., Preuss, A., Gelbard, A. S., Perumal, A. S., and Cooper, A. J. L. 1994. Vitamin E, ascorbate, glutathione, glutathione disulfide, and enzymes of glutathione metabolism in astrocytes and neurons: evidence that astrocytes play an important role in antioxidative processes in the brain. *Journal of Neurochemistry*. 62:45–53.

Mårtensson, J., and Meister, A. 1989. Mitochondrial damage in muscle occurs after marked depletion of glutathione and is prevented by giving glutathione monoester. *Proceedings of the National Academy of Sciences of the U.S.A.* 86:471–5.

Mårtensson, J., Lai, J. C. K., and Meister, A. 1990. A high-affinity transport of glutathione is part of a multicomponent system essential for mitochondrial function. *Proceedings of the National Academy of Sciences of the U.S.A.* 87:7185–9.

Meister, A. 1986. Glutathione: metabolism, transport and effects of selective modifications of cellular glutathione levels. In *Thioredoxin and glutaredoxin systems: structure and function*, ed. A. Holmgren, C.-I. Bründén, H. Jörnvall, and B.-M. Sjöberg, 245–75. Karolinska Nobel Conference Series. New York: Raven.

Messina, J. E., Page, R. H., Hetzel, F. W., and Chopp, M. 1989. Administration of L-2-oxothiazolidine-4-carboxylate increases glutathione levels in rat brain. *Brain Research* 478:181–3.

Minn, A., and Besagni, D. 1983. Uptake of L-glutamine into synaptosomes. Is the γ-glutamyl cycle involved? *Life Sciences* 33:245–96.

Mizui, T., Kinouchi, H., and Chan, P. H. 1992. Depletion of brain glutathione by buthionine sulfoximine enhances cerebral ischemic injury in rats. *American Journal of Physiology* 262:H313–17.

Murphy, T. H., Miyamoto, M., Satre, A., Schnaar, A. L., and Coyle, J. T. 1989. Glutamate toxicity in a neural cell line involves inhibition of cystine transport leading to oxidative stress. *Neuron* 2:1547–58.

Murphy, T. H., Schnarr, R. L., and Coyle, J. T. 1990. Immature cortical neurons are uniquely sensitive to glutamate toxicity by inhibition of cystine uptake. *FASEB Journal* 4:1624–33.

Noble, P. G., Antel, J. P., and Yong, V. W. 1994. Astrocytes and catalase prevent the toxicity of catecholamines to oligodendrocytes. *Brain Research* 633:83–90.

Noguchi, K., Higuchi, S., and Matsui, H. 1989. Effects if glutathione isopropyl ester on glutathione concentration in ischemic brain. *Brain Research* 18:1337–40.

O'Connor, E., Devesa, A., García, C., Puertes, I. R., Pellín, A., and Viña, J. R. 1995. Biosynthesis and maintenance of GSH in primary astrocyte cultures: role of cystine and ascorbate. *Brain Research* 680:157–63.

Ogita, K., Enomoto, R., Nakahara, F., Ishitsubo, N., and Yoneda, Y. 1995. A possible role of glutathione as an endogenous agonist at the N-methyl-D-aspartate recognition domain in rat brain. *Journal of Neurochemistry* 64:1088–96.

Oldendorf, W. H. 1971. Brain uptake of radiolabeled amino acids, amines, and hexoses after arterial injection. *American Journal of Physiology* 221:1629–39.

Oldendorf, W. H., and Szabo, J. 1976. Amino acid assignment to one of three blood–brain amino acid carriers. *American Journal of Physiology* 230:94–8.

Olney, J. W., Ho, O. L., and Rhee, V. 1971. Cytotoxic effects of acidic and sulphur containing amino acids on the infant mouse central nervous system. *Experimental Brain Research* 14:61–70.

Olney, J. W., Zorumski, C., Price, M. T., and Larbruyere, J. 1990. L-Cysteine, a bicarbonate-sensitive endogenous excitotoxin. *Science* 248:596–9.

Orlowski, M., and Meister, A. 1970. The γ-glutamyl cycle: a possible transport system for amino acids. *Proceedings of the National Academy of Sciences of the U.S.A.* 67:1248–55.

Orwar, O., Li, X., Andine, P., Bergström, C. M., Hagberg, H., Folestad, S., and Sandberg, M. 1994. Increased intra- and extracellular concentrations of γ-glutamylglutamate and related dipeptides in the ischemic rat striatum: involvement of γ-glutamyltranspeptidase. *Journal of Neurochemistry* 63:1371–6.

Pan, Z., and Perez-Polo, R. 1993. Role of nerve growth factor in oxidant homeostasis: glutathione metabolism. *Journal of Neurochemistry* 61:1713–21.

Patel, N. J., Fullone, J., and Anders, M. W. 1993. Brain uptake of S-(1,2-dichlorovinyl)glutathione and S-(1,2-dichlorovinyl)-L-cysteine, the glutathione and cysteine conjugates of the neurotoxin dichloroacetylene. *Molecular Brain Research* 17:53–8.

Pardridge, W. M. 1977. Kinetics of competitive inhibitors of neutral amino acid transport across the blood–brain barrier. *Journal of Neurochemistry* 28:103–8.

Petito, C. K. 1986. Transformation of postischemic perineural glial cells. 1. Electron microscopic changes. *Journal of Cerebral Blood Flow and Metabolism* 6:616–24.

Petito, C. K., and Babiak, T. 1982. Early proliferative changes in astrocytes in postischemic noninfarcted rat brain. *Annals of Neurology* 11:510–8.

Philbert, M. A., Beiswinger, C. M., Waters, D. K., Reuhl, K. R., and Lowndes, H. E. 1991. Cellular and regional distribution of reduced glutathione in the nervous system of the rat: histochemical localization by mercury orange and o-phthaldialdehyde. *Toxicology and Applied Pharmacology* 107:215–27.

Pileblad, E., Eriksson, P. S., and Hansson, E. 1991. The presence of glutathione in primary neuronal and astroglial cultures from the rat cerebral cortex and brain stem. *Journal of Neural Transmission* 86: 43–9.

Raps, S. P., Lai, J. C. K., Hertz, L., and Cooper, A. J. L. 1990. Glutathione is present in high concentrations in cultured astrocytes but not in cultured neurons. *Brain Research* 493:398–401.

Rehncrona, S., Folbergrová, J., Smith, D. S., and Siesjö, B. L. 1980. Influence of complete and pronounced incomplete ischemia and subsequent recirculation on cortical concentrations of oxidized and reduced glutathione in the rat. *Journal of Neurochemistry* 34:477–86.

Reichelt, K. L., and Fonnum, F. 1969. Subcellular localization of *N*-acetyl-aspartyl-glutamate, *N*-acetyl-glutamate and glutathione in brain. *Journal of Neurochemistry* 16:1409–16.

Saetre, R., and Rabenstein, D. L. 1978. Determination of cysteine in plasma and urine and homocysteine in plasma by high-pressure liquid chromatography. *Analytical Biochemistry* 90:684–92.

Sagara, J.-I., Miura, K., and Bannai, S. 1993a. Cystine uptake and glutathione levels in fetal brain cells in primary culture and in suspension. *Journal of Neurochemistry* 61:1667–71.

Sagara, J.-I., Miura, K., and Bannai, S. 1993b. Maintenance of neuronal glutathione by glial cells. *Journal of Neurochemistry* 61:1672–6.

Sagara, J.-I., Makino, N., and Bannai, S. 1996. Glutathione efflux from cultured astrocytes. *Journal of Neurochemistry* 66:1876–81.

Schapira, A. H. V., Cooper, J. M., Dexter, D., Clark, J. B., Jenner, P., and Marsden, C. D. 1990. Mitochondrial complex 1 deficiency in Parkinson's disease. *Journal of Neurochemistry* 54:823–7.

Seaton, T. A., Jenner, P., and Marsden, L. D. 1996. Mitochondrial respiratory enzyme function and super-oxide dismutase activity following brain glutathione depletion in the rat. *Biochemical Pharmacology* 52:1657–63.

Semon, B. A., Leung, P. M. B., Rogers, Q. R., and Gietzen, D. W. 1989. Plasma and brain ammonia and amino acids in rats measured after feeding 75% casein or 28% egg white. *Journal of Nutrition* 119:1583–92.

Shank, R. P., and Campbell, G. LeM. 1984. α-Ketoglutarate and malate uptake and metabolism by synaptosomes: further evidence for an astrocyte-to-neuron metabolic shuttle. *Journal of Neurochemistry* 42: 1153–61.

Shine, H. D., Hertz, L., de Vellis, J., and Haber, B. 1981. A fluorometric assay for γ-glutamyltranspeptidase: demonstration of enzymatic activity in cultured cells of neural origin. *Neurochemical Research* 6:435–63.

Slivka, A., and Cohen, G. 1993. Brain ischemia markedly elevates levels of the neurotoxic amino acid cysteine. *Brain Research* 608:33–7.

Slivka, A., Mytilineou, C., and Cohen, G. 1987. Histochemical evaluation of glutathione in brain. *Brain Research* 409:275–84.

Slivka, A., Spina, M. B., and Cohen, G. 1988. Reduced and oxidized glutathione in human and monkey brain. *Neuroscience Letters* 74:112–8.

Spina, M. B., and Cohen, G. 1989. Dopamine turnover and glutathione oxidation: implications for Parkinson disease. *Proceedings of the National Academy of Sciences of the U.S.A.* 86:1398–400.

Tansey, F. A., and Cammer, W. 1991. Depletion of glutathione interferes with induction of glycerolphosphate dehydrogenase in the brains of young rats. *Brain Research* 564:31–6.

Tate, S. S., Ross, L. L., and Meister, A. 1973. The γ-glutamyl cycle in the choroid plexus. Its possible function in amino acid transport. *Proceedings of the National Academy of Sciences of the U.S.A.* 70:1447–9.

Tayarani, I., Lefauconnier, J.-M., Roux, F., and Bourre, J.-M. 1987. Evidence for an alanine, serine, and cysteine system of transport in isolated brain capillaries. *Journal of Cerebral Blood Flow and Metabolism* 7:585–91.

Thompson, G. A., and Meister, A. 1975. Utilization of L-cystine by the γ-glutamyl transpeptidase-γ-glutamyl cyclotransferase pathway. *Proceedings of the National Academy of Sciences of the U.S.A.* 72:1985–8.

Vatassery, G. 1996. Oxidation of vitamin E, vitamin C, and thiols in rat brain synaptosomes by peroxynitrite. *Biochemical Pharmacology* 52:579–86.

Wade, L. A., and Brady, H. M. 1981. Cysteine and cystine transport at the blood–brain barrier. *Journal of Neurochemistry* 37:730–4.

Werner, P., and Cohen, G. 1993. Glutathione disulfide (GSSG) as a marker of oxidative injury in brain mitochondria. *Annals of the New York Academy of Sciences* 679:364–9.

Winkler, B. S., De Santis, N., and Solomon, F. 1986. Multiple NADPH-producing pathways control glutathione (GSH) content in retina. *Experimental Eye Research* 43:829–47.

Wüllner, U., Löschmann, P. A., Schulz, J. B., Schmid, A., Dringen, R., Eblen, F., Turski, L., and Klockgether, T. 1996. Glutathione depletion potentiates MPTP and MPP+ toxicity in nigral dopaminergic neurons. *Neuroreport* 7:921–3.

Yang, C. S., Chou, S. T., Lin, N. N., Liu, L., Tsai, P. J., Kuo, J. S., and Lai, J. S. 1994. Determination of extracellular glutathione in rat brain by microdialysis and high-performance liquid chromatography with fluorescence detection. *Journal of Chromatography. B. Biomedical Applications* 661:231–5.

Yonezawa, M., Back, S. A., Gan, Y., Rosenberg, P. A., and Volpe, J. J. 1996. Cystine deprivation induces oligodendroglial death: rescue by free radical scavengers and by a diffusible glial factor. *Journal of Neurochemistry* 67:566–75.

Yudkoff, M., Pleasure, D., Cregar, L., Lin, Z.-P., Nissim, I., Stern, J., and Nissim, I. 1990. Glutathione turnover in cultured astrocytes: studies with [15N]glutamate. *Journal of Neurochemistry* 55:137–45.

Yudkoff, M., Nissim, I., Hertz, L., Pleasure, D., and Erecinska, M. 1992. Nitrogen metabolism: neuronal-astroglial relationships. *Progress in Brain Research* 94:213–24.

Zängerle, L., Cuénod, M., Winterhalter, K. H., and Do, K. Q. 1992. Depolarization-induced release of glutathione and cysteine from rat brain slices. *Journal of Neurochemistry* 59:181–9.

Zlokovic, B., Mackic, J. B., McComb, J. G., Weiss, M. H., Kaplowitz, N., and Kannan, R. 1994. Evidence for transcapillary transport of reduced glutathione in vascular perfused guinea-pig brain. *Biochemical and Biophysical Research Communications* 201:402–8.

Glutathione in the Nervous System
Edited by Christopher A. Shaw
Copyright © 1998 Taylor & Francis

6

Glutathione and the Regulation of Apoptosis in the Nervous System

Khalequz Zaman and Rajiv R. Ratan

Neurology Laboratories, Harvard Institutes of Medicine, Boston, Massachussetts 02115

ABBREVIATIONS

BHA butylated hydroxyanisole
BSO buthionine sulfoximine
CAT catalase
GSHPX glutathione peroxidase
GR glutathione reductase

GST glutathione-S-transferase
GSH glutathione (reduced form)
GSSG glutathione (oxidized form)
HCA homocysteate
H_2O_2 hydrogen peroxide
LDH lactate dehydrogenase
NAC N-acetyl-L-cysteine
NMDA N-methyl-D-asparate
OTC L-2-oxothiazolidine-4-carboxylate
PC_{12} pheochromocytoma
SOD superoxide dismutase
SV Sindbis virus

1. INTRODUCTION

A critical role for glutathione in opposing the toxic effects of free radicals in neurons and glia is well established, and much of the biochemistry and cell biology linking glutathione to cell survival in the nervous system has been cogently discussed in other chapters of this book (see chapter 5 by Cooper). Less well understood, however, are the precise pathways by which glutathione insufficiency in the brain leads to neuronal loss. Given the putative primary role of glutathione insufficiency in a host of neurodegenerative states, including Parkinson's disease, greater understanding of the molecular targets of radicals under conditions of glutathione insufficiency as well as of mechanisms for compensating for such insufficiency is likely to suggest novel and broadly applicable therapeutic approaches. In this chapter, we will describe an in vitro model of glutathione depletion in mixed neuron–glia cultures. Further we will describe findings using this model linking glutathione depletion to neuronal apoptosis and examine evidence for glutathione-dependent and glutathione-independent paths to apoptotic death.

2. AN IN VITRO MODEL OF GLUTATHIONE-DEPLETION-INDUCED DEATH IN PRIMARY NEURONS AND NEUROBLASTOMA CELLS

Over the past decade, considerable attention has been focused on the role that the excitatory amino acid glutamate plays as the proximate toxin in acute and chronic neurodegenerative states, including hypoxia–ischemia, Huntington's disease, and amyotrophic lateral sclerosis (Choi 1994; Brouillet and Beal 1993; Rothstein 1996). These studies have elucidated several plausible schemes by which glutamate may induce neurotoxicity. Most of these schemes have invoked glutamate-mediated activation of one of three classes of plasma membrane receptors (excitatory glutamate receptors), named, nearly a decade ago, for structural analogs that are specific agonists: quisqualate, kainate, and N-methyl-D-aspartate (NMDA) (Monaghan, Bridges, and Cotman 1989; Robinson and Coyle 1987; Rothman and Olney 1987). However, while investigating glutamate toxicity in the N18-RE-105 neuroblastoma × retina cell line, Murphy, Coyle and coworkers identified a novel type of glutamate neurotoxicity that depends on the ability of glutamate or glutamate analogs, such as quisqualate, to interact with a chloride-dependent, high-affinity antiport uptake site (Murphy et al. 1989; Murphy, De Long, and Coyle 1991) and not plasma membrane receptors. This antiport uptake site, which was initially described in fibroblasts by Bannai and colleagues (Bannai and Kitamura 1980, 1981), is highly specific for cystine and glutamate, and was designated the X_c^- system (Makowske and Christensen 1982).

Evidence that neurotoxicity due to elevated glutamate results from competitive inhibition of cystine transport from the extracellular medium to the cytoplasm through the X_c^- system comes from five observations (Murphy et al. 1989, 1991). First, non-receptor-mediated cell death induced by glutamate is observed on continuous exposure of neurons (for 24 h) to high concentrations of glutamate or its analogs (1–10 mM). In contrast, maximal cell death in excitotoxic paradigms is observed with short exposures (5–30 min) of primary neurons to concentrations of glutamate from 0.1 to 0.5 mM. Second, only glutamate analogs capable of inhibiting radioactive cystine uptake into the cell are toxic to immature cortical neurons and some neuroblastoma cell lines. Exposure to the glutamate analog N-methyl-D-aspartate (an excitotoxin) does not inhibit radioactive cystine uptake and is not lethal to immature cortical neurons. Third, elimination of cystine from the bathing medium of cultured cells results in cell death with the same temporal profile and morphology as exposure to elevated glutamate. Fourth, neuronal death associated with cystine depletion cannot be blocked by competitive or noncompetitive antagonists of the three types of glutamate receptors. Fifth, in immature cortical neurons, toxicity is observed 1–3 days in vitro (DIV), but inward currents in response to glutamate (mediated by cell surface glutamate receptors) are not seen till 4–7 days later (Murphy and Baraban 1990). Cystine deprivation induced by glutamate has been observed in primary cortical cultures, N18-RE-105 cells, some clones of PC12 cells (Froissard and Duval 1994), and primary sympathetic neurons.

Following up on previous studies by Bannai and colleagues, Murphy et al. demonstrated that inhibition of cystine uptake in neuroblastoma cells or primary neuronal cultures leads to cellular depletion of cysteine, the rate-limiting precursor in glutathione synthesis (Murphy et al. 1989, 1991). Many cells in culture require cystine as an essential nutrient because they are unable to synthesize (through the transulfuration pathway) sufficient cysteine (the reduced form of cystine) for survival and growth. Furthermore, when added to the culture medium, cysteine is rapidly oxidized to cystine. Cells take up cystine from the medium through the X_c^- system described above in exchange for glutamate. Intracellularly, cystine is reduced to cysteine, where it is utilized primarily for synthesis of the antioxidant glutathione or the synthesis of protein. The first and rate-limiting step of glutathione synthesis involves the formation of a γ-glutamyl linkage between L-glutamate and L-cysteine and is catalyzed by γ-glutamylcysteine synthetase. Thus, inhibition of cystine transport leads to depletion of intracellular cysteine and the antioxidant glutathione, leading in turn to oxidative-stress-induced death.

2.1 Oxidative Stress, Glutathione, and the Nervous System

Oxidative stress, is, by definition, an imbalance in the cell between oxidant production and antioxidant defenses in which oxidant production is favored (Sies 1985). Intracellular sources of oxidants (e.g., superoxide, peroxide, and hydroxyl radicals) include the mitochondria, where partial reduction of O_2 by electron-rich intermediates in the electron transport chain results in the formation of the superoxide radicals (Boveris and Chance 1973); the plasma membrane, where activation of phospholipase A2 can lead to superoxide production via the subsequent metabolism of its product arachidonic acid (Chan and Fishman 1980); and the synaptic cleft, where peroxide can be produced by the enzyme monoamine oxidase or by the autooxidation of endogenous nervous-system compounds such as catecholamines (Graham et al. 1978). In all subcellular locales, superoxide and peroxide can give rise to the formation of hydroxyl radicals, the most reactive of oxygen species, via interactions with loosely bound or free metals such as iron (Halliwell and Gutteridge 1989). Proposed cytotoxic mechanisms of oxidative stress including lipid peroxidation, oxidation of proteins, DNA damage, reduction in ATP levels, and increases in intracellular Ca^{2+} concentration with concomitant activation of Ca^{2+}-dependent proteases (Halliwell 1992).

To counteract the threat imposed by oxidants as a result of normal and abnormal cell metabolism, neurons have developed several enzymatic and nonenzymatic antioxidant defense systems (Das et al. 1981; Cammer et al. 1989; Shivakumar, Anandatheerthavarada, and Ravindranath 1991; Brannan et al. 1980). The enzymatic system consists of five basic enzymes: superoxide dismutase (SOD), catalase (CAT), and three glutathione dependent enzymes: glutathione peroxidase (GSHPX), glutathione-S-transferase (GST), and glutathione reductase (GR).

There are three forms of SOD, encoded by three separate genes, which are found in distinct subcellular locations. Copper–zinc superoxide dismutase (Cu/Zn SOD) is found in the cytosol; another Cu/Zn SOD is found in the extracellular space; and manganese superoxide dismutase (Mn SOD) is localized in the mitochondrial matrix. All three forms of SOD catalyze the dismutation of superoxide radical to H_2O_2.

Hydrogen peroxide can be detoxified by catalase (in peroxisomes) or by glutathione peroxidase (present in all organelles). Reduced glutathione acting in concert with glutathione peroxidase is the major pathway of hydrogen and lipid peroxide detoxification in many cells (Meister 1983). It is thus important for the protection of membrane lipids against oxidation.

A distinct family of glutathione-dependent enzymes, the glutathione-S-transferases (GST), catalyze the conjugation of GSH with a variety of electrophilic compounds and play a major role in detoxification processes in the cells. This system also provides a secondary defense against protein cross-linking and loss of membrane function after saturation of Se-dependent GSH-peroxidase (Orrenius and Moldeus 1984).

Glutathione reductase (GR) is responsible for maintaining cellular glutathione in the reduced state. It catalyzes the reduction of oxidized glutathione using NADPH as a hydrogen donor. It requires flavin adenine dinucleotide (FAD) as a coenzyme.

In summary, the major functions of glutathione in protection against oxidative stress are believed to be due to three type of reactions. First, reduced glutathione is used by glutathione peroxidases to eliminate toxic peroxides. Second, reduced glutathione (maintained by the action of glutathione reductase and NADPH) functions to reduce oxidized forms of vitamin C back to its functional reduced form. Because vitamin C, in turn, functions to maintain vitamin E in its reduced and functional form, GSH indirectly supports the free-radical reductions and free-radical chain termination functions of vitamin C and vitamin E. Finally, GSH functions through glutathione-S-transferases to detoxify reactive aldehydes that are generated during lipid peroxidation, to deotxifiy xenobiotics, and to detoxify DNA hydroperoxides. These reactions highlight the versatile role that glutathione plays in oxidant homeostasis in cells.

Despite the abundance of evidence linking glutathione depletion to oxidative stress in nonneuronal cells, evidence that glutamate-induced glutathione depletion leads to oxidative stress in neurons is somewhat indirect. The ability of a number of antioxidants, including idebenone (a soluble coenzyme Q analog), vitamin E, butylated hydroxyanisole, and iron chelators to abrogate cystine deprivation and glutathione depletion-induced neuronal death (Murphy, Schnaar, and Coyle 1990; Ratan, Murphy, and Baraban 1994a,b; Ratan, Lee, and Baraban 1996) suggests that exogenously applied antioxidants are balancing the cellular antioxidant debt created by glutathione depletion and reversing oxidative stress. Of course, the existence of oxidative stress in cells depleted with glutathione is consistent with the tripeptide's well-established role as a member of the antioxidant defense team.

Glutathione depletion following glutamate exposure is likely a cause and not a consequence of cell death in neurons. Detailed measurements of total glutathione in primary neuronal cultures reveals that total glutathione is depleted by 3–6 h after exposure to glutamate or its analogs, and yet cells are not irreversibly committed to die until 4–7 h after glutathione is depleted (Ratan et al., unpublished observations). Thus, glutathione depletion occurs well in advance of the irreversible commitment point to cell death. Further, while a number of antioxidants abrogate glutathione-depletion-induced death, they do so without restoring glutathione to control levels. These results indicate that glutathione depletion and cell death can be uncoupled and support the notion that glutathione depletion subsequent to glutamate exposure in immature cortical neurons is a cause and not a consequence of cell death.

Volpe and colleagues have shown that like immature neurons, immature oligodendrocytes are sensitive to glutamate through inhibition of cystine transport leading to glutathione depletion and oxidative stress (Oka et al. 1993). However, in contrast to immature neurons and oligodendrocytes, astrocytes are resistant to cystine-deprivation-induced toxicity. Using specific immunocytochemical markers for astroglial cells (glial fibrillary acid protein) and neurons (neuron specific enolase), Murphy et al. demonstrated that glial cells exposed to low ambient concentrations of cystine (10 μM) or high concentrations of glutamate (10 mM) do not undergo cell death, despite the ability of these treatments to lower intracellular glutathione levels (Murphy et al. 1991). The selective resistance of glia cells was also observed in mixed oligodendrocyte–astroglial cultures exposed to glutamate (Oka et al. 1993). Several possible explanations exist for the selective resistance of glial cells to cystine deprivation and glutathione depletion. First, glial cells in vitro (Raps et al. 1990; Sagara, Miura, and Bannai 1993b) and are known to have significantly higher levels of glutathione than neurons. Thus, while cystine deprivation leads to glial glutathione depletion, the reduced levels of glutathione may remain above the threshold for cell death. Second, glial cells appear to have higher levels of glutathione-independent antioxidant enzymes, and these enzymes may confer protection in the setting of glutathione-depletion. Identification of the precise mechanisms by which glial cells are resistant to glutathione depletion may enlighten novel therapeutic approaches to glutathione-depletion-induced death in neurons and oligodendrocytes and is a fertile area for further study.

Despite the resistance of astrocytes to cystine-deprivation-induced death, Bannai, Sagara and coworkers have proposed that inhibiton of cystine uptake into astrocytes rather than neurons is the pathway by which neurons die in response to elevated glutamate in mixed neuronal cultures (Sagara, Miura, and Bannai 1993a,b). Using nearly pure cultures of neurons and glia, Sagara et al. showed that uptake of radioactive cystine uptake is negligible in neurons as compared to glia. In contrast, cysteine is taken up avidly by neurons and glial cells. Further they demonstrated that cysteine supplemented to neurons increased glutathione levels in these cells, whereas cystine supplementation to isolated neurons had no effect. Finally, they measured the level of cysteine from the media of mixed neuronal–glial cultures and found it to be greater than from the media of isolated neurons or glia. From these observations they proposed a scheme in which cystine is taken up into glial cells and reduced to cysteine. Intracellularly, cysteine is in part converted to glial glutathione. The remaining pool of glial cysteine is exported to the extracellular medium and taken up by neurons for glutathione synthesis. From, this scheme they proposed that depletion of glutathione induced by glutamate in mixed neuronal cultures results from inhibition of cystine uptake into glial cells and not neurons.

While intriguing, the model proposed by Bannai awaits further experimental support. For example, one strong prediction from this model is that inhibitors of neuronal cysteine transport (e.g., high concentrations of serine or alanine) should deplete neuronal glutathione without affecting glial glutathione. Additionally, the model is based on data using currently available methods for measuring glutathione in living cells. These methods are rather primitive and do not allow definition of the potential spatial or temporal heterogeneity that exists in glutathione levels in a single cell or among different cells in the dish. Indeed, studies from our laboratory, using the glutathione-sensitive fluorescent reporter monochlorobimane, suggest that glutathione levels in cocultured neurons and glial cells are significantly more heterogeneous than previously appreciated (Ratan et al. 1994b; see Fig. 1). Recent studies indicate that heterogeneity in neuronal glutathione we have observed in vitro may reflect differences in the cellular origin or maturity of cultured neurons (Beiswanger et al. 1995).

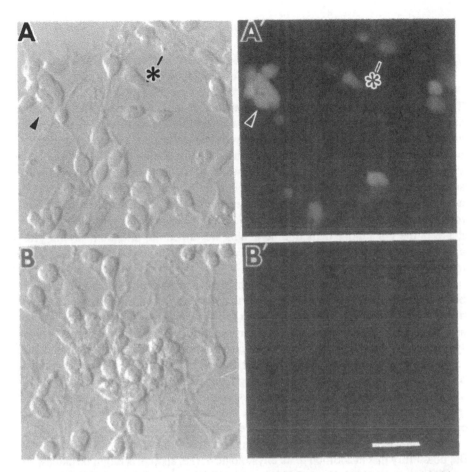

FIG. 1. Staining with the glutathione reporter monochlorobimane demonstrates that glutathione is heterogenously produced in neurons and glia in mixed cortical cultures. Forty-eight hours after plating, cultured neurons were loaded with 5 μM monochlorobimane (dissolved in media) for 10 min at 37°C. Monochlorobimane becomes fluorescent only after enzymatic conjugation to reduced glutathione. Cultures were then rinsed gently in warm PBS and transferred to a buffered salt solution. A: cultured cells visualized under Hoffman microscopy. A': cells depicted in A loaded with monochlorobimane and visualized under fluoresence microscopy. Arrowheads (A and A') point to a cluster of cells with the morphology of neurons. Asterisks (A and A') refer to flat, large cells typical of glial cells. Note that all the cells in the field are labeled. To determine the specificity of the fluorescent signal for reduced glutathione, cultures depicted under Hoffman microscopy in B were treated with diethyl maleate (2 mM) for 15 min prior to labeling with monochlorobimane. Diethylmaleate depletes glutathione by conjugating to it, a process catalyzed by the enzyme glutathione transferase. B': monochlorobimane fluorescence of cells depicted in B after depletion of glutathione. Note the absence of labeling in any cells in the field. (Reprinted by permission from the *Journal of Neuroscience* 14(7):4385–92.)

3. GLUTATHIONE AND APOPTOSIS IN THE NERVOUS SYSTEM

3.1 Glutathione Depletion Leads to Apoptotic Neuronal Death
in Embryonic Cortical Neurons

The precise role that glutamate-induced cystine depletion plays in neurodegenerative states remains uncertain and it is likely that the glutamate-receptor-mediated mechanisms of cytotoxicity, which have received much attention over the past decade, predominate. Nevertheless, glutamate-induced cystine deprivation can serve as an in vitro model to study the mechanisms by which glutathione depletion and oxidative stress lead to neuronal death. Such a model involving a single type of insult has clear advantages over excitotoxic paradigms where multiple serial and interacting pathways to cell death are simultaneously active. To begin to address the mechanism of cell death in the glutathione depletion paradigm, we turned our attention, several years ago, to the question of whether neuronal death under conditions of cystine deprivation is necrotic or apoptotic.

Apoptosis is a morphologically distinct form of cell death that historically has been associated with physiological processes such as nervous-system development, organ morphogenesis, and immunity (Kerr and Harmon 1991). For example, as many as 80 percent of the original pool of neurons are lost in some areas of the nervous system during development. This developmental neuronal loss has been postulated to be an efficient way to match an overabundant neuronal pool to a finite number of targets that secrete a limiting level of growth-factor support (Server and Mobley 1991). According to this scheme, only those neurons that reach their proper target and receive sufficient trophic support survive. In this way, dauntingly complex neuronal circuits that connect targets separated by several feet can be sculpted.

Studies by Kerr, Wyllie, and coworkers in the late 1960s to early 1970s suggested the possibility that in addition to its physiological role, apoptosis might be unleashed, possibly inappropriately, by pathological stimuli (Kerr and Harmon 1991). The original intent of their studies was to characterize the histochemical changes that occurred in lysosomal enzymes following hepatic ischemia. They found that—within hours after ligating the portal vein branches leading to the left and median lobes of the liver—cell lysis, cell swelling, and a robust inflammatory response consistent with coagulative necrosis could be observed around terminal hepatic veins. The periportal area appeared normal, probably because the hepatic artery was still able to supply this area with glucose and oxygen. However, over the ensuing few weeks, the preserved periportal area gradually shrank. During the period of regression, histochemical studies revealed cytoplasmic masses containing nearly normal organelles inside normal hepatocytes. Kerr and Wyllie inferred that these cytoplasmic masses reflected the compartmentalization of cells into multiple membrane-bound vesicles containing normal organelles, which were subsequently phagocytized by neighboring cells. This was clearly a manifestation of cell death, yet it was different from the necrosis seen soon after the onset of ischemia in that there was no cell swelling and no inflammatory response. They called this type of cell death, distinct from necrosis, *apoptosis*. This term is derived from the Greek, "to fall away from," as in leaves falling from a tree. It was coined for contrast with *mitosis*, a biological process where cells divide and organs grow. The seminal studies of Kerr and Wyllie highlighted the ability of apoptosis to be activated by a pathological process such as hepatic ischemia and raised the question whether apoptosis could be activated by a pathological stimulus such as glutathione depletion in neurons.

Although morphological criteria are its most consistent features, no one criterion defines cells as apoptotic. Therefore, to address whether glutathione depletion in embryonic cortical neurons leads to apoptosis or necrosis, several criteria were considered. Necrotic death is characterized by an early loss of membrane integrity and cell and organelle swelling. In vivo, membrane breakdown leads to a release of cytoplasmic contents and a fulminant immune response, including polymorphonuclear lymphocytes. DNA is degraded randomly, leading to a continuous smear upon agarose gel electrophoresis. Apoptosis, on the other hand, is characterized morphologically by chromatin condensation and nuclear and cytoplasmic shrinkage. In vivo, the cell compartmentalizes into many small, membrane-bound vesicles, which can be quietly digested by macrophages or neighboring cells. Loss of membrane integrity is a late event, and DNA is degraded into multiples of oligonucleosomal fragments of approximately 200 base pairs (with some exceptions). In some cases, apoptotic death requires RNA and DNA synthesis. This observation has fueled the notion that this form of death is "active" and requires the synthesis of specific "death proteins."

We used the above criteria to determine whether glutathione depletion and oxidative stress in embryonic cortical neurons leads to necrosis or apoptosis. Nuclear morphology of cells dying from glutathione depletion was assessed by staining control and treated cultures with DNA-intercalating dyes, acridine orange or propidium iodide. Sixteen hours after exposure to 1 mM glutamate or the glutamate analog homocysteate (HCA), a significant increase in cells with the condensed chromatin and fragmented nuclei characteristic of apoptosis was noted (Ratan et al. 1994a). The increase in apoptosis induced by glutamate could be prevented by coapplication of the antioxidant idebenone or butylated hydroxyanisole. These antioxidants not only inhibited the glutamate-induced changes characteristic of apoptosis, but also the glutamate-induced lactate dehydrogenase (LDH) (a measure of loss of membrane integrity) observed 6–8 h after the onset of chromatin condensation. The apoptotic morphologic changes induced by glutamate were associated with ordered cleavage of DNA into multiples of nucleosomal-sized DNA fragments as measured by agarose gel electrophoresis. Finally, as has been observed in paradigmatic neuronal apoptosis induced by growth-factor deprivation (Martin et al. 1988), coadministration of cycloheximide (a protein synthesis inhibitor) or actinomycin-D (an RNA synthesis inhibitor) suppresses oxidative-stress-induced death (Fig. 2). These findings demonstrate that the apoptotic pathway can be triggered by glutathione depletion in cultured neurons and suggest that if apoptosis is triggered in neurological disease states, then glutathione depletion is a candidate trigger for this process.

In addition to triggering apoptosis, glutathione depletion has also been shown to lead to necrosis. Using buthionine sulfoximine (BSO), a specific and essentially

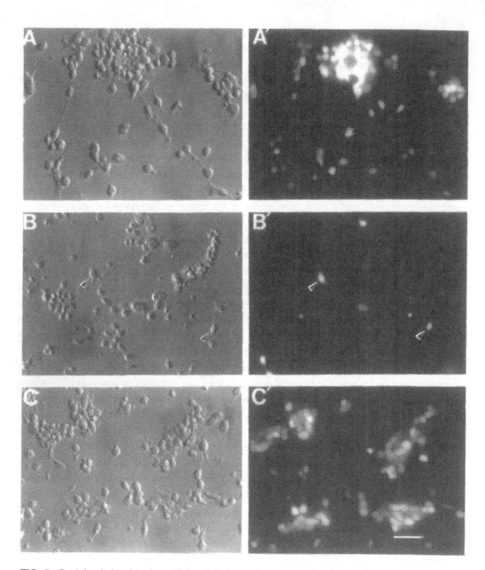

FIG. 2. Cycloheximide blocks cell death induced by cystine deprivation in embryonic cortical cultures. A: Cultured cells visualized under Hoffman microscopy. A': Cells depicted in A loaded with the live-cell stain fluorescein diacetate and visualized under fluorescence microscopy. Note that nearly all the cells in the field are stained. B: Cultured cells visualized as in A after treatment with the glutamate analog, homocysteate (HCA). HCA has been shown to inhibit cystine transport, leading to glutathione depletion and oxidative-stress-induced cell death (Murphy et al. 1990). B': Cells depicted in B loaded with the live-cell stain fluorescein diacetate. Arrows identify two live-cells with the morphological appearance of neurons. C: A parallel cultured treated with the glutamate analog HCA and the protein synthesis inhibitor, cycloheximide. C': Live cell staining of cells in C. Note the dramatic increase in cell viability as compared to B'.

irreversible inhibitor of γ-glutamylcysteine synthetase, to inhibit glutathione synthesis in GT1-7 cells, a hypothalamic neural cell line, Bredesen and colleagues showed that depletion of glutathione leads to death with morphologic and biochemical features of necrosis (Kane et al. 1993). How can these results be reconciled with our observations that glutathione depletion induces apoptosis? Previous studies have shown that low levels of a toxic stimulus in neurons induce apoptosis, whereas high levels induce necrosis (Ankarcrona et al. 1995; Bonfoco et al. 1995). These observations suggest that depletion of glutathione can lead to apoptosis or necrosis, depending on the level of ambient oxidant production or levels of antioxidants, which could, in part, compensate for the absence of glutathione. Such a model is also consistent with the notion that in vivo, apoptosis is the "preferred way to die" (because of the absence of an inflammatory response) and that only when the insult is severe enough to damage the apoptotic machinery will necrosis ensue.

3.2 Macromolecular Synthesis Inhibitors Prevent Glutathione-Depletion-Induced Apoptosis in Embryonic Cortical Neurons by Shunting Cysteine from Protein Synthesis to Glutathione

Inhibitors of macromolecular synthesis have been shown to abrogate neuronal death by a host of stimuli. In a number of these paradigms, protection by inhibitors of macromolecular synthesis has been interpreted as evidence for a gene-directed program of cell suicide. In vertebrate systems, recent data has identified c-jun as one protein fulfilling the definition of a "death protein" (Estus et al. 1994; Ham et al. 1995). However, in considering other possible mechanisms by which protein synthesis inhibitors might abrogate cell death, we noted that in addition to being a vital precursor for glutathione synthesis, intracellular cysteine is also utilized for protein synthesis. We therefore examined whether protection conferred by macromolecular synthesis inhibitors in the presence of glutamate could result from a redistribution of cysteine from protein synthesis into the formation of glutathione. Unexpectedly, we found that RNA and protein synthesis inhibitors increase glutathione levels in glutamate-treated cultures (Ratan et al. 1994b). Studies of radioactive cystine incorporation into acid-soluble (glutathione and its precursors) and acid-precipitable (protein) fractions indicate that sufficient cysteine, derived from cystine, is incorporated into protein to account for the increased glutathione levels in cells treated with glutamate and cycloheximide. Depletion of glutathione in the presence of cycloheximide abrogates the protective effects of protein synthesis inhibition suggesting that increases in glutathione observed in glutamate-treated cultures exposed to cycloheximide is necessary for the survival-promoting effects of these agents.

The ability of protein synthesis inhibitors to prevent cell death by liberating cysteine for use in glutathione synthesis led to a consideration of other agents known to deliver cysteine into the cell as antiapoptotic agents. One agent, which has already been used in humans to treat acetaminophen poisoning, is the drug N-acetyl-L-cysteine (NAC) (Miners, Drew, and Birkett 1984). NAC has been shown to increase cysteine in fibroblasts in the presence of inhibitors of the cystine–glutamate antiporter such as glutamate (Issels et al. 1988; Tayarani et al. 1987). Acetylation of cysteine in NAC is believed to prevent binding of metals to the cysteine moiety and thus to slow oxidation of the amino acid in the medium. NAC is therefore presumed to enter cells through the ASC transport system (alanine, serine, and cysteine) (Issels et al. 1988). Intracellularly, NAC is spontaneously deacetylated to cysteine, which can then be used for glutathione synthesis. Consonant with this model, we showed that 100 μM NAC can prevent apoptosis induced by cystine deprivation in cortical neuronal cultures and that protection is associated with restoration of glutathione levels to 50% above control values.

3.3 Glutathione Depletion and Oxidative Stress: A Common Pathway of Apoptotic Death in the Nervous System?

Our observations with cycloheximide and NAC led us to ask whether glutathione depletion and oxidative stress are part of a common pathway by which neuropathologic stimuli induce apoptotic death. An assumption inherent in this question is that such a common pathway exists. This assumption is based on several lines of evidence.

It is now established that in vitro, a number of pathologic insults—including growth-factor deprivation (Deckwerth and Johnson 1993), calcium overload (Zhong et al. 1993), depletion of the antioxidant enzyme superoxide dismutase (Troy and Shelanski 1994), mitochondrial electron-transport inhibition (DiPasquale, Marini, and Youle 1991), hypoxia (Rosenbaum et al. 1994), and ionizing radiation (Ferrer 1992)—induce the characteristic morphologic and biochemical features that define apoptosis. These observations suggest the coexistence of multiple intracellular signaling pathways that converge upstream of a sequence of events that lead to apoptosis.

Evidence that these pathways converge prior to commitment of cells to die comes from studies with the remarkable gene Bcl-2. Bcl-2 was first cloned from a breakpoint of a translocation present in many B-cell lymphomas (Tsujimoto and Croce 1986). Since its initial cloning, a number of studies have defined a role for Bcl-2 in preventing apoptotic death (Hockenbery et al. 1993). Bcl-2 can prevent neuronal death induced by growth-factor deprivation, the calcium ionophore, A23187, and glutathione depletion (Zhong et al. 1993; Kane et al. 1993). Although Bcl-2 is not protective in all apoptotic cell death paradigms, its broad spectrum of action suggests that the death program triggered by multiple stimuli converge at, or prior to, a step that is blocked by Bcl-2.

From these observations, it seemed reasonable to hypothesize that glutathione depletion is part of a common pathway of apoptotic death. Identification of common pathways of neuronal death may lead to elucidation of novel and broadly applicable strategies to treat neurodegenerative diseases. To address this hypothesis, we wanted a pathological stimulus that fulfilled the following criteria:

1. The stimulus had to be an established inducer of neuronal apoptosis.
2. Apoptosis induced by the stimulus had to be inhibited by Bcl-2 overexpression, suggesting that pathways common to multiple stimuli are engaged.
3. The stimulus had to be one that was not known to involve the generation of glutathione depletion and oxidative stress.

3.4 Sindbis Virus Induces Neuronal Apoptosis
in a Bcl-2-Sensitive Manner

One paradigm of apoptosis that meets all of the above criteria involves a single-stranded RNA virus of positive polarity, Sindbis virus (SV). Infection of mice with SV results in encephalitis, and it has thus been used to model human encephalitides due to alphaviruses (Strauss and Strauss 1994). Encephalitis results primarily from infection of neurons of the central nervous system (Levine et al. 1993; Lewis et al. 1996). Studies of SV infection in primary cultured neurons and cell lines grown in vitro and in whole animals reveal that SV induces all the morphologic characteristics of apoptosis: chromatin condensation, DNA fragmentation, and membrane blebbing (Levine et al. 1993; Lewis et al. 1996). Previous studies have also established that overexpression of Bcl-2 in cultured neurons or the intact brain prevents SV-induced death (Levine et al. 1996).

3.5 Structure and Replication of Sindbis Virus

SV is an enveloped virus containing three structural proteins: a capsid protein and two surface glycoproteins, E1 and E2. The viral genome is a plus-stranded RNA of 11,703 nucleotides. It is a single, continuous strand that is capped on the 5' end and polyadenylated on the 3' end, and it is infectious. The genome is organized into two distinct regions. The 5' two-thirds encodes the nonstructural proteins, which are translated as a polyprotein from full-length 49S RNA. The 3' one-third encodes structural proteins, which are translated as a polyprotein from subgenomic 26S RNA. During replication, the parental RNA is transcribed into a complementary minus strand, which serves as a template for the synthesis of genomic and subgenomic RNA (Strauss and Strauss 1994).

SV's nonstructural proteins include nsP1, an initiator of minus-strand synthesis; nsP2, a cysteine protease; and nsP4, a viral-RNA polymerase. The structural proteins include E1 and E2, transmembrane proteins that associate with each other soon after synthesis. Trimers of E2–E1 heterodimers are linked by E1 and form flower-shaped knobs on the surface of the mature virion. The other structural protein, the capsid, binds viral RNA. In forming a mature virion, the capsid protein and its associated viral RNA bind to the E2 glycoprotein on the plasma membrane surface. Then, by unclear mechanisms, the plasma membrane buds, forming an envelope around the capsid–viral-RNA complex, and virions are released into the media.

3.6 The Antioxidant *N*-Acetylcysteine Blocks
Glutathione-Depletion-Induced Death and Sindvis Virus-Induced Death

We utilized the SV paradigm to begin to address the hypothesis that oxidative stress resulting from glutathione depletion is a common signal utilized by many stimuli to initiate neuronal apoptosis. Our initial strategy was to determine whether antioxidants that elevate glutathione and block glutathione-depletion-induced apoptosis also inhibit SV-induced apoptosis in the N18 neuroblastoma cell lines. We chose to look at the effects of antioxidants in cell lines rather than primary neuronal cultures for several reasons. SV-induced death in cultured cell lines is morphologically indistinct from that induced in primary neurons in vitro and in vivo (Levine et al. 1993; Lewis et al. 1996). Additionally, cultured cell lines offer advantages of homogeneity and biomass that facilitate biochemical and molecular studies.

We first examined whether the antioxidant *N*-acetylcysteine (NAC) could prevent SV-induced apoptosis. We had previously shown that 100 mM NAC completely abrogates glutathione depletion and oxidative-stress-induced apoptosis in primary neurons (Ratan et al. 1994b). The protective effects of NAC were correlated with its ability to increase the antioxidant glutathione (Aruoma et al. 1989; Ratan et al. 1996). We therefore examined whether 100 mM NAC could prevent SV-induced death in cultured cell lines. For these studies, NAC was added 2 h before infection and viability was assessed 48 h after infection. Viability was assessed by trypan-blue exclusion or the release of cytosolic lactate dehydrogenase. Whereas 100 μM NAC had no effect, 30 mM NAC was completely protective (Lin et al. 1995). The protective effects of 30 mM NAC could not be attributed to the ability of this drug to inhibit viral entry or viral replication. Nevertheless, the high levels of NAC required to prevent SV-induced death as compared to glutathione-depletion-induced death led us to examine possible nonspecific effects of the drug that might account for protection, separate from its glutathione-enhancing properties.

3.7 *N*-Acetylcysteine Inhibits Sindbis-Virus-Induced and Growth-Factor-Deprivation-Induced Apoptosis Independent of Glutathione Synthesis

To determine whether NAC's protective action depends on its ability to increase (or prevent a decrease) in intracellular glutathione, we used buthionine sulfoximine (BSO) to buffer glutathione levels as was previously described for our cycloheximide studies in primary neurons. BSO (20 μM) lowered total cellular glutathione (GSSG+GSH) to less than 20% of control levels by 3 h after SV infection, but did not cause cell death in uninfected cells, nor did it affect SV's ability to induce apoptosis. Additionally, when de novo glutathione synthesis was blocked with BSO, NAC still prevented SV-induced death. These results indicate that glutathione synthesis is not required for NAC protection. Further, measures of total glutathione after SV infection but prior to the onset of morphological and biochemical features of apoptosis revealed that levels of the tripeptide increase rather than decrease as in the cystine deprivation model. Altogether, these results suggest that glutathione depletion is not a common final pathway of apoptotic death and that NAC (in the millimolar range) prevents cell death by a glutathione-independent mechanism (Fig. 3). In support of these conclusions, Greene, Ferrari and coworkers showed that PC12 cells and primary sympathetic neurons could be protected from serum deprivation-induced apoptosis

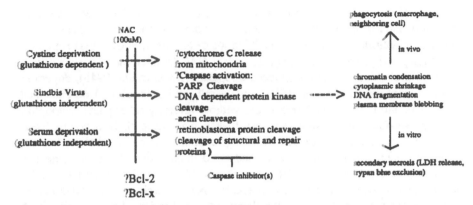

FIG. 3. Glutathione-dependent and -independent pathways to apoptosis in embryonic cortical neurons. Initial steps in the activation of apoptosis are likely to be highly divergent and include glutathione-dependent and -independent pathways, cell-type-specific pathways, and stimulus-specific pathways. These divergent signaling pathways converge on a common pathway to apoptosis, which may include the release of cytochrome c from the mitochondria, leading to the activation of a family of proteases known as caspases. Caspases have been shown cleave a host of proteins, including those involved in cell repair and structure. Cleavage of these proteins disables homeostatic mechanisms and prepares the cell for its controlled demise. Manifestation of the late stages of apoptosis include chromatin and cytoplasmic condensation and compartmentalization of cell constituents into multiple small vesicles. These vesicles can be phagocytized by professional phagocytes or neighboring cells. In vitro, the absence of phagocytic cells leads to secondary necrosis. Secondary necrosis can be assessed by lactate dehydrogenase release or trypan-blue exclusion.

with 60 mM NAC, and that the protection is independent of glutathione synthesis (Ferrari, Yan, and Greene 1995; Yan, Ferrari, and Greene 1995). Moreover, cortical neurons induced to undergo apoptosis by deprivation of growth factors or serum are not protected by concentrations of NAC (100 μM), which raise glutathione and prevent cystine-deprivation-induced death (Ratan et al. 1996). Additionally, cycloheximide prevention of apoptosis induced by staurosporine (a nonselective kinase inhibitor) in cortical neurons (Koh et al. 1995) or glutamate-induced apoptosis in PC12 cells (Froissard, Monrocq, and Duval 1997), in contrast to cystine-deprivation-induced apoptosis (Ratan et al. 1994b), cannot be reversed by the glutathione synthesis inhibitor buthionine sulfoximine. Paradoxically, some groups have presented data suggesting that glutathione depletion induced by BSO actually prevents apoptosis (Castagne and Clarke, unpublished observations). Whether these observations relate to nonspecific effects of BSO remains to be determined, but they suggest that the relationship between glutathione and neuronal apoptosis is not a simple one.

The ability of cells to activate the apoptotic program through glutathione-dependent and -independent pathways is of significant therapeutic import. Indeed, a concern in using agents that target common final pathways of apoptosis present in all cells is that both pathological and physiological apoptosis may be inhibited. The massive inhibition of apoptosis within and outside the central nervous system that would result from chronic treatment with a general apoptosis inhibitor would be expected to inhibit the neurological disease process but might lead to other, unwanted consequences, such as cancer or autoimmunity. Thus, identification of disease-specific paths to apoptosis is of high priority. The ability of agents such as NAC (at micromolar concentrations) to block glutathione-depletion-induced apoptosis (Ratan et al. 1994b), dopamine-induced apoptosis (Gabby et al. 1996), and overall survival in cultured mesencephalic neurons (Colton et al. 1995) selectively without affecting other paths to apoptosis (e.g., growth-factor deprivation) suggests that such agents might be useful in disease states such as Parkinson's disease, where glutathione depletion is believed to be a necessary event for neuronal loss secondary to impairment of energy- generating systems (Zeevalk et al. 1997). However, the established ability of cysteine and its analogs to bind to and activate glutamate receptors and thereby to cause excitotoxicity is a considerable impediment to the therapeutic application of these agents in the nervous system (Olney et al. 1990).

There are several alternatives to cysteine analogs such as NAC. L-2-Oxothiazolidine 4-carboxylate (OTC) is a cysteine prodrug that augments intracellular cysteine levels in brain (Anderson and Meister 1989; Rose et al., 1996), but that may not act as an excitotoxin. Studies from our laboratory indicate that OTC is capable of abrogating cystine-deprivation-induced death in embryonic cortical neurons (Ratan et al., unpublished observations). In addition to cysteine prodrugs, recent evidence suggests that electrophilic agents, some of which are present in the diet, can induce the expression of the cystine transporter in astrocytes and thereby increase glutathione levels (Bannai 1984; Murphy et al. 1991; Prostera et al. 1993; Sagara, Makino, and Bannai 1996; Han, Mytilineou, and Cohen 1996). If neuronal glutathione is influenced by increases in astrocyte cystine transport or if dietary electrophiles such as isothiocyanates induce cystine transport in neurons, then this may be a practical strategy for chronically elevating glutathione in the nervous system. In addition to glutathione-repleting agents, a host of antioxidants, including vitamin E (Murphy et al. 1991), idebenone (Ratan et al. 1994a), and lazaroids (Grasbon-Frodl, Anderson, and Brundin 1996) have been shown to prevent glutathione-depletion-induced death in neurons and are viable candidates for disease therapy.

4. SUMMARY

It is now well established that in vitro glutamate can induce cytotoxicity in immature cortical neurons and oligodendroglia through a non-receptor-mediated mechanism involving the inhibition of cellular cystine uptake leading to depletion of the versatile antioxidant glutathione and oxidative-stress-induced death. In these paradigms, glutathione insufficiency triggers apoptosis, a regulated process of cell destruction in which homeostasis is lost and cellular constituents are packaged into multiple small vesicles, which can be phagocytized by macrophages or neighboring cells without the adverse side effects of an inflammatory response. Studies over the past several years indicate that glutathione insufficiency is only one of a panoply of cellular pathways that can trigger neuronal apoptosis and not part of a common final pathway. These observations suggest that pharmacological strategies for increasing glutathione may be used in chronic neurodegenerative diseases without concerns about inhibiting apoptosis inappropriately inside and outside the nervous system. Future studies will clarify the precise molecular targets altered by glutathione insufficiency leading to the activation of apoptosis. Identification of novel endogenous regulators of apoptosis promises to show the way to novel therapeutic strategies to prevent glutathione depletion induced loss in the central nervous system.

ACKNOWLEDGMENTS

Parts of this chapter were adapted from a previous review by Ratan and Baraban: Apoptotic death in an in vitro model of neuronal oxidative stress, *Clinical and Experimental Pharmacology and Physiology* 22(4):309–10. We would like to thank Tim Murphy, Kuo-I Lin, Jay Baraban, Marie Hardwick, Swu-Hua Lee, and Paul Lee for their contributions to work cited in this manuscript.

REFERENCES

Anderson, M. E., and Meister, A. 1989. Marked increase of cysteine levels in many regions of brain after administration of 2-oxo-thiazolidine-4-carboxylate *FASEB Journal* 3:1632.

Ankarcrona, M., Dypbukt, J. M., Bonfoco, E., Zhivtovsky, B., Orrenius, S., Lipton, S. A., and Nicotera, P. 1995. Glutamate-induced neuronal death: a succession of necrosis or apoptosis depending on mitochondrial function. *Neuron* 15(4):961–73.

Aruoma, O. I., Halliwell, B., Hoey, B. M., and Butler, J. 1989. The antioxidant *N*-acetylcysteine: its reaction with hydrogen peroxide, hydroxyl radical, superoxide, and hypochlorous acid. *Free Radicals in Biology and Medicine* 6:593–7.

Bannai, S. 1984. Induction of cystine and glutamate transport activity in human fibroblasts by diethylmaleate and other electrophilic agents. *Journal of Biological Chemistry* 259(4):2435–40.

Bannai, S., and Kitamura, E. 1980. Transport inhibition of L-cystine and L-glutamate in human diploid fibroblasts in culture. *Journal of Biological Chemistry* 255:2372–6.

Bannai, S., and Kitamura, E. 1981. Role of proton dissociation in the transport of cystine and glutamate in human diploid fibroblasts in culture. *Journal of Biological Chemistry* 256:5770–2.

Beiswanger, C. M., Diegmann, M. H., Novak, R. F., Philbert, M. A., and Graessle, T. L. 1995. Developmental changes in the cellular distribution of glutathione and glutathione-*S*-transferases in the murine nervous system. *Neurotoxicology* 16(3):425–40.

Bonfoco, E., Krainc, D., Ankarcrona, M., Nicotera, P., and Lipton, S. A. 1995. Apoptosis and necrosis: two distinct events induced, respectively, by mild and intense insults with *N*-methyl-D-aspartate or nitric oxide/superoxide in cortical cultures. *Proceedings of the National Academy of Sciences of the U.S.A.* 92(16):7162–6.

Boveris, A., and Chance, B. 1973. The mitochondrial generation of hydrogen peroxidase. General properties and effect of hyperbaric oxygen. *Biochemical Journal* 134:707–16.

Brannan, T. S., Maker, H. S., Weiss, C., and Cohen, G. 1980. Regional distribution of glutathione peroxidase in the adult rat brain. *Journal of Neurochemistry* 36:307–9.

Brouillet, E., and Beal, M. F. 1993. NMDA antagonists partially protect against MPTP induced neurotoxicity in mice. *Neuroreport* 4(4):387–90.

Cammer, W., Tansey, F., Abramovitz, M., Ishigaki, S., and Listowsky, I. 1989. Localization of glutathione-*S*-transferase Yp and Yb subunits in oligodendrocytes and astrocytes of rat brain. *Journal of Neurochemistry* 52:876–83.

Chan, P. H., and Fishman, R. A. 1980. Transient formation of superoxide radicals in poltunsaturated fatty acid-induced brain swelling. *Journal of Neurochemistry* 35:1004–7.

Choi, D. W. 1994. Calcium and excitotoxic neuronal injury. *Annals of the New York Academy of Sciences* 747:162–71.

Colton, C. A., Pagan, F., Snell, J., Colton, J. S., Cummins, A., and Gilbert, D. L. 1995. Protection from oxidation enhances the survival of cultured mesencephalic neurons. *Experimental Neurology* 132(1):54–61.

Das, M., Dixit, R., Set, P. K., and Mukhtar, H. 1981. Glutathione-*S*-transferase activity in the brain, species, sex, regional and age differences. *Journal of Neurochemistry* 36:1439–42.

Deckwerth, T. L., and Johnson, E. M., Jr. 1993. Temporal analysis of events associated with programmed cell death (apoptosis) of sympathetic neurons deprived of nerve growth factor. *Journal of Cell Biology* 123(5):1207–22.

DiPasquale, B., Marini, A. M., and Youle, R. J. 1991. Apoptosis and DNA degradation induced by 1-methyl-4-phenylpyridinium in neurons. *Biochemical and Biophysical Research Communications* 181:1442–8.

Estus, S., Zaks, W. J., Freeman, R. S., Gruda, M., Bravo, R., and Johnson, E. M., Jr. 1994. Altered gene expression in neurons during programmed cell death: identification of c-jun as necessary for neuronal apoptosis. *Journal of Cell Biology* 127:1717–27.

Ferrari, G., Yan, C. Y., and Greene, L. A. 1995. *N*-acetylcysteine (D- and L-stereoisomers) prevents apoptotic death of neuronal cells. *Journal of Neuroscience* 15(4):2857–66.

Ferrer, I. 1992. The effect of cycloheximide on natural and x-ray induced cell death in the developing cortex. *Brain Research* 588:351–7.

Froissard, P., and Duval, D. 1994. Cytotoxic effects of glutamic acid on PC12 cells. *Neurochemistry International* 24(5):485–93.

Froissard, P., Monrocq, H., and Duval, D. 1997. Role of glutathione metabolism in the glutamate-induced programmed cell death of neuronal-like PC12 cells. *European Journal of Pharmacology*. 326(1):93–99.

Gabby, M., Tauber, M., Porat, S., and Simantov, R. 1996. Selective role of glutathione in protecting human neuronal cells from dopamine-induced apoptosis. *Neuropharmacology* 35(5):571–8.

Graham, D. G., Tiffany, S. M., Bell, W. R., and Gutknecht, W. F. 1978. Autoxidation versus covalent binding of quinones as the mechanism of toxicity of dopamine, 6-hydroxydopamine, and related compounds toward C 1300 neuroblastoma cells, in vitro. *Molecular Pharmacology* 14:644–53.

Grasbon-Frodl, E. M., Anderson, A., and Brundin, P. 1996. Lazaroid treatment prevents death of cultured rat mesencephalic neurons following glutathione depletion. *Journal of Neurochemistry*. 67(4):1653–60.

Greenamyre, J. T., Olson, J. M. M., Penney, J. B., and Young, A. B. 1985. Autoradiographic characterization of *N*-methyl-D-aspartate-, quisqualate-, and kainate-sensitive glutamate binding sites. *Journal of Pharmacology and Experimental Therapeutics* 233:254–63.

Halliwell, B., and Gutteridge, J. M. C. 1989. *Free radicals in biology and medicine*, 2nd ed. Oxford, England: Clarendon Press.

Halliwell, B. 1992. Reactive oxygen species and the central nervous system. *Journal of Neurochemistry* 59:1609–23.

Ham, J., Babij, C., Whitfield, J., Pfarr, C. M., Lallemand, D., Yaniv, M., and Rubin, L. L. 1995. A c-jun dominant negative mutant protects sympathetic neurons against programmed cell death. *Neuron* 14:927–39.

Han, S. K., Mytilineou, C., and Cohen, G. 1996. L-Dopa upregulates glutathione and protects mesencephalic cultures against oxidative stress. *Journal of Neurochemistry* 66(2):501–10.

Hockenbery, D. M., Oltivai, Z. N., Yin, X.-M., Millman, C. L., and Korsmeyer, S. J. 1993. Bcl-2 functions in an antioxidant pathway to prevent apoptosis. *Cell* 75:241–51.

Issels, R. D., Nagele, A., Eckert, K. G., and Wilmanns, W. 1988. Promotion of cystine uptake and its utilization for glutathione biosynthesis induced by cysteamine and *N*-acetylcysteine. *Biochemical Pharmacology* 37:881–8.

Kane, D. J., Sarafian, T. A., Anton, R., Hahn, H., Gralla, E. B., Selverstone Valentine, J., Ord, T., and Bredesen, D. E. 1993. Bcl-2 inhibition of neural death: decreased generation of reactive oxygen species. *Science* 262:1274–6.

Kerr, J. F. R., and Harmon, B. V. 1991. Definition and incidence of apoptosis: a historical perspective. In *Apoptosis: the molecular basis of cell death*, ed. L. D. Tomei, and F. O. Cope, 5–29. Cold Spring Harbor, NY: Cold Spring Harbor Laboratory Press.

Koh, J. Y., Wie, M. B., Gwag, B. J., Sensi, S. L., Canzoniero, L. M., Demaro, J., Csernansky, C., and Choi, D. W. 1995. Staurosporine-induced neuronal apoptosis. *Experimental Neurology* 135(2):153–9.

Levine, B., Huang, Q., Isaacs, J. T., Reed, J. C., Griffin, D. E., and Hardwick, J. M. 1993. Conversion of lytic to persistent alphavirus infection by the Bcl-2 cellular oncogene. *Nature* 361:739–41.

Levine, B., Goldman, J. E., Jang, H. H., Griffin, D. E., and Hardwick, J. M. 1996. Bcl-2 protects mice against fatal encephalitis. *Proceedings of the National Academy of Sciences of the U.S.A.* 93:4810–5.

Lewis, J., Wesselingh, S. L., Griffin, D. E., and Hardwick, J. M. 1996. Alphavirus-induced apoptosis in mouse brains correlates with neurovirulence. *Journal of Virology* 70(3):1828–35.

Lin, K. I., Lee, S. H., Narayanan, R., Baraban, J. M., Hardwick, J. M., and Ratan, R. R. 1995. Thiol agents and Bcl-2 identify an alphavirus induced apoptotic pathway that requires activation of the transcription factor, NF-kB. *Journal of Cell Biology* 131(5):1149–61.

Makowske, M., and Christensen, H. N. 1982. Contrasts in transport systems for anionic amino acids in heptaocytes and a hepatoma cell line, HTC. *Journal of Biological Chemistry* 257:5663–70.

Martin, D. P., Schmidt, R. E., DiStefano, P. S., Lowry, O. H., Carter, J. G., and Johnson, E. M., Jr. 1988. Inhibitors of protein synthesis and RNA synthesis prevent neuronal death caused by nerve growth factor deprivation. *Journal of Biology* 106:829–44.

Meister, A. 1983. Selective modification of glutathione metabolism. *Science* 220:472–7.

Miners, J. O., Drew, R., and Birkett, D. J. 1984. Mechanism of action of paracetomol protective agents in mice in vivo. *Biochemical Pharmacology* 33:2995–3000.

Monaghan, D. T., Bridges, R. J., and Cotman, C. W. 1989. The excitatory amino acid receptors: their classes, pharmacology, and distinct properties in the function of the central nervous system. *Annual Review of Pharmacology and Toxicology* 29:365–402.

Murphy, T. H., and Baraban, J. M. 1990. Glutamate toxicity in immature cortical neurons precedes development of glutamate receptor currents. *Developmental Brain Research* 57:146–50.

Murphy, T. H., Miyamato, M., Sastre, A., Schnaar, R. L., and Coyle, J. T. 1989. Glutamate toxicity in a neuronal cell line involves inhibition of cystine transport leading to oxidative stress. *Neuron* 2:1547–58.

Murphy, T. H., Schnaar, R. L., and Coyle, J. T. 1990. Immature cortical neurons are uniquely sensitive to glutamate toxicity by inhibition of cystine uptake. *FASEB Journal* 4:1624–33.

Murphy, T. H., De Long, M. J., and Coyle, J. T. 1991. Enhance NAD(P)H: quinone reductase activity prevents glutamate toxicity produced by oxidative stress. *Journal of Neurochemistry* 56:990–5.

Oka, A., Belliveau, M. J., Rosenberg, P. A., and Volpe, J. J. 1993. Vulnerability of oligodendroglia to glutamate: pharmacology, mechanisms and prevention. *Journal of Neuroscience* 13(4):1441–53.

Olney, J. W., Zorumski, C., Price, M. T., and Labruyere, J. 1990. L-cysteine, a bicarbonate senstive endogenous excitotoxin. *Science* 248:596–9.

Orrenius, S., and Moldeus, P. 1984. The multiple roles of glutathione in drug metabolism. *Trends in Pharmacological Science* 5:432–8.

Prostera, T., Zhang, Y., Spencer, S. R., Wilczak, C. A., and Talalay, P. 1993. The electrophile counterattac response: protection against neoplasia and toxicity. *Advances in Enzyme Regulation* 33:281–96.

Raps, S. P., Lai, J. C. K., Hertz, L., and Cooper, A. J. L. 1990. Glutathione is present in high concentrations in cultured astrocytes but not in cultured neurons. *Brain Research* 493:398–401.

Ratan, R. R., Murphy, T. H., and Baraban, J. M. 1994a. Oxidative stress induces apoptosis in embryonic cortical neurons. *Journal of Neurochemistry* 62:376–9.

Ratan, R. R., Murphy, T. H., and Baraban, J. M. 1994b. Macromolecular synthesis inhibitors prevent oxidative stress-induced apoptosis by shunting cysteine to glutathione. *Journal of Neuroscience* 14:4385–92.

Ratan, R. R., Lee, P. J., and Baraban, J. M. 1996. Serum deprivation inhibits glutathione-depletion induced death in embryonic cortical neurons: evidence against oxidative stress as a common final mediator neuronal apoptosis. *Neurochemistry International* 29(2):153–7.

Robinson, M. B., and Coyle, J. T. 1987. Glutamate and related acidic excitatory neurotransmitters: from basic science to clinical application. *FASEB Journal* 1:446–55.

Rose, M., Hochwald, S. N., Harrison, L. E., and Burt, M. 1996. Selective glutathione repletion with oral oxothiatolidine carboxylate in the radiated tumor-bearing rat. *Journal of Surgical Research* 62(2):224–8.

Rosenbaum, D. M., Michaelson, M., Batter, D. K., Doshi, P., and Kessler, J. A. 1994. Evidence for hypoxia-induced, programmed cell death in neurons. *Annals of Neurology* 36(6):864–70.

Rothman, S. M., and Olney, J. W. 1987. Excitotoxicity and the NMDA receptor. *Trends in Neurology Science* 10:299–302.

Rothstein, J. D. 1996. Excitotoxicity hypothesis. *Neurology* 47(4 Suppl. 2):S19–25.

Sagara, J.-I., Miura, K., and Bannai, S. 1993a. Cystine uptake and glutathione levels in fetal brain cells in primary culture and in suspension. *Journal of Neurochemistry* 61:1667–71.

Sagara, J.-I., Miura, K., and Bannai, S. 1993b. Maintenance of neuronal glutathione by glial cells. *Journal of Neurochemistry* 61:1672–6.

Sagara, J.-I., Makino, N., and Bannai, S. 1996. Glutathione efflux from cultured astrocytes. *Journal of Neurochemistry* 66:1876–81.

Server, A. C., and Mobley, W. C. 1991. Neuronal cell death and the role of apoptosis. In *Apoptosis: the molecular basis of cell death*, ed. L. D. Tomei and F. O. Cope, 263–78. Cold Spring Harbor, NY: Cold Spring Harbor Press.

Shivakumar, B. R., Anandatheerthavarada, H. K., and Ravindranath, V. 1991. Free radical scavenging systems in developing rat brain. *International Journal of Neuroscience* 9:181–5.

Sies, H. 1985. Oxidative stress. In *Oxidants and Antioxidants*, ed. H. Sies, 1–8, London: Academic Press.

Strauss, J. H., and Strauss, E. G. 1994. The alphaviruses: gene expression, replication and evolution. *Microbiological Review* 58:491–562.

Tayarani, I., Lefauconnier, J.-M., Roux, F., and Bourre, J.-M. 1987. Evidence for an alanine, serine, and cysteine uptake system of transport in isolated brain capillaries. *Journal of Cerebral Blood Flow and Metabolism* 7:585–91.

Troy, C. M., and Shelanski, M. L., 1994. Down regulation of copper/zinc superoxide dismutase causes apoptotic death in PC12 neuronal cells. *Proceedings of the National Academy of Sciences, U.S.A.* 91:6384–6387.

Tsujimoto, Y., and Croce, C. M. 1986. Analysis of the structure, transcripts and protein products of Bcl-2, the gene involved in human follicular lymphoma. *Proceedings of the National Academy of Sciences of the U.S.A.* 83:5214–18.

Yan, C. Y., Ferrari, G., and Greene, L. A. 1995. *N*-acetylcysteine-promoted survival of PC12 cells is glutathione-independent but transcription dependent. *Journal of Biological Chemistry* 270(45):26827–32.

Zeevalk, G. D., Bernard, L. P., Albers, D. S., Mirochnitchenko, O., Nicklas, W. J., and Sonsalla, P. K. 1997. Energy stress-induced dopamine loss in glutathione peroxidase-overexpressing transgenic mice and in glutathione-depleted mesencephalic cultures. *Journal Neurochemisty* 68(1):426–9.

Zhong, L., Sarafian, T., Kane, D. J., Charles, A. C., Mah, S. P., Edwards, R. H., and Bredesen, D. E. 1993. Bcl-2 inhibits death of central neural cells induced by multiple agents. *Proceedings of the National Academy of Sciences of the U.S.A.* 90:4533–7.

Glutathione in the Nervous System
Edited by Christopher A. Shaw
Copyright © 1998 Taylor & Francis

7

Possible Modulation by Glutathione of Glutamatergic Neurotransmission

Kiyokazu Ogita, Makoto Shuto, Hiroko Maeda,
Takao Minami, and Yukio Yoneda

Department of Pharmacology, Setsunan University, Hirakata, Osaka, Japan

ABBREVIATIONS

AMPA DL-α-amino-3-hydroxy-5-methylisoxazole-4-propionic acid
BSO L-buthionine-[S,R]-sulfoximine
CGP 39653 DL-(E)-2-amino-4-propyl-5-phosphono-3-pentenoic acid
CGS 19755 *cis*-4-phosphomethyl-2-piperidine carboxylic acid
CPP ($\pm$)-3-(2-carboxypiperazin-4-yl)propyl-1-phosphonic acid
CySH L-cysteine
D-AP5 D-2-amino-5-phosphonovaleric acid

DCKA 5,7-dichlorokynurenic acid
DTE dithioerythritol
DTT dithiothreitol
Glu L-glutamic acid
Gly glycine
GSH reduced glutathione
GSSG oxidized glutathione
IP$_3$ inositol-1,4,5-triphosphate
KA kainic acid
MK-801 5-methyl-10,11-dihydro-5*H*-dibenzo[*a*,*d*]cyclohepten-5,10-imine
MPTP 1-methyl-4-phenyl-1,2,5,6-tetrahydropyridine
NAAG *N*-acetylaspartylglutamic acid
NBQX 2,3-dihydroxy-6-nitro-7-sulfamoylbenzo(*F*)quinoxaline
NMDA *N*-methyl-D-aspartic acid
NO nitric oxide
SHG *S*-hexylglutathione
SMG *S*-methylglutathione
SNP sodium nitroprusside

1. INTRODUCTION

Reduced glutathione (GSH) is a tripeptide that is composed of glycine (Gly), L-cysteine (CySH), and L-glutamic acid (Glu). For years, GSH has been considered to play an important physiological role in the maintenance of homeostasis in various mammalian tissues. For instance, GSH detoxicates xenobiotics through the catalytic action of different glutathione-*S*-transferases to lead to formation of their S-conjugates. GSH also plays a functional role in mechanisms underlying the protection of a variety of cells against the oxidative stress mediated by free radicals and peroxides through the transformation to oxidized glutathione (GSSG) by the action of glutathione peroxidases (Meister and Anderson 1983). In the rat brain, GSH is found at high concentrations such as 1–2 mmol per kilogram of wet tissue and distributed in crude synaptosomal fractions, in addition to cytosol fractions. In the crude synaptosomal fractions, moreover, around 35 percent of total GSH is distributed in synaptosomal fractions, 41 percent in mitochondrial fractions, and 24 percent in myelin fractions, respectively. Therefore, subcellular distribution profiles of GSH are quite similar to those of excitatory amino acids, including Glu and aspartic acid (Reichelt and Fonnum 1969).

On the other hand, very low activities of glutathione *S*-transferases B (Orlowsky and Karkowsky 1976) and glutathione *S*-alkyltransferase (Johnson 1966) are detected in the rat brain. These findings do not argue in favor of the idea that GSH may also participate in the aforementioned detoxification mechanisms for xenobiotics in the central nervous system. In contrast, GSH may play more important roles in antioxidative processes in astrocytes than in neurons. For instance, GSH levels and the ratio of GSSG to total glutathione (GSH+GSSG) are invariably higher in astrocytes than in neurons in cultured chick glial and neuronal cells. The astrocytes have high activities of both GSH reductase and GSH peroxidase compared with neurons (Makar et al. 1994). In PC12 cells, nerve growth factor might maintain oxidant homeostasis by elevating GSH content and GSH peroxidase activity (Pan and Perez-Polo 1993). In contrast, some in vivo experiments have raised the possibility that GSH has no function as an antioxidant in the brain. Namely, levels of GSSG are below 1 percent of total glutathione in any structures of the rat brain. The cortical levels of GSSG are unaltered following cerebral ischemia and postischemic reperfusion, though GSH levels are significantly reduced under these pathological conditions (Rehncrona et al. 1980; Cooper, Pulsinelli, and Duffy 1980; see also Cooper, chapter 5, this volume). In the brain, therefore, GSH could play specific functional roles that are distinct from those in peripheral tissues such as detoxification and antioxidation.

In this chapter, the effects of GSH on ligand binding to Glu receptors (Ogita et al. 1986, 1995) and the presence of specific binding sites of GSH in synaptic membranes of rat brain (Ogita and Yoneda 1987, 1988a; Ogita, Ogawa, and Yoneda 1988; Ogita and Yoneda 1989) are outlined. Based on these findings, possible modulatory roles of GSH in glutamatergic neurotransmission are discussed.

2. GLUTAMATE RECEPTORS

2.1 Classification

Receptors for Glu are divided into two major categories according to their signal transduction systems in the mammalian central nervous system: metabotropic and ionotropic (Table 1; for a review, see Hollmann and Heinemann 1994). The metabotropic Glu receptor family comprises at least eight subtypes (mGluR1–8) so far; it is subdivided into three groups on the basis of the differences in amino acid sequences as well as signal transduction pathways. The group I (mGluR1 and mGluR5) subtype of metabotropic Glu receptors stimulates phospholipase C followed by formation of inositol-1,4,5-triphosphate (IP$_3$) and diacylglycerol, which

TABLE 1. *Classification of Glu receptors in rat brain*

Receptor type	Subunits	Main function
Ionotropic:		
NMDA	NMDAR1, NMDAR2A-D	Ca^{2+} influx ↑
AMPA(AMPA/KA)	GluR1–4	Na^+ influx ↑
KA	GluR5–7 (low affinity) KA1–2 (high affinity)	Na^+ influx ↑
Metabotropic:		
I	mGluR1, 5	IP$_3$/DG formation ↑
II	mGluR2, 3	cAMP formation ↓
III	mGluR4, 6, 7, 8	cAMP formation ↓

then induce Ca^{2+} release from intracellular stores and activation of protein kinase C in cytosol, respectively. The group II (mGluR2 and mGluR3) and III (mGluR4, mGluR6, mGluR7, and mGluR8) subtypes of the metabotropic receptors are negatively coupled to adenylate cyclase to reduce the amount of intracellular cyclic AMP (for reviews, see Nakanishi 1992; Nicoletti et al. 1996; Pin and Duvoison 1995; Riedel 1996).

On the other hand, ionotropic Glu receptors linked to cation channels are pharmacologically classified on the basis of differential sensitivities to the exogenous agonist, *N*-methyl-D-aspartic acid (NMDA). The ionotropic receptors insensitive to NMDA (non-NMDA receptors), which consist of at least nine different subunits (GluR1–7, KA1–2) in rat brain, are further divided into two different subtypes, namely DL-α-amino-3-hydroxy-5-methylisoxazole-4-propioic acid (AMPA) receptors (GluR1–4) and kainic acid (KA) receptors (GluR5–7, KA1–2), according to the differential preference to those exogenous agonists. AMPA receptors are also called AMPA/KA receptors, because the receptors consisting of GluR1–4 subunits are activated by KA as well as AMPA, which is the most potent agonist for them. Two KA binding sites with different affinities are found in rat brain by using membrane binding and autoradiographic techniques. GluR5–7 subunits may form KA receptors with the low-affinity binding sites, which are likely to mediate KA-induced excitotoxicity in particular brain regions, while KA1–2 subunits seem to participate in the formation of the high-affinity KA receptors together with GluR5–7 subunits. Notably, it has been demonstrated in immunoprecipitation studies that the KA2 subunit forms native heteromeric receptor complexes with GluR6 and/or GluR7 subunits (Puchalski et al. 1994). In addition, cDNAs of two KA binding proteins are isolated from frog and chick brains. On the other hand, the NMDA-sensitive subclass is composed of heteromeric assemblies of NMDAR1 and NMDAR2 subunits. The expression of NMDAR1, a common principal subunit, together with any one of the four different modulatory NMDAR2 subunits (A, B, C, and D), results in the construction of ion channels with profiles characteristic of the NMDA receptor . The NMDA channel is highly permeable to Ca^{2+} as well as Na^+ (for reviews, see Lodge and Collingridge 1991; Yoneda and Ogita 1991).

2.2 Glutamate Receptors and Neuropsychiatric Diseases

The NMDA receptor is supposed to take part not only in neuronal plasticity (Moris et al. 1986), but also in delayed neuronal cell death after ischemia (Simon et al. 1984) and hypoglycemia (Wieloch 1985). Moreover, antagonists for the NMDA receptors are proposed to be beneficial for the therapy of epilepsy (Croucher, Collins, and Meldrum 1982) and Parkinson's disease (Klockgether and Turski 1989; Turski et al. 1991). Possible involvement of Glu receptors in Alzheimer's disease (Greenamyre et al. 1987; Maragos et al. 1987), Huntington's disease (Greenamyre et al. 1985; Young et al. 1988), and schizophrenia (Harrison, McLaughlin, and Kerwin 1991; Kurumaji, Ishimaru, and Toru 1992; Akbarian et al. 1996) has also been demonstrated.

Nevertheless, the non-NMDA receptors are also involved in neuronal cell death after ischemia and hypoglycemia, because antagonism against the non-NMDA receptors augments neuroprotective properties of an NMDA antagonist alone against neuronal degeneration by ischemia (Mosinger et al. 1991) and prolonged deprivation of oxygen/glucose (Kaku, Goldberg, and Choi 1991). In animals with global ischemia, hippocampal damage is attenuated by the selective non-NMDA antagonist 2,3-dihydroxy-6-nitro-7-sulfamoyl-benzo (*F*)quinoxaline (NBQX) (Sheardown et al. 1990), which is also shown to posses anti-Parkinsonian properties in rats and monkeys (Klockgether et al. 1991). In addition, [^{3}H]AMPA binding is reduced in the substantia nigra of mice treated with 1-methyl-4-phenyl-1,2,5,6-tetrahydropyridine (MPTP), which induces degeneration of nigrostriatal dopaminergic neurons (Wüllner et al. 1993). Significant alterations are also observed with ligand binding to the non-NMDA receptors in particular hippocampal regions of patients with Alzheimer's disease (Geddes et al. 1992). Moreover, several lines of evidence suggest that the individual metabotropic Glu receptors could mediate neurodegenerative processes in particular situations (for a review, see Nicoletti et al. 1996). Namely, the group I receptors may participate in mechanisms underlying neuronal toxicity through formation of IP$_3$ and/or inhibition of K$^+$ channels, so that antagonism of these receptors may be neuroprotective against cell death in brain ischemia or other acute neuronal degeneration. By contrast, the group II and III receptors could attenuate neuronal toxicity by reducing formation of cAMP and/or Ca^{2+} influx across the voltage-gated Ca^{2+} channels. Thus, all subtypes of Glu receptors could be responsible for molecular mechanisms related to neurodegeneration in a variety of neurological diseases.

3. NEUROLOGICAL DISORDERS ASSOCIATED WITH GSH

There is accumulating evidence in the literature for a positive correlation between neurological diseases and changes in cerebral GSH levels. Patients with inherited GSH synthetase deficiency exhibit mental retardation, spastic tetraparesis, and various signs of cerebellar damages, and γ-glutamylcysteine synthetase–deficient patients show signs of spinocerebellar degradation, absence of lower-limb reflexes, and ataxia (Orlowsky and Karkowsky 1976). Post mortem studies in human being have revealed that GSH levels are decreased with aging in the substantia nigra, and also in the substantia nigra of patients suffering from Parkinson's disease with the degeneration of dopaminergic neurons (Perry, Godin, and Hansen 1982; Bannon, Goedert, and Wolff 1984; Perry and Yong 1986; Riederer et al. 1989). Treatment with MPTP causes a marked decrease in GSH levels in the brain stem of C57B1 mice (Yong, Perry, and Krisman 1986). In addition, degeneration of the nigrostriatal pathway similarly resulted from the depletion of cerebral GSH by the administration of the inhibitor of GSH synthesis, L-buthionine-[*S,R*]-sulfoximine (BSO), in mice (Andersen et al. 1996). Cerebral GSH levels are significantly decreased during convulsive seizures in some experimental models for epilepsy (Berl et al. 1959; Hiramatsu and Mori 1981) and in genetically epileptic mice (Abbott et al. 1990). Furthermore, seizures develop following GSH depletion induced by treatment of BSO in adult rats (Hu et al. 1996). Treatment of BSO also induces definite signs of neurological disturbance in newborn rats, such as lethargy, intermittent tremors, and minor fits (Jain et al. 1991).

Accordingly, neuropsychiatric symptoms associated with abnormally functioning Glu receptors have something in common with neurological abnormalities resulted from changes of cerebral GSH levels. These findings therefore give rise to the proposition that GSH may directly or indirectly modulate glutamatergic neurotransmission in the brain in particular situations.

4. MODULATION OF [^{3}H]GLU BINDING

As mentioned in the preceding paragraph, similar neuropsychiatric symptoms are seen for patients with abnormalities of glutamatergic neurotransmission and of GSH levels in the brain. The fact that glutathione indeed contains Glu in its molecular structure leads us to test the possibility that glutathione may modulate neurotransmission mediated by a particular subtype of the aforementioned Glu receptors through the interaction with specific binding sites in the brain. For this purpose, the effects of in vitro addition of both GSH and GSSG on Na$^+$-dependent binding of [^{3}H]Glu have been evaluated in synaptic membranes from rat brain (Ogita et al. 1986).

The addition of GSH markedly suppresses Na$^+$-dependent binding of [^{3}H]Glu with maximal inhibition of 80–90 percent at both 2°C and 30°C incubation temperature [IC$_{50}$ values (μM): 2°C, 11.5; 30°C, 9.72]. Similarly, GSSG is also effective in inhibiting Na$^+$-dependent binding of [^{3}H]Glu in rat brain synaptic membranes [IC$_{50}$ values (μM): 2°C, 59.2; 30°C, 12.4]. In addition, Na$^+$-dependent [^{3}H]Glu binding is significantly inhibited by the addition of different peptides containing Glu with free α-amino residue under similar experimental conditions (Table 2). These include α-glutamyl-L-alanine, γ-L-glutamyl-L-glutamine, γ-L-glutamyl-L-leucine, γ-glutamyl-L-tyrosine, γ-L-glutamylhistidine, γ-L-glutamyl-L-phenylalanine, α-L-glutamyl-L-valyl-L-phenylalanine, and γ-L-glutamyl–glycyl-L-phenylalanine. However, binding was not significantly inhibited by the addition of N-carbobenzoxyl-Glu, folic acid, and L-pyroglutamyl-L-histidyl-L-prolinamide (thyrotropin-releasing hormone), which all have a Glu molecule with no free α-amino residue. Furthermore, none of SH-containing compounds, including 2-mercaptoethanol, dithiothreitol (DTT), and dithioerythritol (DTE), significantly inhibited Na$^+$-dependent [^{3}H]Glu binding. These data clearly indicate that free α-amino moiety, but not SH residue, is absolutely required for the inhibition of Na$^+$-dependent [^{3}H]Glu binding by Glu-containing peptides in brain synaptic membranes. Therefore, both GSH and GSSG may inhibit Na$^+$-dependent [^{3}H]Glu binding through mechanisms involving their γ-glutamyl moieties but not SH residues.

Several independent lines of evidence suggest that Na$^+$-dependent [^{3}H]Glu binding reflects high-affinity and Na$^+$-dependent transport processes for Glu in synaptic membranes (Vincent and McGeer 1980; Baudry and Lynch 1981; Ogita and Yoneda 1986). These previous findings give rise to the proposal that both GSH and GSSG may modulate neurotransmission mediated by Glu through significant inhibition of

TABLE 2. *Effects of Glu-containing peptides (0.1 mM) on Na⁺-dependent binding of [³H]Glu*

Peptide	Na⁺-dependent [³H]Glu binding (% of control)
GSH	16 ± 2^a
GSSG	20 ± 3^a
α-L-Glutamyl-L-alanine	22 ± 4^a
γ-L-Glutamyl-L-glutamine	22 ± 5^a
γ-L-Glutamyl-L-leucine	38 ± 4^a
γ-L-Glutamyl-tyrosine	39 ± 3^a
γ-L-Glutamyl-histidine	33 ± 9^a
γ-L-Glutamyl-L-phenylalanine	42 ± 1^a
α-L-Glutamyl-L-valyl-L-phenylalanine	19 ± 7^a
γ-L-Glutamyl-glycyl-L-phenylalanine	19 ± 9^a
N-Carbobenzoxyl-Glu	91 ± 4
Folic acid	107 ± 1
L-Pyroglutamyl-L-histidyl-L-prolinamide	92 ± 5

Synaptic membranes were obtained from the brains of male Wistar rats weighing 200–250 g and then washed three times by suspension in 50 mM Tris–acetate buffer (pH 7.4) and subsequent centrifugation at $50,000g$ for 30 min. On the day of the experiment, membranes were further washed twice with buffer. The final pellets were suspended in buffer, and the suspensions were used as "untreated" membranes. An aliquot of "untreated" membranes was incubated with 10 nM [³H]Glu at 30°C for 60 min in 0.5 ml of 50 mM Tris–acetate buffer (pH 7.4) containing 100 mM sodium acetate in either the presence or the absence of the peptide indicated at a concentration of 0.1 mM. Nonspecific binding was determined by the addition of 1 mM unlabeled Glu. Each value represents the mean±SEM obtained from four separate measurements.
$^a p < 0.01$, compared with each control value obtained in the absence of any test peptides. These data are from Ogita et al. (1986).

Na⁺-dependent uptake at synaptic clefts in the brain. As the content of GSSG is below 1 percent of total glutathione in any brain structures, however, it is likely that GSH but not GSSG indeed exerts an inhibitory influence on Glu uptake under physiological conditions. In addition, endogenous glutathione is released from slices of different brain structures upon K⁺ depolarization in a Ca^{2+}-dependent manner, and a large portion (85–100 percent) of glutathione released is in the reduced form (Zängerle et al. 1992). It is thus conceivable that GSH originating from the neuronal compartment may stimulate glutamatergic neurotransmission through the suppression of Glu uptake. The latter has been considered to be an inactivation mechanism at synaptic clefts in the mammalian brain. However, no direct evidence has been shown so far for the inhibition by GSH of Glu uptake.

5. MODULATION OF GLUTAMATE RECEPTORS

5.1 Another Peptide Containing Glu

Glu itself is undoubtedly one of the physiologically important neurotransmitters with agonistic properties for Glu receptors. In addition to Glu, however, an endogenous peptide containing Glu, *N*-acetylaspartylglutamate (NAAG), has also been reported as a possible candidate of agonists at glutamatergic neurons (Zaczek et al. 1983; Koller and Coyle 1984; Ffrench-Mullen et al. 1985). In fact, NAAG acts as an agonist with a low potency at the NMDA receptor on spinal (Westbrook et al. 1986) and olfactory (Trombley and Westbrook 1990) neurons. However, NAAG lacks excitatory action in preparations that easily respond to other Glu agonists (Henderson and Salt 1988; Schneider and Perl 1988; Whittemore and Koerner 1989), suggesting that this peptide may not directly elicit excitatory responses in some cases. The findings that Glu is released from NAAG by a specific peptidase in the brain, furthermore, raises the possibility that NAAG may activate Glu receptors via the liberation of its metabolic product, Glu (Robinson et al. 1987; Williamson and Neale 1992).

Finally, Glu-containing peptides may, at least in part, participate in glutamatergic neurotransmission through modulation at the level of synaptic receptors in the brain. Therefore, a possible physiological role of GSH as a ligand for Glu receptors is discussed in the following paragraphs.

5.2 Non-NMDA Receptors

The non-NMDA receptors are classified into the AMPA and KA receptors on the basis of sensitivity to the exogenous agonists that are used to label the receptors. In order to evaluate the possible roles of glutathione in neurotransmission mediated by the non-NMDA receptors, the effects of both forms of glutathione should be examined on binding of [^{3}H]AMPA and [^{3}H]KA in rat brain synaptic membranes by using membrane binding techniques (Ogita et al. 1995). The effects of glutathione on [^{3}H]AMPA binding are determined in either the presence or the absence of KSCN, since the addition of KSCN not only induces potentiation of [^{3}H]AMPA binding but also changes pharmacological characteristics of the binding sites (Ogita et al. 1994). In the absence of KSCN, GSH is effective in inhibiting [^{3}H]AMPA binding in a concentration-dependent manner at concentrations above 1 μM. Similarly, GSSG inhibits the binding in the absence of KSCN at concentrations above 10 μM, with an IC$_{50}$ value two times higher than that obtained with GSH (Table 3). In the presence of KSCN, in contrast, both GSH and GSSG are similarly potent in inhibiting [^{3}H]AMPA binding, with IC$_{50}$ values of more than 100 μM (Table 3). Thus, GSH has a more potent ability to inhibit [^{3}H]AMPA binding than does GSSG in the absence of KSCN, and the addition of KSCN notably reduces the potencies of GSH and GSSG to inhibit [^{3}H]AMPA binding.

The addition of KSCN has been shown to be effective in significantly reducing the potencies of AMPA antagonists to displace [^{3}H]AMPA binding, without affecting those of agonists (Ogita et al. 1994). These findings give rise to the speculation that GSH/GSSG may play physiological roles as endogenous antagonists at the AMPA receptor. If this is the case, depletion of GSH/GSSG could be at least in part responsible for mechanisms underlying crisis of a variety of neurological disorders associated

TABLE 3. *Potencies of glutathione to displace binding
of various ligands to non-NMDA receptors*

	IC_{50} (μM)	
Binding	GSH	GSSG
[^{3}H]AMPA (−KSCN)	21.0 ± 5.0	55.4 ± 7.4
[^{3}H]AMPA (+KSCN)	>100	>100
[^{3}H]KA	24.2 ± 4.0	26.9 ± 3.4

Synaptic membranes were obtained from the brains of male
Wistar rats and then washed three times with 50 mM Tris–acetate
buffer (pH 7.4). On the day of the experiment, membranes were
treated with 0.08-percent Triton X-100 at 2°C for 10 min. The
treatment was terminated by centrifugation at 50,000*g* for 30 min,
and the resultant pellets were washed once more. The final pellets
obtained were suspended in buffer as Triton-treated membranes
as described elsewhere (Ogita and Yoneda 1988b). An aliquot of
Triton-treated membranes was incubated with 10 nM [^{3}H]AMPA
or 10 nM [^{3}H]KA at 2°C for 30 min in either the presence or
the absence of four different concentrations of GSH or GSSG
at a concentration range of 1 to 0.1 mM. [^{3}H]AMPA binding was
measured by the incubation in either the presence or the absence
of 100 mM KSCN. Each nonspecific binding of [^{3}H]AMPA and
[^{3}H]KA was defined by the addition of 0.1 mM unlabeled Glu.
Values are the mean±SEM. These data are from Ogita et al.
(1995).

with excessive activation of the AMPA receptor through release from the negative
modulation by GSH.

On the other hand, [^{3}H]KA binding is displaced by both GSH and GSSG in a
concentration-dependent manner at concentrations above 0.1 μM. Their IC_{50} values
are shown in Table 3. In contrast to the inhibition of [^{3}H]AMPA binding, GSH and
GSSG inhibited [^{3}H]KA binding with similar potencies. In addition, DTT, CySH,
and GSH have been shown to invariably inhibit [^{3}H]KA binding in porcine striatal
membrane homogenates, due to a significant decrease in the density of binding sites
(Liu and Quirion 1992), suggesting that the negative modulation by GSH of the KA
receptor involves the reducing capacity common to those three different compounds
containing SH residues. However, the inhibition by GSSG (Table 3) could not be
explained by the possible redox regulation of the KA receptor, as shown by Liu and
Quirion (1992). Moreover, the release of [^{3}H]GABA evoked by KA is significantly
enhanced and prolonged by DTT in rat hippocampal slices (Janáky et al. 1994), and the
KA-induced $^{45}Ca^{2+}$ influx is enhanced by DTT but not by GSH in cultured cerebellar
granule cells (Janáky et al. 1994). These data do not support the idea that the KA receptors are solely sensitive to negative modulation by those reducing agents. Therefore, it
seems more likely that both GSH and GSSG directly displace [^{3}K]KA binding in brain
synaptic membranes. Functional roles of glutathione in neurotransmission mediated
by the non-NMDA receptors in the brain, however, should be clarified in future studies.

5.3 NMDA Receptor

The NMDA receptor is proposed to be a receptor ionophore complex consisting of at least three distinct domains having positive stimulatory properties for opening processes of a cation channel within the complex. These are: (i) an NMDA recognition domain with high affinity for Glu, (ii) a Gly recognition domain insensitive to the classic antagonist strychnine, and (iii) a polyamine recognition domain with several indefinite profiles [for reviews see Monaghan, Bridge, and Cotman (1989), Yoneda and Ogita (1991)].

A. NMDA and Gly Recognition Domains

The NMDA recognition domain is labeled by [^{3}H]Glu (Ogita and Yoneda 1988b; Yoneda et al. 1989), [^{3}H]($\pm$)-3-(2-carboxypiperazin-4-yl)propyl-1-phosphonic acid (CPP) (Yoneda et al. 1990), [^{3}H]cis-4-phosphomethyl-2-piperidine carboxylic acid (CGS 19755) (Murphy et al. 1988, and [^{3}H]DL-(E)-2-amino-4-propyl-5-phosphono-3-pentenoic acid (CGP 39653) (Zuo et al. 1993). The Gly recognition domain is labeled by [^{3}H]5,7-dichlorokynurenic acid (DCKA) (Yoneda et al. 1993), which is a specific antagonist for the Gly recognition domain, as well as [^{3}H]Gly (Ogita et al. 1989). Table 4 summarizes K_i values of glutathione, NMDA agonists, and NMDA antagonists for the binding of [^{3}H]Glu, [^{3}H]CPP, and [^{3}H]CGP 39653 to the NMDA recognition domain on the complex in synaptic membranes treated with Triton X-100 (Ogita and Yoneda 1990; Zuo et al. 1993; Ogita et al. 1995). Both GSH and GSSG are effective in displacing binding of [^{3}H]Glu, [^{3}H]CPP, and [^{3}H]CGP 39653 at

TABLE 4. *Potencies of glutathione and of NMDA agonists and antagonists to displace binding of various ligands to NMDA recognition domain*

Substance	K_i (μM)		
	[^{3}H]Glu	[^{3}H]CPP	[^{3}H]CGP 39653
GSH	0.60 ± 0.06	3.79 ± 0.90	4.54 ± 0.47
GSSG	1.11 ± 0.22	1.13 ± 0.39	7.49 ± 1.95
Glu	0.022 ± 0.004	0.057 ± 0.009	0.056 ± 0.007
NMDA	1.70 ± 0.13	5.15 ± 1.21	2.85 ± 0.26
CPP	0.42 ± 0.07	0.095 ± 0.042	0.063 ± 0.007
D-AP5	0.31 ± 0.03	0.083 ± 0.011	0.087 ± 0.006

An aliquot of Triton-treated membranes was incubated with 10 nM [^{3}H]Glu, 10 nM [^{3}H]CPP, or 2 nM [^{3}H]CGP 39653 in either the presence or the absence of four different concentrations of the test compound at a concentration range of 1 nM to 0.1 mM. Binding of [^{3}H]Glu and [^{3}H]CPP was measured by the incubation at 2°C for 10 min, but [^{3}H]CGP 39653 was incubated at 2°C for 60 min in order to obtain equilibrium. Unlabeled compounds used at 0.1 mM to define each nonspecific binding were as follows: NMDA for [^{3}H]Glu, and Glu for [^{3}H]CPP and [^{3}H]CGP 39653. K_i values were calculated according to the equation $K_i = IC_{50}/(1+[L]/K_d)$. Values are the mean±SEM. These data are from Ogita et al. (1995), Yoneda et al. (1990), and Zuo et al. (1993).

concentrations higher than 0.01 μM. Although GSH is less potent in inhibiting binding of all radioligands shown here than Glu, the potencies of GSH are almost the same as those of NMDA. Moreover, both GSH and GSSG are more potent in inhibiting binding of [^{3}H]Glu than binding of [^{3}H]CPP and [^{3}H]CGP 39653, except for the inhibition by GSSG of [^{3}H]CPP binding. Under these experimental conditions, NMDA agonists, including Glu and NMDA, are more potent in displacing [^{3}H]Glu binding than binding of [^{3}H]CPP and [^{3}H]CGP 39653. In contrast, NMDA antagonists, including CPP and D-2-amino-5-phosphonovaleric acid (D-AP5), have greater potencies in displacing binding of [^{3}H]CPP and [^{3}H]CGP 39653 than of [^{3}H]Glu. It is thus likely that NMDA agonists exhibit higher affinity for the agonist-preferring form than for the antagonist-preferring form of the NMDA recognition domain. The fact that both GSH and GSSG have profiles similar to those of NMDA agonists in terms of displacement of binding to agonist- and antagonist-preferring forms raises the possibility that glutathione may in principal have agonistic properties at the NMDA recognition domain on the receptor ionophore complex. As neither GSH and GSSG inhibits binding of either [^{3}H]Gly or [^{3}H]DCKA at concentrations of up to 10 μM (Ogita et al. 1995), on the contrary, it is unlikely that glutathione has affinity for the Gly recognition domain on the complex.

B. NMDA Channel

The NMDA channel is labeled by noncompetitive antagonists such as N-[1-(2-thienyl) cyclohexyl]piperidine and 5-methyl-10,11-dihydro-5H-dibenzo[a,d]cyclohepten-5,10-imine (MK-801). Notably, MK-801 is extremely useful for labeling the NMDA channels because of its high affinity and selectivity (Yoneda and Ogita 1989a,b; Yoneda, Ogita, and Enomoto 1991). To decide whether both GSH and GSSG act as NMDA agonists, the effects of GSH and GSSG on [^{3}H]MK-801 binding have been investigated using brain synaptic membranes treated with Triton X-100 (Ogita et al. 1995). In these membranes, [^{3}H]MK-801 binding is quite sensitive to potentiation by any type of agonists for the NMDA complex as described above. As shown in Fig. 1, the addition of Glu alone markedly leads to potentiation of [^{3}H]MK-801 binding in a concentration-dependent fashion over a concentration range of 0.1 μM to 1 mM. Similarly, GSH potentiates [^{3}H]MK-801 binding in the absence of any added agonists. However, GSH is ineffective in additionally potentiating binding when determined in the presence of Glu at the maximally effective concentration. Further addition of Gly almost doubles the binding in the presence of GSH alone at the highest concentration used. Spermidine remarkably increases binding in the presence of both GSH and Gly at maximally effective concentrations. In addition, GSSG also potentiates [^{3}H]MK-801 binding in a manner similar to the potentiation by GSH. Figure 2 shows the effects of several different antagonists for the NMDA and Gly recognition domains on potentiation by GSH of [^{3}H]MK-801 binding. GSH alone potentiates [^{3}H]MK-801 binding in a bell-shaped manner at a concentration range of 0.1 μM to 10 mM. The potentiation by GSH alone is not only prevented by the further addition of NMDA

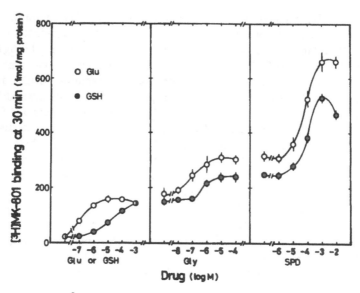

FIG. 1. Potentiation of [^{3}H]MK-801 binding by GSH. An aliquot of Triton-treated membranes was incubated with 5 nM [^{3}H]MK-801 at 30°C for 30 min in 0.5 ml of 50 mM Tris–acetate buffer (pH 7.4) containing different concentrations of agonists as indicated (Yoneda and Ogita, 1989a). Membranes were incubated with [^{3}H]MK-801 in buffer containing five different concentrations of GSH or Glu at concentrations from 100 nM to 1 mM (left panel). Incubation was also carried out in buffer containing five different concentrations of Gly in the presence of either 1 mM GSH or 10 µM Glu (middle panel). In addition, incubation was performed in buffer containing 10 µM Gly and five different concentrations of spermidine in the presence of either 1 mM GSH or 10 µM Glu (right panel). Nonspecific binding of [^{3}H]MK-801 was defined by the addition of both D-AP5 and 7-chlorokynurenic acid at 0.1 mM. Values are the mean±SEM from four independent experiments. These data are from Ogita et al. (1995).

antagonists such as CGP 39653 and D-AP5, but also exacerbated by that of the Gly antagonist DCKA, in a concentration-dependent manner. In addition to these data from membrane binding assays, the effects of GSH and GSSG are investigated on intracellular Ca^{2+} levels using *fura*-2-loaded brain neuronal cells dissociated from newborn rat pups (Leslie et al. 1992). Both GSH and GSSG induce a marked increase in intracellular Ca^{2+} levels in a manner that is prevented by competitive (i.e., D-AP5) and noncompetitve (i.e., MK-801) antagonists for the NMDA receptor complex. All these previous findings support the proposal that GSH and GSSG themselves may function as endogenous agonists selective for the NMDA recognition domain on the NMDA receptor complex. However, evidence against the proposal is also available in the literature. For instance, GSH has no significant effect on nonstimulated basal $^{45}Ca^{2+}$ influx into cultured rat cerebellar granule cells, whereas GSH as well as DTT significantly facilitates the influx induced by Glu and NMDA probably through the redox regulation of NMDA receptor mechanisms (Janáky et al. 1993). Leslie et al. (1992) have also discussed the possibility that binding of Gly and/or CySH of the glutathione molecule to modulatory sites on the NMDA receptor complex may produce

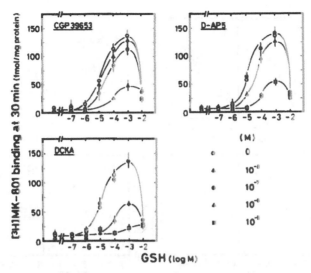

FIG. 2. Effects of several different antagonists on [³H]MK-801 binding in the presence of GSH. An aliquot of Triton-treated membranes was incubated with 5 nM [³H]MK-801 at 30°C for 30 min in buffer containing six different concentrations of GSH at a concentration range of 100 nM to 10 mM in either the presence or absence of the antagonist indicated at concentrations from 10 nM to 10 μM. Values are the mean±SEM from four independent experiments. These data are from Ogita et al. (1995).

an allosteric alteration of the NMDA recognition domain on the receptor complex, for both GSH and GSSG reduce B_{max} values of [³H]CGP 39653 binding, with a Hill coefficient of significantly less than unity.

C. Redox Modulatory Site

Several lines of evidence indicate that the NMDA receptor complex has a redox modulatory site consisting of thiol groups that are sensitive to sulfhydryl-oxidizing agents (for reviews see Lipton 1993; Lipton and Stamler 1994). In primary cultured neurons of rat neocortex, a reducing agent, DTT, potentiates NMDA currents by reducing the redox modulatory site within the NMDA receptor complex in a manner sensitive to abolition by oxidizing agents such as 5,5-dithiobis(2-nitrobenzoic) acid (Aizenman et al. 1989). Nitric oxide (NO) donors, including S-nitrosocysteine, nitroglycerin, and sodium nitroprusside (SNP), not only inhibit Ca^{2+} influx and the depolarization by NMDA, but also protect from cell death due to NMDA, through oxidation of the sulfhydryl residues at the redox modulatory site, independent of the generation of NO radicals (Kiedrowski et al. 1991; East, Batchelar, and Garthwaite 1991; Kiedrowski, Costa, and Wroblewski 1992; Lei et al. 1992). We have also raised the possible regulation by SNP at the redox modulatory site of binding of [³H]MK-801 to the open NMDA channel in rat brain synaptic membranes (Shuto, Ogita, and Yoneda 1997). In

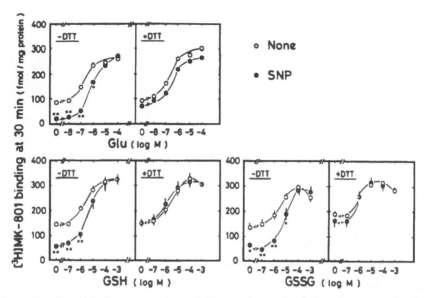

FIG. 3. Effects of SNP on potentiation by Glu, GSH, and GSSG of [³H]MK-801 binding. Synaptic membranes prepared were washed once with 50 mM Tris–acetate buffer (pH 7.4). On the day of the experiment, these membranes were incubated at 30°C for 30 min in either the presence or the absence of 0.1 mM SNP, followed by centrifugation at 50,000g for 30 min. The pellets thus obtained were washed once with buffer, and the resultant pellets were suspended in buffer again. The suspensions were incubated with 5 nM [³H]MK-801 at 30°C for 30 min in buffer containing five different concentrations of Glu, GSH, or GSSG in either the presence or the absence of DTT at 1 mM. Values are the mean±SEM from six independent experiments. *p< 0.05, **p < 0.01, compared with each control value obtained in membranes not treated with SNP.

fact, SNP causes marked inhibition of binding of both [³H]Glu and [³H]CGP 39653 to the NMDA recognition domain, without significantly affecting the binding of either [³H]Gly or [³H]DCKA to the Gly recognition domain. In addition, treatment with SNP leads to marked inhibition of [³H]MK-801 binding in the presence of Glu at low concentrations below 10 μM, without significantly altering that in the presence of Glu at higher concentrations over 1 μM, when determined in the absence of added DTT (Fig. 3). However, the addition of DTT completely abolishes the inhibition of binding by the treatment with SNP, irrespective of the concentration of added Glu. Hence, SNP could decrease the efficiency of NMDA agonists without affecting the maximal potency to potentiate [³H]MK-801 binding. By contrast, other NO donors such as S-nitroso-L-acetylpenicillamine and S-nitrosoglutathione have no inhibitory effect on ligand binding to any domains on the NMDA receptor complex. These results suggest that SNP may inhibit opening processes of the NMDA channel through oxidation of the redox modulatory site at the NMDA recognition domain in a manner independent of the generation of NO radicals, as previously shown (Lipton et al. 1993; Lipton and Stamler 1994; Shuto et al. 1997).

Figure 3 also shows the effects of SNP on the potentiation by GSH and GSSG of [³H]MK-801 binding. As described above, both GSH and GSSG are effective in enhancing binding in a concentration-dependent fashion at concentrations higher than 0.1 μM in membranes not treated with SNP. Treatment with SNP leads to inhibition of the binding detected in the presence of low concentrations of either GSH or GSSG added. The inhibitory effects of SNP are completely prevented by the addition of DTT at 1 mM, as seen with Glu. In the absence of DTT, however, SNP significantly inhibits the binding in the presence of GSSG at 10 μM, without significantly altering the binding in the presence of GSH at 10 μM. In other words, GSH may counteract the inhibition by SNP at a concentration range over 10 μM as a reducing agent, in addition to potentiating binding as an NMDA agonist. These findings further support the proposition that GSH functions as an endogenous agonist for the NMDA recognition domain on the receptor complex. GSH might also act an endogenous modulator for the redox modulatory site on the NMDA receptor ionophore complex. These propositions are supported by the findings that GSH as well as DTT enhances NMDA-induced $^{45}Ca^{2+}$ influx into cultured cerebellar granule cells (Janáky et al. 1993).

6. GSH BINDING SITES

As mentioned above, GSH and GSSG could have common propensities to modulate neurotransmission mediated by Glu in the rodent brain. Hence, it is conceivable that the rodent brain may contain specific sites with abilities to recognize glutathione in both reduced and oxidized forms, and thereby to induce a variety of alterations of central excitabilities associated with glutamatergic transmission. These possibilities prompted us to evaluate labeling by [^{3}H]GSH of the sites in crude synaptic membranes of rat brain.

6.1 Presence of [^{3}H]GSH Binding Sites

Our group was the first to demonstrate the presence of [^{3}H]GSH binding sites in synaptic membranes of rat brain (Ogita et al. 1987). The binding found in the presence of an excess of nonradioactive GSH is considered to be due to nonspecifically bound [^{3}H]GSH and subtracted from the total binding to calculate the specific binding, which occupies about 70% of the total binding. The specific binding is about 2 times higher at 30°C than that found at 2°C. The binding increases linearly with increasing concentrations of membranous proteins used, and also in proportion to the incubation time. The binding reaches equilibrium within 60 min, independently of the incubation temperature. The binding is saturable with increasing concentrations of [^{3}H]GSH at 2°C and 30°C. Scatchard analysis reveals that the binding sites consist of two components with different affinities for the ligand at 30°C ($K_{d1} = 0.56$ μM, $B_{max1} = 2.5$ pmol per milligram of protein; $K_{d2} = 12.6$ μM, $B_{max2} = 28.5$ pmol per milligram of protein). At 2°C, however, only a single component with a K_d of 0.77 μM and a B_{max} of 5.6 pmol per milligram of protein is observed (Fig. 4). Thus, [^{3}H]GSH binding sites have two independent constituents, of high and low affinity, respectively, in rat brain synaptic membranes. High-affinity binding sites

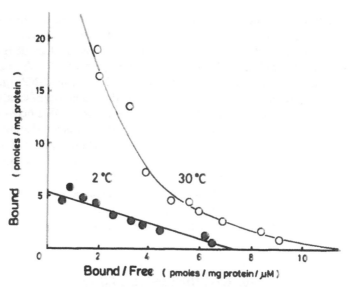

FIG. 4. Scatchard plots of [^{3}H]GSH binding in synaptic membranes from rat brain. An aliquot (350–400 μg) of "untreated" membranes were incubated with 100 or 200 nM [^{3}H]GSH in the presence of varying concentrations of nonradiolabeled GSH to cover the concentration range of 0.1 to 10 μM in 0.5 ml of buffer at 2 or 30°C for 60 min. Incubation was terminated by the addition of 3 ml of ice-cold buffer and subsequent rapid filtration through a Whatman GF/B glass filter under a constant vacuum of 15 mm Hg. Nonspecific binding was defined by the addition of 10 mM unlabeled GSH. Each point indicates the mean of eight independent experiments. At 2°C: $K_d = 0.77 \pm 0.19\ \mu$M, $B_{max} = 5.6 \pm 1.9$ pmol per milligram of protein. At 30°C: $K_{d1} = 0.56 \pm 0.11\ \mu$M, $B_{max1} = 2.5 \pm 0.5$ pmol per milligram of protein, $K_{d2} = 12.6 \pm 3.7\ \mu$M, $B_{max2} = 28.5 \pm 2.8$ pmol per milligram of protein. These data are from Ogita and Yoneda (1988a).

occur in a temperature-independent manner, whereas low-affinity binding sites are entirely depend on incubation temperature. It is unlikely that both types of binding sites reflect an association of GSH with membranous enzymes with affinity for GSH, such as GSH peroxidase, GSH-S-transferase, and γ-glutamyltranspeptidase: Products from the reaction of these enzymes, including Glu, cysteinylglycine, and GSSG, are not detected after the incubation of [^{3}H]GSH with synaptic membrane preparations (Ogita and Yoneda 1987).

The binding of [^{3}H]GSH is displaced by GSH in a concentration-dependent manner (IC$_{50}$ = 2.76 μM). Similarly complete displacement is seen for GSSG and some GSH derivatives without SH moiety, such as S-methylglutathione (SMG) and S-hexylglutathione (SHG) [IC$_{50}$ values (μM): GSSG, 17.6; SMG, 2.18; SHG, 22.5]. Scatchard analysis reveals that binding sites consist of a single component with a K_d of 7.7 μM in the presence of SMG, indicating that high-affinity sites of [^{3}H]GSH binding are completely masked by the addition of SMG. Furthermore, [^{3}H]GSH binding is partially inhibited by different peptides containing Glu, including α-L-glutamyl-Glu, γ-L-glutamyl-Glu, γ-L-glutamyl-Gly, γ-L-glutamyl-L-tyrosine, α-L-glutamyl-L-alanine and α-L-glutamyl-L-valine-L-phenylalanine, but not

TABLE 5. Effects of glutathione derivatives and
Glu-containing peptides (1 mM) on [³H]GSH binding

Substance	[³H]GSH binding (% of control)
GSSG	32.2 ± 5.3[b]
SMG	11.5 ± 2.7[b]
SHG	7.0 ± 1.5[b]
α-L-glutamyl-Glu	79.0 ± 3.3[a]
α-L-glutamyl-L-alanine	57.5 ± 2.6[a]
α-L-glutamyl-L-valyl-L-phenylalanine	60.3 ± 5.5[a]
γ-L-glutamyl-Glu	60.8 ± 4.7[b]
γ-L-glutamyltyrosine	68.4 ± 5.1[a]
γ-L-glutamyl-Gly	63.2 ± 3.8[b]
γ-D-glutamyl-Gly	86.0 ± 5.3

"Untreated" membranes obtained from rat brain were incubated with 100 nM [³H]GSH at 30°C for 60 min in either the presence or the absence of the peptide indicated at a concentration of 1 mM. Each value represents the mean±SEM obtained from four separate measurements.

Significance levels p are in comparison with control values obtained in the absence of any test peptides. These data are from Ogita and Yoneda (1987).

[a] $p < 0.05$.
[b] $p < 0.01$.

γ-D-glutamyl-Glu (Table 5). These data do not give support to the idea that [³H]GSH binding results from the formation of disulfide bonding between the SH residues of membranous proteins and the SH moiety of the ligand. In contrast, the binding seems to be at least in part sensitive to inhibition by the glutamyl residue.

The findings described above all support the proposal that rat synaptic membranes indeed contain structure-selective and saturable binding sites for [³H]GSH with different affinities. The high-affinity sites are completely inhibited by SMG, while the low-affinity sites show temperature dependence. Therefore, the high- and low-affinity binding sites may each play their own independent physiological roles in the brain.

6.2 Potentiation by CySH

The effects of three constituent amino acids of GSH on the binding are next examined for [³H]GSH binding. Both Glu and Gly only inhibit 20–30% of the maximum binding. In contrast, CySH markedly potentiates the binding in a concentration-dependent manner at concentrations above 0.1 mM. However, no significant change is induced by the addition of other SH-containing compounds, including L-cysteinesulfinic acid, L-cysteinic acid, DL-homocysteine, L-homocysteinic acid, DTT, and DTE. These findings suggest that CySH selectively potentiates the binding of [³H]GSH through mechanisms associated with its molecular structure, but not with SH moiety, in rat brain synaptic membranes.

The binding sites consist of two distinct components with high and low affinities in the absence of CySH, as mentioned in section 6.1. The addition of CySH not only induces abolition of the high-affinity sites, but also leads to an increase in the low-affinity sites. In the presence of CySH, therefore, the binding sites consist of a single component with low affinity ($K_d = 8.5 \ \mu$M) and high density ($B_{max} = 105$ pmol per milligram of protein). Extensive washing is ineffective in eliminating the stimulatory effect of CySH on [^{3}H]GSH binding. Hence, L-CySH irreversibly increases the low-affinity binding sites of [^{3}H]GSH in synaptic membranes of the rat brain.

The CySH-induced potentiation is observed in the hippocampus, cerebral cortex, and retina, but not in other brain regions, including the striatum, hypothalamus, midbrain, cerebellum, medulla–pons, and spinal cord. Moreover, CySH fails to significantly potentiate [^{3}H]GSH binding in all peripheral structures examined. In contrast, CySH inhibits the binding in particular peripheral structures such as the pituitary, adrenal, intestinal mucosa, and skeletal muscle. Thus, rat brain could have binding sites of [^{3}H]GSH, which are entirely distinguishable from those in the peripheral tissues with regard to potentiation by CySH. The possible diversity raises the hypothesis that [^{3}H]GSH binding sites may play specific functional roles in the brain, different from those in the peripheral organs.

6.3 Regional Distribution

The binding of [^{3}H]GSH is heterogeneously distributed in the central nervous system (Fig. 5). At 30°C, the retina has the highest binding among the central structures studied, with progressively lower binding in the hypothalamus, striatum, spinal cord, midbrain, medulla–pons, hippocampus, cerebellum, and cerebral cortex. At 2°C the highest binding is also detected in the retina, followed by the hypothalamus, hippocampus, striatum, cerebral cortex, cerebellum, medulla–pons, midbrain, and spinal cord in order of decreasing binding. The binding at 30°C is higher than that at 2°C in all central structures. It is therefore clear that temperature-dependent binding of [^{3}H]GSH is found in all central structures with a distribution profile identical to that determined at 30°. In addition, the binding of [^{3}H]GSH is detected in peripheral tissues with heterogeneous distribution profiles (Fig. 6). The pituitary possesses the highest binding among all structures examined including the central and peripheral tissues. In the peripheral tissues, higher binding is also found in the adrenals, liver, spleen, skeletal muscle, and heart, but no significant binding is detected in the kidney. The liver has a very small portion of the temperature-dependent binding, and the testicular and pulmonary binding are inversely dependent on the incubation temperature.

6.4 Physiological Significance

The physiological roles of [^{3}H]GSH binding sites in the brain are not known. However, the aforementioned findings all support possible differentiation between the functional significance of the low-affinity and high-affinity sites in the maintenance

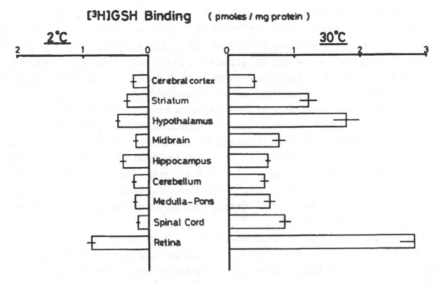

FIG. 5. Regional distribution of [³H]GSH binding in central structures. Membrane preparations were obtained from each central structure of male Wistar rats followed by extensively washing with the buffer as described in the legend of Fig. 4. The suspensions (350–400 μg) were incubated with 100 nM [³H]GSH at 2 or 30°C for 60 min. Data are the mean±SEM obtained from eight independent experiments. These data are from Ogita and Yoneda (1988a).

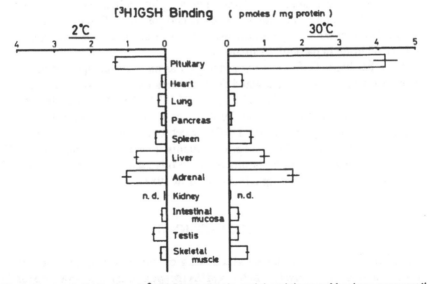

FIG. 6. Regional distribution of [³H]GSH binding in peripheral tissues. Membrane preparations were obtained from each tissue of male Wistar rats followed by extensively washing with the buffer. The suspensions (350–400 μg) were incubated with 100 nM [³H]GSH at 2 or 30°C for 60 min. Data are the mean±SEM obtained from eight independent experiments. n.d., not detectable. These data are from Ogita and Yoneda (1988).

of neural excitation. Furthermore, differential potentiation by CySH suggests that central [^{3}H]GSH binding sites may play specific functional roles different from those in peripheral tissues. For example, the low-affinity sites may at least in part participate in the transport mechanism for glutathione in the brain, which has been shown to be different from that in peripheral tissues such as the liver and kidney (Akerboom and Sies 1990). In contrast, the high-affinity sites could reflect the association of glutathione with membranous sites mediating specific functional alterations in neuronal and/or glial cells. In fact, recent studies have demonstrated that specific binding sites of biotinylated GSH and [^{35}S]GSH exist in the white matter of rat brain and cultured astrocytes (Guo and Shaw 1992; Guo, McIntosh, and Shaw 1992). In cultured astrocytes, moreover, GSH elicits a more than 40% increase in intracellular IP$_3$ levels at micromolar concentrations that occur in vivo (Guo and Shaw 1992). These findings argue in favor of the idea that GSH binding sites may indeed respond to a variety of physiological signals in the brain.

Furthermore, GSH may regulate activities of Glu receptors as described above. Among different types of Glu receptors, the metabotropic Glu receptor (group I) notably stimulates the hydrolysis of phosphatidylinositol followed by the accumulation of intracellular IP$_3$. In addition, NMDA and KA themselves could enhance IP$_3$ formation in a manner sensitive to the respective antagonists in cultured striatal neurons (Sladeczek et al. 1985) and in brain synaptoneurosomes (Récasens et al. 1987). It is thus possible to speculate that GSH binding sites may at least in part overlap a certain type of Glu receptors linked to IP$_3$ formation. In order to clarify the functional roles of GSH binding sites, exact mechanisms underlying intracellular signaling processes mediated by the binding sites should be evaluated further in future studies. Protein phosphorylation as well as gene expression might occur in response to glutathione signals in the brain in particular situations.

On the other hand, GSSG has been identified as an active component of a sleep-promoting substance that is extracted from the brain stems of sleep-deprived rats (Komoda et al. 1990; Honda, Komoda, and Inoué 1994). The sleep-inducing effect of GSSG may result from the activation by itself of the NMDA receptors as described in this chapter. An activation of the NMDA receptors could increase the release of arachidonic acid from membranous phospholipids in neurons (Dumuis et al. 1988; Sanfeliu, Hunt, and Patel 1990), followed by production of prostaglandin D$_2$; the latter has a strong sleep-inducing effect (Hayaishi 1988).

7. CONCLUSIONS

It appears that GSH may modulate glutamatergic neurotransmission in the central nervous system through activation of the NMDA receptor complex as an agonist and/or a redox modulator. The correlation between this modulation and the GSH binding sites demonstrated here, however, is still unclear at present. Elucidation of molecular mechanisms underlying the modulation by GSH of glutamatergic neurotransmission is undoubtedly of great potential benefit for the therapy and treatment of neurological disorders and in a variety of neuropsychiatric symptoms associated with abnormalities of GSH levels in the human being.

ACKNOWLEDGMENTS

This work was in part supported by Grants-in-Aid for Scientific Research to Y.Y. from the Ministry of Education, Science, Sports, and Culture, Japan.

REFERENCES

Abbott, L. C., Nejad, H. H., Bottje, W. G., and Hassan, A. S. 1990. Glutathione levels in specific brain regions of genetically epileptic (tg/tg) mice. *Brain Research Bulletin* 25:629–31.

Aizenman, E., Lipton, S. A., and Loring, R. H. 1989. Selective modulation of NMDA responses by reduction and oxidation. *Neuron* 2:1257–63.

Akbarian, S., Sucher, N. J., Bradley, D., Tafazzoli, A., Trinh, D., Hetrick, W. P., Potkin, S. G., Sandman, C. A., Bunney, Jr, W. E., and Jones, G. 1996. Selective alterations in gene expression for NMDA receptor subunits in prefrontal cortex of schizophrenics. *Journal of Neuroscience* 16:19–30.

Akerboom, T., and Sies, H. 1990. Glutathione transport and its significance in oxidative stress. In *Glutathione metabolism and physiological functions*, ed. Jose Vina, 45–55. CRC Press.

Andersen, J. K., Mo, J. Q., Hom, D. G., Lee, F. Y., Harnish, P., Hamill, R. W., and McNeill, T. H. 1996. Effect of buthionine sulfoximine, a synthesis inhibition of the antioxidant glutathione, on the murine nigrostriatal neurons. *Journal of Neurochemistry* 67:2164–71.

Bannon, M. L., Goedert, M., and Williams, B. 1984. The possible relationship of glutathione, melanin, and 1-methyl-4-phenyl-1,2,5,6-tetrahydropyridine (MPTP) to Parkinson's disease. *Biochemical Pharmacology* 33:2697–8.

Baudry, M., and Lynch, G. 1981. Characterization of two [^{3}H]glutamate binding sites in rat hippocampal membranes. *Journal of Neurochemistry* 36:811–20.

Berl, S., Purpura, D. P., Girado, M., and Waelsch, H. 1959. Amino acid metabolism in epileptogenic and non-epileptogenic lesions of the neocortex (cat). *Journal of Neurochemistry* 4:311–7.

Cooper, A. J. L., Pulsinelli, W. A., and Duffy, T. E. 1980. Glutathione and ascorbate during ischemia and postischemic reperfusion in rat brain. *Journal of Neurochemistry* 35:1242–5.

Croucher, M. J., Collins, J. F., and Meldrum, B. S. 1982. Anticonvulsant action of excitatory amino acid antagonists. *Science* 216:899–901.

Dumuis, A., Sebben, M., Haynes, L., Pin, J.-P., and Bokaert, J. 1988. NMDA receptors activate the arachidonic acid cascade system in striatal neurons. *Nature* 336:68–70.

East, S. J., Batchelar A. M., and Garthwaite, J. 1991. Selective blockade of N-methyl-D-aspartate receptor function by the nitric oxide donor, nitroprusside. *European Journal of Pharmacology* 209: 119–21.

Ffrench-Mullen, J. M. H., Koller, K., Zaczek, R., Coyle, J. T., Hori, H., and Carpenter, D. O. 1985. N-Acetylaspartylglutamate: possible role as the neurotransmitter of the lateral olfactory tract. *Proceedings of the National Academy of Sciences of the U.S.A.* 82:3897–900.

Geddes, J. W., Ulas, J., Brunner, L. C., Choe, W., and Cotman, C. W. 1992. Hippocampus excitatory amino acid receptors in elderly, normal individuals and those with Alzheimer's disease: non-N-methyl-D-aspartate receptors. *Neuroscience* 50:23–34.

Greenamyre, J. T., Penny, J. B., Young, A. B., D'Amato, C. J., Hicks, S. P., and Shoulson, I. 1985. Alterations in L-glutamate binding in Alzheimer's and Huntington's disease. *Science* 227:1496–9.

Greenamyre, J. T., Penny, J. B., D'Amato, C. J., and Young, A. B. 1987. Dementia of Alzheimer's type: changes in hippocampal L-[^{3}H]glutamate binding. *Journal of Neurochemistry* 48:543–51.

Guo, N., and Shaw, C. 1992. Characterization and localization of glutathione binding sites in cultured astrocytes. *Molecular Brain Research* 15:207–15.

Guo, N., McIntosh, C., and Shaw, C. 1992. Glutathione: new candidate neuropeptide in the central nervous system. *Neuroscience* 51:835–42.

Harrison, P. J., McLaughlin, D., and Kerwin, R. W. 1991. Decreased hippocampal expression of a glutamate receptor gene in schizophrenia. *Lancet* 333:450–2.

Hayaishi, O. 1988. Sleep–wake regulation by prostaglandin D_2 and E_2. *Journal of Biological Chemistry* 263:14593–6.

Henderson, Z., and Salt, T. E. 1988. The effects of N-acetylaspartylglutamate (NAAG) and distribution of NAAG-like immunoreactivity in the rat somatosensory thalamus. *Neuroscience* 25:899–906.

Hiramatsu, M., and Mori, A. 1981. Reduced and oxidized glutathione in brain and convulsions. *Neurochemical Research* 6:301–6.

Hollmann, M., and Heinemann, S. 1994. Cloned glutamate receptors. *Annual Review of Neuroscience* 17:31–108.

Honda, K., Komoda, Y., and Inoué, S. 1994. Oxidized glutathione regulates physiological sleep in unrestrained rats. *Brain Research* 636:253–8.

Hu, H. L., Bennett, N., Holton, J. L., Nolan, C. C., Lister, T., and Ray, D. E. 1996. Increased susceptibility of brain to m-dinitrobenzene neurotoxicity by glutathione depletion. *Human Experimental Toxicology* 15:145–55.

Jain, A., Mårtensson, J., Stole, E., Auld, P. A. M., and Meister, A. 1991. *Proceedings of the National Academy of Sciences of the U.S.A.* 88:1913–7.

Janáky, R., Varga, V., Saransaari, P., and Oja, S. S. 1993. Glutathione modulates the N-methyl-D-aspartate receptor-activated calcium influx into cultured rat cerebellar granule cells. *Neuroscience Letters* 156:153–7.

Janáky, R., Varga, V., Oja, S. S., and Saransaari, P. 1994. Release of [^{3}H]GABA evoked by glutamate agonists from hippocampal slices: effects of dithiothreitol and glutathione. *Neurochemical International* 24:575–82.

Johnson, M. K. 1966. Studies on glutathione S-alkyltransferase of the rat. *Biochemical Journal* 98:44–51.

Kaku, D. A., Goldberg, M. P., and Choi, D. W. 1991. Antagonism of non-NMDA receptors augments the neuroprotective effect of NMDA receptor blockade in cortical culture subjected to prolonged deprivation of oxygen and glucose. *Brain Research* 554:344–7.

Kiedrowski, L., Manev, H., Costa, E., and Wroblewski, J. T. 1991. Inhibition of glutamate-induced cell death by sodium nitroprusside is not mediated by nitric oxide. *Neuropharmacology* 30:1241–3.

Kiedrowski, L., Costa, E., and Wroblewski, J. T. 1992. Sodium nitroprusside inhibits N-methyl-D-aspartate-evoked calcium influx via a nitric oxide- and cGMP-independent mechanism. *Molecular Pharmacology* 41:779–84.

Klockgether, T., and Turski, L. 1989. Excitatory amino acids and basal ganglia: implication for the therapy of Parkinson's disease. *Trends in Neurosciences* 2:285–6.

Klockgether, T., Turski, L., Honoré, T., Zhang, Z., Gash, D. M., Kurlan, R., and Greenamyre, J. T. 1991. The AMPA receptor antagonist NBQX has antiparkinsonian effect in monoamine-depleted rats and MPTP-treated monkeys. *Annals of Neurology* 30:717–23.

Koller, K. J., and Coyle, J. T. 1984. Characterization of the interactions of N-acetyl-aspartyl-glutamate with [^{3}H]glutamate receptors. *European Journal of Pharmacology* 98:193–9.

Komoda, Y., Honda, K., and Inoué, S. 1990. SPS-B, a physiological sleep regulator, from the brainstems of sleep-deprived rat, identified as oxidized glutathione. *Chemical & Pharmaceutical Bulletin* 38:2057–9.

Kurumaji, A., Ishimaru, M., and Toru, M. 1992. α-[^{3}H]Amno-3-hydroxy-5-methylisoxazole-4-propionic acid binding to human cerebral cortical membranes: minimal changes in postmortem brains of chronic schizophrenics. *Journal of Neurochemistry* 59:829–37.

Lei, S. Z., Pan, Z.-H., Aggarwal, S. K., Chen, H.-S. V., Hartman, J., Sucher, N. J., and Lipton, S. A. 1992. Effect of nitric oxide production on the redox modulatory site on the NMDA receptor-channel complex. *Neuron* 8:1087–99.

Leslie, S. W., Brown, L. M., Trent, R. D., Lee, Y.-H., Morris, J. L., Jones, T. W., Randall, P. K., Lau, S. S., and Monks, T. J. 1992. Stimulation of N-methyl-D-aspartate receptor-mediated calcium entry into dissociated neurons by reduced and oxidized glutathione. *Molecular Pharmacology* 41:308–14.

Lipton, S. A., 1993. Prospects for clinically tolerated NMDA antagonists: open-channel blockers and alternative redox states of nitric oxide. *Trends in Neurosciences* 16:527–32.

Lipton, S. A., and Stamler, J. S. 1994. Actions of redox-related congeners of nitric oxide at the NMDA receptor. *Neuropharmacology* 33:1229–33.

Lipton, S. A., Choi, Y.-B., Pan, Z.-H., Lei, S. Z., Chen, H.-S. V., Sucher, N. J., Loscalzo, J., Singel, D. J., and Stamler, J. S. 1993. A redox-based mechanism for the neuroprotective and neurodestructive effects of nitric oxide and related nitroso-compounds. *Nature* 364:626–32.

Liu, Y. F., and Quirion, R. 1992. Modulatory role of glutathione on μ-opioid, substance P/neurokinin-1, and kainic acid receptor binding sites. *Journal of Neurochemistry* 59:1024–32.

Lodge, D., and Collingridge, G. 1990. Les agents provocateus: a series on the pharmacology of excitatory amino acids *Trends in Pharmacological Sciences* 11:22–4.

Makar, T. K., Nedergaad, M., Preuss, A., Gelbard, A. S., Perumal, A. S., and Cooper, A. J. L. 1994. Vitamin E, ascorbate, glutathione, glutathione disulfide, and enzymes of glutathione metabolism in cultures of chick astrocytes and neurons: evidence that astrocytes play an important role in antioxidative process in the brain. *Journal of Neurochemistry* 62:45–53.

Maragos, W. B., Chu, D. C. M., Young, A. B., D'Amato, C. J., and Penny, J. B. 1987. Loss of hippocampal [^{3}H]TCP binding in Alzheimer's disease. *Neuroscience Letters* 74:371–6.

Meister, A., and Anderson, M. E. 1983. Glutathione. *Annual Review of Biochemistry* 52:711–60.

Monaghan, D. T., Bridge, R. J., and Cotman, C. W. 1989. The excitatory amino acid receptors: their classes, pharmacology, and distinct properties in the function of the central nervous system. *Annual Review of Pharmacology and Toxicology* 29:365–402.

Moris, R. G. M., Anderson, E., Lynch, G. S., and Baudry, M. 1986. Selective impairment of learning and blockage of long-term potentiation by an *N*-methyl-D-aspartate receptor antagonist, AP5. *Nature* 319:774–6.

Mosinger, J. L., Price, M. T., Bai, H. Y., Xiao, H., Wozniak, D. F., and Olney, J. W. 1991. Blockade of both NMDA and non-NMDA receptors is required for optimal protection against ischemic neuronal degeneration in the in vivo adult mammalian retina. *Experimental Neurology* 113:10–17.

Murphy, D. E., Hutchinson, A. J., Hurt, S. D., Williams, M., and Sills, M. A. 1988. Characterization of the binding of [^{3}H]-CGS 19755: a novel *N*-methyl-D-aspartate antagonist with nanomolar affinity in rat brain. *British Journal of Pharmacology* 95:115–6.

Nakanishi, S. 1992. Molecular diversity of glutamate receptors and implications for brain function. *Science* 258:597–603.

Nicoletti, F., Bruno, V., Copani, A., Casabono, G., and Knöpfel, T. 1996. Metabotropic glutamate receptors: a new target for the therapy of neurodegenerative disorders. *Trends in Neurosciences* 19:267–71.

Ogita, K., and Yoneda, Y. 1986. Characterization of Na$^+$-dependent binding sites of [^{3}H]glutamate in synaptic membranes from rat brain. *Brain Research* 397:137–44.

Ogita, K., and Yoneda, Y. 1987. Possible presence of [^{3}H]glutathione (GSH) binding sites in synaptic membranes from rat brain. *Neuroscience Research* 4:486–96.

Ogita, K., and Yoneda, Y. 1988a. Temperature-dependent and -independent apparent binding activities of [^{3}H]glutathione in brain synaptic membranes. *Brain Research* 463:37–46.

Ogita, K., and Yoneda, Y. 1988b. Disclosure by Triton X-100 of NMDA-sensitive [^{3}H]glutamate binding sites in brain synaptic membranes. *Biochemical and Biophysical Research Communications* 153:510–7.

Ogita, K., and Yoneda, Y. 1989. Selective potentiation by L-cysteine of apparent binding activity of [^{3}H]glutathione in synaptic membranes of rat brain. *Biochemical Pharmacology* 38:1499–505.

Ogita, K., and Yoneda, Y. 1990. Temperature-independent binding of [^{3}H](±)-3-(2-carboxypiperazin-4-yl)propyl-1-phosphonic acid in brain synaptic membranes treated by Triton X-100. *Brain Research* 515:51–6.

Ogita, K., Kitago, T., Nakamuta, H., Fukuda, Y., Koida, M., Ogawa, Y., and Yoneda, Y. 1986. Glutathione-induced inhibition of Na$^+$-independent and dependent binding of L-[^{3}H]glutamate in rat brain. *Life Sciences* 39:2411–8.

Ogita, K., Ogawa, Y., and Yoneda, Y. 1988. Apparent binding activity of [^{3}H]glutathione in rat central and peripheral tissues. *Neurochemistry International* 13:493–7.

Ogita, K., Suzuki, T., and Yoneda, Y. 1989. Strychnine-insensitive binding of [^{3}H]glycine in synaptic membranes in rat brain, treated with Triton X-100. *Neuropharmacology* 28:1263–70.

Ogita, K., Sakamoto, T., Han, D., Azuma, Y., and Yoneda, Y. 1994. Discrimination by added ions of ligands at ionotropic excitatory amino acid receptors insensitive to *N*-methyl-D-aspartate in rat brain using membrane binding techniques. *Neurochemistry International* 24:379–88.

Ogita, K., Enomoto, R., Nakahara, F., Ishitsubo, N., and Yoneda, Y. 1995. A possible role of glutathione as a endogenous agonist at the *N*-methyl-D-aspartate recognition domain in rat brain. *Journal of Neurochemistry* 64:1088–96.

Orlowsky, M., and Karkowsky, A. 1976. Glutathione metabolism and some possible functions of glutathione in the nervous system. *International Reviews of Neurobiology* 19:75–121.

Pan, Z., and Perez-Polo, P. 1993. Role of nerve growth factor in oxidant homeostasis: glutathione metabolism. *Journal of Neurochemistry* 61:1713–21.

Perry, T. L., and Yong, V. W. 1986. Idiopathic Parkinson's disease, progressive supranuclear palsy and glutathione metabolism in the substantia nigra of patients. *Neuroscience Letters* 67:269–74.

Perry, T. L., Godin, D. V., and Hansen, S. 1982. Parkinson's disease, a disorder due to nigral glutathione deficiency? *Neuroscience Letters* 33:305–10.

Pin, J.-P, and Duvoison, R. 1995, The metabotropic glutamate receptors: structure and functions. *Neuropharmacology* 34:1–26.

Puchalski, R. B., Louis, J.-C., Brose, N., Traynelis, S. F., Egebjerg, J., Kukekov, V., Wenthold, R. J., Rogers, S. W., Lin, F., Moran, T., Morrison, J. H., and Heinemann, S. F. 1994. Selective RNA editing and subunit assembly of native glutamate receptors. *Neuron* 13:131–47.

Récasens, M., Sassetti, I., Nourigat, A., Sladeczek, F., and Bockaert, J. 1987. Characterization of subtypes of excitatory amino acid receptors involved in the stimulation of inositol phosphate synthesis in rat brain synaptoneurosomes. *European Journal of Pharmacology* 141:87–93.

Rehncrona, S., Folbergrova, J., Smith, D. S., and Siesjo, B. K. 1980. Influence of complete and pronounced incomplete cerebral ischemia and subsequent recirculation on cortical concentrations of oxidized and reduced glutathione in the rat. *Journal of Neurochemistry* 34:477–86.

Reichelt, K. L., and Fonnum, F. 1969. Subcellular localization of N-acetyl-aspartyl-glutamate, N-acetyl-glutamate and glutathione in brain. *Journal of Neurochemistry* 16:1409–16.

Riedel, G: 1996. Function of metabotropic glutamate receptors in learning and memory. *Trends in Neurosciences* 19:216–24.

Riederer, P., Sofic, E., Rausch, W.-D., Schmidt, B., Reynolds, G. P., Jellinger, K., and Youdium, M. B. H. 1989. Transition metals, ferritin, glutathione, and ascorbic acid in parkinsonian brains. *Journal of Neurochemistry* 52:515–20.

Robinson, M. B., Blakely, R. D., Couto, R., and Coyle, J. T. 1987. Hydrolysis of the brain dipeptide N-acetyl-L-aspartyl-L-glutamate. *Journal of Biological Chemistry* 262:14498–506.

Sanfeliu, C., Hunt, A., and Patel, A. J. 1990. Exposure to N-methyl-D-aspartate increases release of arachidonic acid in primary cultures of rat hippocampal neurons and not in astrocytes. *Brain Research* 526:241–8.

Schneider, S. P., and Perl, E. R. 1988. Comparison of primary afferent and glutamate excitation of neurons in the mammalian spinal dorsal horn. *The Journal of Neuroscience* 8:2062–73.

Sheardown, M. J., Nielson, E. O., Hansen, A. J., Jacobsen, P., and Honoré, T. 1990. 2.3-Dihydroxy-6-nitro-7-sulfamoyl-benzo(F)quinoxaline: a neuroprotectant for cerebral ischemia. *Science* 247:571–4.

Shuto, M., Ogita, K., and Yoneda, Y. 1997. Nitric oxide-independent inhibition by sodium nitroprusside of the N-methyl-D-aspartate recognition domain in rat brain. *Neurochemistry International*. In press.

Simon, R. P., Swan, J. H., Griffiths, T., and Meldrum, B. S. 1984. Blockage of N-methyl-D-aspartate receptors may protect against ischemic damage in the brain. *Science* 226:850–2.

Sladeczek, F., Pin, J. -P., Récasens, M., Bockaert, J., and Weiss, S. 1985. Glutamate stimulates inositol phosphate formation in striatal neurons. *Nature* 317:717–9.

Trombley, P. Q., and Westbrook, G. L. 1990. Excitatory synaptic transmission in cultures of rat olfactory bulb. *Journal of Neurophysiology* 64:598–606.

Turski, L., Bressler, K., Retting, K. L., Loschmann, P. A., and Wachtel, H. 1991. Protection of substantia nigra from MPP+ neurotoxicity by N-methyl-D-aspartate antagonists. *Nature* 349:414–8.

Vincent, S. R., and McGeer, E. G. 1980. A comparison of sodium-dependent glutamate binding with high-affinity glutamate uptake in rat striatum. *Brain Research* 184:99–108.

Westbrook, G. L., Mayer, M. L., Namboodiri, M. A. A., and Neale, J. H. 1986. High concentrations of N-acetylaspartylglutamate (NAAG) selectively activate NMDA receptor on mouse spinal cord neurons in cell culture. *The Journal of Neuroscience* 6:3385–92.

Whittemore, E. R., and Koerner, J. F. 1989. An explanation fro the purported excitation of piriform cortical neurons by N-acetyl-L-aspartyl-L-glutamic acid (NAAG). *Proceedings of the National Academy of Sciences of the U.S.A* 86:9602–5.

Wieloch, T. 1985. Hypoglycemia-induced neuronal damages prevented by an N-methyl-D-aspartate antagonist. *Science* 230:681–3.

Williamson, L. C., and Neale, J. H. 1992. Uptake, metabolism, and release of N-[³H]acetylaspartylglutamate by the avian retina. *Journal of Neurochemistry* 58:2191–9.

Wüllner, U., Brouillet, E., Isacson, O., Young, A. B., and Penney, J. B. 1993. Glutamate receptor binding sites in MPTP-treated mice. *Experimental Neurology* 121:284–7.

Yoneda, Y., and Ogita, K. 1989a. Labeling of NMDA receptor channels by [³H]MK-801 in brain synaptic membranes treated with Triton X-100. *Brain Research* 499:305–14.

Yoneda, Y., and Ogita, K. 1989b. Abolition of the NMDA-mediated responses by a specific glycine antagonist, 6,7-dichloroquinoxaline-2,3-dione (DCQX). *Biochemical and Biophysical Research Communications* 164:841–9.

Yoneda, Y., and Ogita, K. 1991. Neurochemical aspects of the N-methyl-D-aspartate receptor complex. *Neuroscience Research* 10:1–33.

Yoneda, Y., Ogita, K., Ohgaki, T., Uchida, S., and Meguri, H. 1989. N-Methyl-D-aspartate-sensitive [³H]glutamate binding sites in brain synaptic membranes treated with Triton X-100. *Biochimica et Biophysica Acta* 1012:74–80.

Yoneda, Y., Ogita, K., Kouda, T., and Ogawa, Y. 1990. Radioligand labeling of N-methyl-D-aspartic acid (NMDA) receptors by [^{3}H](±)-3-(2-carboxypiperazin-4-yl)propyl-1-phosphonic acid in brain synaptic membranes treated with Triton X-100. *Biochemical Pharmacology* 39:225–8.

Yoneda, Y., Ogita, K., and Enomoto, R. 1991. Characterization of spermidine-dependent [^{3}H](+)-5-methyl-10,11-dihydro-5H-dibenzo[a,d]cyclohepten-5,10-imine (MK-801) binding in brain synaptic membranes treated with Triton X-100. *Journal of Pharmacology and Experimental Therapeutics* 256:1161–72.

Yoneda, Y., Suzuki, T., Ogita, K., and Han, D. 1993. Support for radiolabeling of a glycine domain on the N-methyl-D-aspartate receptor ionophore complex by 5,7-[^{3}H]dichlorokynurenate in rat brain. *Journal of Neurochemistry* 60:634–54.

Yong, V. W., Perry, T. L., and Krisman, A. A. 1986. Depletion of glutathione in brainstem of mice by N-methyl-4-phenyl-1,2,3,6-tetrahydropyridine is prevented by antioxidant treatment. *Neuroscience Letters* 63:56–60.

Young, A. B., Greenamyre, J. T., Hollingsworth, Z., Albin, R., D'Amato, C. J., Shoulson, L., and Penny, J. B. 1988. NMDA receptor losses in putamen form patients with Huntington's disease. *Science* 241:981–3.

Zaczek, R., Koller, K., Cotter, R., Heller, D., and Coyle, J. T. 1983. N-Acetylaspartylglutamate: an endogenous peptide with high affinity for a brain "glutamate" receptor. *Proceedings of the National Academy of Sciences of the U.S.A.* 80:1116–9.

Zängerle, L., Cuénod, M., Winterhalter, K. H., and Do, K. Q. 1992. Screening of thiol compounds: depolarization-induced release of glutathione and cysteine from rat brain slices. *Journal of Neurochemistry* 59:181–9.

Zuo, P., Ogita, K., Suzuki, T., Han, D., and Yoneda, Y. 1993. Further evidence for multiple forms of an N-methyl-D-aspartate recognition domain in rat brain using membrane binding techniques. *Journal of Neurochemistry* 61:1865–73.

Glutathione in the Nervous System
Edited by Christopher A. Shaw
Copyright © 1998 Taylor & Francis

8

Glutathione and Glutathione Derivatives: Possible Modulators of Ionotropic Glutamate Receptors

Réka Janáky

Tampere Brain Research Center, University of Tampere Medical School, Tampere, Finland

Vince Varga and Zsolt Jenei

Tampere Brain Research Center, University of Tampere Medical School, Tampere, Finland and Department of Animal Physiology, Kossuth Lajos University of Science, Debrecen, Hungary

Pirjo Saransaari

Tampere Brain Research Center, University of Tampere Medical School, Tampere, Finland

Simo S. Oja

Tampere Brain Research Center, University of Tampere Medical School, Tampere, Finland and Department of Clinical Physiology, Tampere University Hospital, Tampere, Finland

---

- **Introduction**
- **Glutathione in the CNS**
- **Glutamate Receptors**
- **Glutathione and Glutathione Derivatives as Modulators of Glutamatergic Neurotransmission**
- **Glutathione as a Peptide Ligand of Ionotropic Glutamate Receptors**

ABBREVIATIONS

APH 2-amino-7-phosphonoheptanoate
APV 2-amino-5-phosphonovalerate
AMPA 2-amino-3-hydroxy-5-methyl-4-isoxazolepropionate
cAMP cyclic adenosine $3',5'$-monophosphate
cGMP cyclic guanosine $3',5'$-monophosphate
CGP-39653 DL-(E)-2-amino-4-propyl-5-phosphono-3-pentenoate
CNS central nervous system
CNQX 6-cyano-7-nitroquinoxaline-2,3-dione
CPP 3-(2-carboxypiperazin-4-yl)propyl-1-phosphonate
DCKA 5,7-dichlorokynurenate
DNQX 6,7-dinitroquinoxaline-2,3-dione
DTT dithiothreitol
DTNB $5,5'$-dithio-bis-2-nitrobenzoate
GSA glutathione sulfonate
GSH reduced glutathione
GSSG oxidized glutathione
HA 1-hydroxy-3-amino-2-pyrrolidone
LTP long-term potentiation

NBQX sulfamoylbenzo(*f*)quinoxaline
PCP phencyclidine
SMG *S*-methylglutathione
SEG *S*-ethylglutathione
SPG *S*-propylglutathione
SBG *S*-butylglutathione
SPeG *S*-pentylglutathione
t-ACPD trans-1-aminocyclopentane-1, 3-dicarboxylate

1. INTRODUCTION

γ-Glutamylcysteinylglycine (GSH and GSSG, reduced and oxidized glutathione) is ubiquitous in the organism and the most abundant peptide in the central nervous system (CNS) (Martin and MacIlwain 1959; Orlowski and Karkowsky, 1976). Its redox functions are well known; for example, the molecule can scavenge free radicals as an antioxidant, protecting cell membranes against oxidative stress, DNA against radiation and ultraviolet light, and the whole cell against xenobiotics (Meister and Anderson 1983; see also chapters 1, Shaw, and 2, Cooper, in this volume). The other roles of glutathione, stemming from its ability to serve as a cofactor or substrate for various enzymes, are less well known, for example, in the regulation of the cell cycle and in cellular metabolism (Kosower and Kosower 1978; Meister 1988; Max 1989). Other effects—those on sleep (Honda, Komoda, and Inoué 1994) and in protection against glutamate neurotoxicity (Levy, Sucher, and Lipton 1991), for example—may be explicable by the regulation of homeostasis of intraneuronal calcium (Gilbert, Aizenman, and Reynolds 1991) and of second-messenger functions (Guo, McIntosh, and Shaw 1992).

The three amino acids in the GSH molecule are neuroactive, and two of them, glutamate and glycine, are neurotransmitters in the CNS. Its chemical structure thus tends to make of glutathione another neuroactive compound, which may function as both neurotransmitter and neuromodulator. On one hand, glutathione has been reported to possess specific binding sites in the CNS, which may mediate signal transmission in neurons (Ogita and Yoneda 1987, 1988, 1989; Lanius et al. 1994) and glial cells (Guo et al. 1992; see also chapter 1, Shaw). On the other, glutathione may interact with glutamate receptors and may have a neuromodulatory role in the CNS (see section 4). In this chapter, drawing upon recent findings of our own and those of other investigators, we summarize the present knowledge of the putative neuromodulatory role of glutathione in the CNS.

2. GLUTATHIONE IN THE CNS

Glutathione (Reichelt and Fonnum 1969; Slivka et al. 1987a,b; Kirstein et al. 1991), *S*-methylglutathione (SMG) (Kanazawa et al. 1965), and glutathione sulfonate (GSA) (Li et al. 1993b) are endogenous constituents of brain tissue. The total level of glutathione in the CNS is within the range of 1.4–3.4 mM. The greater part of glutathione (about 95 percent) is in reduced form (Slivka et al. 1987b), but under oxidizing conditions GSSG may be present at higher concentrations (Folbergrova, Rehncrona, and Siesjö 1979). The concentration of SMG is 0.1–0.2 mg per kilogram of fresh brain tissue. SMG was the first example to be recognized of an occurrence of *S*-methylcysteine in a peptide (Kanazawa et al. 1965).

Glutathione is synthesized from its constituent amino acids and broken down via the γ-glutamyl cycle. The synthesis of GSH is catalyzed by γ-glutamylcysteine synthetase and glutathione synthetase, the former being the rate-limiting enzyme. The activity of both enzymes is under substrate and product (feedback) control. GSH is oxidized to GSSG by GSH peroxidase, and GSSG is reduced to GSH by glutathione reductase. The breakdown of glutathione is catalyzed by γ-glutamyl transferase, a membrane-bound enzyme that catalyzes the transfer of the γ-glutamyl moiety to free amino acids. γ-Glutamyl dipeptides are then split by γ-glutamyl cyclotransferase into its constituent amino acids and 5-oxoproline. The latter is further metabolized to glutamate by 5-oxoprolinase (Meister and Anderson 1983; Deneke and Fanburg 1989). The cellular and regional distribution of the enzymes involved in glutathione metabolism parallels that of glutathione (Philbert et al. 1991).

Glutathione is present in both intra- and extracellular spaces (Orlowski and Karkowsky 1976). The concentration of GSH in the cerebrospinal fluid is in the micromolar range (Rehncrona and Siesjö 1979). Cellularly, glutathione has been histochemically located in nonneuronal elements (epithelial and glial cells) and neurons (axons and nerve terminals, but not perikarya except for cerebellar granule cells) (Slivka et al. 1987a; Philbert et al. 1991; Hjelle, Chaudry, and Ottersen 1994; see also Hjelle et al., chapter 4, this volume). Raps et al. (1989) have reported that the glutathione levels are much higher in cultured differentiated astrocytes than in undifferentiated astrocytes or cultured neurons. They suggest that there exists an exchange of intercellular GSH between astrocytes and neurons. In cultured astrocytes the intracellular glutathione concentration is 8–20 mM (Yudkoff et al. 1990). There exist different intracellular pools (mitochondrial and cytoplasmic) of glutathione (Jain et al. 1991). Glutathione is rapidly released from astrocytes, but it is not known whether or not a specific efflux system exists (Yudkoff et al. 1990).

The possible intercellular exchange of glutathione between astrocytes and neurons in vivo (Raps et al. 1989), the fast release of glutathione from astrocytes under normoxic (Yudkoff et al. 1990) and hypoxic conditions (V. Varga et al., unpublished data), and its in vivo release during the early postischemic phase (Andiné et al. 1991) witness the extracellular modulatory action of glutathione. On the other hand, for a molecule with neuromodulatory or transmitter functions, depolarization-evoked release is an essential prerequisite, and glutathione (GSH 85 percent and GSSG 15 percent) is indeed released upon K^+ depolarization in a Ca^{2+}-dependent manner from slices from different regions of the rat brain. This release is most prominent in the mesodiencephalon, cortex, hippocampus, and striatum and lowest in the pons–medulla and cerebellum (Andiné et al. 1991). The Ca^{2+} dependence of the release bespeaks a neuronal origin of GSH. The amounts of GSH released have been found to be within the concentration range (i.e., micromolar) of the binding affinity of GSH to synaptic membranes (see section 5). It has thus been concluded that glutathione may be of physiological relevance in synaptic events (Zängerle et al. 1992). Although these preliminary data are promising, our present knowledge regarding the release of glutathione from neural cells is far from sufficient to support a neuromodulatory or transmitter role of this tripeptide.

3. GLUTAMATE RECEPTORS

L-Glutamate is the major excitatory neurotransmitter in the brain. It is present at high concentrations throughout the CNS. It is also likely that glutamate influences neuronal growth and development of interneuronal connections (Récasens, Mayat, and Vignes 1992) and participates in most CNS functions, including regulation of sensation, motion, motivation, behavior, learning, and memory (Fagg and Foster 1983; Collingridge 1987). On the other hand, glutamate may play an essential role in cellular death associated with epilepsy and in neurodegenerative disorders such as Huntington's, Alzheimer's, and Parkinson's diseases. It has also been implicated in ischemia, brain injury, amyotrophic lateral sclerosis, and AIDS encephalopathy (Lipton and Rosenberg 1994).

The glutamate receptors fall into two families, ionotropic and metabotropic (Nakanishi 1992; Cunningham, Ferkany, and Enna 1994). The ionotropic receptors contain integral cation-specific ion channels. They are classified after the agonists N-methyl-D-aspartate (NMDA), 2-amino-3-hydroxy-5-methyl-4-isoxazolepropionate (AMPA), and kainate (KA), the last two receptor classes being collectively referred to as non-NMDA receptors (Récasens et al. 1992). The metabotropic receptors are coupled to G-proteins and functionally linked either to the turnover of inositol phosphates (Conn and Desai 1991) or to the metabolism of cyclic nucleotides (Nakanishi 1992). The glutamate receptors are probably homo- or heteromeric pentamers, consisting of subunits NR1 and NR2A–D (NMDA receptors), GluR1–4 (AMPA receptors), KA1, 2 and GluR5–7 (kainate receptors), and mGlu1–5 and 7, 8 (metabotropic receptors) (Nakanishi 1992; Stone 1993). One ionotropic receptor subunit has three transmembrane segments (TM1, 3, and 4) and one segment folding as a loop into the membrane from its intracellular surface (TM2). The amino acid side chains of the TM2 segments surrounding and forming the pore determine its ion selectivity. The metabotropic receptors have subunits with seven transmembrane segments (TM1 to 7). The pharmacological properties of the receptors are determined by their subunit composition, the structure of the receptor subunits, and the availability and charges of functional amino acid side chains in the receptor protein.

The ionotropic AMPA receptors are activated by AMPA, quisqualate, and glutamate. Kainate, the most potent activator of the kainate receptors, also cross-reacts with AMPA receptors, though less effectively. The AMPA and kainate receptors are inhibited by 6,7-dinitroquinoxaline-2,3-dione (DNQX), 6-cyano-7-nitroquinoxaline-2,3-dione (CNQX), and sulfamoylbenzo(f)quinoxaline (NBQX) and are coupled to channels highly permeable to Na^+ and K^+ and much less permeable to Ca^{2+}. An activation of these receptors increases the membrane permeability for these ions, accompanied by rapid depolarization (Récasens et al. 1992). They are thus primarily responsible for the fast excitatory glutamate transmission and neuronal–glial interaction. The colocalization of NMDA and AMPA sites (reviewed in Nicoll, Malenka, and Kauer 1990) may be of significance in the repetitive activation of NMDA receptors and in the generation of long-term potentiation (LTP). The kainate receptors may function as autoreceptors, since they are located presynaptically. Their exact physiological role is not however known (Seeburg 1993).

The ionotropic NMDA receptors are receptor–channel complexes with several interacting binding sites for ligands:

1. Sites for agonists (e.g., NMDA, glutamate, and aspartate) and antagonists [e.g., 2-amino-5-phosphonovalerate (APV) and 3-(2-carboxypiperazine-4-yl) propyl-1-phosphonate (CPP)]. Because of the presence of two binding sites with different affinities for CPP (van Amsterdam et al. 1992), the existence of both agonist- and antagonist-preferring sites or conformational states has been suggested (Stone 1993).
2. A glycine coactivatory site blocked by cyclic glycine analogs (e.g., 1-hydroxy-3-amino-2-pyrrolidone, HA-966).
3. An inside-channel Mg^{2+} site involved in the voltage-dependent block of the ionophore.
4. A phencyclidine (PCP) site within the ion channel, which binds dissociative anesthetics with NMDA antagonistic properties—among them PCP itself, ketamine, and dizocilpine (MK-801)—in a use- and voltage-dependent manner.
5. A modulatory site for Zn^{2+}, which has a voltage-independent antagonistic action.
6. Polyamine activatory sites, which are blocked by ifenprodil, a new neuroprotective agent.

In addition to the above agonists and antagonists, alcohols, gangliosides, tricyclic antidepressants (other than desipramine), oxygen free radicals, H^+ ions, and agents that modify the redox state of thiols also affect the functions of NMDA receptors (Récasens et al. 1992).

The NMDA receptors have been claimed to function as molecular coincidence detectors (Seeburg 1993). Glutamate released from the nerve terminal cannot activate a Ca^{2+} flux through the receptor-governed channel unless the postsynaptic membrane is sufficiently depolarized to remove the Mg^{2+} block. Converging repetitive stimulation of AMPA or kainate receptors (tetanic stimulation), parallel stimulation of metabotropic glutamate receptors, or reduction of synaptic inhibition leads to a prolonged depolarization and potentiation of Ca^{2+} entry via the NMDA receptor-gated channels. Hence, the NMDA receptor is a sensor for synchronous activity of pre- and postsynaptic cells and a molecular basis for the generation of some forms of LTP, a process thought to be involved in synaptic plasticity (Collingridge 1987). On the other hand, overexcitation via the NMDA receptors leads to neuronal death. These plastic and toxic events may involve activation of various Ca^{2+}-dependent enzymes and consequent generation of second messengers, resulting in the following positive and negative feedback processes:

1. Activation of tyrosine kinase and nitric oxide synthase results in synthesis of nitrogen monoxide (NO) and cyclic guanosine monophosphate (cGMP) (Garthwaite 1991; Rodriguez et al. 1994). NO, readily diffusing out from the postsynaptic neuron, is believed to be a "retrograde transmitter" acting at the presynaptic terminal and increasing neurotransmitter release (positive feedback) (Bredt and Snyder 1992). The alternative redox states of NO evoke different cellular responses. The reduced form of nitric oxide ($NO^•$) is a free radical that reacts with the superoxide anion and forms neurotoxic peroxylnitrite ($ONOO^-$). In addition to osmotic damage, this reaction is responsible for neuronal death. In contrast, the oxidized form, nitrosonium ion (NO^+), can react with the redox modulatory site of the NMDA receptor and downregulate the receptor activity (negative feedback) (Lipton et al. 1993).
2. Activation of ornithine decarboxylase leads to the formation of polyamines and potentiation of NMDA receptor activity (positive feedback) (Porcella et al. 1992).
3. Stimulation of protein kinase C results in phosphorylation of the receptor and relief of the Mg^{2+} block (positive feedback) (Chen and Huang 1992).
4. Activation of phospholipase A_2 leads to the formation of arachidonic acid (Canonico et al. 1985; Ohmichi et al. 1989; Dumuis et al. 1988), which enhances presynaptic glutamate release and reduces glial reuptake (positive feedback) (Rhoads et al. 1983; Chan, Kerlan, and Fishman 1983; Miller et al. 1992).
5. Stimulation of adenylyl cyclase causes an accumulation of cyclic adenosine monophosphate (cAMP) (Jones, Snell, and Johnson 1987; Smart 1989).

4. GLUTATHIONE AND GLUTATHIONE DERIVATIVES AS MODULATORS OF GLUTAMATERGIC NEUROTRANSMISSION

GSH is composed of three amino acids, all known as transmitters or modulators in glutamatergic neurotransmission. *Glutamate* is an agonist of all classes of glutamate receptors. In spite of the large size and limited flexibility of the glutathione molecule, which may affect its interactions with receptor and transport proteins, GSH and GSSG still remain promising ligands of the glutamate binding sites. *Cysteine*—when it is free—is a neurotoxin (Shen and Dryhurst 1996) which can also cause an overstimulation of NMDA receptors in several ways, for example, via reduction of functional SH groups, formation of complexes with inhibitory Zn^{2+} ions, and elevation of the level of the NMDA receptor agonist cysteine sulfinate (Olney et al. 1990; Varga et al., unpublished observations). Bath cysteine application to cortical slices gives a rapid depolarization, which is blocked by dizocilpine (Pace et al. 1992). Treatment of rats with cysteine alters the affinity and number of NMDA sites in cortical membranes, which indicates an excitotoxic effect (Varga et al., unpublished observations). As a component of the tripeptide, cysteine is nontoxic, and moreover, GSH and GSSG are neuroprotective. In spite of this, GSH may still influence the redox state of glutamate receptors by means of its free thiol group (see section 6). *Glycine*, the main inhibitory transmitter in the spinal cord, is a coagonist at the NMDA receptor ionophores (see section 3). Thus, the following mechanisms may participate in the modulation of glutamatergic neurotransmission:

1. The γ-glutamyl moiety may act at the agonist or competitive antagonist site(s) of glutamate receptors.
2. The glycine moiety may act at the glycine coactivatory site.
3. The cysteine moiety modulates the uptake, reuptake, or binding sites of glutamate by (i) direct interaction with functional thiol groups in the transport or receptor proteins (Kiskin et al. 1986; Terramani et al. 1988; Aizenman, Lipton, and Loring 1989; Lazarewicz et al. 1989; Sucher, Wong, and Lipton 1990; Tang and Aizenman 1993), (ii) regulation of the level of other redox agents, for example, NO (Aizenman, Hartnett, and Reynolds 1990; Manzoni et al. 1992) and ascorbate, or (iii) formation of complexes with inhibitory metal ions, e.g., Zn^{2+} and Mg^{2+}.
4. The functions of ionophores are modulated, with a consequent change in membrane polarization (Gilbert et al. 1991; Ruppersberg et al. 1991).

Glutathione may interact with glutamatergic neurotransmission at the following steps: (i) release of excitatory amino acids, (ii) neuronal or glial uptake of excitatory amino acids, (iii) post- and presynaptic binding of glutamate and its selective analogs to different glutamate receptors, (iv) generation of second messengers, and (v) release of other transmitters evoked by the activation of glutamate receptors.

5. GLUTATHIONE AS A PEPTIDE LIGAND OF IONOTROPIC GLUTAMATE RECEPTORS

The *γ-glutamyl moiety* is the major functional group in the γ-glutamyl dipeptides that affects the different steps of glutamatergic neurotransmission (Varga et al. 1994a). γ-Glutamylglutamate, -aspartate, and -glycine enhance the release of L-glutamate and D-aspartate (a nonmetabolized analog of L-glutamate) from cerebral cortical slices (Varga et al. 1994a). Several γ-glutamyl peptides also inhibit the uptake of L-glutamate and D-aspartate and the Na^+-independent binding of glutamate and ligands selective for different ionotropic glutamate receptors (Davies and Watkins 1981; Ungerer et al. 1988; Varga et al. 1989, 1994a,b). γ-Glutamylglutamate acts as an agonist at the NMDA receptors, increasing the neuronal uptake and the intraneuronal level of Ca^{2+} (Li et al. 1993a; Varga et al. 1992, 1995); γ-glutamylaspartate and -taurine inhibit the glutamate- and glutamate-agonist-evoked neuronal Ca^{2+} and cyclic nucleotide (cAMP and cGMP) accumulation (Varga et al. 1990,1992, 1994a, b). These data imply that the γ-glutamyl moieties are also important for the action(s) of GSH and GSSG.

5.1 Binding of Glutamate

GSH and GSSG inhibit the Na^+- and temperature-independent binding of [³H]glutamate to synaptic membranes in a concentration-dependent manner within the concentration range prevailing in vivo (Ogita et al. 1986; Oja et al. 1988, 1994, 1995;

TABLE 1. *Inhibition of Na⁺-independent binding of tritiated glutamate receptor ligands to rat brain synaptic membranes by γ-glutamyl oligopeptides*

Displacer	IC_{50} (95% confidence intervals), μM			
	Glutamate	AMPA	APH	Kainate
GSH	9.6 (6.4–11.2)	8.4 (2.3–31.3)	103 (41.3–257)	180 (121–267)
GSSG	14.4 (12.5–16.3)	3.5 (1.7–7.1)	78.3 (31.3–195)	184 (132–281)
γ-L-Glutamylamino-methylsulphonate	178 (124–255)	639 (462–884)	42.4 (33.0–54.5)	400 (186–861)
γ-L-Glutamyltaurine	864 (663–1080)	689 (645–736)	694 (586–821)	>1000
γ-L-Glutamylamino-propylsulphonate	307 (172–549)	302 (211–431)	>1000	162 (108–241)
γ-L-Glutamylcysteine-sulphonate	290 (261–322)	416 (259–670)	>1000	701 (589–835)
γ-L-Glutamylglycine	61.7 (24.7–154)	48.8 (27.9–85.1)	124 (89.9–171)	24.5 (12.3–49.0)
γ-L-Glutamylglutamate	3.8 (2.8–5.2)	4.2 (3.0–5.9)	13.0 (8.5–20.0)	15.9 (14.3–17.6)
γ-L-Glutamylaspartate	112 (97.8–128)	49.0 (37.1–64.7)	104 (50.3–213)	9.7 (8.6–10.9)
L-Glutamate	1.7 (1.0–2.8)	1.2 (1.0–1.4)	18.8 (10.3–34.3)	0.7 (0.6–0.8)

Data from Varga et al. (1989) and Varga et al. (1994a). APH = 2-amino-7-phosphonoheptanoate.

Varga et al. 1989, 1994a). GSH and GSSG belong to the γ-glutamyl peptides, which displace glutamate from its postsynaptic binding sites with a medium affinity. No significant difference has been found between the effects of GSH and GSSG (Table 1). A number of γ-glutamyl dipeptides also displace [³H]glutamate binding. Of these, γ-L-glutamylglutamate is the most effective. Similarly to GSH and GSSG, glutathione sulfonate and S-alkyl derivatives of glutathione interact with the binding of glutamate (Table 2). Their efficacies are comparable to those of glutamate, GSH, and GSSG.

5.2 Binding of AMPA and Kainate

Both GSH and GSSG show some selectivity towards AMPA sites in rat forebrain membranes. The affinity of these sites for GSH and GSSG exceeds that of other γ-glutamyl

TABLE 2. *Inhibition of binding of tritiated glutamate receptor ligands to mouse brain synaptic membranes by glutathione and glutathione derivatives*

Displacer	IC_{50} (95% confidence intervals), μM			
	Glutamate	AMPA	CPP	Kainate
GSH	7.3 (0.2–25.1)	15.9 (4.5–55.5)	6.9 (4.1–11.6)	514 (82–3246)
GSSG	12.0 (3.0–47.4)	3.1 (0.9–10.3)	2.0 (0.4–9.2)	41.1 (5.0–332)
SMG	6.6 (1.0–42.3)	2.6 (1.6–4.5)	1.7 (0.2–14.2)	>1000
SEG	0.6 (0.0–39.9)	1.7 (0.6–5.0)	2.4 (0.1–54.4)	>1000
SPG	1.6 (0.3–9.6)	1.3 (0.4–4.2)	3.1 (0.5–16.7)	>1000
SBG	2.0 (0.5–7.3)	4.0 (1.1–13.9)	2.7 (1.0–6.7)	>1000
SPeG	56.4 (3.1–113)	0.8 (0.1–7.1)	3.0 (0.4–9.2)	>1000
GSA	19.8 (5.2–75.7)	1.5 (0.5–4.5)	21.2 (5.0–89.8)	544 (214–1385)

Data from Jenei et al. (1997).

peptides (except γ-L-glutamylglutamate) and approaches that for glutamate (Table 1). GSH; GSSG; S-methyl-, S-ethyl- (SEG); S-propyl- (SPG), S-butyl- (SBG), and S-pentylglutathione (SpeG); and glutathione sulfonate all strongly displace [^{3}H]AMPA binding (Table 2). All estimated IC$_{50}$ values have been in the low micromolar range. The compounds significantly reduce the B_{max} constants for binding, and some of them also tend to diminish K_D (Jenei et al. 1997). There are no marked differences in the efficacy of any of the compounds tested. GSSG and GSA also diminish the low-affinity binding of AMPA, in contrast to the other compounds (Oja et al. 1995).

[^{3}H]Kainate binding is affected by GSH and GSSG much less than the binding of other ligands of ionotropic glutamate receptors (Tables 1 and 2; Varga et al. 1997). In most cases the binding to kainate receptors has been inhibited less by the peptides than the binding to AMPA receptors. Only γ-L-glutamylaspartate displaces kainate more effectively than AMPA. However, both GSH and GSSG are still more effective inhibitors of kainate binding than cysteine or dithiothreitol (DTT), which also reduce the number of binding sites without affecting K_D (Liu and Quirion 1992). Even though GSA is slightly active, the S-alkylation of glutathione seems generally to prevent the inhibition of kainate binding to receptors (Table 2).

The above results have been the impetus for our hypothesis that glutathione and glutathione derivatives mainly interact with the AMPA class of non-NMDA receptors. At the AMPA sites, the redox modulation of sulfhydryl groups in the receptor complex is apparently not the mechanism of their action. This inference is in concert with the observed ineffectiveness of DTT and 5,5'-dithio-bis-2-nitrobenzoate (DTNB) in modulating AMPA binding and AMPA receptor-mediated responses (Aizenman et al. 1989). Furthermore, the efficacies of GSH and GSSG have proved to be about the same in AMPA binding and tended to be enhanced when the sulfhydryl group in the cysteine residue was S-alkylated. There is no indication that the mechanism of inhibition is pure competition at the ligand binding site, since none of the tested compounds increased K_D for the high-affinity binding of AMPA (Jenei et al. 1997). The mechanism of action is apparently complex, possibly involving allosteric effects.

5.3 Binding of NMDA Receptor Ligands

Micromolar concentrations of GSH and GSSG displace the binding of the ^{3}H-labeled *competitive NMDA antagonists* 2-amino-7-phosphonoheptanoate (APH) (Table 1), 3-(2-carboxypiperazin-4-yl)propyl-1-phosphonoate (CPP) (Table 2), and DL-(E)-2-amino-4-propyl-5-phosphono-3-pentenoate (CGP-39653) from the NMDA recognition site (Yoneda et al. 1990; Ogita et al. 1995; for a review, see Yoneda and Ogita 1991 and Ogita et al., chapter 7, this volume). In our recent study on rat brain membranes, GSH and GSSG also strongly inhibited the binding of CPP. GSH was somewhat more effective than GSSG, IC$_{50}$ with 95% confidence interval being 3.0 (2.3–3.7) and 7.0 (5.0–9.8) μM, respectively (Varga et al. 1997). Neither GSH and GSSG nor glutathione analogs altered K_D in CPP binding, but, with the exception of SBG, they all lowered B_{max}.

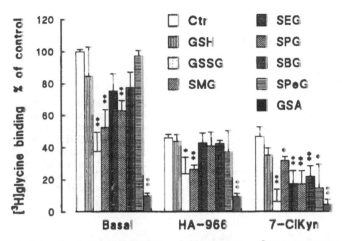

FIG. 1. Effects of glutathione and its derivatives on the binding of [^{3}H]glycine in the presence of glycine antagonists. Peptides (1 and 0.5 mM in the case of SPeG) were first preincubated with the mouse brain membranes together with 50 μM 3-amino-1-hydroxypyrrolidin-2-one (HA-966) or 50 μM 7-chlorokynurenate (7-ClKyn), for 10 min and then for 20 min with the antagonists and 20 nM [^{3}H]glycine. Mean values (±SEM) of three triplicate experiments as a percentage of the control binding without any effectors. Significantly different from the controls: * $p < 0.05$; ** $p < 0.01$. Reproduced from Jenei et al. (1997) by permission.

The strychnine-insensitive binding of [^{3}H]*glycine* to NMDA receptors is significantly affected only by GSSG and GSA. GSH and *S*-alkyl derivatives of glutathione have failed to affect glycine binding at micromolar concentration (Jenei et al. 1997). Some of the peptides slightly inhibit glycine binding at a 1 mM concentration. This effect is additive to inhibition by 7-chlorokynurenate and HA-966 (Fig. 1).

Both GSH and GSSG enhance the binding of [^{3}H]*dizocilpine* in a time- and concentration-dependent manner. The effect of GSH is clearly biphasic. It enhances the binding at low concentrations, but at higher concentrations the enhancement is gradually more and more attenuated (Fig. 2). The activating effects of GSH and GSSG are additive to the effect of glycine, but not to that of saturating concentrations of glutamate or glutamate plus glycine. Similar results were obtained with GSA and *S*-alkyl derivatives of glutathione (Fig. 3). These results are in agreement with those of Ogita et al. (1995), who found that both glycine and spermidine potentiated the binding of dizocilpine even in the presence of a maximally effective concentration of GSH. However, the effect was smaller than in the presence of a maximally effective concentration of glutamate.

The activation of dizocilpine binding by GSH and GSSG is prevented by the competitive NMDA and glycine antagonists APV and 7-chlorokynurenate (Fig. 4). These results are again in concert with those of Ogita et al. (1995), who used CGP-39653 and 5,7-dichlorokynureate (DCKA). On the other hand, GSH diminished the inhibitions by CGP-39653, APV, DCKA, and DNQX in dizocilpine binding in the presence of glutamate and glycine (Ogita et al. 1995). Pretreatment with glutamate dehydrogenase,

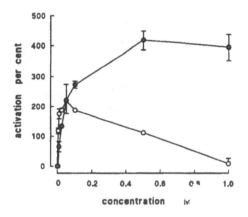

FIG. 2. Activation of [³H]dizocilpine binding to synaptosomal membranes by different concentrations of reduced (GSH) and oxidized (GSSG) glutathione. The results show the activation of binding as a percentage of the control binding without any effectors. Three experiments in triplicate, ±SEM. o = GSH, ● = GSSG. Reproduced from Varga et al. (1997).

an NAD-dependent enzyme, which catalyzes the breakdown of glutamate, induced a marked rightward shift of the concentration–response curve for glutamate but failed to affect the potentiation by GSH or GSSG (Ogita et al. 1995).

The above data indicate that GSH, GSSG, and SMG are endogenous ligands of the NMDA receptors, binding preferably to the glutamate recognition site via their γ-glutamyl moieties. They affect the glycine coactivatory site with considerably less affinity, which effect may not be of great functional importance.

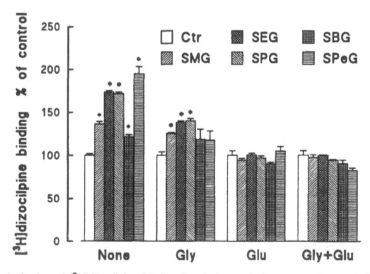

FIG. 3. Activation of [³H]dizocilpine binding by glycine and glutamate: effects of glutathione derivatives. The peptides and glycine (10 μM) and/or glutamate (50 μM) were present during the final incubation (20 min) with 1 nM [³H]dizocilpine. Mean values (±SEM) of three triplicate experiments as a percentage of the corresponding controls, designated 100 percent, in the absence of peptides. Significantly different from the corresponding control: * $p < 0.01$. Reproduced from Jenei et al. (1997).

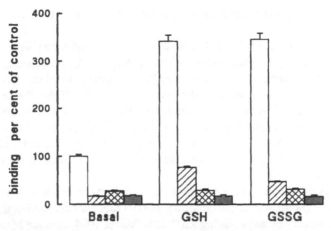

FIG. 4. Inhibition of the basal and GSH- and GSSG-activated (both 50 μM) binding of [³H]dizocilpine by 100 μM 2-amino-5-phosphonovalerate and 50 μM 7-chlorokynurenate. The antagonists were applied together with GSH or GSSG. The results (±SEM) of three triplicate experiments as a percentage of control binding without any effectors. White bars: control; hatched: APV; double-hatched: 7-chlorokynurenate; black: APV plus 7-chlorokynurenate. The antagonists significantly ($p < 0.01$) inhibited the binding in all cases. Reproduced from Varga et al. (1997).

6. GLUTATHIONE AS A REDOX MODULATOR OF NMDA RECEPTORS

The activation of NMDA receptors is regulated by the redox potential, as both [³H]dizocilpine binding and NMDA-induced Ca^{2+} fluxes are enhanced under reducing and inhibited under oxidizing conditions (Aizenman et al. 1989; Reynolds, Rush, and Aizenman 1990; Sucher et al. 1990). The redox modulation of NMDA receptor activity is a complex process involving regulation of several (at least three) extracellular sites (Sullivan et al. 1994; Gozlan and Ben-Ari 1995). The native NMDA receptor generally fluctuates between the fully reduced and fully oxidized state. Since all modulatory functions of the NMDA receptor (i.e., potentiation by glycine and blocks by magnesium and zinc) manifest themselves in all redox states, physiologically induced reduction and oxidation may provide an overall control on the receptor functions (Aizenman et al. 1989). Interestingly, the fully oxidized NMDA receptors are still functional, indicating that redox modulation does not work as an all-or-nothing switch, but rather as a buffer of NMDA receptor hyperactivity (Gozlan and Ben-Ari 1995).

Glutathione, which constitutes an antioxidant system maintaining the redox state of membrane thiols via reversible thiol–disulfide exchange reactions and by scavenging free radicals (Gilbert 1982; Walters and Gilbert 1986; Cappel and Gilbert 1988), has also been proposed as a putative endogenous redox modulator of NMDA receptor activity (Gozlan and Ben-Ari 1995). On the other hand, the discrepancies between the activation of NMDA receptors and the neurotoxicity evoked by DTT and cysteine (Tolliver and Pellmar 1987, 1988; Olney et al. 1990) and the neuroprotection via NMDA receptor inhibition by glutathione (Levy et al., 1991) clearly indicate that the mechanism of glutathione action is different from that of DTT and other thiol compounds.

Both extra- and intracellular GSH may form nitrosoglutathione (Hogg, Singh, and Kalyanaraman 1996; see also Cuénod and Do, chapter 12, this volume) and hence serve as a buffer of intra- and extracellular NO. S-Nitrosoglutathione may react with GSH and generate nitrous oxide and GSSG (Hogg et al. 1996). The reduction of the latter is inhibited by S-nitrosoglutathione (Becker, Gui, and Schirmer 1995). Both release of nitrous oxide and accumulation of GSSG may result in a short- and long-term downregulation of the NMDA receptor function (Manzoni et al. 1992; Janáky et al. 1993a; Lipton et al. 1993). However, more experimental work is needed to prove this hypothesis.

GSH and cysteine form complexes with Zn^{2+} (Li, Gawron, and Basuas 1954; Li and Manning 1955; Sato, Frazier, and Goldberg 1984). GSH may thus indirectly influence the NMDA receptor by sequestering Zn^{2+} ions (Maret 1994; also see Maret, chapter 12, this volume). Such a mechanism of action has been proved only for cysteine, which activates the NMDA receptor-mediated influx of Ca^{2+} and promotes NMDA-evoked cytotoxicity by forming complexes with the inhibitory Zn^{2+} ions (Olney et al. 1990; Varga et al., unpublished results).

7. INDIRECT MODULATION OF NMDA RECEPTOR ACTIVATION VIA REGULATION OF ION CHANNELS

Both extra- and intracellular GSH and GSSG may interact with ion channels, regulate their ion permeability, cause changes in the membrane potential, and hence modulate NMDA receptor activity. The fast-inactivating K^+ currents mediated by the cloned K^+ channel subunits derived from the mammalian brain and expressed in *Xenopus oocytes* are regulated by intracellular glutathione. In the oxidized state, the ball domain of the channel is fixed by a disulfide bridge, and it can no longer cork (i.e., inactivate) the channel (Ruppersberg et al. 1991). A consequent leakage of K^+ ions may lead to hyperpolarization. Such a hyperpolarization is absent when the channel is corked by the freely moving ball domain in the presence of GSH. GSSG may also oxidize and thereby inhibit neuronal voltage-sensitive Ca^{2+} channels, since it inhibits depolarization-induced changes in intracellular Ca^{2+} in cultured rat forebrain neurons and DTT reverses this effect (Gilbert et al. 1991). The nimodipine-sensitive part of the neuronal NMDA-evoked Ca^{2+} response is probably manifested via voltage-sensitive Ca^{2+} channels and reversibly inhibited by GSSG (Gilbert et al. 1991).

8. INTERACTION WITH THE RELEASE OF GLUTAMATE

GSH and cysteine (both at 1 mM) evoke release of glutamate from cultured cerebellar granule cells [138.6 ± 1.9 and 315.8 ± 10.1 percent of control, respectively (mean ± SEM]. GSH does not interfere with the release evoked by cysteine (Janáky et al., unpublished results). This indicates that the mechanism of glutathione action differs from redox modulation, although an inhibition of glutamate uptake may also contribute (Volterra et al. 1994). We assume that GSH may inhibit presynaptic glutamate autoreceptors.

9. INTERACTION WITH THE UPTAKE OF GLUTAMATE

In a crude synaptic membrane preparation, GSH and GSSG inhibit the Cl^-/Ca^{2+}-dependent, temperature-sensitive binding of glutamate (Varga et al. 1994a), which event has been considered to represent glutamate transport (Pin, Bockaert, and Récasens 1984). However, in synaptosomes GSH and GSSG (both at 1 mM) enhance the uptake of L-glutamate [176.4 ± 9.4 and 155.1 ± 4.6 percent of the control, respectively (mean $\pm$ SEM)]. One-millimolar GSSG (but not GSH) inhibits by 29.0 ± 8.0 percent the uptake of D-aspartate. L- and D-aspartate and L-glutamate inhibit both uptake processes, whereas glycine is without effect. The enhancement of synaptosomal glutamate uptake is unique for glutathione, since other γ-glutamyl peptides lacking the free thiol group, including γ-L-glutamylglycine, were generally inhibitory (Varga et al. 1994a). D-Aspartate, the unmetabolized analog of L-glutamate, has been thought to mix with the transmitter pool of endogenous glutamate and mimic its behavior (Drejer, Larsson, and Schousboe 1983; Wheeler 1984), but it is a poor substrate for vesicular transport systems of glutamate in nerve endings (Naito and Ueda 1985; Maycox et al. 1988). It is likely therefore that the vesicular, mitochondrial, and other transport systems for glutamate are differentially modulated by glutathione.

The results on glial uptake of glutamate are likewise contradictory. Volterra et al. (1994) report that extracellular GSH prevents inhibition of the glutamate uptake of astrocytes caused by oxygen free radicals. On the other hand, it has been postulated by Albrecht et al. (1993) that extracellular GSH does not interfere with glutamate uptake in astrocytes, because the SH groups critical for this process are located within the membrane.

10. INTERACTIONS WITH THE NEURONAL INFLUX AND INTRANEURONAL LEVELS OF Ca^{2+}

Neuronal excitation elevates cytoplasmic free Ca^{2+}, which triggers a number of physiological responses. A sustained depolarization and an increase in free intracellular Ca^{2+} are also causative factors in neuronal death (Orrenius et al. 1989). Three major mechanisms are involved in the increase in Ca^{2+} (Mayer and Miller 1990). Ca^{2+} ions are released into the cytosol from intracellular pools emptied by inositol triphosphate (Berridge and Irvine 1989), extracellular Ca^{2+} may enter into cells through depolarization-activated Ca^{2+} channels located in plasma membranes, and Ca^{2+} entry may also occur via agonist-activated ionotropic glutamate receptor-gated channels. Among these, the channels associated with the NMDA receptors exhibit the highest selectivity for Ca^{2+} and a voltage-dependent Mg^{2+} block, which is possibly alleviated by other cations or intracellular Ca^{2+} (Récasens et al. 1992).

Since reduction of disulfide bonds activates NMDA receptors (Lazarewicz et al. 1989; Aizenman et al. 1989, 1990; Reynolds et al. 1990; Sucher et al. 1990; see section 6), it is not surprising that thiol-reducing agents, for example, DTT and L-cysteine, may elicit epileptiform firing in hippocampal neurons (Tolliver and Pellmar 1987) and long-term potentiation (Tauck and Ashbeck 1990), and aid glutamate in killing cultured retinal ganglion cells (Levy, Sucher, and Lipton 1990; Olney et al. 1990). On the other hand, cysteine and DTT have been shown to suppress the cytotoxicity evoked by 10 mM glutamate in a neuronal cell line (Miyamoto et al. 1989; Gilbert et al. 1991). Moreover, if GSH acts as a simple thiol-reducing agent, why does it prevent NMDA receptor-mediated neurotoxicity (Levy et al. 1991)?

We have therefore studied the effects of GSH, GSSG, GSA, and S-alkyl derivatives of glutathione on the influx of Ca^{2+} into cultured cerebellar granule cells induced by glutamate and its agonists. For comparison, the possible interference of DTT and cysteine with the effects of GSH and GSSG was also subjected to study (Figs. 5 and 6, Table 3). GSH, GSSG and the glutathione derivatives tested did not markedly affect the basal influx of $^{45}Ca^{2+}$ into cultured cerebellar granule cells, DTT alone being moderately activating. In agreement with some previous data (Levy et al. 1990; Lazarewicz et al. 1989; Sucher et al. 1990), DTT strongly potentiated the enhancement of Ca^{2+} influx by glutamate and NMDA. The kainate- and quisqualate-evoked influxes were significantly less enhanced (Table 3). The higher the concentration of glutamate, the greater was the potentiation of the enhanced influx of Ca^{2+} by DTT (Fig. 6). This finding indicates that in the activated receptor complexes DTT may preferentially act at accessible disulfide bonds, probably not located within the binding site for glutamate

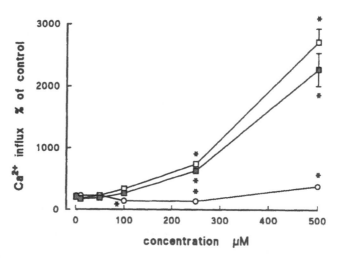

FIG. 5. Effects of various concentrations of reduced glutathione and dithiothreitol on $^{45}Ca^{2+}$ influx into cultured rat cerebellar granule cells enhanced by 1 mM glutamate. Mean values (±SEM) of three triplicate experiments. GSH (o), DTT (□), GSH + DTT (■). Statistical significance of the differences from the glutamate-activated influx without the other effectors: * $p < 0.01$. Reproduced from Janáky et al. (1993a).

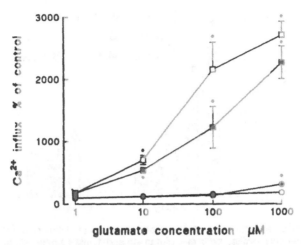

FIG. 6. Effects of reduced and oxidized glutathione and dithiothreitol on $^{45}Ca^{2+}$ influx into cultured rat cerebellar granule cells enhanced by glutamate at various concentrations. Glutamate (o); glutamate + 0.5 mM GSH (•); glutamate + 0.5 mM DTT (□); glutamate + 0.5 mM GSH + 0.5 mM DTT (■). The graph shows mean values ± SEM of triplicate experiments. Statistical significance of the differences from the glutamate-stimulated influx at each corresponding concentration: * p <0.01. Reproduced from Janáky et al. (1993a).

(Reynolds et al. 1990). Such a synergism between endogenous glutamate agonists and DTT may readily lead to a lethal perturbation of Ca^{2+} influx into neural cells.

The above results show that the perturbation of neuronal Ca^{2+} influx can be attenuated by glutathione (Figs. 5 and 6), although high (millimolar) concentrations of GSH may also slightly increase the influx evoked by millimolar (but not smaller) concentrations of glutamate (Table 3, Figs. 5, 6, and 7). On the other hand, micromolar (i.e., the physiological extracellular) concentrations of GSH slightly reduce the influx evoked by 1 mM glutamate and are already sufficient to reduce significantly the potentiation by similar concentrations of DTT (Fig. 5). GSSG (0.5 mM, Table 3), SMG

TABLE 3. *Effects of glutamate agonists, glutathione, and dithiothreitol on $^{45}Ca^{2+}$ influx into cultured rat cerebellar granule cells*

Agonist	$^{45}Ca^{2+}$ influx (% of control)				
	No effectors	GSH	DTT	GSH + DTT	GSSG
None	100 ± 9 (23)	117 ± 15 (5)	153 ± 9 (7)[a]	165 ± 29 (5)	102 ± 11 (11)
Glutamate	197 ± 11 (15)	381 ± 28 (7)[a]	2713 ± 215 (8)[b]	2271 ± 263 (10)[b]	112 ± 4 (4)[a]
Kainate	324 ± 16 (3)	351 ± 3 (3)	799 ± 27 (4)[b]	341 ± 8 (3)	179 ± 4 (4)[b]
Quisqualate	184 ± 7 (3)	193 ± 8 (3)	795 ± 21 (4)[b]	333 ± 29 (3)[a]	94 ± 3 (4)[b]
NMDA	152 ± 8 (3)	233 ± 13 (3)[a]	2496 ± 67 (4)[b]	1894 ± 41 (3)[b]	107 ± 4 (4)[a]

Data from Janáky et al. (1993a).
[a]Statistical significance of the effects of reducing agents: $p < 0.05$.
[b]$p < 0.01$.

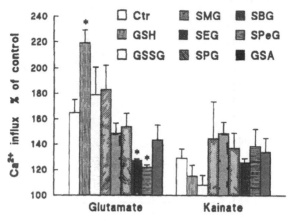

FIG. 7. Effects of glutathione and glutathione derivatives on the influx of Ca^{2+} into cultured rat cerebellar granule cells evoked by 1 mM glutamate and 1 mM kainate. Mean values (±SEM) of three triplicate experiments as the percentage enhancement of influx without any effectors. Significantly different from the corresponding control: * $p < 0.01$. Reproduced from Jenei et al. (1997).

(1 mM), and SPeG (0.5 mM) inhibit the Ca^{2+} influxes evoked by glutamate (Fig. 7). GSSG inhibits the enhancements by glutamate agonists (Table 3). As predicted by binding studies, S-alkyl derivatives of glutathione have no effect on kainate-evoked Ca^{2+} influx (Fig. 7). At a 0.5 mM concentration, GSH almost completely abolishes the DTT potentiation of the kainate-evoked and considerably attenuates that of the glutamate-, quisqualate-, and NMDA-evoked neuronal Ca^{2+} influxes (Table 3). The enhancement of NMDA-stimulated Ca^{2+} influx by 0.5 mM cysteine is likewise significantly attenuated by 0.5 mM GSH: 28 ± 7 and 23 ± 3 percent (mean ± SEM) at 1.0 and 0.1 mM NMDA, respectively (Janáky et al. 1993a). Similar inhibition by GSSG is predictable, but was not tested for because of the obvious thiol–disulfide interaction with DTT molecules. Instead, we tested the effect of SMG (0.5 mM), which significantly attenuates the enhancement of 1 mM NMDA–evoked responses by DTT and cysteine: 40 ± 8 and 25 ± 6 percent, respectively.

We conclude that GSH can bind to the recognition sites in the NMDA receptors as a γ-glutamyl compound (Varga et al. 1989, 1992), displace glutamate, and hence diminish the number of accessible thiol groups. Such an inference is in keeping with the above results and the assumption that glutathione modulates glutamatergic neurotransmission by displacing glutamate from its binding sites (Ogita et al. 1986; Oja et al. 1988; Varga et al. 1989; see also Ogita et al., chapter 7, this volume). GSH itself also causes a slight activation by reducing disulfide bonds in the vicinity of the γ-glutamyl moiety of the bound GSH molecule, where it competes with other more effective disulfide-reducing agents. This is consistent with the finding that GSH mimics the DTT effects in NMDA receptors with NR1-NR2A (but not NR1-NR2B,C or D) subunit composition, in which DTT rapidly potentiates glutamate-activated whole-cell currents and reduces the time course of desensitization and reactivation (Köhr et al. 1994). GSSG (Oja et al. 1988, 1994; Varga et al. 1989, 1997) and S-alkyl deriva-tives of glutathione (Oja et al. 1995; Jenei et al. 1997) may also displace glutamate and diminish receptor activation, but have no enhancing effects because of the lacking free SH group. The contribution of such a displacement of agonists to the inhibition of glutamate-evoked influx of Ca^{2+} into neurons is corroborated by the finding that DTNB, a sulfhydryl-oxidizing agent, has only a slight inhibitory effect (Janáky et al. 1993a). Nevertheless, GSSG may also oxidize neighboring thiol groups. This assump-tion is in concert with the results of Gilbert et al. (1991) obtained on primary cultures of forebrain neurons. In their experiments, GSSG attenuated the NMDA-evoked in-crease in intracellular free Ca^{2+}. This effect was still apparent after oxidation of the cells by DTNB, but completely reversed only by addition of DTT. Nor can we exclude the possibility of intracellular regulatory actions of GSSG and GSH [e.g., interaction with the intracellular regulatory sites of ion channels (Gilbert et al. 1991; Ruppersberg et al. 1991); see section 7], even though GSH transport has been hitherto demonstrated only in certain nonneural cells (Boobis, Fawthrop, and Davies 1989; Mårtensson et al. 1989; Albrecht et al. 1993; Kannan et al., chapter 3, this volume).

Although the results on the use-dependent binding of dizocilpine may indicate an NMDA agonistic action of GSH, GSSG (Ogita et al. 1995; Varga et al. 1997) and glutathione derivatives (Jenei et al. 1997), their effects on the neuronal influx and intraneuronal level of Ca^{2+} support rather a biphasic concentration-dependent action of GSH and an inhibitory action of GSSG and S-alkyl derivatives of glutathione. The activation of dizocilpine binding by glutathione derivatives may represent a change in the receptor conformation not necessarily involving any enhanced Ca^{2+} influx and intracellular elevation of free Ca^{2+}. We would be inclined to infer that the tested glutathione analogs are antagonistic in nature.

11. INTERACTIONS WITH THE Ca^{2+}-DEPENDENT EVOKED RELEASE OF OTHER NEUROTRANSMITTERS

Glutamate affects the release of several other neurotransmitters. This action may also be subject to modulation by glutathione. Although the mechanisms of the redox modulation have not been completely elucidated (see section 6), the addition of DTT to cortical, hippocampal, and striatal slices results in an increase in the NMDA-evoked release of noradrenaline and dopamine. These effects are associated with an enhancement in the activity of the NMDA receptor through its redox modulatory sites (Woodward and Blair 1991; Woodward 1994). Glutathione may thus also interact with the effects of DTT by means of competition at the NMDA redox sites.

11.1 Modulation of the Glutamate-Evoked Release of GABA in the Hippocampus

Both Ca^{2+}-dependent and seemingly Ca^{2+}-independent releases of GABA can be induced by glutamate and its agonists in the hippocampus (Janáky, Saransaari, and Oja

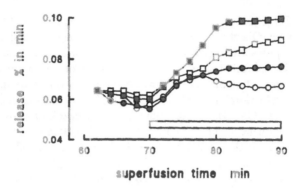

FIG. 8. Time course of the quisqualate-evoked release of [³H]GABA from rat hippocampal slices. The slices were superfused with 0.1 mM Mg²⁺ medium, supplemented during the last 20 min by 1.0 mM quisqualate in the absence of DTT and GSH (●), and in the presence of 1 mM GSH (o), 1 mM DTT (■), or 1 mM GSH and 1 mM DTT (□), as indicated by the bar. The release, mean values of 6–13 experiments, is shown as a percentage of the total radioactivity remaining in the slices at each time point. SEM varied from 5 to 10 percent of the mean. Reproduced from Janáky et al. (1994b).

1993b). Activation of both ionotropic NMDA and non-NMDA receptors is involved in the transient Ca²⁺-dependent release of GABA from hippocampal slices, which can contribute to physiological feedback and feedforward inhibition. Quisqualate evokes a sustained release of GABA via activation of metabotropic glutamate receptors (Janáky 1994; Janáky et al. 1994a), which can contribute to disinhibition and be involved in the generation of LTP.

DTT, GSH, and GSSG have no effect on the basal release of [³H]GABA from hippocampal slices. Coadministration of DTT with glutamate agonists enhances and prolongs the evoked release. The effect of DTT is dose-dependent. A 10-min pre-treatment with 1 mM DTT followed by washing has no effect on the evoked release. The effect of DTT is also discernible in the absence of external Ca²⁺. It is insensitive to 2-amino-3-phosphonopropionate, an NMDA antagonist, but blocked by MK-801 and Mg²⁺ (Janáky et al. 1994b). In contrast to DTT, GSH, and GSSG do not enhance the release. Moreover, they generally attenuate the enhancement by DTT (Fig. 8). GSH attenuates the release evoked by 1 mM glutamate and glycine, and GSSG the release evoked by quisqualate, glycine, and NMDA together with glycine. The effect of DTT on the kainate-evoked release of GABA is not modified by GSH or GSSG (Janáky et al. 1994b).

11.2 Modulation of the Glutamate-Evoked Release of Dopamine in the Striatum

The excitatory glutamatergic input from the cerebral cortex to the neostriatum (DiChiara and Morelli 1993) controls dopamine release (Chéramy et al. 1994; Morari et al. 1996) in a complex manner. A triple regulation has been proposed to exist in vivo (Leviel, Gobert, and Guibert 1990). The cortical glutamatergic pathway exerts tonic inhibition on GABAergic interneurons mediated by activation of the NMDA receptors. Low concentrations of glutamate temporarily activate dopamine release via AMPA/kainate receptors at dopaminergic nerve endings, and high concentrations of glutamate evoke complex inhibition, possibly via GABAergic interneurons. An indirect facilitation mediated by somatostatinergic and cholinergic interneurons has also been proposed. The glutamatergic regulation of striatal dopamine release thus involves both direct and indirect facilitation and indirect inhibition by glutamate receptors of different types of striatal interneurons and different receptors at nigrostriatal dopaminergic synaptic terminals (Leviel et al. 1990; Krebs et al. 1991; Chéramy et al. 1994; O'Connor, Drew, and Ungerstedt 1995). Such a triple regulation forms the basis for the exact timing of dopamine release from the nigrostriatal nerve endings.

A variety of interactions of glutathione, both direct and indirect, with the striatal dopamine release regulated by glutamate are possible. In our in vitro studies glutamate, kainate, and AMPA evoked [^{3}H]dopamine release in a concentration- and time-dependent manner, kainate and AMPA (0.5 mM) being significantly more effective than glutamate (1 mM). NMDA, glycine, and the metabotropic glutamate receptor agonist *trans*-1-aminocyclopentane-1,3-dicarboxylate (tACPD) (0.5 mM) were without effect in standard Krebs–Ringer–Hepes solution (Table 4). γ-Glutamylcysteine, GSH, GSSG, DTT, DTNB, and L-cysteine (all 1 mM) were without any marked effect. Likewise, the antagonists of different ionotropic glutamate receptors, for example, CNQX, DNQX, NBQX, APV, and dizocilpine (all 0.1 mM), failed to affect the basal release (Janáky et al. 1997). The release evoked by 1 mM glutamate was strongly enhanced by 1 mM GSH (Fig. 9, Table 5) and less effectively by 1 mM L-cysteine.

TABLE 4. Effects of glutamate receptor agonists on the release of [^{3}H]dopamine from the mouse striatum

Effector, mM	Efflux rate constants ± SD (n), % of k_1 (60–70 min)	
	k_2 (72–80 min)	k_3 (80–90 min)
None (control)	66 ± 17 (25)	49 ± 13 (24)
Glutamate, 1	99 ± 22 (24)[b]	90 ± 25 (22)[b]
NMDA, 1	76 ± 34 (8)	60 ± 29 (8)
Kainate, 0.1	89 ± 11 (7)[b]	63 ± 6 (7)[b]
Kainate, 0.5	131 ± 17 (7)[b]	95 ± 13 (7)[b]
Kainate, 1	168 ± 14 (7)[b]	102 ± 9 (7)[b]
AMPA, 0.1	83 ± 21 (7)[a]	70 ± 18 (7)[b]
AMPA, 0.5	141 ± 26 (7)[b]	100 ± 9 (6)[b]
t-ACPD, 0.1	60 ± 11 (4)	49 ± 8 (4)
t-ACPD, 0.5	72 ± 52 (8)	61 ± 36 (8)

The fractional efflux rate constants k_2 and k_3 for the superfusion periods of 72–80 min (early stimulation phase) and 80–90 min (late stimulation phase) in percent of the constant k_1 (60–70 min, prestimulation phase). Data from Janáky et al. (1997).
[a]Statistical significance of the effects: $p < 0.05$.
[b]$p < 0.01$.

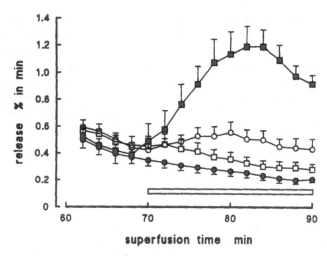

FIG. 9. Glutamate-evoked release of [³H]dopamine from the mouse striatum: effect of reduced glutathione. The results (means ± SEM) of eight experiments are shown as percentages of the total radioactivity released per minute. (●) Control, (○) 1 mM glutamate, (□) 1 mM GSH, (■) 1 mM glutamate + 1 mM GSH. The effectors were present as indicated by the bar. The experiments were carried out as described in Janáky et al. (1997).

GSSG, glycine, DTT, DTNB, and L-cystine were without effect. γ-Glutamylcysteine is a weak inhibitor (Janáky et al. 1997).

The slight enhancing effect (20 ± 4 percent, mean $\pm$ SD, $n = 12$) of NMDA observed in 0.1 mM Mg^{2+} medium in the presence of $50\ \mu$M glycine is not discernible in standard Krebs–Ringer–Hepes solution (Table 4). This block is relieved in the presence of GSH, but not GSSG (Janáky et al., unpublished results). The release evoked by 1 mM kainate is enhanced by GSSG (Table 5), but only during the late stimulation phase by GSH. It is inhibited by CNQX and DNQX, NBQX being without effect (Janáky et al. 1997). The release evoked by 0.5 mM AMPA is enhanced by GSSG, GSH being ineffective (Table 5). t-ACPD fails to influence the release of dopamine in all conditions (Janáky et al. 1997).

TABLE 5. Effects of glutathione on the release of [³H]dopamine evoked by 1 mM glutamate, NMDA, kainate, and 0.5 mM AMPA from the mouse striatum

| Effector | Efflux rate constants ± SD (n): k_2 (72–80 min), % of k_1 (60–70 min) | | |
	None (control)	GSH	GSSG
Glutamate	99 ± 17 (24)	229 ± 30 (9)[b]	90 ± 16 (7)
NMDA	76 ± 34 (8)	121 ± 19 (6)[a]	91 ± 5 (8)
Kainate	168 ± 14 (7)	158 ± 14 (8)	200 ± 28 (8)[a]
AMPA	141 ± 26 (7)	156 ± 15 (6)	178 ± 14 (7)[b]

Data from Janáky et al. (1997).
[a]Significance of differences from control: $p < 0.05$.
[b]$p < 0.01$.

The studies cited above indicate that striatal dopamine release is regulated by activation of both NMDA and non-NMDA classes of glutamate receptors. This is in concert with both direct and indirect facilitation of dopaminergic nerve terminals or interneurons via glutamate receptors. The results show that striatal dopamine release is enhanced by both reduced and oxidized glutathione, albeit probably by means of different mechanisms and mediated by different glutamate receptors. GSH may act in several ways:

1. Directly via the NMDA receptors on dopamine nerve endings, enhancing the activity solely as a thiol reagent. This is less probable, because DTT fails to facilitate the NMDA effect.
2. Directly as an agonist at the non-NMDA receptors on dopamine nerve endings with a consequent relief of the Mg^{2+} block of NMDA receptors.
3. Indirectly as an agonist at the NMDA receptors on facilitatory somatostatinergic and cholinergic interneurons.
4. Indirectly as a γ-glutamyl competitive antagonist at the NMDA receptors on inhibitory GABAergic interneurons.

Although we cannot at present exclude any of the above mechanisms, the fact that S-ethyl, -propyl, and -butyl derivatives of glutathione are inhibitors of striatal dopamine release indicates the possible importance of both γ-glutamyl and cysteinyl moieties in the enhancement of glutamate-evoked dopamine release. On the other hand, the presence of the γ-glutamyl moiety is necessary for the agonistic action of GSSG exerted at the non-NMDA receptors on the dopaminergic terminals.

11.3 Modulation of the K^+-Evoked Release of Dopamine in the Striatum

In mouse striatal slices, K^+ ions evoke dopamine release in a concentration- and time-dependent manner. The release evoked by 25 mM K^+ is inhibited by 1 mM GSA, GSH, GSSG, and S-butyl, -pentyl, -propyl, and -ethyl derivatives of glutathione in decreasing order of potency. γ-Glutamylcysteine, glycine, DTNB, and L-cystine (all 1 mM) exhibit a moderate efficacy, while SMG and L-cysteine are without effect. The thiol antioxidant DTT is a strong activator (Table 6). A 10-min preperfusion of the striata with 50 μM $ZnCl_2$ has no effect on the K^+-evoked release, but prevents activation by 1 mM DTT applied for the subsequent 10 min (Janáky et al., unpublished results).

It is surprising that, opposite to the physiological upregulation via glutamate receptors (see earlier), both GSH and GSSG inhibit the depolarization-evoked release of dopamine from the striatum. This inhibition may be a consequence of glutathione–membrane protein interaction(s), or inhibition of voltage-dependent Ca^{2+} channels (Gilbert et al. 1991). In any case, the mechanism of this inhibition is different from that of DTT, namely, it is not thiol redox modulation and formation of complexes with Zn^{2+} ions, which even enhance the K^+-evoked dopamine release. Intracellular redox modulation of K^+ channels (Ruppersberg et al. 1991) is likewise not probable, because GSH and GSSG have both been seen to be inhibitory and glutathione (opposite to DTT) does not penetrate across the plasma membrane.

TABLE 6. *Potassium-evoked (25 mM) release of [³H]dopamine from the mouse striatum: effects of glutathione, glutathione derivatives, and thiol redox agents*

Effector, 1 mM	Efflux rate constants ± SD (n), % of k_1 (60–70 min)	
	k_2 (72–80 min)	k_3 (80–90 min)
None (control)	637 ± 72 (21)	487 ± 61 (21)
GSH	353 ± 37 (7)[b]	358 ± 32 (7)[b]
GSSG	358 ± 47 (7)[b]	289 ± 76 (7)[b]
SMG	673 ± 193 (16)	479 ± 89 (16)
SEG	554 ± 150 (15)[a]	467 ± 110 (15)
SPG	457 ± 10 (7)[a]	360 ± 103 (6)[b]
SBG	360 ± 111 (4)[b]	258 ± 47 (4)[b]
SPeG	411 ± 47 (4)[b]	230 ± 10 (4)[b]
GSA	335 ± 72 (4)[b]	186 ± 16 (4)[b]
γ-L-Glutamylcysteine	486 ± 135 (4)[b]	422 ± 83 (4)
Glycine	471 ± 69 (8)[b]	319 ± 78 (8)[b]
DTT	799 ± 252 (16)[b]	679 ± 114 (16)[b]
DTNB	483 ± 92 (11)[b]	375 ± 74 (11)[b]
L-Cysteine	707 ± 145 (8)	506 ± 89 (8)
L-Cystine	506 ± 66 (8)[b]	372 ± 47 (8)[b]

Results from unpublished experiments carried out as in Janáky et al. (1997).
[a]Statistical significance of differences from the corresponding control: $p < 0.05$.
[b]$p < 0.01$.

12. GLUTATHIONE: NEUROTRANSMITTER OR NEUROMODULATOR?

There is now evidence to suggest that GSH itself might be involved in synaptic transmission as a transmitter (see Pasqualotto et al., chapter 9, this volume). Glutathione is bound to synaptic membranes. S-Methyl-GSH and S-hexyl-GSH completely abolish this binding (Ogita and Yoneda 1987, 1988, 1989). As dissociation of bound GSH from its binding site is slow, GSH could exert a long-term effect on synaptic transmission. In keeping with such a concept, the inhibitory effects of SMG and GSA on CPP binding are likewise only slowly reversible (Jenei et al. 1997). GSH also possesses saturable binding sites in the dorsal and ventral horns of the mouse and human spinal cord. The bound GSH is displaceable by GSH, cysteine, and S-hexyl-GSH, but not by glutamate, NMDA, or glycine. The binding is enhanced by activation of protein kinase C. These findings corroborate the assumption of the presence of specific GSH receptors not identical to any known glutamate binding sites (Wagey et al. 1993; Lanius et al. 1993, 1994). As neurotransmitters, GSH and GSSG may stimulate phospholipase C activity and inositol triphosphate metabolism (Guo et al. 1992). The possible existence of glial GSH receptors (Guo and Shaw 1992; Guo et al. 1992) might also afford GSH a role in neuronal–glial communication. On the other hand, the data summarized above witness unequivocally that glutathione belongs to the endogenous γ-glutamyl peptides, which are ligands of ionotropic glutamate receptors. They may regulate the presynaptic release and uptake of excitatory amino acids, neuronal Ca^{2+} responses, and release of other neurotransmitters evoked by these receptors. We therefore infer that GSH and GSSG are likely to be both neurotransmitters and neuromodulators acting at different receptors and affecting different cellular second-messenger systems.

13. THE ROLE OF GLUTATHIONE IN THE REGULATION OF PHYSIOLOGICAL PROCESSES IN THE CNS

There is little definite knowledge regarding the role of glutathione in the regulation of physiological processes in the CNS. Glutathione—as a neuromodulator acting mainly at the AMPA and NMDA receptors—may affect wakefulness (Honda et al. 1994). Our results indicate, however, that released glutathione may also regulate neural signal transduction in the vicinity of its release site. The coincidence-detecting function of NMDA receptors (Seeburg 1993) may be one crucial point of this glutathione intervention, since glutathione displaces glutamate from both NMDA and AMPA sites and inhibits the neuronal Ca^{2+} responses. We assume therefore that glutathione at micromolar concentrations in the extracellular space, due to depolarization or hypoxia-evoked release (Andiné et al. 1991; Zängerle et al. 1992) or leakage from glial cells (Yudkoff et al. 1990), may exert a long-lasting or even almost permanent suppression of the pre- and postsynaptic AMPA and NMDA receptors, which must be overcome by glutamate. As a consequence, the elimination of glutamate from the synaptic cleft may not be the only means of abolishing glutamatergic excitation. The possible block of presynaptic receptors and the consequent sensitization of glutamate release together with an increase in the postsynaptic threshold for NMDA receptor activation could increase the specificity of the postsynaptic responses. At the same time, extracellular glutathione—similarly to Mg^{2+} ions—would provide a defense against glutamate excitotoxicity. The regional and age-dependent changes in the extracellular concentrations of glutathione may thus determine both the plasticity and the vulnerability of neurons to excitotoxic agents.

14. NEW ASPECTS OF NEUROPROTECTION BY GLUTATHIONE

14.1 Glutamate Neurotoxicity and Neurological Diseases

Under pathological conditions (e.g., in hypoxia and hypoglycemia) and in certain neurodegenerative diseases (e.g., in epilepsies and amyotrophic lateral sclerosis—see Lipton and Rosenberg 1994), the release of glutamate is generally increased (Globus et al. 1991) and/or the uptake decreased (Nicholls and Attwell 1990). The consequent converging and long-lasting overactivation of non-NMDA and metabotropic glutamate receptors sensitize and overstimulate NMDA-governed ionophores, leading to harmful positive feedback mechanisms (e.g., in the second-messenger systems after NMDA receptor activation) and to Ca^{2+}-dependent and osmotic neuronal destruction (reviewed in Janáky 1994). In addition to this, extracellular glutamate competitively inhibits the high-affinity uptake of cystine, leading to glutathione depletion (Sagara,

Miura, and Bannai 1993), which kills neurons (Murphy, Schnaar, and Coyle 1990) and oligodendroglial cells (Oka et al. 1993). Such a loss of GSH has recently been thought to be an important mechanism in glutamate toxicity (Murphy et al. 1990). It has been suggested that not only acute but also chronic GSH depletion may lead to neurodegeneration [e.g., during normal aging (Ravindranath, Shivakumar, and Anandatheerthavarada 1989; Benzi et al. 1992; Benzi and Moretti 1995; see also Benji and Moreth, chapter 11, this volume)] and may be a pathogenic factor in neuropsychiatric diseases [e.g., in Parkinson's disease (Maker et al. 1981; Perry, Godin, and Hansen 1982; Spencer, Jenner, and Halliwell 1995; Owen et al. 1996)]. The neuroprotection by intracellular GSH is thus no longer a matter of conjecture.

In addition to the above considerations, Levy et al. (1991) reported that extracellular glutathione provides protection against NMDA receptor-mediated neurotoxicity without penetrating across the cell membrane (Albrecht et al. 1993). However, in their study nonphysiologically high concentrations of GSH (10 mM) and GSSG (5 mM) were applied to prevent neuronal death. We assume that extracellular GSH and GSSG may also be neuroprotective at lower concentrations. Besides its general action mechanisms discussed above, glutathione exerts other beneficial effects on neurotransmission in some pathological processes.

14.2 Glutathione and GABA Release in Pathological States

In the hippocampus—possibly the most vulnerable part of the brain—overexcitation of glutamate receptors evokes massive release of GABA preceding neuronal death (Margaill et al. 1992). This GABA release may cause disinhibition, that is, it may exempt principal hippocampal excitatory neurons from GABAergic inhibition because of desensitization of presynaptic $GABA_A$ receptors. The feedforward excitation, which elevates intracellular Ca^{2+} in principal cells, contributes to this desensitization. This may lead to seizures, further energy depletion, and neuronal death. Our results indicate that reducing agents elicit a sustained, both Ca^{2+}-dependent and apparently Ca^{2+}-independent, release of GABA from hippocampal slices (Janáky et al. 1994b). Ca^{2+} ions needed for Ca^{2+}-dependent release may come from the extracellular space through overactivated NMDA-gated ionophores. The apparently Ca^{2+}-independent release may in fact be due to Ca^{2+} released from intracellular stores, for example, from dysfunctional mitochondria. The sustained GABA release is obviously nonphysiological and could in vivo cause harmful disinhibition and excitotoxicity. This may explain the epileptogenic and neurotoxic effects of DTT and cysteine. GSH and GSSG inhibit such a sustained release of GABA and can consequently disrupt these harmful disinhibition processes. If glutathione is depleted by an excess of extracellular glutamate (see section 14.1), the accessibility and redox state of the NMDA receptors is not balanced and the increased hippocampal aminoacidergic transmission escapes normal control. We suggest therefore that extracellular GSH and GSSG as neuromodulators balance the excitatory–inhibitory input of hippocampal neurons and determine both their plasticity and their vulnerability.

14.3 Glutathione and Striatal Dopamine Release: Implications in Extrapyramidal Neurodegenerative Diseases

Parkinson's disease is an adult-onset neurodegenerative disorder with progressive motor decline. It is characterized by a selective loss of dopaminergic neurons in the substantia nigra. Our results cited above seem to indicate that the enhancement of striatal dopamine release by glutathione may be of physiological importance in the regulation of extrapyramidal motor functions. Glutathione may act as an anti-Parkinsonian compound enhancing the striatal release of dopamine (Janáky et al. 1997). On the other hand, there is no striatal GSH deficiency in Parkinson's disease (Owen et al. 1996), which finding is not in keeping with the assumption that such a process participates in the pathogenesis of this disease. GSSG, the level of which is decreased to 50 percent in the caudate nucleus in patients with Huntington's disease (Sian et al. 1994), also potentiates the glutamate-evoked release of dopamine. The significance of this finding needs further evaluation.

14.4 Glutathione and Spreading Depolarization

Spreading depolarization is crucial for neuronal survival, being the last "cry for help" sign of neuronal and glial exhaustion caused by the excessive release of K^+ ions from neurons. In this case, the capacity of astrocytes is no longer sufficient for effective reuptake of K^+ ions. This sequence of events evokes more release of neurotransmitters from the neighboring neurons, followed by further depolarization and neural exhaustion. The mechanism of this harmful cascade is unknown, and its prevention remains an open question (Rader and Lanthorn 1989). Since the extracellular K^+ concentration rises to 3–15 mM in anoxic depolarization, application of 25 mM K^+ as in our experiments (see section 11.3) may serve as a model of this pathological process. Here we have presented evidence that GSH, GSSG, and glutathione derivatives inhibit the release of dopamine evoked by high concentrations of potassium. If this is also the case in vivo and with the release of other transmitters, then glutathione released in hypoxic conditions (Andiné et al. 1991) and upon depolarization (Zängerle et al. 1992) may forestall the final neuronal depression (dumbness), a lethal consequence of extreme transmitter release.

15. CONCLUSIONS AND PERSPECTIVES

We suggest that glutathione acts not only as an antioxidant or disulfide-reducing agent but also as a neurotransmitter and neuromodulator in the CNS. Glutathione and the endogenous glutathione analogs SMG and GSA regulate the functions of AMPA and NMDA receptors. In view of the colocalization of these receptors, we are of the of opinion that glutathione and its endogenous derivatives may affect the coincidence-detecting function of NMDA receptors possibly involved in neuronal information processing. The glutamate-antagonist activity of glutathione provides the basis for the sharpening of neuronal responses to glutamate and for defense against glutamate toxicity. The other nontoxic S-alkyl derivatives of glutathione are compounds that can be used as synthetic ligands of the AMPA and NMDA classes of ionotropic glutamate receptors. Since they are more hydrophobic than glutathione itself and hence readily cross the blood–brain barrier, they may prove to be antiepileptic and neuroprotective in the stroke and neurodegenerative diseases involving glutamate neurotoxicity and pathological overexcitation of NMDA receptors.

In the future, the role of extracellular GSH should be taken into account in more sophisticated studies of mammalian CNS, especially in Parkinson's disease, ischemia, and mental retardation. The possible interaction of glutathione with metabotropic glutamate receptors and the involvement of second messengers other than Ca^{2+} in its action are also provocative questions still awaiting answers. The evaluation of interactions of glutathione as a neuromodulator with the release of other neurotransmitters, ontogenetic and phylogenetic follow-up studies of GSH binding, behavioral studies on glutathione actions, design and characterization of active glutathione analogs and evaluation of their possible therapeutic potencies in diseases connected with glutamate or cysteine toxicity are further challenges for investigators.

ACKNOWLEDGMENTS

The skillful technical assistance of Ms. Oili Pääkkönen, Ms. Irma Rantamaa, and Ms. Sari Perttunen and the financial support of the Emil Aaltonen Foundation and the Medical Research Fund of Tampere University Hospital are gratefully acknowledged.

REFERENCES

Aizenman, E., Lipton, S. A., and Loring, R. H. 1989. Selective modulation of NMDA responses by reduction and oxidation. *Neuron* 2:1257–63.

Aizenman, E., Hartnett, K. A., and Reynolds, I. J. 1990. Oxygen free radicals regulate NMDA receptor function via a redox modulatory site. *Neuron* 5:841–6.

Albrecht, J., Talbot, M., Kimelberg, H. K., and Ascher, M. 1993. The role of sulfhydryl groups and calcium in the mercuric chloride–induced inhibition of glutamate uptake in rat primary astrocyte cultures. *Brain Research* 607:249–54.

Andiné, P., Orwar, O., Jacobson, I., Sandberg, M., and Hagberg, H. 1991. Extracellular acidic sulfur-containing amino acids and γ-glutamyl peptides in global ischemia: postischemic recovery of neuronal activity is paralleled by a tetrodotoxin-sensitive increase in cysteine sulfinate in the CA_1 of the rat hippocampus. *Journal of Neurochemistry* 57:230–6.

Becker, K., Gui, M., and Schirmer, R. H. 1995. Inhibition of human glutathione reductase by S-nitrosoglutathione. *European Journal of Biochemistry* 234:472–8.

Benzi, G., and Moretti, A. 1995. Age- and peroxidative stress–related modifications of the cerebral enzymatic activities linked to mitochondria and the glutathione system. *Free Radical Biology and Medicine* 19:77–101.

Benzi, G., Pastoris, O., Marzatico, F., Villa, R. F., Dagani, F., and Curti, D. 1992. The mitochondrial electron transfer alteration as a factor involved in the brain aging. *Neurobiology of Aging* 13:361–8.

Berridge, M. J., and Irvine, R. F. 1989. Inositol phosphates and cell signalling. *Nature* 341:197–205.

Boobis, A. R., Fawthrop, D. J., and Davies, D. S. 1989. Mechanisms of cell death. *Trends in Pharmacological Sciences* 10:275–80.

Bredt, D. S., and Snyder, S. H. 1992. Nitric oxide, a novel neuronal messenger. *Neuron* 8:3–11.

Canonico, P. L., Judd, A. M., Koike, K., Valdenergo, C. A., and MacLeod, R. M. 1985. Arachidonate stimulates prolactin release in vitro: a role for the fatty acid and its metabolites as intracellular regulator(s) in mammotrophs. *Endocrinology* 116:218–25.

Cappel, R. E., and Gilbert, H. F. 1988. Thiol/disulfide exchange between 3-hydroxy-3-methylglutaryl-CoA reductase and glutathione. A thermodynamically facile dithiol oxidation. *Journal of Biological Chemistry* 263:12204–12.

Chan, P. H., Kerlan, R., and Fishman, R. A. 1983. Reduction of γ-aminobutyric acid and glutamate uptake and (Na^+-K^+)-ATPase activity in brain slices and synaptosomes by arachidonic acid. *Journal of Neurochemistry* 40:309–316.

Chen, L., and Huang, L. Y. 1992. Protein kinase C reduces Mg^{2+} block of NMDA-receptor channels as a mechanism of modulation. *Nature* 356:521–3.

Chéramy, A., Desce, J. M., Godeheu, G., and Glowinski, J. 1994. Presynaptic control of dopamine synthesis and release by excitatory amino acids in rat striatal synaptosomes. *Neurochemistry International* 25:145–54.

Collingridge, G. 1987. Synaptic plasticity. The role of NMDA receptors in learning and memory. *Nature* 330:604–5.

Conn, P. J., and Desai, M. A. 1991. Pharmacology and physiology of metabotropic glutamate receptors in mammalian central nervous system. *Drug Development Research* 24:207–29.

Cunningham, M. D., Ferkany, J. W., and Enna, S. J. 1994. Excitatory amino acid receptors: a gallery of new targets for pharmacological intervention. *Life Sciences* 54:135–48.

Davies, J., and Watkins, J. C. 1981. Differentiation of kainate and quisqualate receptors in the cat spinal cord by selective antagonism with γ-D (and L) glutamylglycine. *Brain Research* 206:172–7.

Deneke, S. M., and Fanburg, B. L. 1989. Regulation of cellular glutathione. *American Journal of Physiology* 257:L163–73.

DiChiara, G., and Morelli, M. 1993. Dopamine–acetylcholine–glutamate interactions in the striatum: a working hypothesis. *Advances in Neurology* 60:102–6.

Drejer, J., Larsson, O. M., and Schousboe, A. 1983. Characterization of uptake and release processes for D- and L-aspartate in primary cultures of astrocytes and cerebellar granule cells. *Neurochemical Research* 8:231–43.

Dumuis, A., Sebben, M., Haynes, L., Pin, J. P., and Bockaert, J. 1988. NMDA receptors activate the arachidonic acid cascade system in striatal neurons. *Nature* 336:68–70.

Fagg, G. E., and Foster, A. C. 1983. Amino acid neurotransmitters and their pathways in the mammalian central nervous system. *Neuroscience* 9:701–19.

Folbergrova, J., Rehncrona, S., and Siesjö, B. K. 1979. Oxidized and reduced glutathione in the rat brain under normoxic and hypoxic conditions. *Journal of Neurochemistry* 32:1621–70.

Garthwaite, J. 1991. Glutamate, nitric oxide and cell–cell signalling in the nervous system. *Trends in Neuroscience* 14:60–7.

Gilbert, H. F. 1982. Biological disulfides: the third messenger? Modulation of phosphofructokinase activity by thiol/disulfide exhange. *Journal of Biological Chemistry* 257:12086–91.

Gilbert, K. R., Aizenman, E., and Reynolds, I. J. 1991. Oxidized glutathione modulates N-methyl-D-aspartate- and depolarization-induced increases in intracellular Ca^{2+} in cultured rat forebrain neurons. *Neuroscience Letters* 133:11–4.

Globus, M. Y., Busto, R., Martinez, E., Valdes, I., Dietrich, W. D., and Ginsberg, M. D. 1991. Comparative effect of transient global ischemia on extracellular levels of glutamate, glycine and γ-aminobutyric acid in vulnerable and nonvulnerable brain regions in the rat. *Journal of Neurochemistry* 57:470–8.

Gozlan, H., and Ben-Ari, Y. 1995. NMDA receptor redox sites: are they targets for selective neuronal protection? *Trends in Pharmacological Sciences* 16:368–74.

Guo, N., and Shaw, C. 1992. Characterization and localization of glutathione binding sites on cultured astrocytes. *Molecular Brain Research* 15:207–15.

Guo, N., McIntosh, C., and Shaw, C. 1992. Glutathione: new candidate neuropeptide in the central nervous system. *Neuroscience* 51:835–42.

Hjelle, O. P., Chaudry, F. A., and Ottersen, O. P. 1994. Antisera to glutathione: characterization and immunocytochemical application to the rat cerebellum. *European Journal of Neuroscience* 6:793–804.

Hogg, N., Singh, R. J., and Kalyanaraman, B. 1996. The role of glutathione in the transport and catabolism of nitric oxide. *FEBS Letters* 382:223–8.

Honda, K., Komoda, Y., and Inoué, S. 1994. Oxidized glutathione regulates physiological sleep in unrestrained rats. *Brain Research* 636:253–8.

Jain, A., Mårtensson, J., Stole, E., Auld, P. A., and Meister, A. 1991. Glutathione deficiency leads to mitochondrial damage in brain. *Proceedings of the National Academy of Sciences of the U.S.A.* 88: 1913–7.

Jankáky, R. 1994. Glutamate agonist-evoked calcium influx in the cerebellum and release of GABA in the hippocampus. Modulation of the receptor functions by thiol-modulating agents. *Acta Universitatis Tamperensis, Series A* 403:1–84.

Jankáky, R., Varga, V., Saransaari, P., and Oja, S. S. 1993a. Glutathione modulates the N-methyl-D-aspartate receptor-activated calcium influx into cultured rat cerebellar granule cells. *Neuroscience Letters* 156:153–7.

Jankáky, R., Saransaari, P., and Oja, S. S. 1993b. Release of GABA from rat hippocampal slices: involvement of quisqualate/N-methyl-D-aspartate-gated ionophores and extracellular magnesium. *Neuroscience* 53:779–85.

Jankáky, R., Varga, V., Saransaari, P., and Oja, S. S. 1994a. Glutamate agonists and [^{3}H]GABA release from rat hippocampal slices: involvement of metabotropic glutamate receptors in the quisqualate-evoked release. *Neurochemical Research* 19:729–34.

Jankáky, R., Varga, V., Oja, S. S., and Saransaari, P. 1994b. Release of [^{3}H]GABA evoked by glutamate agonists from hippocampal slices: effects of dithiothreitol and glutathione. *Neurochemistry International* 24:575–82.

Jankáky, R., Varga, V., Hermann, A., Serfőző, Z., Dohovics, R., Saransaari, P., and Oja, S. S. 1997. Effects of glutathione on [^{3}H]dopamine release from the mouse striatum evoked by glutamate receptor agonists. In *Proceedings of the 11th ESN Meeting*, Groningen. In press.

Jenei, Zs., Jankáky, R., Varga, V., Saransaari, P., and Oja, S. S. 1997. Interference of S-alkyl derivatives of glutathione with brain ionotropic glutamate receptors. *Neurochemical Research*. In press.

Jones, S. M., Snell, L. D., and Johnson, K. M. 1987. Phencyclidine selectively inhibits N-methyl-D-aspartate-induced hippocampal [^{3}H]noradrenaline release. *Journal of Pharmacology and Experimental Therapeutics* 240:492–7.

Kanazawa, A., Kakimoto, Y., Nakajima, T., and Sano, I. 1965. Identification of γ-glutamylserine, γ-glutamylalanine, γ-glutamylvaline and S-methylglutathione of bovine brain. *Biochimica et Biophysica Acta* 111:90–5.

Kirstein, C. L., Coopersmith, R., Bridges, R. J., and Leon, M. 1991. Glutathione levels in olfactory and non-olfactory neural structures of rats. *Brain Research* 543:341–6.

Kiskin, N. I., Kristal, O. A., Tsyndrenko, A. Ya., and Akaike, N. 1986. Are sulfhydryl groups essential for function of the glutamate-operated receptor–ionophore complex? *Neuroscience Letters* 66:305–10.

Köhr, G., Eckardt, S., Lüddens, H., Monyer, H., and Seeburg, P. H. 1994. NMDA receptor channels: subunit-specific potentiation by reducing agents. *Neuron* 12:1031–40.

Kosower, N. S., and Kosower, E. M. 1978. The glutathione status of cells. *International Review of Cytology* 54:109–60.

Krebs, M. O., Desce, J. M., Kemel, M. L., Gauchy, C., Godeheu, G., Cheramy, A., and Glowinski, J. 1991. Glutamatergic control of dopamine release in the rat striatum: evidence for presynaptic N-methyl-D-aspartate receptors on dopaminergic nerve terminals. *Journal of Neurochemistry* 56:81–5.

Lanius, R. A., Krieger, C., Wagey, R., and Shaw, C. A. 1993. Increased [^{35}S]glutathione binding sites in spinal cords from patients with sporadic amyotrophic lateral sclerosis. *Neuroscience Letters* 163:89–92.

Lanius, R. A., Shaw, C. A., Wagey, R., and Krieger, C. 1994. Characterization, distribution, and protein kinase C-mediated regulation of [^{35}S]glutathione binding sites in mouse and human spinal cord. *Journal of Neurochemistry* 63:155–60.

Lazarewicz, J. W., Wroblewski, J. T., Palmer, M. E., and Costa, E. 1989. Reduction of disulfide bonds activates NMDA-sensitive glutamate receptors in primary cultures of cerebellar granule cells. *Neuroscience Research Communications* 4:91–7.

Leviel, V., Gobert, A., and Guibert, B. 1990. The glutamate-mediated release of dopamine in the rat striatum: further characterization of the dual excitatory–inhibitory function. *Neuroscience* 39:305–12.

Levy, D.I., Sucher, N. J., and Lipton, S. A. 1990. Redox modulation of NMDA receptor-mediated toxicity in mammalian central neurons. *Neuroscience Letters* 110:291–6.

Levy, D. I., Sucher, N. J., and Lipton, S. A. 1991. Glutathione prevents N-methyl-D-aspartate receptor-mediated neurotoxicity. *Neuroreport* 2:345–7.

Li, N. C., and Manning, R. A. 1955. Some metal complexes of sulfur-containing amino acids. *Journal of the American Chemical Society* 77:5225–8.

Li, N. C., Gawron, O., and Basuas, G. 1954. Stability of zinc complexes with glutathione and oxidized glutathione. *Journal of the American Chemical Society* 76:225–9.

Li, X., Orwar, O., Persson, J., Sandberg, M., and Jacobson, I. 1993a. γ-L-Glutamyl-L-glutamate is an endogenous dipeptide in the rat olfactory bulb which activates *N*-methyl-D-aspartate receptors. *Neuroscience Letters* 155:42–6.

Li, X., Orwar, O., Varga, V., and Sandberg, M. 1993b. Determination of acidic sulfur-containing amino acids, β-aspartyl and γ-glutamyl dipeptides in rat brain. *Journal of Neurochemistry, Suppl.* 61:S89A.

Lipton, S. A., and Rosenberg, P. A. 1994. Excitatory amino acids as a final common pathway for neurologic disorders. *New England Journal of Medicine* 330:613–22.

Lipton, S. A., Choi, Y. B., Pan, Z. H., Lei, S. Z., Chen, H. S., Sucher, N. J., Loscalzo, J., Singel, D. J., and Stamler, J. S. 1993. A redox-based mechanism for the neuroprotective and neurodestructive effects of nitric oxide and related nitroso compounds. *Nature* 364:626–32.

Liu, Y. F., and Quirion, R. 1992. Modulatory role of glutathione on mu-opioid, substance P/neurokinin-1, and kainic acid receptor binding sites. *Journal of Neurochemistry* 59:1024–32.

Maker, H. S., Weiss, C., Silides, D. J., and Cohen, G. 1981. Coupling of dopamine oxidation (monoamine oxidase activity) to glutathione oxidation via the generation of hydrogen peroxide in rat brain homogenates. *Journal of Neurochemistry* 36:589–93.

Manzoni, O., Prezeau, L., Martin, P., Deshager, S., Bockaert, J., and Fagni, L. 1992. Nitric oxide–induced blockade of NMDA receptors. *Neuron* 8:653–62.

Maret, W. 1994. Oxidative metal release from metallothionein via zinc-thiol/disulfide interchange. *Proceedings of the National Academy of Sciences of the U.S.A.* 91:237–41.

Margaill, I., Miquet, J. M., Doble, A., Blanchard, J. C., and Boireau, A. 1992. K$^+$ ATP channels modulate GABA release in hippocampal slices in the absence of glucose. *Fundamentals in Clinical Pharmacology* 6:295–300.

Mårtensson, J., Jain, A., Frayer, W., and Meister, A. 1989. Glutathione metabolism in the lung: inhibition of its synthesis leads to lamellar body and mitochondrial defects. *Proceedings of the National Academy of Sciences of the U.S.A.* 86:5296–300.

Martin, H., and MacIlwain, H. 1959. Glutathione, oxidized and reduced, in the brain and isolated cerebral tissue. *Biochemical Journal* 71:275–80.

Max, B. 1989. This and that: the war on drugs and the evolution of sulfur. *Trends in Pharmacological Sciences* 10:483–6.

Maycox, P. R., Deckwerth, T., Hell, J. W., and Jahn, R. 1988. Glutamate uptake by brain synaptic vesicles. Energy dependence of transport and functional reconstitution in proteoliposomes. *Journal of Biological Chemistry* 263:15423–8.

Mayer, M. L., and Miller, R. J. 1990. Excitatory amino acid receptors, second messengers and regulation of intracellular Ca^{2+} in mammalian neurons. *Trends in Pharmacological Sciences* 11:254–60.

Meister, A. 1988. Glutathione metabolism and its selective modification. *Journal of Biological Chemistry* 263:17205–8.

Meister, A., and Anderson, M. E. 1983. Glutathione. *Annual Review of Biochemistry* 52:711–25.

Miller, B., Sarantis, M., Traynelis, S. F., and Attwell, D. 1992. Potentiation of NMDA receptor currents by arachidonic acid. *Nature* 355:722–5.

Miyamoto, M., Murphy, T. H., Schnaar, R.L., and Coyle, J. T. 1989. Antioxidants protect against glutamate-induced cytotoxicity in a neuronal cell line. *Journal of Pharmacology and Experimental Therapeutics* 250:1132–40.

Morari, M., O'Connor, W. T., Darvelid, M., Ungerstedt, U., Bianchi, C., and Fuxe, K. 1996. Functional neuroanatomy of the nigrostriatal and striatonigral pathways as studied with dual probe microdialysis in the awake rat—I. Effects of perfusion with tetrodotoxin and low-calcium medium. *Neuroscience* 72:79–87.

Murphy, T. H., Schnaar, R. L., and Coyle, J. T. 1990. Immature cortical neurons are uniquely sensitive to glutamate toxicity by inhibition of cystine uptake. *FASEB Journal* 4:1624–33.

Naito, S., and Ueda, T. 1985. Characterization of glutamate uptake into synaptic vesicles. *Journal of Neurochemistry* 44:99–109.

Nakanishi, S. 1992. Molecular diversity of glutamate receptors and implications for brain function. *Science* 258:597–603.

Nicholls, D., and Attwell, D. 1990. The release and uptake of excitatory amino acids. *Trends in Pharmacology Sciences* 11:462–8.

Nicoll, R. A., Malenka, R. C., and Kauer, J. A. 1990. Functional comparison of neurotransmitter receptor subtypes in mammalian central nervous system. *Physiological Reviews* 70:513–65.

O'Connor, W. T., Drew, K. L., and Ungerstedt, U. 1995. Differential cholinergic regulation of dopamine release in the dorsal and ventral neostriatum of the rat: an in vivo microdialysis study. *Journal of Neuroscience* 15:8353–61.

Ogita, K., and Yoneda, Y. 1987. Possible presence of [³H]glutathione (GSH) binding sites in synaptic membranes from rat brain. Neuroscience Research 4:486–96.

Ogita, K., and Yoneda, Y. 1988. Temperature-dependent and -independent apparent binding activities of [³H]glutathione in brain synaptic membranes. Brain Research 463:37–46.

Ogita, K., and Yoneda, Y. 1989. Selective potentiation by L-cysteine of apparent binding activity of [³H]glutathione in synaptic membranes of rat brain. Biochemical Pharmacology 38:1499–505.

Ogita, K., Kitago, T., Nakamuta, H., Fukuda, Y., Koida, M., Ogawa, Y., and Yoneda, Y. 1986. Glutathione-induced inhibition of Na⁺-independent and -dependent bindings of L-[³H]glutamate in rat brain. Life Sciences 39:2411–8.

Ogita, K., Enomoto, R., Nakahara, F., Ishitsubo, N., and Yoneda, Y. 1995. A possible role of glutathione as an endogenous agonist at the N-methyl-D-aspartate recognition domain in rat brain. Journal of Neurochemistry 64:1088–96.

Ohmichi, M., Hirota, K., Koike, K., Kadowaki, K., Miyake, A., Kiyama, H., Tohyama, M., and Tanizawa, O. 1989. Involvement of extracellular calcium and arachidonate in [³H]dopamine release from rat tuberoinfundibular neurons. Neuroendocrinology 50:481–7.

Oja, S. S., Varga, V., Janáky, R., Kontro, P., Aarnio, T., and Marnela, K.-M. 1988. Glutathione and glutamatergic neurotransmission in the brain. In Frontiers in excitatory amino acid research, ed. E. A. Cavalheiro, J. Lehmann, and L. Turski, 75–78. New York: Alan R. Liss.

Oja, S. S., Jenei, Zs., Janáky, R., Saransaari, P., and Varga, V. 1994. Thiol reagents and brain glutamate receptors. Proceedings of the Western Pharmacology Society 37:59–62.

Oja, S. S., Janáky, R., Saransaari, P., Jenei, Zs., and Varga, V. 1995. Interaction of glutathione derivatives with brain 2-amino-3-hydroxy-5-methyl-4-isoxazolepropionate (AMPA) receptors. Proceedings of the Western Pharmacology Society 38:9–11.

Oka, A., Belliveau, M. J., Rosenberg, P. A., and Volpe, J. J. 1993. Vulnerability of oligodendroglia to glutamate: pharmacology, mechanisms and prevention. Journal of Neuroscience 13:1441–53.

Olney, J. W., Zorumski, C., Price, M. T., and Labruyere, J. 1990. L-Cysteine, a bicarbonate-sensitive endogenous excitotoxin. Science 248:596.

Orlowski, M., and Karkowsky, A. 1976. Glutathione metabolism and some possible functions of glutathione in the nervous system. International Review of Neurobiology 19:75–121.

Orrenius, S., McConkey, D. J., Bellomo, G., and Nicotera, P. 1989. Role of Ca²⁺ in toxic cell killing. Trends in Pharmacological Sciences 10:281–5.

Owen, A. D., Schapira, A. H., Jenner, P., and Marsden, C. D. 1996. Oxidative stress and Parkinson's disease. Annals of the New York Academy of Sciences 786:217–23.

Pace, J. R., Martin, B. M., Paul, S. M., and Rogawski, M. A. 1992. High concentrations of neutral amino acids activate NMDA receptor currents in rat hippocampal neurons. Neuroscience Letters 141:97–100.

Perry, T. L., Godin, D. V., and Hansen, S. 1982. Parkinson's disease: a disorder due to nigral glutathione deficiency? Neuroscience Letters 33:305–10.

Philbert, M. A., Beiswanger, C. M., Waters, D. K., Reuhl, K. R., and Lowndes, H. E. 1991. Cellular and regional distribution of reduced glutathione in the nervous system of the rat: histochemical localization by mercury orange and o-phthaldialdehyde-induced histofluorescence. Toxicology and Applied Pharmacology 107:215–27.

Pin, J. P., Bockaert, J., and Récasens, M. 1984. The Ca²⁺/Cl⁻-dependent L-[³H]glutamate binding: a new receptor or a particular transport process? FEBS Letters 175:31–6.

Porcella, A., Fage, D., Voltz, C., Bourdiol, F., Benavides, J., Scatton, B., and Carter, C. 1992. Implication of the polyamines in the neurotoxic effects of N-methyl-D-aspartate. Neurological Research 14:S181–3.

Rader, R. K., and Lanthorn, T. H. 1989. Experimental ischemia induces a persistent depolarization blocked by decreased calcium and NMDA antagonists. Neuroscience Letters 99:125–30.

Raps, S. P., Lai, J. C., Hertz, L., and Cooper, A. J. 1989. Glutathione is present in high concentrations in cultured astrocytes but not in cultured neurons. Brain Research 493:398–401.

Ravindranath, V., Shivakumar, B. R., and Anandatheerthavarada, H. K. 1989. Low glutathione levels in brain regions of aged rats. Neuroscience Letters 101:187–90.

Récasens, M., Mayat, E., and Vignes, M. 1992. The multiple excitatory amino acid receptor subtypes and their putative interactions. Molecular Neuropharmacology 2:15–31.

Rehncrona, S., and Siesjö, B. K. 1979. Cortical and cerebrospinal fluid concentrations of reduced and oxidized glutathione during and after cerebral ischemia. Advances in Neurology 26:285–6.

Reichelt, K. L., and Fonnum, F. 1969. Subcellular localization of N-acetyl-aspartyl-glutamate, N-acetyl-glutamate and glutathione in brain. Journal of Neurochemistry 16:1409–16.

Reynolds, I. J., Rush, E. A., and Aizenman, E. 1990. Reduction of NMDA receptors with dithiothreitol increases [³H]MK–801 binding and NMDA-induced Ca²⁺ fluxes. *British Journal of Pharmacology* 101:178–82.

Rhoads, D. E., Osburn, L. D., Peterson, N. A., and Raghupathy, E. 1983. Release of neurotransmitter amino acids from synaptosomes: enhancement of calcium-independent efflux by oleic and arachidonic acids. *Journal of Neurochemistry* 41:531–7.

Rodriguez, J., Quignard, J.-F., Fagni, L., Lafon-Cazal, M., and Bockaert, J. 1994. Blockade of nitric oxide synthesis by tyrosine kinase inhibitors in neurones. *Neuropharmacology* 33:1267–74.

Ruppersberg, J. P., Stocker, M., Pongs, O., Heinemann, S. H., Frank, R., and Koenen, M. 1991. Regulation of fast inactivation of cloned mammalian I_k(A) channels by cysteine oxidation. *Nature* 352: 711–4.

Sagara, J., Miura, K., and Bannai, S. 1993. Cystine uptake and glutathione level in fetal brain cells in primary culture and in suspension. *Journal of Neurochemistry* 61:1667–71.

Sato, S. M., Frazier, J. M., and Goldberg, A. M. 1984. A kinetic study of the in vivo incorporation of ⁶⁵Zn into the rat hippocampus. *Journal of Neuroscience* 4:1671–5.

Seeburg, P. H. 1993. The TINS/TIPS lecture. The molecular biology of mammalian glutamate receptor channels. *Trends in Neuroscience* 16:359–65.

Shen, X. M., and Dryhurst, G. 1996. Oxidation chemistry of (−)-norepinephrine in the presence of L-cysteine. *Journal of Medical Chemistry* 39:2018–29.

Sian, J., Dexter, D. T., Lees, A. J., Daniel, S., Agid, Y., Javoy-Agid, F., Jenner, P., and Marsden, C. D. 1994. Alterations in glutathione levels in Parkinson's disease and other neurodegenerative disorders affecting basal ganglia. *Annals of Neurology* 36:348–55.

Slivka, A., Mytilineou, C., and Cohen, G. 1987a. Histochemical evaluation of glutathione in brain. *Brain Research* 409:275–84.

Slivka, A., Spina, M. B., and Cohen, G. 1987b. Reduced and oxidized glutathione in human and monkey brain. *Neuroscience Letters* 74:112–8.

Smart, T. G. 1989. Excitatory amino acids: the involvement of second messengers in the signal transduction process. *Cellular and Molecular Neurobiology* 9:193–206.

Spencer, J. P., Jenner, P., and Halliwell, B. 1995. Superoxide-dependent depletion of reduced glutathione by L-dopa and dopamine. Relevance to Parkinson's disease. *Neuroreport* 6:1480–4.

Stone, T. W. 1993. Subtypes of NMDA receptors. *General Pharmacology* 24:825–32.

Sucher, N. J., Wong, L. A., and Lipton, S. A. 1990. Redox modulation of NMDA receptor-mediated Ca²⁺ flux in mammalian central neurons. *Neuroreport* 1:29–32.

Sullivan, J. M., Traynelis, S. F., Chen, H. S., Escobar, W., Heinemann, S. F., and Lipton, S. A. 1994. Identification of two cysteine residues that are required for redox modulation of the NMDA subtype of glutamate receptor. *Neuron* 13:929–36.

Tang, L.-H., and Aizenman, E. 1993. The modulation of N-methyl-D-aspartate receptors by redox and alkylating reagents in rat cortical neurones in vitro. *Journal of Physiology* (Lond.) 465:303–23.

Tauck, D. L., and Ashbeck, G. A. 1990. Glycine synergistically potentiates the enhancement of LTP induced by a sulfhydryl reducing agent. *Brain Research* 519:129–32.

Terramani, T., Kessler, M., Lynch, G., and Baudry, M. 1988. Effects of thiol-reagents on [³H]α-amino-3-hydroxy-5-methylisoxazole-4-propionic acid binding to rat telencephalic membranes. *Molecular Pharmacology* 34:117–23.

Tolliver, J. M., and Pellmar, T. C. 1987. Dithiothreitol elicits epileptiform activity in CA₁ of the guinea pig hippocampal slice. *Brain Research* 404:133–41.

Tolliver, J. M., and Pellmar, T. C. 1988. Effects of dithiothreitol, a sulfhydryl reducing agent, on CA₁ pyramidal cells of the guinea pig hippocampus in vitro. *Brain Research* 456:49–56.

Ungerer, A., Schmitz-Bourgeois, M., Mélan, C., Boulanger, Y., Reinbolt, J., Amiri, I., and deBarry, J. 1988. γ-L-Glutamyl-L-aspartate induces specific deficits in long-term memory and inhibits [³H]glutamate binding on hippocampal membranes. *Brain Research* 446:205–11.

van Amsterdam, F. T., Giberti, A., Mugnaini, M., and Ratti, E. 1992. 3-[(±)-2-Carboxypiperazin-4-yl]propyl-1-phosphonic acid recognizes two N-methyl-D-aspartate binding sites in rat cerebral cortex membranes. *Journal of Neurochemistry* 59:1850–6.

Varga, V., Janáky, R., Marnela, K.-M., Gulyás, J., Kontro, P., and Oja, S. S. 1989. Displacement of excitatory amino acid receptor ligands by acidic oligopeptides. *Neurochemical Research* 14:1223–7.

Varga, V., Janáky, R., Holopainen, I., Kontro, P., and Oja, S. S. 1990. Effect of glutamyltaurine on calcium influx in cultured cerebellar granule cells. In *Taurine: functional neurochemistry, physiology, and cardiology*, ed. H. Pasantes-Morales, D. L. Martin, W. Shain, and R. Martín del Río, 141–5. New York: Wiley-Liss.

Varga, V., Janáky, R., and Oja, S. S. 1992. Modulation of glutamate agonist-induced influx of calcium into neurons by γ-L-glutamyl and β-L-aspartyl dipeptides. *Neuroscience Letters* 138:270–4.

Varga, V., Janáky, R., Saransaari, P., and Oja, S. S. 1994a. Endogenous γ-L-glutamyl and β-L-aspartyl peptides and excitatory aminoacidergic neurotransmission in the brain. *Neuropeptides* 27:19–26.

Varga, V., Janáky, R., Marnela, K.-M., Saransaari, P., and Oja, S. S. 1994b. Interactions of γ-L-glutamyltaurine with excitatory aminoacidergic neurotransmission. *Neurochemical Research* 19:243–8.

Varga, V., Janáky, R., Holopainen, I., Oja, S. S., and Åkerman, K. E. O. 1995. Endogenous γ-L-glutamylglutamate is a partial agonist at the N-methyl-D-aspartate receptors in cultured cerebellar granule cells. *Neurochemical Research* 20:1471–6.

Varga, V., Jenei, Zs., Janáky, R., and Oja, S. S. 1997. Glutathione is an endogenous ligand of rat brain N-methyl-D-aspartate (NMDA) and 2-amino-3-hydroxy-5-methyl-4-isoxazolepropionate (AMPA) receptors. *Neurochemical Research* 22:1165–71.

Volterra, A., Trotti, D., Tromba, C., Floridi, S., and Racagni, G. 1994. Glutamate uptake inhibition by oxygen free radicals in rat cortical astrocytes. *Journal of Neuroscience* 14:2924–32.

Wagey, R., Lanius, R. A., Krieger, C., and Shaw, C. A. 1993. Glutathione receptors in human spinal cord: a potential role in normal and abnormal function? *Society for Neuroscience Abstracts* 19:984.

Walters, D. W., and Gilbert, H. F. 1986. Thiol/disulfide exchange between rabbit muscle phosphofructokinase and glutathione. Kinetics and thermodynamics of enzyme oxidation. *Journal of Biological Chemistry* 261:15372–7.

Wheeler, D. D. 1984. Kinetics of D-aspartic acid release from rat cortical synaptosomes. *Neurochemical Research* 9:1599–1614.

Woodward, J. J. 1994. The effects of thiol reduction and oxidation on the inhibition of NMDA-stimulated neurotransmitter release by ethanol. *Neuropharmacology* 33:635–40.

Woodward, J. J., and Blair, R. 1991. Redox modulation of N-methyl-D-aspartate-stimulated neurotransmitter release from rat brain slices. *Journal of Neurochemistry* 57:2059–64.

Yoneda, Y., and Ogita, K. 1991. Neurochemical aspects of the N-methyl-D-aspartate receptor complex. *Neuroscience Research* 10:1–33.

Yoneda, Y., Ogita, K., Kouda, T., and Ogawa, Y. 1990. Radioligand labeling of N-methyl-D-aspartic acid (NMDA) receptors by [^{3}H](±)-3-(2-carboxypiperazin-4-yl)propyl-1-phosphonic acid in brain synaptic membranes treated with Triton X-100. *Biochemical Pharmacology* 39:225–8.

Yudkoff, M., Pleasure, D., Cregar, L., Lin, Z.-P., Nissim, I., Stern, J., and Nissim, I. 1990. Glutathione turnover in cultured astrocytes: studies with [^{15}N]glutamate. *Journal of Neurochemistry* 55:137–45.

Zängerle, L., Cuénod, M., Winterhalter, K. H., and Do, K. Q. 1992. Screening of thiol compounds: depolarization-induced release of glutathione and cysteine from rat brain slices. *Journal of Neurochemistry* 59:181–9.

Glutathione in the Nervous System
Edited by Christopher A. Shaw
Copyright © 1998 Taylor & Francis

9

Excitatory Actions of GSH on Neocortex

Bryce A. Pasqualotto

*Department of Physiology, University of British Columbia,
Vancouver, British Columbia, Canada V6T 1Z3*

Kenneth Curry

Precision Biochemicals Inc., Vancouver, British Columbia, Canada V6T 1Z3

Christopher A. Shaw

*Departments of Ophthalmology and Physiology and Neuroscience Program, University of
British Columbia, Vancouver, British Columbia, Canada V6T 1Z3*

ABBREVIATIONS

AMPA α-amino-3-hydroxy-5-methyl-4-isoxazolepropionate
AP5 2-amino-5-phosphonovalerate
C-A L-cysteinyl-L-alanine

197

C-G γ-L-Cys-Gly
Cys L-cysteine
D-C L-aspartyl-L-cysteine
DNQX 6,7-dinitroquinoxaline-2,3-dione
E-C γ-L-Glu-L-Cys
efp's excitatory field potentials
E-G γ-L-Glu-Gly
Glu L-glutamate
Gly glycine
GSA glutathione sulfonic acid
GSH reduced glutathione
GSSG oxidized glutathione
MK801 (5S,10R)-(--)-5-methyl-10,11-dihydro-5H-dibenzo[a,d]cyclohepten-
   5,10-imine
NMDA N-methyl-D-aspartate
SMG S-methylglutathione

1. GLUTATHIONE IN THE NERVOUS SYSTEM

The tripeptide γ-L-glutamyl-L-cysteinylglycine (glutathione) is widely distributed throughout the animal kingdom. In its reduced form (GSH) glutathione is best known for its role in protecting cells from oxidative stress through its actions as a free-radical scavenger, but also has a number of other functions in cell survival and metabolism, including detoxification of xenobiotics, amino acid transport, maintenance of protein redox state, and more (Meister and Anderson 1983; Meister 1995). Glutathione is a monomer in its reduced form (GSH) and acts as a powerful reducing agent by virtue of the thiol group of its cysteine residue. In the oxidized form (GSSG), it is a dimer that is recycled to GSH by the actions of the enzyme GSSG disulfide reductase and the reducing power of NADPH. The evolution of GSH is closely associated with the evolution of aerobic metabolism in eukaryotes (Fahey and Sundquist 1991) and probably arose out of the need for an antioxidant defense mechanism to protect cells from reactive oxygen species produced as a result of oxidative phosphorylation. The ratio of GSH to GSSG is thought to be a key determinant of cell survival (Kosower and Kosower 1978). Since its discovery in 1888 by De Rey-Pailhade, the number of functions attributed to glutathione has grown steadily (Meister 1989).

Glutathione is ubiquitously distributed throughout the body and is present in the central nervous system at tissue concentrations in the millimolar range (Reichelt and Fonnum 1969; Slivka et al. 1987a,b; Raps et al. 1989; Kudo et al. 1990) with the majority (95–99 percent) in the reduced form (Slivka et al. 1987a,b; Sofic et al. 1992). The cellular localization of glutathione in the CNS has been an issue of some debate. Several authors have reported glutathione to be localized primarily to glial cells (Raps et al. 1989; Pow and Crook 1995) with little evidence of neuronal glutathione. However, immunocytochemical and histochemical methods have been used to demonstrate the presence of glutathione in both neurons and glia (Pileblad, Eriksson, and Hansson 1991; Philbert et al. 1991; Amara, Coussemacq, and Geffard 1994; Hjelle, Chaudry, and Ottersen 1994; Hjelle et al., chapter 4, this volume). Recent immunocytochemical studies using antibodies specific for glutathione (both GSH and GSSG) demonstrate the heterogeneity of glutathione distribution in the CNS (Hjelle et al. 1994; Hjelle et al., chapter 4, this volume). In these studies, a considerable fraction of glutathione immunoreactivity was found to reside in neurons, frequently appearing as labeling of the axon terminals in addition to other neuronal compartments. In cerebellum and spinal cord, glutathione immunoreactivity is present in putative glutamatergic terminals. The widespread appearance of glutathione in the nervous system is well documented; however, the multiple roles of glutathione in nervous-system function are still being elucidated.

2. GLUTATHIONE AND NEUROPATHOLOGY

Perturbations of glutathione levels have been observed in various neurodegenerative and neuropathological processes. Alterations of total glutathione [GSH + (2×GSSG)] or of the GSH/GSSG ratio have been implicated in experimental models of neuropathology, including glutamate-induced cytotoxicity (Kato et al. 1992; Levy, Sucher, and Lipton 1991; Miyamoto et al. 1989; Murphy et al. 1989), ischemic brain damage (Cooper, Pulsinelli, and Duffy 1980; Rehncrona et al. 1980; Andine et al. 1991; see also Cooper, chapter 5, this volume), MPTP toxicity (Wullner et al. 1996), and animal models of epileptogenesis (Hiramatsu and Mori 1981). GSH depletion is associated with brain damage in developing animals (Jain et al. 1991). One prominent theory with regard to the etiology of several neurodegenerative diseases includes the proliferation of injurious free radicals and subsequent cell death resulting from the highly reactive nature of these molecules (Evans 1993; Fahn and Cohen 1992; Di Monte, Chan, and Sandy 1992). Although this theory is not universally accepted (Calne 1992), there is sufficient evidence supporting this type of mechanism in several neurodegenerative diseases to warrant further examination. As this topic is the focus of several chapters in this book, it will not be dealt with further here.

3. GLUTATHIONE AND SIGNAL TRANSDUCTION

3.1 Glutathione Receptors of Lower Animals

Another aspect of glutathione function in the nervous system was first identified in the early 1950s by Loomis in his studies of the feeding behavior of the coelenterate *Hydra*. In a somewhat serendipitous series of events, Loomis identified GSH as a specific trigger of the *Hydra* feeding response (Loomis 1955; Lenhoff, chapter 2, this volume). Loomis demonstrated that GSH released from the punctured body wall of prey species impaled by the nematocysts of *Hydra* acts as a chemical messenger to stimulate the feeding response. These data provided the first evidence of a role for GSH in nervous-system signal transduction.

The existence of a GSH receptor in membrane fractions of *Hydra* homogenates has been demonstrated in radioligand binding studies (Koizumi and Kijima 1980; Bellis, Kass-Simon, and Rhoads 1992; Venturini 1987; Grosvenor et al. 1992; Bellis et al. 1991), although efforts to purify the *Hydra* GSH receptor are still in early stages. Antibodies against sepharose-binding proteins of *Hydra* inhibit specific components of the feeding response of *Hydra japonica* induced by S-methyl-glutathione (SMG), and analysis of the antigens for these antibodies was used to identify a 220-kDa protein from membrane fractions of *Hydra* tentacles (Ohta et al. 1992). This 220-kDa protein may represent the putative GSH receptor. Using an affinity column of immobilized GSH, Bellis, Laux, and Rhoads (1994) purified several peptides (20–48 kDa) from solubilized membrane proteins of *Hydra attenuata*. Antiserum raised against some of these peptides antagonized the feeding response stimulated by GSH, suggesting a role for some of these peptides in *Hydra* chemoreception. However, the 220-kDa protein reported by Ohta, Hanai, and Morita (1992) was not detected in this study. Methodological and species differences between these two reports may account for the disparity of the results.

Radioligand binding studies and behavioral assays of the *Hydra* feeding response have been used to define the pharmacology of the *Hydra* GSH receptor. Equilibrium and kinetic binding parameters calculated in membrane homogenates of *Hydra* using [^{35}S]GSH correspond closely to values calculated in behavioral studies of GSH-induced feeding response (see Bellis et al. 1992). SMG also stimulates *Hydra* feeding behavior (Ohta et al. 1992), and incorporation of S-[^{14}C]methyl-GSH into nematocyst-rich fractions of *Hydra* has been documented (Koizumi and Kijima 1980). Analysis of glutathione analogs and derivatives has yielded information on the active sites of the molecule that are required for receptor activation. The γ-glutamyl residue of GSH is critical for activation of the feeding response, as is the side chain of the second residue, although the presence of the thiol group of cysteine is not necessary (Cobb et al. 1982). Glutamate inhibits the feeding response (Lenhoff and Boivard 1961), but there is some doubt if this occurs by competitive inhibition at the GSH receptor. Competition of radiolabeled GSH binding by glutamate has been reported (Grosvenor et al. 1992). Following inhibition of the cell surface enzyme γ-glutamyl transpeptidase in soluble fractions of *Hydra* membrane homogenates, no competitive effects of glutamate on specific [^{35}S]GSH binding were observed (Bellis et al. 1992). These results indicate that inhibition of the GSH-induced feeding response by glutamate is not mediated by competitive inhibition of the GSH receptor (Bellis et al. 1992). Radiolabeled glutamate binding has been used as a tool to study the *Hydra* GSH receptor (Venturini 1987; Bellis et al. 1991); however, it appears that only a portion of glutamate binding is displaced by GSH (Bellis et al. 1991). [^{3}H]Glutamate binding in this preparation consists of a GSH-sensitive and a GSH-insensitive component, suggesting that, in addition to GSH receptors, *Hydra* also possess glutamate receptors (Bellis et al. 1991).

Despite the progress that has been made in the identification and characterization of the *Hydra* GSH receptor, the signal transduction mechanism upon activation of this receptor is still unknown. Studies of the chemoreceptors on lobster walking legs reveal that GSH receptors are not unique to *Hydra* or to coelenterates in general. Extracellular recordings from nerve bundles of the leg nerves of the lobster (*Homarus americanus*) show that GSH is excitatory and stimulates spiking (i.e., action-potential generation) in these cells (Derby and Atema 1982). GSH receptors thus appear to play an important role as targets of GSH acting as a chemoexcitant in the prey selection and feeding responses of a variety of lower animals, including coelenterates and crustaceans (see Cobb et al. 1982).

3.2 GSH Receptors in the Mammalian Nervous System

Glutathione is a particularly intriguing molecule from the perspective of information transfer in the mammalian nervous system. All three of its constituent amino acids are neuroactive. L-Glutamate is thought to be the endogenous neurotransmitter that activates a family of receptors (collectively referred to as glutamate receptors). L-Glutamate activates at least three ligand-gated ion channels, which are named on the basis of their preferred agonist as follows: N-methyl-D-aspartate receptors (NMDA receptors), α-amino-3-hydroxy-5-methyl-1,4-isoxazole propionate receptors (AMPA receptors), and high-affinity kainate receptors (Monaghan, Bridges, and Cotman 1989). In addition, glutamate also activates multiple G-protein-coupled receptors known collectively as glutamate metabotropic receptors (Bockaert, Pin, and Fagni 1993; Prezeau et al. 1994). Ionotropic glutamate receptors (i.e., glutamate-gated ion channels) are ubiquitous mediators of rapid synaptic excitation in the mammalian nervous system. L-Cysteine is an agonist at NMDA receptors (Pace et al. 1992; Olney et al. 1990) and is known to have excitotoxic properties (Olney et al. 1990). In the brain, glycine is a coagonist at NMDA receptors; on its own it does not activate these receptors, but it must be present for L-glutamate to cause channel opening (Johnson and Ascher 1987). In the spinal cord, glycine binds to and activates its own receptors (glycine receptors), which mediate rapid synaptic inhibition by virtue of the Cl^- currents they gate (Kuhse, Betz, and Kirsch 1995; Betz et al. 1994). Thus, glutathione's components have the potential to activate a wide variety of synaptic receptors, each with unique distributions and biophysical properties within the mammalian CNS.

Specific binding sites for GSH have been described in various CNS tissue preparations. High-affinity ($k_d = 2\text{--}12$ nM) [^{35}S]GSH binding sites have been characterized in primary cultures of rat cortical astrocytes (Guo and Shaw 1992) and in thin sections of mouse and human spinal cord (Lanius et al. 1993, 1994) and rat and cat visual cortex (Bowlsby and Shaw 1991). Autoradiograms of [^{35}S]GSH labeling in human and murine spinal cord show labeling largely restricted to gray matter with fairly uniform density (substantia gelatinosa > dorsal horn > ventral horn: Lanius et al. 1993, 1994). In rat brain sections, dense [^{3}H]GSH binding is observed in the dentate gyrus, habenula, hypothalamus, and retina, and lower density of labeling is observed in other brain regions (Bains et al. 1997).

Ogita and colleagues have extensively characterized the binding of radiolabeled GSH to membrane fractions of rat brain homogenates and the interactions of GSH with glutamate receptors (see Ogita et al., chapter 7, this volume). [^{3}H]GSH binds saturably and reversibly to at least two sites with equilibrium dissociation constants of $k_{d1} = 560$–760 nM and $k_{d2} = 3.7$–12.6 μM (Ogita and Yoneda 1987, 1988, 1989). Additionally, GSH competes for binding of glutamate receptor ligands, particularly NMDA receptor ligands, and is hypothesized by these authors to have a role as an endogenous neurotransmitter at these receptors (Ogita et al. 1995). In support of this hypothesis, both GSH and GSSG stimulate Ca^{2+} influx into dissociated neurons from embryonic rat brains in a manner that is antagonized by NMDA receptor antagonists (Leslie et al. 1992). Furthermore, Zangerle et al. (1992) have demonstrated the release of GSH from hippocampal slices of rat brain following depolarization by elevated extracellular K^+. Of particular interest is the Ca^{2+}-dependent nature of this release, which is suggestive of a neuronal origin. Glutathione (both GSH and GSSG) may also play an indirect, modulatory role in neurotransmission. Endogenous reducing and oxidizing agents such as GSH and GSSG stimulate changes in the redox state of receptor proteins and concomitant modulation of receptor function (Gozlan and Ben-Ari 1995; Pan et al. 1995; Sucher and Lipton 1991; Gilbert, Aizenman, and Reynolds 1991; see also Ogita et al. and Janáky et al., chapters 7 and 8 in this volume, for review).

4. EXCITATORY ACTIONS OF GSH IN NEOCORTEX

Despite the abundance of evidence in favor of a role for glutathione in neurotransmission, much remains to be learned about the actions of glutathione in the CNS. Although radioligand binding studies reveal the presence of specific GSH binding sites, there is little evidence of a direct effect of GSH on neural cells. The studies of the chemoreception of lower animals indicate an excitatory role for GSH, which appears to be mediated by specific GSH receptors (see section 3.1). In primary cultures of rat cortical astrocytes, GSH induces an increase in IP$_3$ accumulation (Guo and Shaw 1992), but the precise pharmacology and physiological significance of this result is unknown. Using extracellular electrophysiological recording methods (Harrison and Simmonds 1985) and radioligand binding, we have sought to address the physiological significance of GSH receptors in the mammalian nervous system and the interactions between GSH and ionotropic glutamate receptors.

Application of GSH to slices of adult rat cortex produces a depolarizing field potential shift (Shaw, Pasqualotto, and Curry 1996; Pasqualotto, Curry, and Shaw 1995, 1996). This GSH-induced excitation is concentration-dependent (EC$_{50}$ = 4.86±2.09 mM), and effective concentrations of GSH ($\geq$50 μM) produce large depolarizing excitatory potentials (efps) with rapid onset and offset characteristics (Fig. 1). Reducing agents such as GSH are known to give rise to quantifiable potentials as a result of the flow of electrons in oxidation–reduction reactions, termed redox potentials. However, there are several reasons why the GSH-induced potentials we described are not likely to be redox potentials. The concentrations of GSH that elicit depolarizing field potentials in cortical slices are lower than the calculated concentration required to

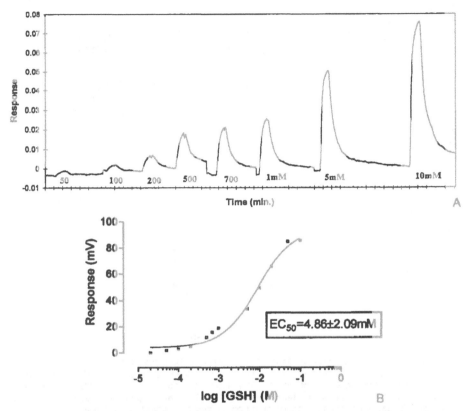

FIG. 1. Concentration dependence of GSH excitatory field potentials in rat cortical slices. A: Field potentials elicited by increasing concentrations of GSH. Concentrations are in micromolar unless otherwise indicated. Slices were perfused for 2 min at each concentration of GSH to produce these responses. Responses are given in inverse volts. The methods employed have been published previously (Harrison and Simmonds 1985; Shaw et al. 1996). B: Dose–response plot of the data represented in A. Peak amplitudes of each trace were measured and plotted against GSH concentration. Data shown are from one representative experiment; the calculated EC_{50} is the mean ± s.e. from three experiments. Curve fitting performed using GraphPad Prism.

produce a redox potential of similar magnitude. Inclusion of oxidizing agents such as GSSG would be expected to diminish the putative GSH redox potential when coapplied and to induce a potential of opposite direction if applied on its own. GSSG has no effect when applied alone or when coapplied with GSH in our preparation. When slices are killed under carefully controlled conditions (hypo- or hyperosmotic shock) or the recording configuration changed, no GSH potentials are observed (data not shown).

It was at first thought that GSH might produce its excitatory effects by breakdown to its constituent amino acids, each of which is known to act at glutamate-gated ion channels to produce excitatory currents. To control for this possibility, GSH was perfused

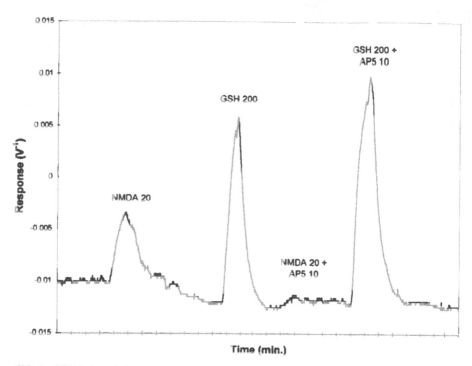

FIG. 2. GSH-induced depolarization is not mediated by glutamate ionotropic receptors. Perfusion of slices with the NMDA receptor antagonist AP5 blocks responses to NMDA but not to GSH. Similar results were obtained for the NMDA antagonist MK801 and the non-NMDA antagonist DNQX (not shown). In this experiment, responses to GSH were slightly potentiated in the presence of AP5, although this potentiation is not consistently observed. Concentrations given are in micromolar.

onto slices in the presence of standard glutamate receptor antagonists. Blockade of NMDA receptors by AP5 or MK801 had no effect on GSH-induced responses, nor did antagonism of non-NMDA receptors (both AMPA and kainate receptors) by DNQX (Fig. 2, Table 1). Complete inhibition of ionotropic glutamate receptor responses by a coperfusion of AP5 and DNQX had no effect on GSH responses. These results stand in contrast to the hypothesis that GSH is an endogenous neurotransmitter at NMDA receptors (Leslie et al. 1992; Ogita et al. 1995; Ogita et al., chapter 7, this volume).

Of the constituent amino acids of GSH, only Cys is an effective competitor in radioligand binding experiments of radiolabeled GSH to crude synaptic membrane fractions of adult rat neocortex (approximately 91-percent inhibition of total binding at 1 mM; see Fig. 3A). L-glu and gly show only weak competition at high concentrations (approximately 15–20-percent inhibition at 1 mM). To control for the possibility that the excitatory effects of GSH are mediated by free Cys (perhaps liberated by the actions of GSH metabolizing enzymes), responses to Cys were studied. Perfusion of cortical slices with Cys also produces efps in a concentration-dependent manner. These responses are qualitatively similar to GSH efps with respect to onset and offset

TABLE 1. *Summary table of cortical-wedge recording data. N/A represents not applicable. Antagonism refers to significant reduction in peak response amplitude by DNQX (5 μM), AP5 (10 μM), or MK801 (2 μM)*

Compound	Response	Antagonism			Classification
		DNQX	AP5	MK801	
Glutathione:					
GSH	Yes	No	No	No	GSH-like
GSSG	No	N/A	N/A	N/A	N/A
Constituents:					
L-glutamate	Yes	Partial	Partial	Partial	Mixed agonist
L-cysteine	Yes	No	Yes	Yes	NMDA-like
glycine	No	N/A	N/A	N/A	N/A
Derivatives:					
SMG	Yes	No	Yes	Yes	NMDA-like
GSA	Yes	No	Yes	Yes	NMDA-like
Dipeptides:					
E-C	Yes	No	No	No	GSH-like
C-G	Yes	No	No	No	GSH-like
G-G	No	N/A	N/A	N/A	N/A
D-C	Yes	No	No	No	GSH-like
C-A	Yes	No	No	No	GSH-like

rates and peak amplitude. Unlike GSH efps, Cys efps are partially antagonized by AP5 (Fig. 4, Table 1). This result is consistent with previous reports of agonist activity of Cys at NMDA receptors (Olney et al. 1990; Pace et al. 1992) but also demonstrates an excitatory effect of Cys that is independent of NMDA receptor activation. Further evidence in favor of unmetabolized GSH as the excitatory stimuli as opposed to one of its constituents is derived from comparison of [^{35}S] cysteine-labeled GSH and [^{3}H] glycine-labeled GSH. The two ligands show very similar characteristics in steady-state-binding assays, suggesting that GSH is not rapidly metabolized under the conditions employed in binding experiments (data not shown).

Ion replacement experiments were performed to study the mechanism of GSH-induced depolarization. Substitution of Na$^+$ in the perfusion buffer eliminated GSH-induced responses but substitution of Ca^{2+} did not (Shaw et al. 1996). Antagonism of K$^+$ currents by the K$^+$-channel toxin TEA or of voltage-dependent Ca^{2+} currents by inclusion of Mg^{2+} in the perfusion buffer did not effect GSH-induced responses. Na$^+$ dependence may be suggestive of an uptake mechanism, and Na$^+$-dependent GSH transport has been demonstrated (Kannan et al., chapter 3, this volume). It seems unlikely that the excitatory effects of GSH are a result of GSH-transport processes, considering that the response kinetics for GSH efps are similar to those observed for activation of ionotropic glutamate receptors. Furthermore, the specific binding of radiolabeled GSH has been characterized using Tris acetate buffers, and thus binding does not appear to be Na$^+$-dependent. Despite the presence of Na$^+$-independent GSH transporters in brain (Kannan et al., chapter 3, this volume), which might be expected to contribute significantly to high nonspecific binding values, over 90 percent of total [^{35}S or ^{3}H]GSH is displaceable by an excess of unlabeled GSH. Thus, the binding

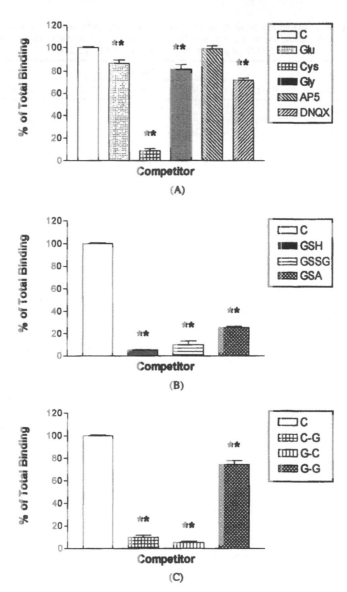

FIG. 3. Competition of [³H]GSH binding by glutathione to crude synaptic membrane fractions of adult rat cortex. Membrane homogenates were prepared as previously described (Zukin, Young, and Snyder 1974) and incubated with 2 nM [³H]GSH (specific activity 44.7 Ci/mmol) for 150 min at 37°C in the presence of the various test compounds at 100 μM (AP5 and DNQX) or 1 mM (all other compounds). Incubations were terminated by vacuum filtration using Whatman GF/B filter disks and rinsing of the homogenates with ice-cold buffer (100 mM Tris acetate, pH 7.4, two rinses of 5 mL each). Radioactivity bound to membrane homogenates was determined by scintillation counting, and the raw data were analyzed by one-way ANOVA with Dunnett's posttest, using GraphPad Prism to determine significance. Each bar represents a minimum of four determinations. Data are plotted as mean percentage of control ±s.e.; ∗∗ represents $p < 0.01$.

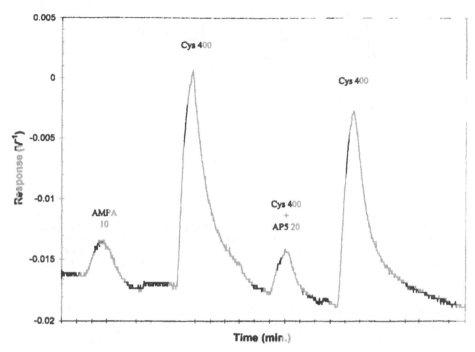

FIG. 4. Excitatory field potentials elicited by L-Cys are partially antagonized by AP5. A residual response to L-Cys in the presence of AP5 is apparent. A template response to AMPA is shown for comparison.

of radiolabeled GSH appears to represent specific labeling of putative GSH receptors that are distinct from GSH transporters, and GSH efps are unlikely to result from rapid Na^+-dependent GSH transport. These results demonstrate a novel role for GSH in mammalian CNS neurotransmission and provide the first direct evidence of an excitatory effect of GSH on mammalian neural cells, which appears to be mediated by a GSH-gated, Na^+-permeable channel.

5. PHARMACOLOGY OF THE PUTATIVE GSH RECEPTOR OF RAT NEOCORTEX

Saturation experiments of [³H]GSH reveal the presence of two binding sites ($k_{d1} = 760 \pm 200$ nM, $K_{d2} = 11.0 \pm 3.1$ μM: Ogita and Yoneda 1987), whereas homologous competition experiments are generally best fitted by a one-site binding model ($IC_{50} = 2.32 \pm 0.34$ μM) (Fig. 5). Although these two methods of analysis are mathematically equivalent, for technical reasons saturation analysis is generally more sensitive for detecting multiple site binding systems (De Blasi, O'Reilly, and Motulsky 1989). In two of seven homologous-competition experiments the curves

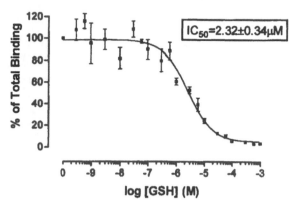

FIG. 5. Homologous competition of [³H]GSH binding. Curve fitting was performed by nonlinear least-squares regression analysis (GraphPad Prism). Data were best fitted by a one-site binding model in most cases (five of seven experiments). Data shown are from a single experiment performed in triplicate. The calculated IC₅₀ represents the mean ± s.e. from five experiments. In two experiments, data were best fitted by a two-site binding model (see section 5 for discussion).

were best fitted by a two-site binding model with k_ds very similar to those reported in saturation experiments (Pasqualotto et al. 1995). Thus it is probable that radiolabeled GSH binds specifically to a high-affinity site ($k_{d1} = 688.5 \pm 206.1$ nM) and a low-affinity site ($k_{d2} = 10.1 \pm 2.2$ µM). Preliminary competition experiments with Cys are also best fitted by a two-site binding model, although the affinity of Cys for these sites is markedly different than the affinity of GSH (data not shown). A significant concern in all binding studies is the specificity of the radioligand employed. Radiolabeled GSH could potentially bind to nonreceptor GSH binding sites in the membrane such as glutathione transporters, or to cell surface enzymes such as glutathione-S-transferase. Since binding studies were performed using a Tris acetate buffer, GSH binding does not appear to require the presence of ionic cofactors, which provide the driving force for transmembrane transport of many amino acids, neurotransmitters, etc. To control for the potential contribution of GSH-S-transferases to observed GSH binding, competition experiments using glutathione analogs that act as inhibitors of GSH-S-transferases (see Ouwerkerk-Mahadevan et al. 1995) were performed. None of the three GSH-S-transferase inhibitors tested displayed competition for GSH binding (Fig. 6). These experiments suggest that radiolabeled GSH binding represents labeling of the putative GSH receptor described in this and other reports (Ogita and Yoneda 1987, 1988, 1989; see also Ogita et al., chapter 7, this volume) under the conditions employed. The presence of two binding sites for GSH may represent different affinity states for the putative receptor as has been observed in binding studies of ionotropic glutamate receptors (Hall, Kessler, and Lynch 1992) or may result from the cross-reactivity of GSH (and Cys) with glutamate binding sites, NMDA receptors in particular, as reported by other authors (Ogita et al. 1995).

The structure–activity relationships of the GSH molecule have been investigated by assaying the competitive effects of various GSH derivatives and metabolites on

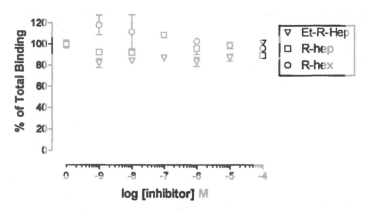

FIG. 6. GSH-*S*-transferase inhibitors do not compete for GSH binding. No competition by the GSH-*S*-transferase inhibitors (*R*)-5-ethyloxycarbonyl-2-γ-(*S*)-glutamylamino-*N*-hexylpentamide (Et-*R*-Hep), (*R*)-5-carboxy-2-γ-(*S*)-glutamylamino-*N*-2-heptylpentamide (*R*-Hep), or (*R*)-5-carboxy-2-γ-(*S*)-glutamylamino-*N*-hexylpentamide (*R*-Hex) was observed in two experiments.

radiolabeled GSH binding. Although GSSG has no apparent effect in electrophysiological recording assays, it is an effective competitor of GSH binding (Fig. 3 B, $IC_{50} = 32.67 \pm 12.35$ μM). Glutathione sulfonic acid (GSA) is also an effective competitor of GSH binding (25.48 ± 1.22 percent of control binding at 1 mM GSA: Fig. 3 B), indicating that the thiol group of GSH is not essential for binding activity. Dipeptide combinations of GSH constituent amino acids also serve as good competitors of GSH binding. Preliminary experiments show that γ-Glu-Cys (E-C) and γ-Cys-Gly (C-G) strongly inhibit GSH binding (reducing binding to 5.34 ± 1.03 and 10.06 ± 1.87 percent of total binding, respectively, at 1 mM), whereas γ-glu-gly (E-G) is a poor competitor (74.57 ± 3.33 percent of total binding at 1 mM E-G), suggesting that the γ-glutamyl residue is also not essential for binding activity, but the presence of the cysteinyl residue appears to be important (Fig. 3 C).

When these compounds were tested in the cortical wedge recording preparation, all were found to have excitatory activity with the exception of GSSG and γ-Glu-Gly. Both SMG and GSA produced concentration-dependent efps that were susceptible to inhibition by NMDA receptor antagonists (Fig. 7). In contrast, the Cys containing dipeptides E-C and C-G also stimulated efps in a concentration-dependent manner, but these responses were insensitive to glutamate-receptor antagonists (Fig. 8). Based on these results we hypothesize that the critical determinant of biological activity in these molecules is the presence of the cysteinyl residue and the state of the thiol group of this residue. In the –SH form, as in the parent molecule GSH or the dipeptide metabolites E-C and C-G, this residue confers specificity for the GSH receptor we have described. In chemically modified forms such as in the GSH derivatives SMG and GSA, specificity for activation of NMDA receptors is conferred. However, in the S–S dimerized form that is present in GSSG, no excitatory effects are observed at either receptor (Table 1; see also Curry, chapter 10, this volume).

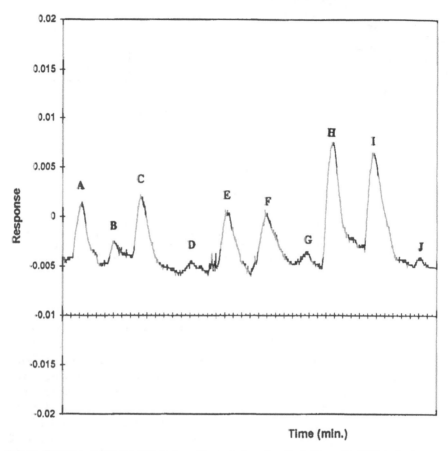

FIG. 7. Excitatory field potentials induced by perfusion of cortical slices with GSH derivatives are mediated by NMDA receptors. Responses are labeled as follows: A: AMPA (10 μM); B: AMPA (10 μM) + DNQX (5 μM); C: NMDA (20 μM); D: NMDA (20 μM) + AP5 (10 μM); E: SMG (400 μM); F: SMG + DNQX; G: SMG + AP5; H: GSA (400 μM); I: GSA + DNQX; J: GSA + AP5. Experiments with MK801 (2 μM) yield similar results to those shown for AP5.

It is also clear that the presence of Cys is not sufficient to confer specificity for the putative GSH receptor. As discussed previously, a significant fraction of the Cys efp is antagonized by NMDA receptor antagonists, although the residual response in the presence of NMDA receptor inhibitors may be mediated by activation of GSH receptors. Thus, we have speculated that the amino acid residues linked to the carboxyl and/or amino terminus of Cys are also important in determining the specific targeting of these compounds. To this end we have assayed the effects of the synthetic Cys containing dipeptides L-aspartylcysteine (D-C) and L-cysteinylalanine (C-A). Both D-C and C-A produce GSH-like responses in cortical slices that are not affected by NMDA receptor antagonists (Table 1). These efps resemble those observed for GSH in their magnitude, rate of onset and offset, concentration dependence, and insensitivity

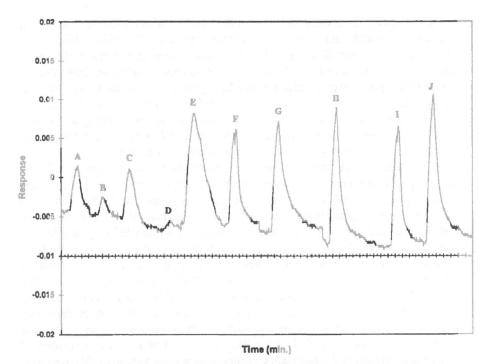

FIG. 8. Excitatory field potentials induced by perfusion of cortical slices with GSH fragments are not inhibited by glutamate receptor antagonists. Responses are labeled as follows: A: AMPA (10 μM); B: AMPA (10 μM) + DNQX (5 μM); C: NMDA (20 μM); D: NMDA (20 μM) + AP5 (10 μM); E: E-C (400 μM); F: E-C + DNQX; G: E-C + AP5; H: C-G (400 μM); I: C-G + DNQX; J: C-G + AP5. Experiments with MK801 (2 μM) yield similar results to those shown for AP5.

to NMDA receptor antagonists. It appears that the addition of amino acid residues to either end of the Cys molecule is sufficient to confer specificity for the putative GSH receptor we have described. We have not yet determined whether any amino acid substitution is sufficient or whether specific amino acids are required to confer specificity. Experiments are currently underway to address this issue.

6. GSH AS A NEUROTRANSMITTER

Although much remains to be done to satisfy the criteria for GSH as a new neuro-transmitter in the mammalian nervous system (see Shaw, chapter 1, this volume), there is a significant body of evidence to suggest that GSH does indeed serve this function. The presence of GSH in brain in relatively high concentrations (Reichelt and Fonnum 1969; Slivka et al. 1987a,b; Raps et al. 1989; Kudo et al. 1990) and its localization in neuropil and specifically in axon terminals (Hjelle et al., chapter 4, this volume) demonstrates that GSH is appropriately located to serve as a synaptic chemical messenger. The identification of specific GSH binding sites in lower animals

(Bellis et al. 1991, 1992; Grosvenor et al. 1992) and in various mammalian CNS cell preparations (Bowlsby and Shaw 1991; Ogita and Yoneda 1987, 1988, 1989; Lanius et al. 1993, 1994; Pasqualotto et al. 1995) indicates the existence of a postsynaptic protein complex that enables cells to sense and respond to released GSH. The release of GSH upon cell depolarization in a Ca^{2+}-dependent fashion is suggestive of a synaptic release mechanism (Zangerle et al. 1992). Finally, the excitatory actions of GSH on the nervous systems of lower animals (Loomis 1955; Derby and Atema 1982) and on cortical slices of rat brain (Shaw et al. 1996; see preceding section) demonstrates the physiological consequences of GSH on neural cells.

The presence of GSH immunoreactivity in glutamatergic nerve terminals (Hjelle et al., chapter 4, this volume) is particularly interesting in light of some of the other functions ascribed to GSH. Potential co-release of glutamate and GSH would add significant complexity to the signal transduction mechanisms at such synapses. Both GSH and glutamate could activate their own receptors while GSH may additionally fine-tune glutamatergic transmission by redox modulation of glutamate receptors.

Our results do not support the hypothesis that GSH (or GSSG) acts as an NMDA-receptor agonist (Ogita et al. 1995; Leslie et al. 1992; Ogita et al., chapter 7, this volume). It is possible that a small proportion of the GSH response is mediated by NMDA receptors. Cys appears to bridge the gap between selectivity of these two receptors according to two-site binding fits and the presence of a small residual response to Cys when coapplied with AP5. Since Cys is an effective competitor of GSH binding, the residual response to Cys may represent activation of GSH receptors. However, binding to GSH receptors is not effected by NMDA receptor antagonists (Fig. 3 A), nor is the GSH response diminished in the presence of such drugs. The GSH receptor we have identified also appears to be different from the *Hydra* GSH receptor, in view of their pharmacological specificities. In *Hydra*, the γ-glutamyl residue is an essential component of the bioactivity of GSH (Cobb et al. 1982), but in rat neocortex we have reported essentially full activity in the absence of this residue. Our results identify two critical features of the GSH molecule: the cysteine residue and the presence of a second residue of unspecified features either 5' or 3' to Cys. The specificity of GSH appears to reside largely with the state of the thiol group of Cys. In the reduced form, specificity for GSH receptors in conferred. When the thiol group is modified, as in SMG or GSA, specificity for NMDA receptors is displayed, and when Cys is absent, as in γ-Glu-Gly, no response is detected. Although the –SH group is necessary for activation of putative GSH receptors in rat cortex, it does not appear to be necessary for binding-site recognition, because GSSG, GSA, and SMG all display competitive inhibition of [^{3}H]GSH binding.

Conclusive proof of the existence of a novel, GSH-gated ionotropic receptor and of a neurotransmitter role for GSH awaits the cloning of this putative protein complex. Multiple subunits have been cloned for glutamate ionotropic receptor subunit proteins (Hollman and Heineman 1994). Included in the glutamate ionotropic receptor super-family are two subunits cloned by sequence homology to subunits encoding other glutamate ionotropic receptor subunits. These subunits have been labeled $\delta 1$ and $\delta 2$ (Yamazaki et al. 1992; Lomeli et al. 1993). $\delta 1$ and $\delta 2$ subunits have not yet been found to form functional channels and do not appear to coassemble with other glutamate ionotropic receptor subunits (Mayat et al. 1995). Neither do they show any affinity for binding of AMPA, kainate, or glutamate (Lomeli et al. 1993; Mayat et al. 1995), thus, they are alternatively referred to as orphan subunits, as no known agonist exists. However, these subunits appear to have a molecular structure very similar that of other glutamate-gated ion channels and are localized in postsynaptic elements (Mayat et al. 1995), suggesting that they do form ionotropic receptors. Because of the ability of GSH to bind to some glutamate ionotropic receptor subtypes (Janáky et al., chapter 8, this volume; Ogita et el., chapter 7), it is possible, albeit highly speculative at this point, that these subunits represent constituents of synaptic GSH receptors.

7. CONCLUSIONS

More than a century has passed since the discovery of glutathione. In this time much has been learned about the important and varied roles of glutathione in cell survival and metabolism. In recent years, evidence for a new role for GSH in nervous-system function has begun to emerge. Although still speculative at this point, we have provided evidence of an excitatory role for GSH on neocortex, which appears to be mediated by activation of a novel, Na^+-permeable ion channel. The chemoexcitant properties of GSH on the chemosensory systems of lower animals and the ubiquitous presence of GSH in the animal kingdom suggest that GSH may be an evolutionarily conserved neuroexcitant of ancient origin. The validity of this hypothesis remains to be determined, and future investigations of the multifunctional roles GSH in nervous-system function promise to be rewarding.

Note Added in Proof

In preliminary experiments, we have used intracellular recording to examine the effects of GSH application to single neurons. Recordings were made from neurons in layer 2/3 of sensory cortex or layer 5/6 of prefrontal cortex in 4 week old rat cortical slices. GSH application leads to a depolarization from resting potential, a decrease in input resistance, and an apparent shift in threshold. More comprehensive experiments are in progress.

ACKNOWLEDGMENTS

This work was supported by an NSERC operating grant to CAS. We wish to thank J. Seamans and Dr. C. Yang for assistance with intracellular recording experiments.

REFERENCES

Amara, A., Coussemacq, M., and Geffard, M. 1994. Antibodies to reduced glutathione. *Brain Research* 659:237–42.

Andine, P., Orwar, O., Jacobson, I., Sandberg, M., and Hagberg, H. 1991. Extracellular acidic sulfur-containing amino acids and γ-glutamyl peptides in global ischemia: postischemic recovery of neuronal activity is paralleled by a tetrodotoxin-sensitive increase in cysteine sulfinate in the CA1 of the rat hippocampus. *Journal of Neurochemistry* 57:230–6.

Bains, J. S., Pasqualotto, B. A., Shaw, C. A., and Curry, K. 1997. Localization of glutathione binding sites in the central nervous system of rat. *Society for Neuroscience Abstracts* 23. In press.

Bellis, S. L., Grosvenor, W., Kass-Simon, G., and Rhoads, D. E. 1991. Chemoreception in *Hydra vulgaris* (attenuata): initial characterization of two distinct binding sites for L-glutamic acid. *Biochimica et Biophysica Acta* 1061:89–94.

Bellis, S. L., Kass-Simon, G., and Rhoads, D. E. 1992. Partial characterization and detergent solubilization of the putative glutathione chemoreceptor from *Hydra*. *Biochemistry* 31:9838-43.

Bellis, S. L., Laux, D., and Rhoads, D. E. 1994. Affinity purification of *Hydra* glutathione binding proteins. *FEBS Letters* 354:320–4.

Betz, H., Kuhse, J., Fischer, M., Schmieden, V., Laube, B., Kuryatov, A., Langosch, D., Meyer, G., Bormann, J., Rundstrom, N., Matzenbach, J., Kirsch, J., and Ramming, M. 1994. Structure, diversity and synaptic localization of inhibitory glycine receptors. *Journal de Physiologie (Paris)* 88:243–8.

Bockaert, J., Pin, J., and Fagni, L. 1993. Metabotropic glutamate receptors: an original family of G protein-coupled receptors. *Fundamental and Clinical Pharmacology* 7:473–85.

Bowlsby, S., and Shaw, C. 1991. Characteristics and distribution of GSH receptors and GSH uptake sites in visual cortex. *Society for Neuroscience Abstracts* 12:75.

Calne, D. B. 1992. The free radical hypothesis in idiopathic Parkinsonism: evidence against it. *Annals of Neurology* 32:799–803.

Cobb, M. H., Heagy, W., Danner, J., Lenhoff, H. M., and Marshall, G. R. 1982. Structural and conformational properties of peptides interacting with the glutathione receptor of *Hydra*. *Molecular Pharmacology* 21:629-36.

Cooper, A. J. L., Pulsinelli, W. A., and Duffy, T. E. 1980. Glutathione and ascorbate during ischemia and postischemic reperfusion in rat brain. *Journal of Neurochemistry* 35:1242–5.

DeBlasi, A., O'Reilly, K., and Motulsky, H. J. 1989. Calculating receptor number from binding experiments using same compound as radioligand and competitor. *TIPS* 10:227–9.

Derby, C. D., and Atema, J. 1982. Chemosensitivity of walking legs of the lobster *Homarus americanus*: neurophysiological response spectrum and thresholds. *Journal of Experimental Biology* 98:303–15.

Di Monte, D., Chan, P., and Sandy, M. S. 1992. Glutathione in Parkinson's disease: a link between oxidative stress and mitochondrial damage? *Annals of Neurology* 32:S111–15.

Evans, P. H. 1993. Free radicals in brain metabolism and pathology. *British Medical Bulletin* 49:577–87.

Fahey, R. C., and Sundquist, A. R. 1991. Evolution of glutathione metabolism. *Advances in Enzymology and Molecular Biology* 64:1–53.

Fahn, S., and Cohen, G. 1992. The oxidant stress hypothesis in Parkinson's disease: evidence supporting it. *Annals of Neurology* 32:804–12.

Gilbert, K. R., Aizenman, E., and Reynolds, I. J. 1991. Oxidized glutathione modulates N-methyl-D-aspartate- and depolarization-induced increases in intracellular Ca^{2+} in cultured rat forebrain neurons. *Neuroscience Letters* 133:11–14.

Gozlan, H., and Ben-Ari, Y. 1995. NMDA receptor redox sites: are they targets for selective neuronal protection? *TIPS* 16:368–74.

Grosvenor, W., Bellis, S. L., Kass-Simon, G., and Rhoads, D. E. 1992. Chemoreception in *Hydra*: specific binding of glutathione to a membrane fraction. *Biochimica et Biophysica Acta* 1117:120–5.

Guo, N., and Shaw, C. 1992. Characterization and localization of glutathione binding sites on cultured astrocytes. *Molecular Brain Research* 15:207–15.

Hall, R. A., Kessler, M., and Lynch, G. 1992. Evidence that high- and low-affinity DL-alpha-amino-3-hydroxy-5-methylisoxazole-4-propionic acid (AMPA) binding sites reflect membrane-dependent states of a single receptor. *Journal of Neurochemistry* 59:1997–2004.

Harrison, N. L., and Simmonds, M. A. 1985. Quantitative studies on some antagonists of N-methyl-D-aspartate in slices of rat cerebral cortex. *British Journal Pharmacology* 84:381–91.

Hiramatsu, M., and Mori, A. 1981. Reduced and oxidized glutathione in brain and convulsions. *Neurochemical Research* 6:301–6.

Hjelle, O. P., Chaudry, F. A., and Ottersen, O. P. 1994. Antisera to glutathione: characterization and immunocytochemical application to the rat cerebellum. *European Journal of Neuroscience* 6:793–804.

Hollmann, M., and Heinemann, S. 1994. Cloned glutamate receptors. *Annual Review of Neuroscience* 17:31–108.

Jain, A., Martensson, J., Stole, E., Auld, P. A. M., and Meister, A. 1991. Glutathione deficiency leads to mitochondrial damage in brain. *Proceedings of the National Academy of Sciences of the U.S.A.* 88:1913–17.

Johnson, J. W., and Ascher, P. 1987. Glycine potentiates the NMDA response in cultured mouse brain neurons. *Nature* 325:529–31.

Kato, S., Negishi, K., Mawatari, K., and Kuo, C.-H. 1992. A mechanism for glutamate toxicity in the C6 glioma cells involving inhibition of cystine uptake leading to glutathione depletion. *Neuroscience* 48:903–14.

Koizumi, O., and Kijima, H. 1980. Specific S-methylglutathione incorporation into a nematocyst-rich fraction of *Hydra. Biochimica et Biophysica Acta* 629:338–48.

Kosower, N. S., and Kosower, E. M. 1978. The glutathione status of cells. *International Review of Cytology* 54:109–160.

Kudo, H., Kokunai, T., Kondoh, T., Tamaki, N., and Matsumoto, S. 1990. Quantitative analysis of glutathione in rat central nervous system: comparison of GSH in infant brain with that in adult brain. *Brain Research* 511:326–8.

Kuhse, J., Betz, H., and Kirsch, J. 1995. The inhibitory glycine receptor: architecture, synaptic localization and molecular pathology of a postsynaptic ion-channel complex. *Current Opinion in Neurobiology* 5:318–23.

Lanius, R. A., Krieger, C., Wagey, R., and Shaw, C. A. 1993. Increased [^{35}S]glutathione binding sites in spinal cords from patients with sporadic amyotrophic lateral sclerosis. *Neuroscience Letters* 163:89–92.

Lanius, R. A., Shaw, C. A., Wagey, R., and Krieger, C. 1994. Characterization, distribution, and protein kinase C-mediated regulation of [^{35}S]glutathione binding sites in mouse and human spinal cord. *Journal of Neurochemistry* 63:155–160.

Lenhoff, H. M., and Boivard, J. 1961. Action of glutamic acid and glutathione analogs on the *Hydra* glutathione receptor. *Nature* 189:486–7.

Leslie, S. W., Brown, L. M., Trent, R. D., Lee, Y.-H., Morris, J. L., Jones, T. W., Randall, P. K., Lau, S. S., and Monks, T. J. 1992. Stimulation of N-methyl-D-aspartate receptor-mediated calcium entry into dissociated neurons by reduced and oxidized glutathione. *Molecular Pharmacology* 41:308–14.

Lomeli, H., Sprengle, R., Laurie, D. J., Kohr, G., Her, A., Seeburg, P. H., and Wisden, W. 1993. The rat delta-1 and delta-2 subunits extend the excitatory amino acid receptor family. *FEBS Letters* 315:318–22.

Loomis, W. F. 1955. Glutathione control of the specific feeding reactions of *Hydra. Annals of the New York Academy of Sciences* 62:209–28.

Mayat, E., Petralia, R. S., Wang, X. Y., and Wenthold, R. J. 1995. Immunoprecipitation, immunoblotting, and immunocytochemistry studies suggest that glutamate receptor delta subunits form novel postsynaptic receptor complexes. *Journal of Neuroscience* 15:2533–46.

Meister, A. 1989. A brief history of glutathione and a survey of its metabolism and functions. In *Glutathione: chemical, biochemical, and medical aspects, part A*, ed. D. Dolphin, R. Poulson, and O. Avramovic, 1–48. New York: Wiley.

Meister, A. 1995. Glutathione metabolism. *Methods of Enzymology* 251:3–7.

Meister, A., and Anderson, M. E. 1983. Glutathione. *Annual Review of Biochemistry* 52:711–60.

Miyamoto, M., Murphy, T. H., Schnaar, R. L., and Coyle, J. T. 1989. Antioxidants protect against glutamate-induced cytotoxicity in a cell neuronal cell line. *Journal of Pharmacology and Experimental Toxicology* 250:1132–40.

Monaghan, D. T., Bridges, R. J., and Cotman, C. 1989. The excitatory amino acid receptors: their classes and distinct properties in the function of the central nervous system. *Annual Review of Pharmacology and Toxicology* 29:365–402.

Murphy, T. H., Miyamoto, M., Sastre, A., Schnaar, R. L., and Coyle, J. T. 1989. Glutamate toxicity in a neuronal cell line involves inhibition of cystine transport leading to oxidative stress. *Neuron* 2:1547–58.

Ogita, K., and Yoneda, Y. 1987. Possible presence of [^{3}H]glutathione (GSH) binding sites in synaptic membranes from rat brain. *Neuroscience Research* 4:486–96.

Ogita, K., and Yoneda, Y. 1988. Temperature-dependent and -independent apparent binding activities of [^{3}H]glutathione in brain synaptic membranes. *Brain Research* 463:37–46.

Ogita, K., and Yoneda, Y. 1989. Selective potentiation by L-Cysteine of apparent binding activity of [^{3}H]glutathione in synaptic membranes of rat brain. *Biochemical Pharmacology* 38:1499–1505.

Ogita, K., Enomoto, R., Nakahara, F., Ishitsubo, N., and Yoneda, Y. 1995. A possible role of glutathione as an endogenous agonist at the N-methyl-D-aspartate recognition domain in rat brain. *Journal of Neurochemistry* 64:1088–96.

Ohta, K., Hanai, K., and Morita, H. 1992. Glutathione-binding proteins identified by monoclonal antibodies which depress the behavioral responses evoked by glutathione in *Hydra. Biochimica et Biophysica Acta* 1117:136–42.

Olney, J. W., Zorumski, C., Price, M. T., and Labruyere, J. 1990. L-Cysteine, a bicarbonate-sensitive endogenous excitotoxin. *Science* 251:596–9.

Ouwerkerk-Mahadevan, S., van Boom, J. H., Dreef-Tromp, M. C., Ploemen, J. H. T. M., Meyer, D. J., and Mulder, G. J. 1995. Glutathione analogues as novel inhibitors of rat and human glutathione *S*-transferase isozymes, as well as of glutathione conjugation in isolated rat hepatocytes and in the rat *in vivo. Biochemical Journal* 308:283–90.

Pace, J. R., Martin, B. M., Paul, S. P., and Rogawski, M. A. 1992. High concentrations of neutral amino acids activate NMDA receptor currents in rat hippocampal slices. *Neuroscience Letters* 141:97–100.

Pan, Z.-H., Bahring, R., Grantyn, R., and Lipton, S. A. 1995. Differential modulation by sulfhydryl redox agents and glutathione of GABA- and glycine-evoked currents in rat retinal ganglion cells. *Journal of Neuroscience* 15:1384–91.

Pasqualotto, B. A., Curry, K., and Shaw, C. A. 1995. GSH as a neuroactive peptide. *Society for Neuroscience Abstracts* 21:1600.

Pasqualotto, B. A., Curry, K., and Shaw, C. A. 1996. Excitatory actions of GSH in neocortex. *Society for Neuroscience Abstracts* 22:1544.

Philbert, M. A., Beiswanger, C. M., Waters, D. K., Reuhl, K. R., and Lowndes, H. E. 1991. Cellular and regional distribution of reduced glutathione in the nervous system of the rat: histochemical localization by mercury orange and o-phthaldialdehyde-induced histofluorescence. *Toxicology and Applied Pharmacology* 107:215–27.

Pileblad, E., Eriksson, P. S., and Hansson, E. 1991. The presence of glutathione in primary neuronal and astroglial cultures from rat cerebral cortex and brain stem. *Journal of Neural Transmission* 86:43–9.

Pow, D. V., and Crook, D. K. 1995. Immunocytochemical evidence for the presence of high levels of reduced glutathione in radial glial cells and horizontal cells in the rabbit retina. *Neuroscience Letters* 193:25–8.

Prezeau, L., Carrette, J., Helpap, B., Curry, K., Pin, J. P., and Bockaert, J. 1994. Pharmacological characterization of metabotropic glutamate receptors in several types of brain cells in primary cultures. *Molecular Pharmacology* 45:570–7.

Raps, S. P., Lai, J. C. K., Hertz, L., and Cooper, A. J. L. 1989. Glutathione is present in high concentrations in cultured astrocytes but not in cultured neurons. *Brain Research* 493:398–401.

Rehncrona, S., Folbergrova, J., Smith, D. S., and Siesjo, B. K. 1980. Influence of complete and pronounced incomplete cerebral ischemia and subsequent recirculation on cortical concentrations of oxidized and reduced glutathione in the rat. *Journal of Neurochemistry* 34:477–86.

Reichelt, K. L., and Fonnum, F. 1969. Subcellular localization of *N*-acetyl-aspartyl-glutamate, *N*-acetyl-glutamate and glutathione in brain. *Journal of Neurochemistry* 16:1409–16.

Shaw, C. A., Pasqualotto, B. A., and Curry, K. 1996. Glutathione-induced sodium currents in neocortex. *NeuroReport* 7:1149–52.

Slivka, A., Mytilineou, C., and Cohen, G. 1987a. Histochemical evaluation of glutathione in brain. *Brain Research* 409:275–84.

Slivka, A., Spina, M. B., and Cohen, G. 1987b. Reduced and oxidized glutathione in human and monkey brain. *Neuroscience Letters* 74:112–8.

Sofic, E., Lange, K. W., Jellinger, K., and Riederer, P. 1992. Reduced and oxidized glutathione in the substantia nigra of patients with Parkinson's disease. *Neuroscience Letters* 142:128–30.

Sucher, N. J., and Lipton, S. A. 1991. Redox modulatory site of the NMDA receptor-channel complex: regulation by oxidized glutathione. *Journal of Neuroscience Research* 30:582–91.

Venturini, G. 1987. The *Hydra* GSH receptor. Pharmacological and radioligand binding studies. *Comparative Biochemistry and Physiology* 87C: 321–4.

Wullner, U., Loschmann, P. A., Schulz, J. B., Schmid, A., Dringen, R., Eblen, F., Turski, L., and Klockgether, T. 1996. Glutathione depletion potentiates MPTP and MPP+ toxicity in nigral dopaminergic neurones. *NeuroReport* 7:921–3.

Yamazaki, M., Araki, K., Shibata, A., and Mishina, M. 1992. Molecular cloning of a cDNA encoding a novel member of the mouse glutamate receptor channel family. *Biochemical and Biophysical Research Communications* 183:886–92.

Zangerle, L., Cuenod, M., Winterhalter, K. H., and Do, K. Q. 1992. Screening of thiol compounds: depolarization-induced release of glutathione and cysteine from rat brain slices. *Journal of Neurochemistry* 59:181–9.

Zukin, S. R., Young, A. B., and Snyder, S. H. 1974. Gamma-aminobutyric acid binding to receptor sites in the rat central nervous system. *Proceedings of the National Academy of Sciences of the U.S.A.* 71:4802–7.

Glutathione in the Nervous System
Edited by Christopher A. Shaw
Copyright © 1998 Taylor & Francis

10

Medicinal Chemistry of Glutathione and Glutathione Analogs in the Mammalian Central Nervous System

Kenneth Curry

Precision Biochemicals Inc., Vancouver, British Columbia, Canada V6T 1Z3

ABBREVIATIONS

AMPA α-amino-3-hydroxy-5-methylisoxazole-4-propionic acid
AP5 2-amino-5-phoshoponopropionic acid
DNQX 6,7-dinitroquinoxaline-2,3-dione
KA kainic acid
NMDA *N*-methyl-D-aspartic acid

1. INTRODUCTION

The structure, occurrence, and actions of the tripeptide glutathione (GSH) have been studied for over a century (Meister 1989). Many symposia and book reviews have been written. The importance of GSH as a free-radical scavenger cannot be understated.

In these studies we will investigate the possible role of GSH in the central nervous system (CNS) as a neurotransmitter candidate. This work is a consequence of our recent report on the ability of GSH to produce depolarizing field potentials in the rat cortical-wedge preparation. We will present here our latest findings on the initial structure–activity relationships (SARs) for the system and propose future directions based on the results found.

2. STRUCTURE OF GSH

The structure of GSH was elucidated following observation of its action (de Rey-Pailhade 1888a,b), degradation (Harrington 1935), and finally synthesis (du Vigneaud and Miller 1936). Chemically it is a tripeptide composed of the amino acids glutamate, cysteine, and glycine, and it is called γ-glutamylcysteinylglycine (γ-Glu-Cys-Gly) because the Glu–Cys peptidic linkage is at the terminal or γ carboxyl group of Glu (Fig. 1). Because of the γ linkage it contains acidic and basic functional groups of well-known pK_a values (Glu α NH_3^+ = 9.5, SH = 9.2, Glu α CO_2^- = 2.5, and Gly CO_2^- = 3.7), and in particular it contains a thiol group capable of undergoing one- and two-electron redox reactions. It is the latter functions that have generated the most attention in that this capability allows GSH to be oxidized to its dimer GSSG and readily reduced back to two molecules of GSH (Fig. 2). The importance of this simple transformation lies in the route by which it happens.

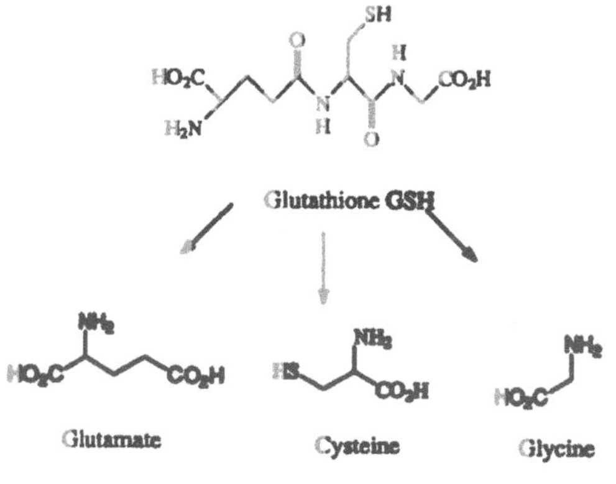

FIG. 1. GSH and its constituent amino acid residues.

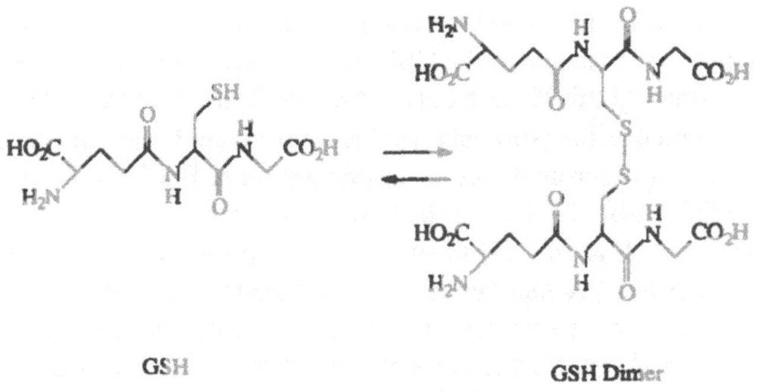

FIG. 2. Relationship between GSH and its dimer GSSG.

2.1 Free Radicals

The generation of free radicals in the mammalian system is continuous and occurs as a result of both normal and abnormal cellular activity and also environmental perturbations. Free radicals are highly reactive chemical species containing, in the outermost or bonding orbital, a single unpaired electron. As a consequence of this they have a very short half-life and are annihilated by a variety of mechanisms.

Simply, a free radical may collide with another free radical to create a stable electron pair and thus a more stable chemical:

$$R\bullet + R'\bullet \rightarrow R\!-\!R'$$

This process is termed annihilation and is the ultimate fate of the free radical.

On another route the free radical may obtain the transfer of another electron from a compound more susceptible to electron donation, thus eliminating the original free radical but replacing it with another possessing perhaps different properties:

$$R\bullet + R' \rightarrow R + R'\bullet$$

This is termed perpetuation and prolongs the life of the free-radical species.

Because of the reactivity of free radicals, they can react with other important structural and functional molecules in the body, particularly the polyunsaturated fatty acids contained in cell walls, which can undergo a form of cascade reaction (Cheeseman 1993; Cheeseman and Slater 1993). This may lead to damage and dysfunction of the normal system and eventual cellular and genetic damage.

GSH itself is capable of forming free radicals; however, the GSH free radical is considered to be relatively unreactive and often will only annihilate on contact with another GSH radical or other proteinous substances containing thiols. This vital function of GSH has been much studied (Kosower and Kosower 1978; Meister 1989) and is well understood. The distribution and concentrations of GSH are also well known (Kosower and Kosower 1978) and it may be present as high as 9 mM in some tissue types. Less well understood or studied is the role of GSH in the CNS, although it has been noted that a change in the GSH *status* of cells is seen to be associated with certain disease states (Jaffe 1970; Konrad et al. 1972; Meister 1974; Sian et al. 1994). GSH status is defined as the ratio between the total concentration of intracellular GSH and the total of all the forms it can be associated with. For instance, the total forms will include GSH itself, GSSG, and the GSS proteins.

Recently some analogs of GSH have been prepared and tested as neuroprotectants against ischemic insult (Yamamoto et al. 1993; Sagara et al. 1993; Saija et al. 1994). The actual mechanism of this action remains unclear, but is proposed to involve inhibition of the lipid peroxidative process. In addition, the direct action of GSH on hydra mobility has been observed (Lenhoff 1961; see also Lenhoff, chapter 2, this volume), lending some credence to the idea that GSH may serve a neurotransmitter role in some species.

2.2 GSH as a Putative Neurotransmitter

The suspicion that GSH may have neurotransmitter actions in the mammalian system is due to a similarity in structure to the acidic amino acid class of neurotransmitters based on glutamic acid. The structure of GSH has a free terminal carboxyl group, and because of the γ-glutamate linkage it also contains a free α amino carboxylic acid group much in the same way as glutamate itself. GSH has been identified as playing some modulatory role in the interaction of KA (Saija et al. 1994) and NMDA (Weaver et al. 1993) with their receptors and also having some effects on their excitotoxic actions in the CNS. In addition it is known that each of the amino acid residues of GSH has neurotransmitter or neuromodulatory actions of its own. Glutamate is the major excitatory neurotransmitter in the CNS, cysteine has excitatory effects at one of the Glu subreceptor types (as will be discussed in detail later), and finally glycine has neuromodulatory actions at one of the Glu subreceptor types at the cortical level, while acting also as an inhibitory neurotransmitter at the spinal level. Overall, to date, the involvement of GSH in neurotransmitter function, and the subsequent effects of the excitotoxicity associated with glutamate-like compounds, have been attributed to the antioxidant or redox effects of GSH (Zhang and Dryhurst 1994; Vanella et al. 1993). The direct actions of GSH as a generator of depolarizing field potentials is much less well known and understood, although there is some evidence to link an increase in intracellular Ca^{2+} to application of GSH (Weaver et al. 1993).

3. GLUTAMATE AS A NEUROTRANSMITTER

3.1 Glutamate Receptor Divisions

It is now widely accepted that glutamate is the major excitatory neurotransmitter in the central nervous system. The heterogeneous population of the family of glutamate receptors has been fairly well characterized, both by molecular pharmacological techniques and from drug–receptor interactions. Glutamate receptors can be divided into two broad classes having distinct actions.

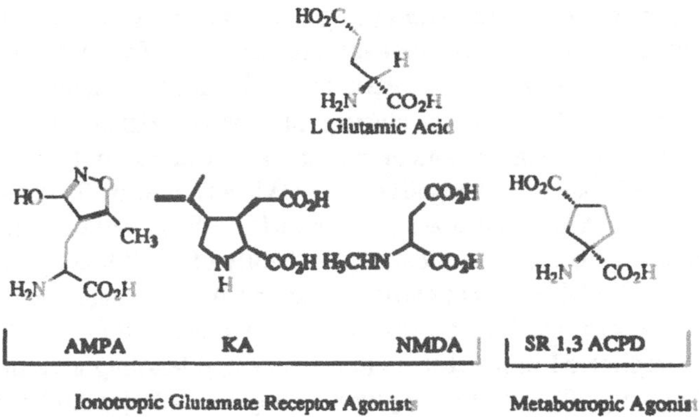

FIG. 3. Representative structures for glutamate and the compounds defining the subreceptor system.

3.2 Ionotropic Glutamate Receptors

The first (and first to be discovered) are the ionotropic receptors, now referred to as the GluR family. This set of receptors are characterized as comprising a pentameric array of membrane-bound protein subunits. The array may be transmembrane and forms an ion channel to allow the passage of sodium or calcium ions; in addition, an extracellular binding domain is present but not necessarily an intracellular domain. These receptors have been classified as associated with the pharmacological agents, which demonstrate high specificity of action at them. Thus classes of GluRs are named for N-methyl-D-aspartic acid (NMDA), kainic acid (KA), and α-amino-3-hydroxy-5-methylisoxazole-4-propionic acid (AMPA), and each is further subdivided into groups as defined by molecular pharmacology (Fig. 3). These receptors are by now well characterized and many reviews are available (for example see Hansen and Krogsgaard-Larsen 1990). The major function of this group of receptors appears to be in the fast type of synaptic transmission and in some cases presynaptic neuromodulation.

3.3 Metabotropic Glutamate Receptors

The second group of GluRs are termed the metabotropic glutamate receptors, or mGluRs. These receptors have much in common with other second-messenger-linked receptor–effector systems. They are composed of a heptameric array of membrane-bound subunits possessing a transmembrane structural region, an extracellular domain of varying complexity containing the binding site, and an intracellular domain associated with a G-protein and subsequently an effector system, which may be either phospholipase C (PLC), phospholipase D (PLD), or adenylate cyclase (AC). Activation of these receptors leads to G-protein-linked activation or inhibition of a second-messenger effect via the effector molecules cited above. These receptors are not named for the pharmacological agents that act at them, although initially the specific agonist SR 1-amino-1,3-cyclopentanedicarboxylic acid (SR 1,3 ACPD, Fig. 3) was proposed as the archetypal agent (Curry et al. 1997, 1988; Manzoni et al. 1992). mGluRs mediate several important intracellular mechanisms in the CNS. Activation of types 1 and 5 leads to the cleavage of phosphatydal inositol and eventually to activation of protein kinases and an increase in the concentration of intracellular Ca^{2+}. Types 2 and 3 are linked negatively to AC via a G_i type of G-protein and inhibit the production of cAMP. These are the two major effects, although much study is currently underway to elucidate the complete family of mGluR actions.

The function of mGluRs is not entirely clear, and these receptors are currently undergoing intense scrutiny. They do however lead to longer-term intracellular changes and thus are implicated in development, memory and learning and the modulation of neuronal activity. The number of pharmacological tools for studying mGluRs is growing rapidly, and soon a complete pharmacological description will be available. For a review see Schoepp and Conn (1993).

4. BIOLOGICAL TEST PREPARATION

Because of the interest in this large family of receptors, we have investigated the hypothesis that synergy exists between the tripeptide GSH and one or more of the GluR receptors. In order to study the effect of GSH on neuronal systems, we focused initially on the ionotropic receptors and the cortical-wedge preparation for both electropharmacology and binding studies. The cortical wedge offers the advantage of being a more intact preparation than others; it also lends itself well to rapid screening and thus is ideally suited for the type of initial SAR studies we are undertaking.

The approach we have used is to compare the test compounds with the effects of known pharmacological agents in the system. Thus NMDA, KA, and AMPA are used as agonists, each producing dose-dependent depolarizing actions. These depolarizations are attenuated by the concurrent application of known specific antagonists. NMDA actions are antagonized by 2-amino-5-phosphonopentanoic acid (AP5) (Stone et al. 1981; Davies et al. 1980), and AMPA and KA are antagonized by 6,7-dinitroquinoxaline-2,3-dione (DNQX) (Honoré et al. 1988).

Initially we observed the effect of GSH on application of increasing concentrations of the compound. We discovered that GSH produced depolarizing field potentials, which appeared to be dose-dependent. This effect could not be reversed by AP5 or DNQX or a combination of the two (Shaw, Pasqualotto, and Curry 1996). Application of GSH at submaximal doses produced additive depolarizations when applied concurrently with NMDA, AMPA, and KA. These results indicated that GSH was acting through a separate receptor system. In order to discover the channel type, a series of channel-blocking and ion-replacement experiments were performed. The only action found to affect the actions of GSH was the replacement of all sodium and so pointed to the involvement of sodium flux via an as yet unidentified receptor–channel system. (It should be noted that depolarizations due to application of GSH were not attenuated by concurrent application of tetrodotoxin.)

5. STRUCTURE–ACTIVITY RELATIONSHIP STUDIES

5.1 Initial Studies

Initial SAR studies are targeted to finding the pharmacophore or functional groups essential for the activity of GSH. The methodology often used with peptides is to test fragments of the peptide to find the active region. GSH contains only three residues and therefore presents a much simpler analysis in that we are able to look at the actions of the individual amino acids and some of the dipeptide combinations. GSH may be regarded as a longer analog of Glu, the backbone chain length being 11 atoms for the diacid of GSH as opposed to 5 for Glu (Fig. 4). The comparison is complicated, however, by the presence of the peptide linkages, which add polarity and the capacity for hydrogen bonding to the molecule. Finally, GSH contains the thiol group, which (as we shall see) appears essential to the actions of GSH as a neuromodulator. The actions of the individual amino acid residues are known, as are those of a variety of dipeptides. Glutamic acid itself is a potent excitant in the CNS. Cysteine is less well studied; however, in our investigations this compound produces NMDA-like depolarizations that are reversibly attenuated by AP5 (Table 1). Glycine is known to act at a site in the NMDA channel in various regions of the CNS and is necessary for activation of this channel by NMDA; glycine is known also to be an inhibitory neurotransmitter in the spinal cord.

5.2 Dipeptide Actions

Several dipeptides based on the GSH structure were tested in the wedge preparation. As with the mentioned Cys-containing dipeptides, no attenuation of the depolarizations was found in the presence of AP5 or DNQX. Therefore, all of the

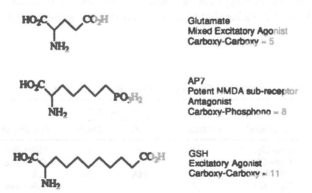

Glutamate
Mixed Excitatory Agonist
Carboxy-Carboxy = 5

AP7
Potent NMDA sub-receptor
Antagonist
Carboxy-Phosphono = 8

GSH
Excitatory Agonist
Carboxy-Carboxy = 11

FIG. 4. Carbon skeleton structures of the different-chain-length acidic amino acids and their pharmacological actions.

TABLE 1. *Comparison of standard agonist responses to GSH and GSH amino acid residues*

| Compound | Response | Antagonism | | Properties |
		AP5	DNQX	
GSH	✓	✗	✗	GSH-like
Glutamate	✓	Partial	Partial	Mixed agonist
AMPA	✓	✗	✓	Non-NMDA
KA	✓	✗	✓	Non-NMDA
NMDA	✓	✓	✗	NMDA-like
Cysteine	✓	✓	✗	NMDA-like
glycine	✗	NA	NA	NA

Cys-containing dipeptides tested to date have displayed characteristics associated with GSH (Table 2).

5.3 Manipulation of the Thiol Function

In an effort to test the limits of substitution at the cysteine thiol functional group, some substituted cysteine-containing residues were tested. Both *S*-methyl-GSH and GSH sulphonic acid produce NMDA-like activity, whereas oxidation to the dimer GSSG makes the compound inactive (Table 3). The inference then is that while substitution at the carboxyl or amino terminals of cysteine turns the NMDA-like cysteine into GSH-like agonists, substitution of the thiol with more oxidized or substituted thiols transforms GSH activity into NMDA-like activity. The one limit we have so far discovered is that the dimerization of GSH to its bulkier counterpart GSSG eliminates activity at neuronal receptors. What is not known is how much manipulation it is possible to perform on the cysteine molecule in order to produce GSH-like activity. The pharmacophore is clearly based on cysteine that has been substituted or protected at the carboxyl and/or amino groups, and it is necessary to produce and test analogs of cysteine that have been protected or substituted at the functional groups.

TABLE 2. *Dipeptide responses in cortical wedge*

| Compound | Response | Antagonism | | Properties |
		AP5	DNQX	
Glu–Cys	✓	✗	✗	GSH-like
Cys–Gly	✓	✗	✗	GSH-like
Glu–Gly	✗	NA	NA	NA
Cys–Ala	✓	✗	✗	GSH-like
Asp–Cys	✓	✗	✗	GSH-like

TABLE 3. *GSH analog responses in rat brain cortical wedge*

Compound	Response	Antagonism		Properties
		AP5	DNQX	
GSSG	×	NA	NA	NA
S-methyl-GSH	√	√	×	NMDA-like
GSH-sulphonic acid	√	√	×	NMDA-like

6. GSH ANTAGONIST DEVELOPMENT

No further development of the study of GSH action can be made until specific and reversible antagonists are found. To date none of the compounds that are structurally related to GSH and unable to produce field potentials have proved to be competitive with GSH in their ability to attenuate GSH depolarizations. GSSG was a good candidate in that there are other systems where the antagonist is closely linked to the agonist. For example, quinolinic acid is an agonist at the NMDA receptor and is a naturally occurring compound in the tryptophan pathway. The actions of quinolinic acid can be antagonized by kynurenic acid, a related substance in the same pathway. However, GSSG does not have antagonist activity, so the answer must lie with some other analog of cysteine, or perhaps a dipeptide containing cysteine and another amino acid. Taking an example from the excitatory amino acid field, it is possible to incorporate non-naturally-occurring amino acid residues into the dipeptide. Phosphono-containing amino acids have been successful in producing antagonists in both the Glu and GABA systems and may prove of value in this emerging field. Likewise, it is not clear that the γ-linkage of glutamate is necessary, and we must investigate the α-substituted di- and tripeptides.

7. FUTURE DIRECTIONS

GSH does not fulfill many of the classical criteria for neurotransmitter function. It is ubiquitous in the mammalian system, abundant in the extreme for a neurotransmitter, and heavily involved in cellular metabolism; it also has an known role as a free-radical scavenger. These drawbacks are not insurmountable; for we have a very good example of a neurotransmitter for which the same is true. Glutamate is ubiquitous, abundant, and involved in metabolism. It has no inactivating enzyme, and every cell seems to respond to it (Curtis and Watkins 1960). At the beginning it was doubted that glutamate could be a neurotransmitter, but time has proven otherwise. This may well prove to be the case for GSH. In its favor, it has depolarizing actions, it is compartmentalized, and it is present in nerve terminals (Hjelle et al., chapter 4, this volume). Finally, it is readily and rapidly inactivated through oxidation to the dimer.

We now have a clearer idea of the parameters that can be manipulated to create other agonists and perhaps antagonists. The evidence is that this peptide is linked to a receptor system that is more abundant than glutamate itself. This may represent a parallel series of neurotransmitter or neuromodulator functions hitherto unknown, and may be of great importance to understanding the mechanisms of brain function.

REFERENCES

Cheeseman, K. H. 1993. Mechanisms and effects of lipid peroxidation. *Molecular Aspects of Medicine* 14(3):191–7.

Cheeseman, K. H., and Slater, T. F. 1993. An introduction to free radical biochemistry. *British Medical Bulletin* 49:481–93.

Curry, K., Magnuson, D. S. K., McLennan, H., and Peet, M. J. 1987. Excitation of rat hippocampal neurones by the stereoisomers of cis- and trans-1-amino-1,3-cyclopentane dicarboxylate. *Canadian Journal of Physiological Pharmacology* 65(11):2196–200.

Curry, K., Peet, M. J., McLennan, H., and Magnuson, D. S. K 1988. Synthesis resolution and absolute configuration of the isomers of the neuronal excitant 1-amino-1,3-cyclopentane dicarboxylic acid. *Journal of Medical Chemistry* 31:864–7.

Curtis, D. R., and Watkins, J. C. 1960. Chemical excitation of spinal neurones by certain acidic amino acids. *Journal of Physiology (London)* 150:656–82.

Davies, J., Francis, A. A., Jones, A. W., and Watkins, J. C. 1980. 2-aminophosphonovalerate (2APV) a highly potent and specific antagonist at spinal NMDA receptors. *British Journal of Pharmacology* 70:52–3.

de Rey-Pailhade, J. 1888a. Sur un corps d'origine organique hydrogénant le soufre à froid. *Comptes rendus de l'Académie des Sciences* 106:1683–94.

de Rey-Pailhade, J. 1888b. Nouvelle researche physiologique sur la substance organique hydrogénant le soufre à froid. *Comptes rendus de l'Académie des Sciences* 107:43–4.

du Vigneaud, V., and Miller, G. L. 1936. A synthesis of glutathione. *Journal of Biological Chemistry* 116:469–76.

Hansen, J. J., and Krogsgaard-Larsen, P. 1990. Structural, conformational and stereochemical requirements of central excitatory amino acid receptors. *Medical Research Review* 10(1):55–94.

Harrington, C. R. 1935. The synthesis of glutathione. *Biochemical Journal* 29:1602–11.

Honoré, T., Davies, S. N., Drejer, J., Fletcher, E. J., Jacobsen, P., Lodge, D., and Nielsen, F. E. 1988. Quinoxalinediones: potent competitive non-NMDA glutamate receptor antagonists. *Science* 241:701–10.

Jaffe, E. R. 1970. Reduction of oxidized glutathione in normal and glucose-6-phosphate-deficient erythrocytes and hemolysates. *Blood* 35:166–72.

Konrad, P. N., Richards, F., Valentine, W. N., and Paglia, D. E. 1972. Glutamyl–cysteine synthetase deficiency. A cause of hereditary hemolytic anemia. *New England Journal of Medicine* 286:557–61.

Kosower, N. S., and Kosower, E. M. 1978. The glutathione status of cells. *International Review of Cytology* 54:109–60.

Lenhoff, H. M. 1961. Activation of the feeding reflex in *Hydra literalis*. 1. Role played by reduced glutathione in quantitative assay of the feeding reflex. *Journal of General Physiology* 45:331–44.

Manzoni, O., Rassendren, F. A., Prézeau, L., Sladeczek, F., Curry, K., and Bockeart, J. 1992. Both enantiomers of 1-aminocyclopentyl-1,3-dicarboxylate are full agonists of metabotropic glutamate receptors coupled to phospholipase C. *Molecular Pharmacology* 42:322–7.

Meister, A. 1974. On the synthesis and utilization of glutamine. *Harvey Lectures* 63:139–78.

Meister, A. 1989. On the biochemistry of glutathione. In *Glutathione Centennial: Molecular Perspectives and Clinical Implications*, ed. N. Taniguchi, T. Higashi, Y. Sakamoto, and A. Meister, 3–22. New York: Academic Press.

Sagara, J., Miura, K., and Bannai, S. 1993. Cystine uptake and glutathione level in fetal brain cells in primary culture and in suspension. *Journal of Neurochemistry* 61:1667–71.

Saija, A., Princi, P., Lanza, M., Scalese, M., Aramnejad, E., Ceserani, R., and Costa, G. 1994. Protective effect of glutathione on kainic acid-induced neuropathological changes in the rat brain. *General Pharmacology* 25(1):97–102.

Schoepp, D. D., and Conn, P. J. 1993. Metabotropic glutamate receptors in brain function and pathology. *TIPS* 14:13–20.

Shaw, C. A., Pasqualotto, B. A., and Curry, K. 1996. Glutathione-induced sodium currents in neocortex. *Neuroreport* 11:1149–52.

Sian, J., Dexter, D. T., Lees, A. J., Daniel, S., Jenner, P., and Marsden, C. D. 1994. Glutathione-related enzymes in brain in Parkinson's disease. *Annals of Neurology* 36:356–61.

Stone, T. W., Perkins, M. N., Collins, J. F., and Curry, K. 1981. Activity of the enantiomers of 2-amino-5-phosphonovaleric acid as specific antagonists of excitatory aminoacids. *Neuroscience* 6:2249–52.

Vanella, A., DiGiacomo, C., Sorrenti, V., Russo, A., Castorini, C., Campisi, A., Renis, M., and Perez-Polo, J. R. 1993. Free radical scavenger depletion in post-ischemic reperfusion brain damage. *Neurochemical Research* 18(12):1337–40.

Weaver, M. S., Lee, Y.-H., Morris, J. L., Randall, P. K., Schaleert, T., and Leslie, S. W. 1993. Effects of in vitro ethanol and fetal ethanol exposure on glutathione-stimulation of *N*-methyl-D-aspartate receptor function. *Alcoholism: Clinical and Experimental Research* 17(3):643–50.

Yamamoto, M., Sakamoto, N., Iwai, A., Yatsugi, S., Hidaka, K., Noguchi, K., and Yusasa, T. 1993. Protective actions of YM737, a new glutathione analogue, against cerebral ischemia in rats. *Research Communications in Chemical Pathology and Pharmacology* 81(2):221–7.

Zhang, F., and Dryhurst, G. 1994. Effects of L-cysteine on the oxidation chemistry of dopamine: new reaction pathways of potential relevance to idiopathic Parkinson's disease. *Journal of Medical Chemistry* 37:1084–98.

PART III

Glutathione and Neurological Disease

Glutathione in the Nervous System
Edited by Christopher A. Shaw
Copyright © 1998 Taylor & Francis

11

Glutathione in Brain Aging and Neurodegenerative Disorders

Gianni Benzi and Antonio Moretti

Institute of Pharmacology, Faculty of Science, University of Pavia, Italy

ABBREVIATIONS

Aβ β-amyloid protein
AD Alzheimer's disease
APP amyloid protein precursor
BSO L-buthionine sulfoximine
t-BuOOH *tert*-butylhydroperoxide
CNS central nervous system
cyclo 2-cyclohexene-1-one
DA dopamine
DDC diethyl dithiocarbamate
L-dopa L-dihydrophenylalanine
DEM diethyl maleate
ETC electron transfer chain
γ-GPT γ-glutamyltranspeptidase
GRI glutathione redox index
GSH reduced glutathione

GSSG oxidized glutathione
GSH-PX glutathione peroxidase (glutathione : hydrogen peroxide oxidoreductase, EC 1.11.1.9)
MDA malondialdehyde
MAO monoamine oxidase [amine : oxygen oxidoreductase (deaminating; flavin-containing), EC 1.4.3.4]
MPP$^+$ 1-methyl-4-phenylpyridinium ion
MPTP 1-methyl-4-phenyl-1,2,3,6-tetrahydropyridine
NO nitric oxide
6-OH-DA 6-hydroxydopamine
PD Parkinson's disease
PPP pentose phosphate pathway
ROS reactive oxygen species
SOD superoxide dismutase

1. INTRODUCTION

This review focuses on changes of the GSH system in aging both in basic conditions and after an exogenous oxidative stress. The possible role of the alterations of the GSH system in the pathogenesis of neurodegenerative diseases (mainly Parkinson's and Alzheimer's diseases) will also be examined.

The biochemistry of the GSH system, the transport of GSH across the blood–brain barrier, and its cellular and subcellular localization in the CNS are described in other chapters of this book. Therefore, mention of these basic aspects will be made only when relevant to the object of the review. In this context, the role of the mitochondrial GSH compartment is especially worth dealing with.

2. MITOCHONDRIAL GLUTATHIONE

Since mitochondria are the site of high oxidative metabolism, they are under physiological oxidative stress. Indeed, it was estimated that about 2 percent of mitochondrial O_2 consumption generates ROS. The mitochondrial electron transfer chain (ETC) is one of the main sources of ROS in aerobic cells, due to electron leakage from energy-transducing sequences leading to the formation of superoxide radicals $O_2^{\bullet-}$ and H_2O_2 (reviewed by Ames, Shigenaga, and Hagen 1995; Benzi and Moretti 1997; Boveris and Chance 1973; Götz et al. 1994; Halliwell 1992; Marzatico and Café 1993; Nohl 1994; Sohal et al. 1994).

Superoxide radicals are then dismutated to H_2O_2 by the action of SOD:

$$2O_2^{\bullet-} + 2H^+ \leftrightarrow H_2O_2 + O_2$$

There are two types of SOD: the mitochondrial Mn SOD interacts with superoxide radicals derived from the ETC, whereas the cytosolic Cu–Zn SOD has a more generic catalytic function. The mitochondria-located MAOs also contribute to H_2O_2

generation during the oxidative deamination of monoamines:

$$RCH_2NH_2 + O_2 + H_2O \rightarrow RCHO + NH_3 + H_2O_2$$

Hydrogen peroxide is potentially deleterious because of: (i) its high membrane permeability and the possibility of migration from the site of formation to other compartments, and (ii) its interaction with superoxides in the presence of Fe^{2+} or Cu^+, giving rise to the highly dangerous hydroxyl radicals $\cdot OH$:

$$O_2^{\bullet-} + H_2O_2 \overset{M^{n+}}{\longleftrightarrow} O_2 + \cdot OH + OH^-$$

where M^{n+} are the metal ions. Indeed, neurons are particularly vulnerable to H_2O_2 (Behl et al. 1994; Buckman, Sutphin, and Mitrovic 1993; Desagher, Glowinski, and Premont 1996; Hinshaw et al. 1993; Whittemore, Loo, and Cotman 1994).

These processes take place even in the brain of young animals, but are enhanced in physiological aging, and even more so in neuropathological diseases, such as Parkinson's and Alzheimer's diseases (reviewed by Beal 1995; Benzi 1990; Benzi and Moretti 1995a; Götz et al. 1994; Simonian and Coyle 1996).

ROS are known to impair ETC components and activity and damage mitochondrial membranes (Benzi et al. 1992; Hillered and Ernster 1983; Sohal and Dubey 1994; Zhang et al. 1990), especially the inner membrane, where they are produced and which holds several iron- and copper-containing enzyme complexes capable of catalyzing the formation of hydroxyl radicals after ion release.

A key role in the protection against the above-described chain of events is played by the H_2O_2 disposition. Neuronal and glial cell cultures and cocultures proved to be very useful for investigating these processes in detail.

At cellular level, astrocytes are less vulnerable to ROS than neurons, and appear to protect them against ROS (Lagenveld et al. 1995). This observation was attributed to the prevalent astrocyte content of GSH and GSH-PX (Damier et al. 1993; Desagher et al. 1996; Geremia et al. 1990; Makar et al. 1994; Savolainen 1978). The CNS has a high concentration of glutathione (1–3 mM, 90 percent of which is in the reduced form), but its localization is controversial. Some biochemical–histochemical data pointed to a prevalent presence in astrocytes (Bolaños et al. 1995; Makar et al. 1994; Raps et al. 1989; Slivka, Mytilineou, and Cohen 1987), whereas others found significant levels also in neurons (Hjelle, Chaudhry, and Ottersen 1994; Hjelle et al., chapter 4, this volume; Pileblad, Eriksson, and Hansson 1991; Sagara, Miura, and Bannai 1993a). Moreover, the astrocyte membranes have glutathione receptors, as shown by studies on binding sites (Guo and Shaw 1992; Guo, McIntosh, and Shaw 1992).

Overall, these observations suggest the existence of glial–neuronal coupling in the oxidant defense system. Astrocytes could act as a source of cysteine for the neuronal GSH synthesis (Sagara et al. 1993a,b). Alternatively, astrocytes may protect neurons by removing H_2O_2 from the external medium through the action of catalase, which is facilitated by the high surface/volume ratio of astrocytes and the consequent rapid diffusion of H_2O_2 inside them. According to this view, neuronal GSH-PX and GSH could also contribute to neuron protection, but to a less extent than astrocytes (Desagher et al. 1996).

A crucial role in H_2O_2 detoxification is played by GSH regeneration through the action of NADPH-dependent glutathione reductase (GSSG-R) and the pentose phosphate pathway (PPP) enzymes, glucose-6-phosphate reductase and 6-phosphogluconate dehydrogenase. Monitoring PPP activity is a useful tool to investigate the metabolic responses to oxidative stress in cortical cultures. Using this procedure, it was found (Ben-Yoseph, Boxer, and Ross 1996) that: (i) exposure to H_2O_2 enhanced PPP in astrocytes more than neurons, and this effect inversely correlated with the sensitivity to H_2O_2 citotoxicity; (ii) partial inhibition of GSH-PX or depletion of GSH decreased the PPP stimulation and exacerbated H_2O_2 neuronal injury. These findings again emphasize the pivotal role of GSH and GSH-PX in neuronal protection from oxidative stress.

As regards the subcellular distribution, mitochondria have about 10–15 percent of the total intracellular concentration of GSH, but do not possess the enzymes for its synthesis. Moreover, at variance with the cytosol, which has a GSSG efflux mechanism, mitochondria are unable to release GSSG and export GSH.

However, mitochondria have a remarkable ability to import GSH from the cytosol and retain it, even when the cytosolic levels are reduced. Two mechanisms come into operation. On the one hand, when the external concentration is lower than 1 mM, GSH is transported into the mitochondrial matrix by a high-affinity, ATP-dependent mechanism (K_m about 60 μM, $V_{max} = 0.5$ nmol/min per milligram of protein), which is saturated at levels of 1–2 mM. On the other hand, the efflux mechanism is very slow when the extramitochondrial GSH levels are low (Meister 1991).

Since catalase is located in peroxisomes, the GSH-PX is the main defense against mitochondrial ROS. The GSH-PX and GSSG-R have both cytosolic and mitochondrial localization. In the mitochondria, GSH-PX is chiefly present in the matrix and in the inner membrane. A lower fraction is located in the contact sites between the inner and outer membranes (Panfili, Sandri, and Ernster 1991; Vitorica, Machado, and Satrùstegui 1984). This location suggests a strategic role in eliminating the peroxides generated by SOD and MAO in inner and outer membranes and the phospholipid hydroperoxides formed in the inner membrane, where lipid peroxidation can take place. Therefore, the mitochondrial GSH system appears to be crucial in order to maintain a normal cellular function and viability.

The protective role of GSH for mitochondria was first described in a number of studies for hepatocytes showing that cytotoxicity occurred only if the intracellular concentration of GSH fell below 10–15 percent, which corresponds to the mitochondrial concentration (reviewed by Reed 1990). The protection by mitochondrial GSH appears to be related to its ability to maintain calcium homeostasis. Mitochondria can take up and retain Ca^{2+}, thus contributing to the regulation of its levels in the cytosol.

Oxidative stress leads to a shift in the redox status and generates high levels of GSSG, which can damage cell integrity and metabolism. Because of their inability to export GSSG, mitochondria are more sensitive than the rest of the cell to the shift in the redox status. In these conditions, the oxidation and depletion of mitochondrial GSH appears to increase the permeability of the inner membrane to Ca^{2+} (through the oxidation of protein thiol groups), an event that is believed to be an initial step in cell injury.

In the CNS, a number of in vivo and in vitro studies were carried out on the effects of manipulating glutathione on mitochondria and ETC.

In vivo, a severe depletion of GSH levels, induced in newborn rats by L-buthionine sulfoximine (BSO), an inhibitor of GSH synthesis, caused striking mitochondrial enlargement and degeneration. Both the decreased GSH and mitochondrial damage were prevented by giving glutathione monoethyl ester (Jain et al. 1991; Meister 1991).

GSH depletion also led to altered mitochondrial ETC (complex IV) (Heales et al. 1995). Unfortunately, because of the use of whole brain, this study did not provide information on possible regional differences.[1]

In an in vitro model of oxidative stress, exposure of rat whole brain mitochondria to low concentrations of *tert*-butylhydroperoxide (*t*-BuOOH) resulted in rapid and irreversible decrease of GSH accompanied by the formation of protein–SSG mixed disulfide.

Brain mitochondria appeared to be more susceptible to oxidative stress than those from the liver, which at low concentrations of *t*-BuOOH were still able to reduce GSSG back to GSH (instead of depleting GSH), and at high concentrations accumulated GSSG (instead of protein mixed disulfide) (Ravindranath and Reed 1990).

A number of recent investigations were made in primary neuron and astrocyte cultures. The sensitivity of neuronal mitochondria to cellular antioxidant status was confirmed in neuronal cultures, where the BSO-induced 93 percent GSH depletion resulted in marked decrease in all ETC complexes assayed in cell lysates: complex I, −34 percent; complex II–III, −60 percent; complex IV, −41 percent (Bolaños et al. 1996).

Other studies dealt with the mechanism(s) of nitric oxide (NO) neurotoxicity, with particular reference to mitochondria and the involvement of GSH. NO has many physiological functions. In the CNS it can act as a neuronal messenger, but, if produced in excess, it may become neurotoxic, probably because of its reaction with the superoxide anion, which generates the potent oxidizing compound peroxynitrite ONOO⁻ (Lipton et al. 1993).

Exposure to NO caused inhibition of ETC (mainly complexes II–III and IV) determined in lysates of neuron and astrocyte cultures, but, whereas the former cells died, the latter did not. A crucial factor in determining the vulnerability of ETC to NO was the availability of GSH (Bolaños et al. 1994, 1995). On the one hand, NO and ONOO⁻ depleted the GSH, inhibited the GSH-PX [probably by nitrosylation of the Se-cysteine catalytic center of the enzyme (Asahi et al. 1995) and GSSG-R (Barker et al. 1996)]. On the other hand, the BSO-induced depletion of GSH either in the brain in vivo or in cultured neurons led to a 55–65-percent increase in NO-synthase activity and cell death (Heales, Bolaños, and Clark 1996).

Neurons are in juxtaposition to astrocytes, which, as mentioned before, provide GSH to neurons, possibly by supplying cysteine. Therefore, in an attempt to reproduce a more physiological condition, the same authors cocultured neurons with astrocytes,

---

[1] The various brain areas have been known for a long time to be biochemically and functionally heterogeneous. It is therefore surprising that this and other studies were made on the whole brain.

either normal or releasing NO. When in culture with normal astrocytes, neurons had 1.7 tinces as much intracellular GSH as neurons cultured alone (22.6 vs. 13.4 nmol/g). When coincubated with NO-releasing astrocytes, neurons had the same high GSH content (22.7 nmol/g) with only a modest (14 percent) decrease in ATP concentration, an insignificant loss of complex IV, and no changes in the other ETC complexes (complexes were determined in cell lysates). Decreases in both ATP and complex IV were prevented by inhibiting nitric oxide synthase (Bolaños et al. 1996). Despite the limited significance of ATP determination alone (in comparison with the energy charge potential), these results are interesting because (i) they confirm the role of cultured astrocytes in maintaining neuronal GSH level, and (ii) they suggest that GSH status is a key factor in determining the susceptibility of cultured neurons to ROS-mediated mitochondrial damage.

However, due to some experimental limitations, it is not easy to tell whether the above-mentioned investigations in cell cultures are predictive of the in vivo situation. For example, mitochondrial ETC complexes in cell lysates might be affected by the presence of other organelles, such as lysosomes and peroxisomes. That the complexes in purified mitochondria behave differently from those in homogenates is indicated by their different sensitivity to the BSO in vivo, despite similar GSH depletion (Heales et al. 1995). Complex activities in cell lysates were also quite different from those in brain mitochondria (Heales et al. 1995; see also the activities reported in other studies, e.g., Battino et al. 1991). Moreover, neurons and astrocytes were obtained from different sources (fetuses and newborn rats, respectively).

3. THE EFFECTS OF PROOXIDANTS ON THE GLUTATHIONE SYSTEM

In the context of the protective role of the glutathione system against oxidative stress, it is interesting to see whether this stress affects GSH function. Table 1 illustrates the effects of prooxidants on GSH and GSSG in the rodent brain.

The *tert*-butylhydroperoxide (*t*-BuOOH) is a lipophilic organic hydroperoxide, which permeates throughout the brain when injected intraventricularly (ivc) and is toxic to the brain (Adams et al. 1993). Cyclohexene-1-one (cyclo) and diethyl maleate (DEM) bear an electrophilic site and are conjugated with GSH, therefore decreasing its level (Masukawa, Sai, and Tochino 1989). The 1-methyl-4-phenyl-1,2,3,6-tetrahydropyridine (MPTP) and its metabolite 1-methyl-4-phenylpyridinium ion (MPP$^+$) are parkinsogenic neurotoxins that damage dopaminergic neurons in the midbrain and nerve terminals in the striatum. These effects appear to involve at least two mechanisms: inhibition of mitochondrial ETC and induction of oxidative stress (Adams and Odunze 1991; Johannessen et al. 1986). The 6-hydroxydopamine (6-OHDA) is another neurotoxin, which selectively lesions nigrostriatal dopamine pathway and induces ROS generation (Cohen and Heikkila 1974).

Despite some discrepancies, most of the above-mentioned compounds affect the glutathione system either by lowering GSH, GSSG-R, GSH-PX, or increasing GSSG.

TABLE 1. *Effect of oxidative stress on brain GSH system in young adult rodents*

Peroxidative compound[a]	Species and cerebral region	Dose (mg/kg), route, and time of assay after treatment	Parameter assayed	Changes vs. controls	Ref.[b]
t-BuOOH	Mouse: Cortex Striatum Thalamus Hippocampus	116, icv, 20 min	GSSG-R	−30% −50% −47% −37%	(a)
	Cortex Striatum Midbrain	93, icv, 5 min	GSSG/GSH	×168 ×159 ×7	(b)
Cyclo	Rat: Forebrain	100 × 1, ip, 18 h 100 × 2, ip, 18 h 100 × 3, ip, 18 h	GSH	−19% −23% −32%	(c)
	Striatum	0.1 ml/kg, sc, 24 h	GSH	−45%	(d)
Cyclo + BSO	Rat: Striatum	0.1 ml/kg, sc, 24 h	GSH	−80%	(d)
DEM	Mouse: Brain	1 g/kg, ip, 3 h	GSH	−58%	(e)
	Rat: Striatum	0.8 ml/kg, sc, 24 h	GSH	−42%	(d)
DEM + BSO	Rat: Striatum	0.8 ml/kg, sc, 24 h	GSH	−78%	(d)
MPTP	Mouse: Striatum S. nigra	30 × 5, ip, 24 h	GSH	Unchanged −34%	(f)
	Striatum Brain stem	40, ip, 1 and 24 h 40, ip, 1 h 40, ip, 24 h		Unchanged +13% −26%	(g)
	Cortex Striatum Midbrain	25 × 4, ip, 2 h	GSSG/GSH	+440% +178% +74%	(h)
	Rat: Striatum (3 mo) Striatum (6 mo) Brain stem (3 mo) Brain stem (6 mo)	35, ip, 1 h	GSH	+10% −14% Unchanged −23%	(i)
6-OHDA	Rat: Striatum	24 h	GSH GSH-PX	−23% −30%	(l)

[a]*t*-BuOOH = *tert*-butylhydroperoxide; Cyclo = 2-cyclohexene-1-one; BSO = L-buthionine sulfoxime; DEM = diethyl maleate; MPTP = 1-methyl-4-phenyl-1,2,3,6-tetrahydropyridine; 6-OHDA = 6-hydroxydopamine.

[b](a) Chang et al. (1995); (b) Adams et al. (1993); (c) Benzi et al. (1991a); (d) Pileblad and Magnusson (1990); (e) Masukawa et al. (1989); (f) Ferraro et al. (1986); (g) Yong et al. (1986); (h) Odunze et al. (1990); (i) Desole et al. (1995); (l) Kumar et al. (1995).

Of note are the following observations:

1. The GSH depletion induced by *cyclohexene-1-one* resulted in alterations of mitochondrial ETC, as shown by the rapid decrease in cytochrome oxidase (complex IV), followed by that in cytochrome *c* and subsequently in complexes I and II (Benzi et al. 1991a). This finding further supports the protective activity of GSH towards mitochondrial function (section 2).

2. The GSH depletion induced by *cyclohexene-1-one* and *diethyl maleate* was aggravated by inhibiting GSH synthesis with BSO (Pileblad and Magnusson 1990).

3. *MPTP* did not greatly affect the GSH concentration (except in the substantia nigra), but caused GSSG accumulation, indicating impaired regeneration (Odunze, Klaidman, and Adams 1990). Moreover, the BSO-induced depletion of GSH potentiated MPTP and MPP$^+$ toxicity in the dopaminergic neurons of substantia nigra (Wüllner et al. 1996). The antioxidants α-tocopherol and β-carotene, which in previous studies had offered partial protection against MPTP-induced damage of nigrostriatal neurons, prevented the loss of GSH in the brain stem (Yong, Perry, and Krisman 1986).

4. The dopamine-depleting effect of 6-*OHDA* in the striatum was also potentiated by inhibiting GSH synthesis with BSO (Pileblad, Magnusson, and Fornstedt 1989).

4. BRAIN AGING AND THE GLUTATHIONE SYSTEM

Glutathione influences and maintains the overall redox cellular potential and therefore has a crucial role in preventing oxidative damage, preserving protein –SH groups, stabilizing cell membranes, and detoxifying xenobiotics. Because of its function as a cofactor for GSH-PX, it is also involved in the detoxification of hydrogen peroxide and lipid peroxides. Since these functions have been implicated in health and longevity, the level of glutathione was suggested to affect the health of individuals at molecular, cellular and organ levels.

Initial work on rodents and mosquitos indicated that glutathione declined with aging due to decreased synthesis (Hazelton and Lang 1980, 1983). A causal role of glutathione was inferred from the observation that correction of its deficiency by administration of its precursors enhanced longevity in the mosquito (Richie, Mills, and Lang 1987).

Some experimental findings in rodents indicated that the blood compartment could give a picture of overall glutathione status in a subject. Therefore, other studies investigated the age-related changes of glutathione in human blood (where it is prevalently localized in erythrocytes). These studies led to mixed, inconclusive findings, though suggesting, as a whole, an association of high glutathione levels with good health in aging (Julius et al. 1994; Lang et al. 1992; Richie and Lang 1986). In contrast, recent papers have reported no age-related changes in the overall antioxidant status of rat serum and brain (Cao, Giovanoni, and Prior 1996), or in the antioxidant enzymes SOD and catalase in human blood (de la Torre et al. 1996).

The findings obtained in such a complex compartment as serum/blood are obviously to be considered with great caution, even more so if an overall parameter, such as the total antioxidant capacity, has been examined. This statement does not mean that determinations of single, well-defined parameters in an organ or tissue are absolutely reliable. The data reported in this and in the following sections as well as those described in a previous review (Benzi and Moretti 1995a), clearly demonstrate the discrepancies between different investigations of each parameter.

TABLE 2. Age-related changes of reduced and oxidized glutathione in the rodent brain

Species	Brain regions	Comparison between ages (months)	GSH (%)	GSSG (%)	GRI[a] (%)	GSSG/total (%)	References
Rat	Forebrain	15 vs. 5	+26	Unchanged	+18		Benzi et al. (1988)
		35 vs. 15	-40	+27	-53		
		35 vs. 5	-24	+37	-44		
	Cortex	14 vs. 5	Unchanged	Unchanged	Unchanged		Iantomasi et al. (1993)
		27 vs. 14	-30	+66	-55		
		27 vs. 5	-30	+60	-50		
	Striatum	18 vs. 3	-23				Desole et al. (1993)
	Cortex	30 vs. 3	-42				Ravindranath et al. (1989)
	Striatum		-35				
	Hippocampus		-43				
	Thalamus		-39				
	Cerebellum		-43				
	Brain stem		Unchanged				
Mouse	Cortex	26 vs. 6	Unchanged				Chen et al. (1989)
	Hippocampus		Unchanged				
	Brain stem		Unchanged				
	Cortex	31 vs. 26	-40				
	Hippocampus		-45				
	Brain stem		-50				
	Cortex	8 vs. 2	-21	+26	-37		Adams et al. (1993)
	Striatum		-21	+157	-69		
	Midbrain		-23	+78	-56		
	Whole brain	24 vs. 12		+10			Mo et al. (1995)
Gerbil	Cortex	15 vs. 3				+30	Zhang et al. (1993)
		24 vs. 15				Unchanged	
	Hippocampus	15 vs. 3				+26	
		24 vs. 15				+12	
	Striatum	15 vs. 3				+22	
		24 vs. 15				+13	

[a]Glutathione redox index, $([GSH] + 2[GSSG])/(2[GSSG] \times 100)$.

TABLE 3. *Age-related changes of reduced and oxidized glutathione in rodent brain mitochondria*

| Parameter | Comparison between ages (months) | | |
| | Synaptic mitochondria[a] | Mitochondria in toto[b] | |
	18 vs. 6	24 vs. 6	24 vs. 3
GSH	+21%	Unchanged	Unchanged
GSSG		+83%	+460%
GSSG/GSH		+110%	+500%
Species	Mouse	Mouse	Rat

[a]Martínez et al. (1994).
[b]de la Asuncion et al. (1996).

The following reasons may account for discordant results: different species, strain, brain regions, cellular and subcellular preparations, methods of assay, etc. Another cause for discrepancies is the age of the animals. In most cases, only two age groups (young and old) were examined, and longitudinal investigations were rare. The absence of intermediate groups made it difficult to determine whether the changes observed were really due to aging or to maturity effects. In some studies, the age ranges (e.g., 2- vs. 8-month-old mice, as in Adams et al. 1993, or 3- vs. 6-month-old rats, as in Desole et al. 1995) did not provide satisfactory information on aged animals.

Nevertheless, despite conflicting results, some reasonable conclusions can be inferred from the available literature on the GSH system.

As shown in Table 2, the *level of GSH* was lower in the brains of aged rodents, especially in comparison with mature animals (Adams et al. 1993; Benzi et al. 1988; Chen, Richie, and Lang 1989; Desole et al. 1993; Iantomasi et al. 1993; Ravindranath, Shivakumar, and Anandatheerthavarada 1989). Since GSSG was increased proportionally to GSH decline, the resulting glutathione redox index (GRI) was lower in the aging brain (Adams et al. 1993; Benzi et al. 1988; Iantomasi et al. 1993).

As regards the *mitochondrial GSH* (Table 3), a significant 21-percent increase in the concentration of total glutathione in synaptic mitochondria of old compared with adult mice was reported, concomitantly with a parallel 31-percent decrease in cytochrome oxidase activity (complex IV of the ETC) and a 17-percent decline in lipid peroxide (assessed as MDA) (Martínez et al. 1994). Another study found that the total glutathione and GSH were unchanged in brain mitochondria of mice, and were insignificantly changed in rats. In both species, GSSG accumulated, resulting in an increment of 110 percent (mice) or 500 percent (rats) of the GSSG/GSH ratio (de la Asuncion et al. 1996). The rise in this ratio correlated with that in the amount of 8-hydroxy-2'-deoxyguanosine (a biomarker of oxidative mtDNA damage). Moreover, oral antioxidant treatment protected both against glutathione oxidation and mtDNA damage.

As a whole, these two studies: (i) confirm the age-related alterations of the ETC and the concomitant oxidative stress; (ii) further prove that, at least in physiological

TABLE 4. *Age-related changes of glutathione peroxidase activity in the rodent brain*

Cerebral region	Comparison between ages (months)							
	35 vs. 5	18 vs. 3	30 vs. 8	22 vs. 3	12 vs. 3	24 vs. 12	30 vs. 0.25	26 vs. 8
Whole brain							Unchanged	Unchanged
Cortex	+25%	+31%	Unchanged	Unchanged	Unchanged	Unchanged		
Hippocampus	Unchanged	Unchanged	Unchanged		Unchanged	Unchanged		
Substantia nigra	Unchanged	Unchanged	Unchanged	Unchanged	Unchanged	Unchanged		
Striatum	+18%	+12%	Unchanged					
Thalamus	-13%	-11%		Unchanged	Unchanged	Unchanged		
Cerebellum		Unchanged	Unchanged	Unchanged	-30%	Unchanged		
Medulla				Unchanged				
Cervical cord				Unchanged				
Species	Rat	Rat	Rat	Rat	Rat	Rat	Mouse	Rat
References	Benzi et al. (1989a)	Mizuno and Ohta (1986)	Carrillo et al. (1992)	Ansari et al. (1989)	Ciriolo et al. (1991)	Ciriolo et al. (1991)	de Haan et al. (1992)	Matsuo et al. (1992)

Cerebral region	Comparison between ages (months)						
	26 vs. 6	24 vs. 12	22–23 vs. 4–5	24 vs. 4	24 vs. 4	28 vs. 9	25 vs. 3[a]
Whole brain	Unchanged	Unchanged	Unchanged	Decreased[b]	Unchanged	Unchanged	Unchanged
Cortex							
Species	Rat	Mouse	Mouse	Rat	Rat	Rat	Rat
References	Rao et al. (1990)	Mo et al. (1995)	Cristiano et al. (1995); de Haan et al. (1995)	Zhang et al. (1989)	Cand and Verdetti (1989)	Barja de Quiroga et al. (1990)	Geremia et al. (1990)

[a] In both neurons and glia.
[b] Data not provided.

conditions, mitochondria are able to maintain their GSH level, but GSSG can accumulate; (iii) underline the importance of preserving a reduced glutathione status to protect mitochondria against oxidative damage of key molecules like DNA.

Table 4 summarizes the numerous investigations on changes of *glutathione peroxidase activity* in the aging rodent brain. Most studies point to unchanged activity, although a slight increase was also reported. The GSH-PX gene expression was unaffected by aging (de Haan, Newman, and Cola 1992; de Haan et al. 1995; Rao, Xia, and Richardson 1990). The constant GSH-PX activity in the aged murine brain is even more significant because of enhanced Cu/Zn SOD activity and expression. Conceivably, the changes in these two antioxidant enzymes during aging lead to alteration in the balance between generation and disposition of H_2O_2, and this may explain the increased lipid peroxidation, which is a clear index of oxidative stress (Cristiano et al. 1995; de Haan et al. 1995). The observation that SOD levels and the resulting SOD/GSH-PX ratio were elevated in various organs in Down's syndrome (de Haan et al. 1995) further supports the hypothesis that these alterations may be responsible for the premature aging of these individuals.

The activity of *glutathione reductase* (Table 5) seemed to be somewhat lower in the brain of old than of adult rodents. Thus, the resulting redox index of the glutathione enzyme system (Benzi et al. 1989a) declined in later life, though with some region-related differences, since the cortex and striatum are affected earlier. Conversely, the activities of the PPP enzymes, *glucose-6-phosphate reductase* and *6-phosphogluconate dehydrogenase*, were unchanged by aging, or increased slightly (Table 6). Therefore, whereas the reducing equivalents of NADPH were provided to the aging brain to the same extent as in the adult one, the reduction of GSSG to GSH by the NADPH-dependent glutathione reductase was decreased.

Interestingly, the glutathione reductase activity was significantly greater in the brain of mice fed on antioxidants, such as α-tocopherol or propyl gallate (Khanna, Garg, and Sharma 1992). This finding must be considered in the light of the free radical hypothesis of aging (Harman 1992) and the demonstration that treatment with

TABLE 5. *Age-dependent changes of glutathione reductase activity in the rodent brain*

Cerebral region	Comparison between ages (months)					
	35 vs. 5	27 vs. 3	28 vs. 9	23 vs. 8	25 vs. 3[a]	24 vs. 1
Whole brain			−20%	−35%		−11%
Cortex	−41%	−30%			Unchanged	
Substantia nigra	−28%					
Caudate–putamen	−18%					
Thalamus	−35%					
Species	Rat	Rat	Rat	Mouse	Rat	Mouse
Reference	Benzi et al. (1989a)	Iantomasi et al. (1993)	Barja de Quiroga et al. (1990)	Khanna et al. (1992)	Geremia et al. (1990)	Mo et a (199!

[a]In both neurons and glia.

TABLE 6. *Age-dependent changes of the activity of glucose-6-phosphate dehydrogenase and 6-phosphogluconate dehydrogenase in the rat brain*

	Comparison between ages (months)				
	Glucose-6-phosphate dehydrogenase			6-phosphogluconate dehydrogenase	
Cerebral region	18 vs. 3	35 vs. 5	27 vs. 3	18 vs. 3	35 vs. 5
ortex	+22%	+30%	Unchanged	+19%	+25%
ippocampus	Unchanged			+35%	
ubstantia nigra	Unchanged	+10%		Unchanged	+12%
triatum	Unchanged	+10%		+11%	+21%
halamus	Unchanged	Unchanged		+25%	+25%
mygdala	+22%			Unchanged	
ypothalamus	+38%			Unchanged	
erebellum	+16%			Unchanged	
eferences	Mizuno and Ohta (1986)	Benzi et al. (1989a)	Iantomasi et al. (1993)	Mizuno and Ohta (1986)	Benzi et al. (1989a)

exogenous antioxidants, or induction of endogenous tissue antioxidants, increased the mean animal life span (Harman 1995a; Sohal et al. 1984) and improved the cognitive performance of aged rats (Socci, Crandall, and Arendash 1995).

5. THE EFFECTS OF PROOXIDANTS ON THE GLUTATHIONE SYSTEM IN BRAIN AGING

As discussed in section 3, various prooxidants cause critical alteration of the glutathione system. The findings summarized in section 4 indicate that, with increasing aging, the brain becomes less capable of protecting itself against oxidative stress, as shown by increased GSSG/GSH ratio and decreased glutathione redox index. The oxidation of GSH may conceivably result in neuron damage owing to their low GSH levels.

It was therefore interesting to examine whether aging further affects the response of the glutathione system to exogenous oxidative stress. Table 7 illustrates the effects of a number of prooxidants on the GSH system in the aged rat brain.

In addition to *cyclohexene-1-one* and *MPTP* (whose mechanisms of action were described in section 3), the following compounds were studied: *diethyl dithiocarbamate* (DDC) and *paraquat*, two agents that raise the superoxide concentration by inactivating SOD and interfering with NADH-diaphorase, respectively; *aminotriazole*, which inhibits catalase; H_2O_2; and *diamide*, a specific glutathione oxidant.

In young adult animals, these agents affected glutathione differently. Some enhanced GSH levels (DDC, paraquat, aminotriazole), but others (diamide) did not. The GSSG level was unchanged or increased (aminotriazole, diamide). The resulting glutathione redox index was mostly unchanged or even increased (DDC). These effects were slight (10–20 percent) and rather similar in young, mature, and old animals.

TABLE 7. Effect of prooxidants on reduced and oxidized glutathione in the forebrain of rats of different ages

Prooxidant	Age (months)	GSH	GSSG	GRI[a]	Reference
DDC	5	+19%	Unchanged	+13%	Benzi et al. (1989
	15	+17%	Unchanged	+10%	
	25	+22%	Unchanged	+17%	
Paraquat	5	+13%	Unchanged	Unchanged	Benzi et al. (1990
	15	+15%	Unchanged	Unchanged	
	25	+22%	Unchanged	+19%	
Aminotriazole	5	+20%	+15%	Unchanged	Benzi et al. (1990
	15	+12%	+15%	Unchanged	
	25	+18%	+15%	Unchanged	
H_2O_2	5	−14%	+36%	−33%	Benzi et al. (1989
	15	−23%	+36%	−41%	
	25	−30%	+62%	−56%	
Diamide	5	Unchanged	+14%	Unchanged	Benzi et al. (1990
	15	Unchanged	+17%	Unchanged	
	25	Unchanged	+25%	−27%	
Cyclohexene-1-one	5	−30%			Benzi et al. (199-
[9 h after dosing	15	−43%			
vs. $t = 0$]	25	−51%			
MPTP	3	Unchanged			Desole et al. (19
(in brain stem)	18	−17%			

[a]Glutathione redox index, ([GSH] + 2[GSSG])/(2[GSSG] ×100).

The rise in GSH might be related to a stress-induced compensatory mechanism similar to that described in fibroblasts and endothelial cells, where oxidant compounds stimulated the incorporation of precursor aminoacids into GSH (Deneke et al. 1989; Ochi 1993).

In contrast to the above-mentioned compounds, H_2O_2 decreased GSH and raised GSSG, therefore lowering the glutathione redox index. As expected (section 3), the level of GSH was also depleted by cyclohexene-1-one.

The most important result obtained with GSH-lowering compounds is that their effects were potentiated by aging (H_2O_2, cyclohexene-1-one) or appeared in the aging brain (MPTP). Diamide also lowered the glutathione redox index in old, but not in young or adult, rats.

In the case of MPTP, this finding is particularly relevant because of the observations that the aging brain is more susceptible to this toxin than the young brain in terms of (i) alteration of the dopaminergic system (Desole et al. 1993), (ii) neurotoxicity and lethality in rodents (Ali, David, and Newport 1993; Gupta et al. 1986; Jarvis and Wagner 1985; Ricaurte et al. 1987) and monkeys (Ovadia, Zhang, and Gash 1995; Rose et al. 1993), and (iii) Parkinsonian features (Gupta et al. 1986; Ovadia et al. 1995; Rose et al. 1993). At first glance, these results could be simply attributed to the age-related increase in the activity of glial MAO B, the enzyme responsible for the transformation of MPTP into the active metabolite MPP^+ (pharmacokinetic interpretation) (Irwin et al. 1992). However, the increase in MAO B was found in aged rodents, but not confirmed in aged monkeys (Langston 1996), a species where the higher age-dependent sensitivity to MPTP is also present. Indeed, a significant inverse correlation was found between age and the dose of MPTP needed to produce stable, moderate Parkinsonian symptoms. This observation points to an increased vulnerability of the nigrostriatal system to insult, "whether it be of endogenous or exogenous origins" (Langston 1996). These considerations are especially relevant in understanding the pathogenesis of Parkinson's disease, of which MPTP is thought to be a model.

6. THE GLUTATHIONE SYSTEM IN PARKINSON'S AND ALZHEIMER'S DISEASES

6.1 Parkinson's Disease

Parkinson's disease (PD) is caused by the loss of dopaminergic cells in the substantia nigra. Its pathogenesis is probably multifactorial. So far, at least two mechanisms appear to be mainly involved and probably intercorrelated, alteration of mitochondrial ETC and oxidative stress, which in turn may raise intracellular calcium.

A specific defect of complex I was reported in the substantia nigra of PD patients (reviewed by Schapira 1994). However, these observations were made in homogenates, and the data were corrected for citrate synthase. Appropriate studies on mitochondrial populations are needed before a definite conclusion on the involvement of complex I in PD is obtained.

The involvement of Ca^{2+} in PD is probable, though the literature is scant (De Erausquin, Costa, and Hanbauer 1994). An increased expression of m-calpain specific for the mesencephalon of PD patients has been recently described and ascribed to raised intracellular Ca^{2+} concentration (Mouatt-Prigent et al. 1996).

The age dependence of PD has given rise to the assumption that oxidative stress may have a role in its pathogenesis. This theory is supported by the following observations, mostly made in the substantia nigra, and in some cases also in the putamen and caudate nucleus, of PD patients compared with age-matched subjects (reviewed by Fahn and Cohen 1992; Gerlach et al. 1994; Jenner and Olanow 1996):

1. Increased formation of H_2O_2 in the mitochondria due to enhanced Mn SOD and MAO B, the enzyme involved in the oxidation of dopamine.
2. Altered metabolism of iron. Despite some disputes regarding the form and origin of iron in PD substantia nigra (Jenner and Olanow 1996), the above-mentioned results suggest an increased rate of •OH synthesis.
3. Increased density of NO-synthase-containing cells (Hunot et al. 1996) and raised concentration of nitrite, a metabolite of NO, in the cerebrospinal fluid of PD subjects (Qureshi et al. 1995).

The concept of oxidative stress implies that antioxidant defenses may be inadequate. Indeed, the available data showed:

4. Lower GSH and higher GSSG concentrations (Perry, Godin, and Hansen 1982; Sian et al. 1994a; Sofic et al. 1992). Interestingly, these alterations were specific for

PD and were not found in other neurodegenerative disorders, also characterized by profound nigral cell loss, namely, multiple-system atrophy and progressive supranuclear palsy (Sian et al. 1994a).

5. Doubled activity of γ-glutamyltranspeptidase (γ-GTP), the membrane enzyme involved in glutathione degradation and translocation. In contrast, the activity of the synthetic enzyme γ-glutamylcysteine synthetase was unaffected (Sian et al. 1994b)

6. Reduced catalase activity (Ambani, Van Woert, and Murphy 1975). However, the findings concerning GSH-PX activity are controversial, both decrease (Ambani et al. 1975; Kish, Morito, and Hornykiewicz 1985) and no change (Marttila, Lorentz, and Rinne 1988; Sian et al. 1994b) having been described. An immunohistochemical study on GSH-PX cellular localization in the mesencephalon showed that: (i) the glial cells in substantia nigra pars compacta (the most vulnerable to PD) of control subjects had the lowest density of GSH-PX immunoreactivity, indicating lower protection of dopaminergic neurons surrounded by these cells; (ii) the density of GSH-PX-positive cells in PD was increased in relation to the severity of the disease, probably as a reaction to excess formation of ROS and as an attempt to protect the surviving dopaminergic neurons (Damier et al. 1993). Conceivably, the discrepancy between immunohistochemical and biochemical data was due to the different entities measured: cells containing the enzyme protein (whether active or inactive) versus enzyme activity, which is critical for the protection against H_2O_2, but was assayed in homogenates and provides average measurement on a heterogeneous cell population.

7. Indirect evidence for oxidative stress: increased lipid hydroperoxide and peroxide, coupled with reduced levels of polyunsaturated fatty acids (the substrate for lipid peroxidation) (Dexter et al. 1989, 1994a; Yoritaka et al. 1996); raised level of 8-hydroxy-2′-deoxyguanosine, an index of oxidative-stress-induced DNA damage (Sanchez-Ramos, Övervik, and Ames 1994).

The importance of oxidative stress in PD is also suggested by the following findings: (i) the surviving neurons of nigrostriatal tract in the PD brain are characterized by enhanced dopamine (DA) turnover, an apparent compensatory response to the loss of neurons; (ii) normal dopaminergic neurons are rich in DA and L-dopa, which can undergo autoxidation, as demonstrated in vitro and in vivo. Both the enzymatic and spontaneous DA oxidations lead to the formation of ROS (H_2O_2, superoxide, hydroxyl radicals) and to neurotoxicity.

The catechol autoxidation also generates quinones and semiquinones. The electron-deficient quinones attract nucleophiles, like the thiol group on cysteine and GSH, giving rise to the formation of 5-S-cysteinyl adducts. Therefore, oxidation of DA (as well as L-dopa) could lead to GSH depletion.

An additional mechanism by which DA and L-dopa may contribute to GSH depletion is their reaction with superoxide anion: incubation in vitro of GSH with superoxide and H_2O_2 induced limited GSH depletion, due to the slow reaction of GSH with ROS. However, the loss of GSH was greatly accelerated when L-dopa or DA was also present (Spencer, Jenner, and Halliwell 1995). The role of GSH in protecting against

DA-induced oxidative stress and neurotoxicity is further emphasized by the finding that GSH depletion aggravated this toxicity in neuronal cultures and that cotreatment with *N*-acetylcysteine or GSH provided partial protection (Offen et al. 1996).

On the whole, the above-mentioned results suggest that glutathione depletion in substantia nigra is involved in the pathogenesis of PD. In this connection, the following comments can be made:

1. In addition to neurons, it is likely that GSH loss mainly affects glia (its prevalent localization), but whether the cytosolic or the mitochondrial compartment is more affected remains unknown (Jenner and Olanow 1996).
2. Oxidative stress could deplete GSH (the GSH-PX was not reduced, but unchanged or even enhanced). The decreased GSH could also be due to increased efflux out of the cell. Growing conversion to GSSG may contribute, because the excess GSSG can be removed from cells, via enhanced γ-GTP (Sian et al. 1994b). All these events might occur in the pathological stage of the disease. They could be preceded by some age-related oxidative-stress-inducing alterations, for example, increased DA enzymatic oxidation and autoxidation (Fornstedt, Pileblad, and Carlsson 1990).
3. In a study on individuals with incidental Lewy bodies (presymptomatic stage of the disease), the decrease in nigral GSH (-35 percent) was accompanied by a statistically insignificant reduction in complex I (-29 percent) and unchanged levels and turnover of DA, as well as normal iron and ferritin levels (Dexter et al. 1994b). These results were interpreted as showing GSH decrease as the early component of PD, an implication of great significance from the point of view of pathology and that of potential therapeutical use. However, the difference between the effects on GSH and complex I was small, and a concomitance might also be envisaged. Further experiments with more subjects and more specific techniques regarding mitochondrial isolations (see section 2) appear to be necessary before we can be certain about the above-mentioned conclusion.
4. A clue to understanding these events has been provided by experiments on BSO-induced GSH depletion in the rat brain. A decline of 40–60 percent in basal ganglia GSH (similar to that occurring in PD) did not reduce the number of dopaminergic cells in the substantia nigra (Wüllner et al. 1996) or affect brain mitochondrial complex I (Heales et al. 1995). Only with greater GSH loss was the mitochondrial ETC altered and did the mitochondrial damage appear (section 1). Whether these experimental, acute conditions provide a relevant model of PD remains to be established. Nevertheless, the GSH loss could be high in the vicinity of specific neurons (Jenner and Olanow, 1996).

In conclusion, as pointed out by Di Monte, Chan, and Sandy (1992), a number of links exist between oxidative stress and mitochondrial damage in PD, and glutathione may represent one such link. Regardless of the initial insult, a sequence of events might be triggered involving both processes and contributing to cell damage. "In this respect, a rigid distinction between oxidative stress and the mitochondrial hypothesis of neurodegeneration in PD may only be artificial" (Di Monte et al. 1992).

6.2 Alzheimer's Disease

Alzheimer's disease (AD), the major cause of dementia, is characterized by the presence of extracellular deposits of amyloid β (Aβ) in plaques, abnormal intracellular cytoskeletal filaments in neurofibrillar tangles, and neuron and synapse loss (Selkoe 1991). In contrast with PD, in which the substantia nigra is specifically damaged, the neuropathological hallmarks of AD are distributed in various cortical and subcortical areas.

Several potential pathogenetic mechanisms are currently debated. In the context of a multifactorial pathogenesis, the involvement of the markedly neurotoxic Aβ and its precursor APP and their relationship with the neurofibrillar pathology are central issues (reviewed by Cummings et al. 1996; Iversen et al. 1995; Joachim and Selkoe 1992; and Yankner 1996).

Since aging is an uncontested risk factor for AD (as for PD), oxidative stress has been implicated in its pathogenesis. The evidence for increased oxidative stress in the AD brain comes from a number of observations (reviewed by Benzi and Moretti 1995b; Harman 1995b; and Yankner 1996):

1. Alteration of mitochondrial function (decrease in complex IV and cytochrome oxidase activity), which is likely to lead to electron leakage in the ETC and the consequent formation of superoxide radicals.
2. High activity of SOD and MAO B, resulting in more H_2O_2 formation.
3. Alteration of iron homeostasis, which stimulates the generation of hydroxyl radicals from H_2O_2 and superoxide.
4. Sparing of NO-synthase-containing neurons in a cerebral area (the hippocampus) that is severely affected in AD (Hyman et al. 1992). This may lead to an imbalance of input to remaining neurons and contribute to further neuronal loss. The elevation of NO-synthase activity in brain microvessels (Dorheim et al. 1994) also suggests increased formation of potentially neurotoxic mediator in AD cerebral microcirculation.
5. Interaction between ROS and Aβ: On the one hand, ROS promote Aβ aggregation, a process that is essential for the deposition of amyloid fibrils in senile plaques. On the other hand, Aβ causes increased production of H_2O_2 and undergoes fragmentation and radicalization, therefore generating free-radical neurotoxic peptides.
6. Inflammatory reaction triggered by microglial activation.
7. Increased lipid peroxidation and membrane alteration, protein oxidation, and oxidative damage to DNA, especially the mtDNA.

Surprisingly, little is known about the status of antioxidant defenses in AD. An increased SOD activity not balanced by changes in catalase and glutathione was reported in the blood of probable AD patients (Famulari et al. 1996). In the brain, both normal levels of total glutathione (Perry et al. 1987) and compensatory increase of GSH and GSSG (Adams et al. 1991) were shown. In the latter study, the GSH/GSSG ratio was lower in the hippocampus (-39 percent) and caudate nucleus (-20 percent) of AD patients than in age-matched healthy subjects.

Interestingly, both GSSG and the thiyl radical GS• inhibit the binding of a specific ligand ([^{3}H]QNB) to muscarinic cholinergic receptors (Frey et al. 1996). Cholinergic transmission is crucial to memory and learning, and its alteration is considered as one of the main causes of cognitive disorders in AD.

These scant reports only indicate that the antioxidant defenses may be inadequate to cope with the oxidant stress which is clearly present in the AD brain. In particular, if the activity of SOD is increased without a concomitant enhancement of the activity of the enzymes which dispose of H_2O_2 (mainly GSH-PX) and of GSH levels, then H_2O_2 accumulates and reacts with superoxide and Fe^{2+} to form the hydroxyl radical $\bullet OH$. Thus, the imbalance between SOD and H_2O_2-converting enzymes may result in toxic effects toward lipids, proteins, and DNA, especially in membranes and mitochondria.

7. CONCLUSIONS

In the rodent brain, the age-induced depletion of GSH and the concomitant accumulation of GSSG are probably related to an imbalance between GSH utilization and formation. The GSH-PX activity is unaffected by aging, whereas GSSG-R activity is diminished.

At first glance, the decline of GSH does not seem to have serious repercussions, since, in any case, its concentration does not fall below 0.7–1 mM, a figure that compares well with the V_{max} of GSH-dependent enzymes, in particular the GSH-PX. However, the picture may not be so simple, and the following questions merit further consideration:

1. Potential differences between animals and humans.
2. Imbalance between H_2O_2-producing and -metabolizing enzymes.
3. Possible variations among different subsets of neurons, even in the same cerebral region, regarding the various constituents of the glutathione system and the sensitivity to oxidative stress.
4. The effects of prooxidant compounds. They may be present in the human brain and aggravate preexisting age-related alterations. That this might be the case is shown by the MPTP neurotoxicity, in which the decline of GSH plays a critical role.

To our knowledge, there are no studies on glutathione system in the "physiological" aging human brain, but, as discussed in section 6.1, various investigations have been made in the PD brain (especially in the substantia nigra). They showed further GSH fall and GSSG accumulation, which probably superimpose on age-induced changes.

PD is an age-related disease. Thus, it is conceivable that even mild modifications caused by oxidative stress and accumulated over decades in the substantia nigra may eventually result in the pathological and clinical signs of the disease. Among these alterations, the continuous formation in some neurons of substantia nigra pars compacta of ROS during DA oxidation, not counterbalanced by enough antioxidant defenses, may be of paramount importance.

In the other age-related neurodegenerative disease (AD), evidence was provided for the presence of oxidative stress, but available data are too limited to reach a logical conclusion regarding the possible role of impaired antioxidant defenses.

ACKNOWLEDGMENTS

The authors thank Mrs. G. Giglio for revision of the English and Mrs. G. Corbellini for secretarial work.

REFERENCES

Adams, J. D., Jr., and Odunze, I. N. 1991. Biochemical mechanisms of 1-methyl-4-phenyl-1,2,3,6-tetrahydropyridine toxicity. Could oxidative stress be involved in the brain? *Biochemical Pharmacology* 41:1099–105.

Adams, J. D., Jr., Klaidman, L. K., Odunze, I. N., Shen, H. C., and Miller, C. A. 1991. Alzheimer's and Parkinson's disease. Brain levels of glutathione, glutathione disulfide and vitamin E. *Molecular and Chemical Neuropathology* 14:213–26.

Adams, J. D., Jr., Wang, B., Klaidman, L. K., LeBel, C. P., Odunze, I. N., and Shah, D. 1993. New aspects of brain oxidative stress induced by tert-butylhydroperoxide. *Free Radicals Biology and Medicine* 15:195–202.

Ali, S. F., David, S. N., and Newport, G. D. 1993. Age-related susceptibility to MPTP-induced neurotoxicity in mice. *Neuro toxicology* 14:29–34.

Ambani, L. M., Van Woert, M. H., and Murphy, S. 1975. Brain peroxidase and catalase in Parkinson disease. *Archives of Neurology* 32:114–18.

Ames, B. N., Shigenaga, M. K., and Hagen, T. M. 1995. Mitochondrial decay in aging. *Biochimica et Biophysica Acta* 1271:165–70.

Ansari, K. A., Kaplan, E., and Shoeman, D. 1989. Age-related changes in lipid peroxidation and protective enzymes in the central nervous system. *Growth Development and Aging* 53:117–21.

Asahi, M, Fujii, J., Suzuki, K., Seo, H. G., Kuzuya, T., Hori, M., Tada, M., Fujii, S., and Taniguchi, N. 1995. Inactivation of glutathione peroxidase by nitric oxide. Implication for cytotoxicity. *The Journal of Biological Chemistry* 270:21035–9.

Barja de Quiroga, G., Pérez-Campo, R., and López-Torres, M. 1990. Anti-oxidant defenses and peroxidation in liver and brain of aged rats. *Biochemical Journal* 272:247–50.

Barker, J. E., Heales, S. J. R., Cassidy, A., Bolaños, J. P., Land, J. M., and Clark, J. B. 1996. Depletion of brain glutathione results in a decrease of glutathione reductase activity; an enzyme susceptible to oxidative damage. *Brain Research* 716:118–22.

Battino, M., Bertoli, E., Formiggini, G., Sassi, S., Gorini, A., Villa, R. F., and Lenaz, G. 1991. Structural and functional aspects of the respiratory chain of synaptic and nonsynaptic mitochondria derived from selected brain regions. *Journal of Bioenergetics and Biomembranes* 23:345–63.

Beal, M. F. 1995. Aging, energy, and oxidative stress in neurodegenerative diseases. *Annals of Neurology* 38:357–66.

Behl, C., Davis, J. B., Lesley, R., and Schubert, D. 1994. Hydrogen peroxide mediates amyloid β protein toxicity. *Cell* 77:817–27.

Ben-Yoseph, O., Boxer, P. A., and Ross, B. D. 1996. Assessment of the role of the glutathione and pentose phosphate pathways in the protection of primary cerebrocortical cultures from oxidative stress. *Journal of Neurochemistry* 66:2329–37.

Benzi, G. 1990. *Peroxidation, energy transduction and mitochondria during aging.* Paris: John Libbey Eurotext.

Benzi, G., and Moretti, A. 1995a. Age- and peroxidative stress-related modifications of the cerebral enzymatic activities linked to mitochondria and the glutathione system. *Free Radical Biology and Medicine* 19:77–101.

Benzi, G., and Moretti, A. 1995b. Are reactive oxygen species involved in Alzheimer's disease? *Neurobiology of Aging* 16:661–74.

Benzi, G., and Moretti, A. 1997. Contribution of mitochondrial alterations to brain aging. In *The Aging Brain*, ed. M. P. Mattson, and J. W. Geddes, Advances in cell aging and gerontology, vol. 2, pp. 129–60. JAI Press Inc.

Benzi, G., Pastoris, O., Marzatico, F., and Villa, R. F. 1988. Influence of aging and drug treatment on the cerebral glutathione system. *Neurobiology of Aging* 9:371–5.

Benzi, G., Pastoris, O., Marzatico, F., and Villa, R. F. 1989a. Cerebral enzyme antioxidant system. Influence of aging and phosphatidylcholine. *Journal of Cerebral Blood Flow and Metabolism* 9:373–80.

Benzi, G., Pastoris, O., Marzatico, F., and Villa, R. F. 1989b. Age-related effect induced by oxidative stress on the cerebral glutathione system. *Neurochemical Research* 14:473–81.

Benzi, G., Marzatico, F., Pastoris, O., and Villa, R. F. 1990. Influence of oxidative stress on the age-linked alterations of the cerebral glutathione system. *Journal of Neuroscience Research* 26:120–8.

Benzi, G., Curti, D., Pastoris, O., Marzatico, F., Villa, R. F., and Dagani, F. 1991a. Sequential damage in mitochondrial complexes by peroxidative stress. *Neurochemical Research* 16:1295–302.

Benzi, G., Pastoris, O., Gorini, A., Marzatico, F., Villa, R. F., and Curti, D. 1991b. Influence of aging on the acute depletion of reduced glutathione induced by electrophilic agents. *Neurobiology of Aging* 12:227–31.

Benzi, G., Pastoris, O., Marzatico, F., Villa, R. F., Dagani, F., and Curti, D. 1992. The mitochondrial electron transfer alteration as a factor involved in the brain aging. *Neurobiology of Aging* 13:361–8.

Bolaños, J. P., Peuchen, S., Heales, S. J. R., Land, J. M., and Clark, J. B. 1994. Nitric oxide–mediated inhibition of the mitochondrial respiratory chain in cultured astrocytes. *Journal of Neurochemistry* 63:910–16.

Bolaños, J. P., Heales, S. J. R., Land, J. M., and Clark, J. B. 1995. Effect of peroxynitrite on the mitochondrial respiratory chain: differential susceptibility of neurons and astrocytes in primary cultures. *Journal of Neurochemistry* 64:1965–72.

Bolaños, J. P., Heales, S. J. R., Peuchen, S., Barker, J. E., Land, J. M., and Clark, J. B. 1996. Nitric oxide–mediated mitochondrial damage: a potential neuroprotective role for glutathione. *Free Radical Biology and Medicine* 21:995–1001.

Boveris, A., and Chance, B. 1973. The mitochondrial generation of hydrogen peroxide. General properties and effect of hyperbaric oxygen. *Biochemical Journal* 134:707–16.

Buckman, T. D., Sutphin, M. S., and Mitrovic, B. 1993. Oxidative stress in a clonal cell line of neuronal origin: effects of antioxidant enzyme modulation. *Journal of Neurochemistry* 60:2046–58.

Cand, F., and Verdetti, J. 1989. Superoxide dismutase, glutathione peroxidase, catalase, and lipid peroxidation in the major organs of the aging rats. *Free Radical Biology and Medicine* 7:59–63.

Cao, G., Giovanoni, M., and Prior, R. L. 1996. Antioxidant capacity decreases during growth but not aging in rat serum and brain. *Archives of Gerontology Geriatrics* 22:27–37.

Carrillo, M.-C., Kanai, S., Sato, Y., and Kitani, K. 1992. Age-related changes in antioxidant enzyme activities are region and organ, as well as sex, selective in the rat. *Mechanisms of Ageing and Development* 65:187–98.

Chang, M. L., Klaidman, L., and Adams, J. D. Jr. 1995. Age-dependent effects of *t*-BuOOH on glutathione disulfide reductase, glutathione peroxidase, and malondialdehyde in the brain. *Molecular and Chemical Neuropathology* 26:95–106.

Chen, T. S., Richie, J. P., Jr., and Lang, C. A. 1989. The effect of aging on glutathione and cysteine levels in different regions of the mouse brain. *Proceedings of the Society for Experimental Biology and Medicine* 190:399–402.

Ciriolo, M. R., Fiskin, K., De Martino, A., Corasaniti, M. T., Nisticò, G., and Rotilio, G. 1991. Age-related changes in Cu,Zn superoxide dismutase, Se-dependent and -independent glutathione peroxidase and catalase activities in specific areas of rat brain. *Mechanisms of Ageing and Development* 61:287–97.

Cohen, G., and Heikkila, R. E. 1974. The generation of hydrogen peroxide, superoxide radical, and hydroxyl radical by 6-hydroxydopamine, dialuric acid, and related cytotoxic agents. *Journal of Biological Chemistry* 249:2447–52.

Cristiano, F., de Haan, J. B., Iannello, R. C., and Kola, I. 1995. Changes in the levels of enzymes which modulate the antioxidant balance occur during aging and correlate with cellular damage. *Mechanisms of Ageing and Development* 80:93–105.

Cummings, B. J., Pike, C. J., Shankle, R., and Cotman, C. W. 1996. β-amyloid deposition and other measures of neuropathology predict cognitive status in Alzheimer's disease. *Neurobiology of Aging* 17:921–33.

Damier, P., Hirsch, E. C., Zhang, P., Agid, Y., and Javoy-Agid, F. 1993. Glutathione peroxidase, glial cells and Parkinson's disease. *Neuroscience* 52:1–6.

De Erausquin, G. A., Costa, E., and Hanbauer, I. 1994. Calcium homeostasis, free radical formation, and trophic factor dependence mechanisms in Parkinson's disease. *Pharmacological Review* 46:467–82.

de Haan, J. B., Newman, J. D., and Kola, I. 1992. Cu/Zn superoxide dismutase mRNA and enzyme activity, and susceptibility to lipid peroxidation, increases with aging in murine brains. *Molecular Brain Research* 13:179–87.

de Haan, J. B., Cristiano, F., Iannello, R. C., and Kola, I. 1995. Cu/Zn-superoxide dismutase and glutathione peroxidase during aging. *Biochemistry and Molecular Biology International* 35:1281–97.

de la Asuncion, J. G., Millan, A., Pla, R., Bruseghini, L., Esteras, A., Pallardo, F. V., Sastre, J., and Viña, J. 1996. Mitochondrial glutathione oxidation correlates with age-associated oxidative damage to mitochondrial DNA. *FASEB Journal* 10:333–8.

de la Torre, M. R., Casado, A., López-Fernández, M. E., Carrascosa, D., Casado, M. C., Venarucci, D., and Venarucci, V. 1996. Human aging brain disorders: role of antioxidant enzymes. *Neurochemical Research* 21:885–8.

Deneke, S. M., Baxter, D. F., Phelps, D. T., and Fanburg, R. L. 1989. Increase in endothelial cell glutathione and precursor amino acids uptake by diethyl-maleate and hyperoxia. *American Journal of Physiology* 257:L265–71.

Desagher, S., Glowinski, J., and Premont, J. 1996. Astrocytes protect neurons from hydrogen peroxide toxicity. *Journal of Neuroscience* 16:2553–62.

Desole, M. S., Esposito, G., Enrico, P., Miele, M., Fresu, L., De Natale, G., Miele, E., and Grella, G. 1993. Effects of aging on 1-methyl-4-phenyl-1,2,3,6-tetrahydropyridine (MPTP) neurotoxic effects on striatum and brainstem in the rat. *Neuroscience Letters* 159:143–6.

Desole, M. S., Miele, M., Esposito, G., Fresu, L. G., Migheli, R., Zangani, D., Sircana, S., Grella, G., and Miele, E. 1995. Neuronal antioxidant system and MPTP-induced oxidative stress in the striatum and brain stem of the rat. *Pharmacology, Biochemistry, and Behavior* 51:581–92.

Dexter, D. T., Carter, C. J., Wells, F. R., Javoy-Agid, F., Agid, Y., Lees, A., Jenner, P., and Marsden, C. D. 1989. Basal lipid peroxidation in substantia nigra is increased in Parkinson's disease. *Journal of Neurochemistry* 52:381–9.

Dexter, D. T., Holley, A. E., Flitter, W. D., Slater, T. F., Wells, F. R., Daniel, S. E., Lees, A. J., Jenner, P., and Marsden, C. D. 1994a. Increased levels of lipid hydroperoxides in the Parkinsonian substantia nigra: an HPLC and ESR study. *Movement Disorders* 9:92–7.

Dexter, D. T., Sian, J., Rose, S., Hindmarsh, J. G., Mann, V. M., Cooper, J. M., Wells, F. R., Daniel, S. E., Lees, A. J., Schapira, A. H. V., Jenner, P., and Marsden, C. D. 1994b. Indices of oxidative stress and mitochondrial function in individuals with incidental Lewy body disease. *Annals of Neurology* 35: 38–44.

Di Monte, D. A., Chan, P., and Sandy, M. S. 1992. Glutathione in Parkinson's disease: a link between oxidative stress and mitochondrial damage? *Annals of Neurology* 32:S111–15.

Dorheim, M.-A., Tracey, W. R., Pollock, J. S., and Grammas, P. 1994. Nitric oxide synthase activity is elevated in brain microvessels in Alzheimer's disease. *Biochemical and Biophysical Research Communications* 205:659–65.

Fahn, S., and Cohen, G. 1992. The oxidant stress hypothesis in Parkinson's disease: evidence supporting it. *Annals of Neurology* 32:804–12.

Famulari, A. L., Marschoff, E. R., Llesuy, S. F., Kohan, S., Serra, J. A., Dominguez, R. O., Repetto, M., Reides, C., and de Lustig, E. S. 1996. The antioxidant enzymatic blood profile in Alzheimer's and vascular diseases. Their association and a possible assay to differentiate demented subjects and controls. *Journal of the Neurological Science* 141:69–78.

Ferraro, T. N., Golden, G. T., DeMattei, M., Hare, T. A., and Fariello, R. G. 1986. Effect of 1-methyl-4-phenyl-1,2,3,6-tetrahydropyridine (MPTP) on levels of glutathione in the extrapyramidal system of the mouse. *Neuropharmacology* 25:1071–74.

Fornstedt, B., Pileblad, E., and Carlsson, A. 1990. In vivo autoxidation of dopamine in guinea pig striatum increases with age. *Journal of Neurochemistry* 55:655–9.

Frey, W. H. II, Najarian, M. M., Kumar, K. S., Emory, C. R., Menning, P. M., Frank, J. C., Johnson, M. N., and Ala, T. A. 1996. Endogenous Alzheimer's brain factor and oxidized glutathione inhibit antagonist binding to muscarinic receptor. *Brain Research* 714:87–94.

Geremia, E., Baratta, D., Zafarana, S., Giordano, R., Pinizzotto, M. R., La Rosa, M. G., and Garozzo, A. 1990. Antioxidant enzymatic systems in neuronal and glial cell-enriched fractions of rat brain during aging. *Neurochemical Research* 15:719–23.

Gerlach, M., Ben-Shachar, D., Riederer, P., and Youdim, M. B. H. 1994. Altered brain metabolism of iron as a cause of neurodegenerative diseases? *Journal of Neurochemistry* 63:793–807.

Götz, M. E., Künig, G., Riederer, P., and Youdim, M. B. H. 1994. Oxidative stress: free radical production in neural degeneration. *Pharmacology and Therapeutics* 63:37–122.

Guo, N., and Shaw, C. 1992. Characterization and localization of glutathione binding sites on cultured astrocytes. *Molecular Brain Research* 15:207–15.

Guo, N., McIntosh, C., and Shaw, C. 1992. Glutathione: new candidate neuropeptide in the central nervous system. *Neuroscience* 51:835–42.

Gupta, M., Gupta, B. K., Thomas, R., Bruemmer, V., Sladek, J. R., Jr., and Felten, D. L. 1986. Aged mice are more sensitive to 1-methyl-4-phenyl-1,2,3,6-tetrahydropyridine treatment than young adults. *Neuroscience Letters* 70:326–31.

Halliwell, B. 1992. Reactive oxygen species and the central nervous system. *Journal of Neurochemistry* 59:1609–23.

Harman, D. 1992. Free radical theory of aging. *Mutation Research* 275:257–66.

Harman, D. 1995a. Role of antioxidant nutrients in aging: overview. *Age* 18:51–62.

Harman, D. 1995b. Free radical theory of aging: Alzheimer's disease pathogenesis. *Age* 18:97–119.

Hazelton, G. A., and Lang, C. A. 1980. Glutathione contents of tissues in the aging mouse. *Biochemical Journal* 188:25–30.

Hazelton, G. A., and Lang, C. A. 1983. Glutathione biosynthesis in the aging adult yellow-fever mosquito [*Aedes aegypti* (Louisville)]. *Biochemical Journal* 210:289–95.

Heales, S. J. R., Davies, S. E. C., Bates, T. E., and Clark, J. B. 1995. Depletion of brain glutathione is accompanied by impaired mitochondrial function and decreased *N*-acetyl aspartate concentration. *Neurochemical Research* 20:31–8.

Heales, S. J. R., Bolaños, J. P., and Clark, J. B. 1996. Glutathione depletion is accompanied by increased neuronal nitric oxide synthase activity. *Neurochemical Research* 21:35–9.

Hillered, L., and Ernster, L. 1983. Respiratory activity of isolated rat brain mitochondria following in vitro exposure to oxygen radicals. *Journal of Cerebral Blood Flow and Metabolism* 3:207–14.

Hinshaw, D. B., Miller, M. T., Omann, G. M., Beals, T. F., and Hyslop, P. A. 1993. A cellular model of oxidant-mediated neuronal injury. *Brain Research* 615:13–26.

Hjelle, O. P., Chaudhry, F. A., and Ottersen, O. P. 1994. Antisera to glutathione: characterization and immunocytochemical application to the rat cerebellum. *European Journal of Neuroscience* 6:793–804.

Hunot, S., Boissière, F., Faucheux, B., Brugg, B., Mouatt-Prigent, A., Agid, Y., and Hirsch, E. C. 1996. Nitric oxide synthase and neuronal vulnerability in Parkinson's disease. *Neuroscience* 72:355–63.

Hyman, B. T., Marzloff, K., Wenninger, J. J., Dawson, T. M., Bredt, D. S., and Snyder, S. H. 1992. Relative sparing of nitric oxide synthase–containing neurons in the hippocampal formation in Alzheimer's disease. *Annals of Neurology* 32:818–20.

Iantomasi, T., Favilli, F., Marraccini, P., Stio, M., Treves, C., Quattrone, A., Capaccioli, S., and Vincenzini, M. T. 1993. Age and GSH metabolism in rat cerebral cortex, as related to oxidative and energy parameters. *Mechanisms of Ageing and Development* 70:65–82.

Irwin, I., Finnegan, K. T., Delanney, L. E., Di Monte, D., and Langston, J. W. 1992. The relationships between aging, monoamine oxidase, striatal dopamine and the effects of MPTP in C57BL/6 mice: a critical reassessment. *Brain Research* 572:224–31.

Iversen, L. L., Mortishire-Smith, R. J., Pollack, S. J., and Shearman, M. S. 1995. The toxicity in vitro of β-amyloid protein. *Biochemical Journal* 311:1–16.

Jain, A., Mårtensson, J., Stole, E., Auld, P. A. M., and Meister, A. 1991. Glutathione deficiency leads to mitochondrial damage in brain. *Proceedings of the National Academy of Sciences of the U.S.A.* 88:1913–17.

Jarvis, M. F., and Wagner, G. C. 1985. Age-dependent effects of 1-methyl-4-phenyl-1,2,5,6-tetrahydropyridine (MPTP). *Neuropharmacology* 24:581–3.

Joachim, C. L., and Selkoe, D. J. 1992. The seminal role of β-amyloid in the pathogenesis of Alzheimer disease. *Alzheimer Disease and Associated Disorders* 6:7–34.

Jenner, P., and Olanow, C. W. 1996. Oxidative stress and the pathogenesis of Parkinson's disease. *Neurology* 47(suppl.3):S161–70.

Johannessen, J. N., Adams, J. D., Schuller, H. M., Bacon, J. P., and Markey, S. P. 1986. 1-Methyl-4-phenylpyridine (MPP+) induces oxidative stress in the rodent. *Life Sciences* 38:743–9.

Julius, M., Lang, C. A., Gleiberman, L., Harburg, E., DiFranceisco, W., and Schork, A. 1994. Glutathione and morbidity in a community-based sample of elderly. *Journal of Clinical Epidemiology* 47:1021–6.

Khanna, S. C., Garg, S. K., and Sharma, S. P. 1992. Antioxidant-influenced alterations in glutathione reductase activity in different age groups of male mice. *Gerontology* 38:9–12.

Kish, S. J., Morito, C., and Hornykiewicz, O. 1985. Glutathione peroxidase activity in Parkinson's disease brain. *Neuroscience Letters* 58:343–6.

Kumar, R., Agarwal, A. K., and Seth, P. K. 1995. Free radical–generated neurotoxicity of 6-hydroxydopamine. *Journal of Neurochemistry* 64:1703–7.

Lagenveld, C. H., Jongenelen, C. A. M., Schepens, E., Stoof, J. C., Bast, A., and Drukarch, B. 1995. Cultured rat striatal and cortical astrocytes protect mesencephalic dopaminergic neurons against hydrogen peroxide toxicity independent of their effect on neuronal development. *Neuroscience Letters* 192:13–16.

Lang, C. A., Naryshkin, S., Schneider, D. L., Mills, B. J., and Lindeman, R. D. 1992. Low blood glutathione levels in healthy aging adults. *Journal of Laboratory and Clinical Medicine* 120:720–5.

Langston, J. W. 1996. The etiology of Parkinson's disease with emphasis on the MPTP story. *Neurology* 47(Suppl.3):S153–60.

Lipton, S. A., Choi, Y.-B., Pan, Z.-H., Lei, S. Z., Chen, H.-S. V., Sucher, N. J., Loscalzo, J., Singel, D. J., and Stamler, J. S. 1993. A redox-based mechanism for the neuroprotective and neurodestructive effects of nitric oxide and related nitroso-compounds. *Nature* 364:626–32.

Makar, T. K., Nedergaard, M., Preuss, A., Gelbard, A. S., Perumal, A. S., and Cooper, A. J. L. 1994. Vitamin E, ascorbate, glutathione, glutathione disulfide, and enzymes of glutathione metabolism in cultures of chick astrocytes and neurons: evidence that astrocytes play an important role in antioxidative processes in the brain. *Journal of Neurochemistry* 62:45–53.

Martínez, M., Ferrándiz, M. L., De Juan, E., and Miquel, J. 1994. Age-related changes in glutathione and lipid peroxide content in mouse synaptic mitochondria: relationship to cytochrome c oxidase decline. *Neuroscience Letters* 170:121–4.

Marttila, R. J., Lorentz, H., and Rinne, U. K. 1988. Oxygen toxicity protecting enzymes in Parkinson's disease. Increase of superoxide dismutase-like activity in the substantia nigra and basal nucleus. *Journal of Neurological Science* 86:321–31.

Marzatico, F., and Café, C. 1993. Oxygen radicals and other toxic oxygen metabolites as key mediators of the central nervous system tissue injury. *Functional Neurology* 8:51–66.

Masukawa, T., Sai, M., and Tochino, Y. 1989. Methods for depleting brain glutathione. *Life Sciences* 44:417–24.

Matsuo, M., Gomi, F., and Dooley, M. M. 1992. Age-related alterations in antioxidant capacity and lipid peroxidation in brain, liver, and lung homogenates of normal and vitamin E-deficient rats. *Mechanisms of Ageing and Development* 64:273–92.

Meister, A. 1991. Glutathione deficiency produced by inhibition of its synthesis, and its reversal; applications in research and therapy. *Pharmacology and Therapeutics* 51:155–94.

Mizuno, Y., and Ohta, K. 1986. Regional distributions of thiobarbituric acid-reactive products, activities of enzymes regulating the metabolism of oxygen free radicals, and some of the related enzymes in adult and aged rat brains. *Journal of Neurochemistry* 46:1344–52.

Mo, J. Q., Hom, D. G., and Andersen, J. K. 1995. Decrease in protective enzymes correlates with increased oxidative damage in the aging mouse brain. *Mechanisms of Ageing and Development* 81:73–82.

Mouatt-Prigent, A., Karlsson, J. O., Agid, Y., and Hirsch, E. C. 1996. Increased m-calpain expression in the mesencephalon of patients with Parkinson's disease but not in other neurodegenerative disorders involving the mesencephalon: a role in nerve cell death? *Neuroscience* 73:979–87.

Nohl, H. 1994. Generation of superoxide radicals as byproduct of cellular respiration. *Annales de Biologie Clinique* 52:199–204.

Ochi, T. 1993. Mechanism for the changes in levels of glutathione upon exposure of cultured mammalian cells to tertiary-butylhydroperoxide and diamide. *Archives of Toxicology* 67:401–10.

Odunze, I. N., Klaidman, L. K., and Adams, J. D., Jr. 1990. MPTP toxicity: differential effects in the striatum, cerebral cortex and midbrain on glutathione, glutathione disulfide and protein sulfhydryl levels. *Research Communications on Substance Abuse* 11:123–34.

Offen, D., Ziv, I., Sternin, H., Melamed, E., and Hochman, A. 1996. Prevention of dopamine-induced cell death by thiol antioxidants: possible implications for treatment of Parkinson's disease. *Experimental Neurolology* 141:32–9.

Ovadia, A., Zhang, Z., and Gash, D. M. 1995. Increased susceptibility to MPTP toxicity in middle-aged rhesus monkeys. *Neurobiology of Aging* 16:931–7.

Panfili, E., Sandri, G., and Ernster, L. 1991. Distribution of glutathione peroxidases and glutathione reductase in rat brain mitochondria. *FEBS Letters* 290:35–7.

Perry, T. L., Godin, D. V., and Hansen, S. 1982. Parkinson's disease: a disorder due to nigral glutathione deficiency? *Neuroscience Letters* 33:305–10.

Perry, T. L., Yong, V. W., Bergeron, C., Hansen, S., and Jones, K. 1987. Amino acids, glutathione, and glutathione transferase activity in the brains of patients with Alzheimer's disease. *Annals of Neurology* 21:331–6.

Pileblad, E., and Magnusson, T. 1990. Effective depletion of glutathione in rat striatum and substantia nigra by L-buthionine sulfoximine in combination with 2-cyclohexene-1-one. *Life Sciences* 47:2333–42.

Pileblad, E., Magnusson, T., and Fornstedt, B. 1989. Reduction of brain glutathione by L-buthionine sulfoximine potentiates the dopamine-depleting action of 6-hydroxydopamine in rat striatum. *Journal of Neurochemistry* 52:978–80.

Pileblad, E., Eriksson, P. S., and Hansson, E. 1991. The presence of glutathione in primary neuronal and astroglial cultures from rat cerebral cortex and brain stem. *Journal of Neural Transmission* 86: 43–9.

Qureshi, G. A., Baig, S., Bednar, I., Södersten, P., Forsberg, G., and Siden, Å. 1995. Increased cerebrospinal fluid concentration of nitrite in Parkinson's disease. *Neuroreport* 6:1642–4.

Rao, G., Xia, E., and Richardson, A. 1990. Effect of age on the expression of antioxidant enzymes in male Fischer F344 rats. *Mechanisms of Ageing and Development* 53:49–60.

Raps, S. P., Lai, J. C. K., Hertz, L., and Cooper, A. J. L. 1989. Glutathione is present in high concentrations in cultured astrocytes, but not in cultured neurons. *Brain Research* 493:398–401.

Ravindranath, V., and Reed, D. J. 1990. Glutathione depletion and formation of glutathione–protein mixed disulfide following exposure of brain mitochondria to oxidative stress. *Biochemical and Biophysical Research Communications* 169:1075–9.

Ravindranath, V., Shivakumar, B. R., and Anandatheerthavarada, H. K. 1989. Low glutathione levels in brain regions of aged rats. *Neuroscience Letters* 101:187–90.

Reed, D. J. 1990. Glutathione: toxicological implications. *Annual Review of Pharmacology and Toxicology* 30:603–31.

Ricaurte, G. A., Irwin, I., Forno, L. S., DeLanney, L. E., Langston, E., and Langston, J. W. 1987. Aging and 1-methyl-4-phenyl-1,2,3,6-tetrahydropyridine-induced degeneration of dopaminergic neurons in the substantia nigra. *Brain Research* 403:43–51.

Richie, J. P., Jr., and Lang, C. A. 1986. The maintenance of high glutathione levels in healthy, very old women. *Gerontologist* 26:80A.

Richie, J. P., Jr., Mills, B. J., and Lang, C. A. 1987. Correction of a glutathione deficiency in the aging mosquito increases its longevity. *Proceedings of the Society for Experimental Biology and Medicine* 184:113–17.

Rose, S., Nomoto, M., Jackson, E. A., Gibb, W. R. G., Jaehnig, P., Jenner, P., and Marsden, C. D. 1993. Age-related effects of 1-methyl-4-phenyl-1,2,3,6-tetrahydropyridine treatment of common marmosets. *European Journal of Pharmacology* 230:177–85.

Sagara, J., Miura, K., and Bannai, S. 1993a. Maintenance of neuronal glutathione by glial cells. *Journal of Neurochemistry* 61:1672–6.

Sagara, J., Miura, K., and Bannai, S. 1993b. Cystine uptake and glutathione level in fetal brain cells in primary culture and in suspension. *Journal of Neurochemistry* 61:1667–71.

Sanchez-Ramos, J. R., Övervik, E., and Ames, B. N. 1994. A marker of oxyradical-mediated DNA damage (8-hydroxy-2'-deoxyguanosine) is increased in nigro-striatum of Parkinson's disease brain. *Neurodegeneration* 3:197–204.

Savolainen, H. 1978. Superoxide dismutase and glutathione peroxidase activities in rat brain. *Research Communications on Chemical Pathology and Pharmacology* 21:173–6.

Schapira, A. H. V. 1994. Evidence for mitochondrial dysfunction in Parkinson's disease—a critical appraisal. *Movement Disorders* 9:125–38.

Selkoe, D. J. 1991. The molecular pathology of Alzheimer's disease. *Neuron* 6:487–98.

Sian, J., Dexter, D. T., Lees, A. J., Daniel, S., Agid, Y., Javoy-Agid, F., Jenner, P., and Marsden, C. D. 1994a. Alterations in glutathione levels in Parkinson's disease and other neurodegenerative disorders affecting basal ganglia. *Annals of Neurology* 36:348–55.

Sian, J., Dexter, D. T., Lees, A. J., Daniel, S., Jenner, P., and Marsden, C. D. 1994b. Glutathione-related enzymes in brain in Parkinson's disease. *Annals of Neurology* 36:356–61.

Simonian, N. A., and Coyle, J. T. 1996. Oxidative stress in neurodegenerative diseases. *Annual Review of Pharmacology and Toxicology* 36:83–106.

Slivka, A., Mytilineou, C., and Cohen, G. 1987. Histochemical evaluation of glutathione in brain. *Brain Research* 409:275–84.

Socci, D. J., Crandall, B. M., and Arendash, G. W. 1995. Chronic antioxidant treatment improves the cognitive performance of aged rats. *Brain Research* 693:88–94.

Sofic, E., Lange, K. W., Jellinger, K., and Riederer, P. 1992. Reduced and oxidized glutathione in the substantia nigra of patients with Parkinson's disease. *Neuroscience Letters* 142:128–30.

Sohal, R. S., and Dubey, A. 1994. Mitochondrial oxidative damage, hydrogen peroxide release, and aging. *Free Radical Biology and Medicine* 16:621–6.

Sohal, R. S., Farmer, K. J., Allen, R. G., and Ragland, S. S. 1984. Effects of diethyldithiocarbamate on lifespan, metabolic rate, superoxide dismutase, catalase, inorganic peroxides and glutathione in the adult male housefly *Musca domestica*. *Mechanisms of Ageing and Development* 24:175–83.

Sohal, R. S., Ku, H.-H., Agarwal, S., Forster, M. J., and Lal, H. 1994. Oxidative damage, mitochondrial oxidant generation and antioxidant defenses during aging and in response to food restriction in the mouse. *Mechanisms of Ageing and Development* 74:121–33.

Spencer, J. P. E., Jenner, P., and Halliwell, B. 1995. Superoxide-dependent depletion of reduced glutathione by L-DOPA and dopamine. Relevance to Parkinson's disease. *Neuroreport* 6:1480–4.

Vitorica, J., Machado, A., and Satrústegui, J. 1984. Age-dependent variations in peroxide-utilizing enzymes from rat brain mitochondria and cytoplasm. *Journal of Neurochemistry* 42:351–6.

Whittemore, E. R., Loo, D. T., and Cotman, C. W. 1994. Exposure to hydrogen peroxide induces cell death via apoptosis in cultural rat cortical neurons. *Neuroreport* 5:1485–8.

Wüllner, U., Löschmann, P.-A., Schulz, J. B., Eblen, F., Turski, L., and Klockgether, T. 1996. Glutathione depletion potentiates MPTP and MPP+ toxicity in nigral dopaminergic neurons. *Neuroreport* 7:921–3.

Yankner, B. A. 1996. Mechanisms of neuronal degeneration in Alzheimer's disease. *Neuron* 16:921–32.

Yong, V. W., Perry, T. L., and Krisman, A. A. 1986. Depletion of glutathione in brainstem of mice caused by *N*-methyl-4-phenyl-1,2,3,6-tetrahydropyridine is prevented by antioxidant pretreatment. *Neuroscience Letters* 63:56–60.

Yoritaka, A., Hattori, N., Uchida, K., Tanaka, M., Stadtman, E. R., and Mizuno, Y. 1996. Immunohistochemical detection of 4-hydroxynonenal protein adducts in Parkinson disease. *Proceedings of the National Academy of Sciences of the U.S.A.* 93:2696–701.

Zhang, L., Maiorino, M., Roveri, A., and Ursini, F. 1989. Phospholipid hydroxyperoxide glutathione peroxidase: specific activity in tissues of rats of different age and comparison with other glutathione peroxidases. *Biochimica et Biophysica Acta* 1006:140–3.

Zhang, Y., Marcillat, O., Giulivi, C., Ernster, L., and Davies, K. J. A. 1990. The oxidative inactivation of mitochondrial electron transport chain components and ATPase. *Journal of Biological Chemistry* 265:16330–6.

Zhang, J.-R., Andrus, P. K., and Hall, E. D. 1993. Age-related regional changes in hydroxyl radical stress and antioxidants in gerbil brain. *Journal of Neurochemistry* 61:1640–7.

Glutathione in the Nervous System
Edited by Christopher A. Shaw
Copyright © 1998 Taylor & Francis

12

The Glutathione Redox State and Zinc Mobilization from Metallothionein and Other Proteins with Zinc–Sulfur Coordination Sites

Wolfgang Maret

Center for Biochemical and Biophysical Sciences and Medicine, Harvard Medical School, Boston, Massachusetts 02115

- Introduction
- Biology of Zinc and of the Zinc-Binding Protein Metallothionein
- Glutathione, Glutathione Disulfide, and the Thiol–Disulfide Redox State
- Zinc–Sulfur Coordination Bonds and Their Reactivity in Biology
- A Link between the Cellular Thiol–Disulfide Redox State and Mobilization of Zinc from Zinc–Sulfur Coordination Sites
- Specific Aspects Pertaining to the Central Nervous System and to Neurodegenerative Diseases
- Conclusions
- References

ABBREVIATIONS

GIF growth inhibitory factor
GSH reduced glutathione
GSSG oxidized glutathione (glutathione disulfide)
MT metallothionein

1. INTRODUCTION

Reactive oxygen species are formed constantly during aerobic metabolism. These can pose a serious threat to a cell, particularly in the presence of prooxidants, if the supply of antioxidants is inadequate, or if enzymatic protection systems for maintaining a proper balance are not intact. Both a tendency toward an oxidizing state, *oxidative stress*, and a tendency toward a reducing state, *reductive stress*, can be deleterious and cause disease. Examples for redox stress include inter alia vitamin deficiencies (ascorbate, tocopherols) or diabetes, where hyperglycemia increases metabolic flux through the polyol pathway while tipping the NAD^+–NADH redox balance (Williamson et al. 1993). The general importance of redox homeostasis is also apparent in aging, where a weakened antioxidant defense is thought to be a major factor (Sohal and Weindruch 1996). It is, however, also becoming increasingly apparent that in redox-sensitive signaling pathways or in redox regulation of gene expression the cell utilizes modulation of the cellular redox state and reactive oxygen species in order to control critical physiological events (Burdon 1995; Monteiro and Stern 1996; Sun and Oberley 1996; Sen and Packer 1996). The determination of cell fate in proliferation and apoptosis of T-lymphocytes is an example where this has been investigated rigorously (Buttke and Sandstrom 1995). There is ample reason, therefore, why definitions of two aspects are becoming focal points of research: (i) the molecular components that regulate the cellular redox state, and (ii) the quantification of redox stress that exceeds the capacity of the homeostatic system and plays a role in the etiology and pathogenesis of diseases.

The thiol–disulfide redox state, which is determined largely by the glutathione–glutathione disulfide couple, is one component in the cellular redox balance. Since this couple and thiolate sulfur from the side chain of cysteine in proteins are metal ligands, metal binding must affect the thiol–disulfide redox state, or vice versa changes in the thiol–disulfide redox state must affect metal binding. This relationship has received almost no attention. This pertains to zinc in particular, since it is now recognized that thiolate sulfur is an essential part of the coordination sites of hundreds of zinc finger proteins and enzymes. In this chapter, therefore, I address the relation between the cellular redox state and the chemistry of the zinc–sulfur interaction in proteins, which I believe to be significant for cellular utilization and distribution of zinc. To this end, I shall first introduce salient points of the biology of zinc and of a major zinc-binding protein, namely metallothionein, whose function is unknown thus far; of glutathione; and of the glutathione–glutathione disulfide redox state. Then I shall discuss the possible intersection of these two areas in an emerging paradigm that suggests that zinc–sulfur coordination sites are redox-sensitive.

2. BIOLOGY OF ZINC AND OF THE ZINC-BINDING PROTEIN METALLOTHIONEIN

The amount of zinc (2–3 g) and its daily requirement (2–7 mg, depending on its availability in the diet) in a 70-kg adult human are close to those of iron (4–5 g and about 10 mg, respectively). It is also true that whereas the functions of iron in oxidoreduction (cytochromes) and oxygen transport and storage (hemoglobin and myoglobin) are specific and well defined, zinc-dependent functions are much more widespread. Thus, zinc is an essential element of paramount importance for growth and development, metabolism, the immune response, gene expression, and neurotransmission, among others. Over the past 10 years, expanding progress in the recognition of zinc-dependent functions (Vallee and Falchuk 1993) and in the elucidation of the molecular structure of many different motifs of zinc sites in proteins (Vallee and Auld 1993) has stimulated interest in this element in almost all branches of the biomedical sciences. Virtually all roles of zinc so far are linked to function and structure of proteins. In fact, zinc-containing proteins form a much larger group than any of the other metalloproteins. More than 200 zinc enzymes have been recognized, in all seven of the classes defined by the commission for enzyme nomenclature of the International Union of Biochemistry and Molecular Biology (Vallee and Falchuk 1993), and an estimated 500 zinc finger genes of just one of the many classes exist in humans (Becker et al. 1995).

It is not understood how the priorities among different systems potentially competing for zinc are set, and which molecules supply zinc and signal or trigger its mobilization. In view of the considerable number and variety of motifs for zinc-binding sites in enzymes and other proteins, it has become a pressing issue to define and recognize efficient zinc regulatory and chemical mechanisms that must safeguard distribution of this element to and among all these proteins in a timely and coordinated manner. Once again in contrast to iron metabolism, where transport systems (e.g., transferrins), storage systems (e.g., ferritins), and regulatory systems (e.g., iron-regulatory element-binding proteins) are well established, we note that analogous proteins are unknown for zinc. There are a series of questions that remain unanswered, viz.: How is zinc transferred from the intestine to the blood and then to tissues? Where and when is zinc acquired in the course of translation? Are the mechanisms that pertain to cytosolic proteins and those that pertain to proteins destined for the export pathway identical? When zinc proteins are degraded, what is the fate of the metal? At present, there are no answers, and the distribution of zinc in the cell basically remains an open question.

One school of thought maintains that a zinc distribution system may not be required, and the amount of *free* zinc may determine its availability. However, zinc is bound very tightly to cellular proteins with picomolar dissociation constants, and the cellular concentration of *free* zinc has been shown for some systems to be exceedingly low (in the picomolar range), (Peck and Ray 1971; Simons 1991; Atar et al. 1995). If *free* zinc were to reach upper nanomolar or micromolar concentrations, biological processes that depend on other metal ions such as calcium (Csermely et al. 1989) and many enzyme systems would be compromised. Therefore, in the cell zinc must be bound tightly, while remaining available. Metallothionein (MT) is a molecule that fulfills this requirement. It binds zinc with high thermodynamic stability [$K_D = 5 \times 10^{-13}$ M at pH 7, and 10^{-18} M at higher pH (Kägi 1993)], while demonstrating remarkable chemical reactivity (Otvos, Petering, and Shaw 1989). MT has long been implicated in cellular zinc (and copper) homeostasis, and it is the only protein whose role in cellular zinc distribution can now be discussed in terms of its characteristic zinc coordination chemistry.

MT is a small protein of about 60 amino acids, 20 of which are invariably cysteines in the mammalian proteins. In humans, at least fifteen genes and eight expressed proteins (viz., MT-I isoforms and MT-II, -III, and -IV) have been identified. The

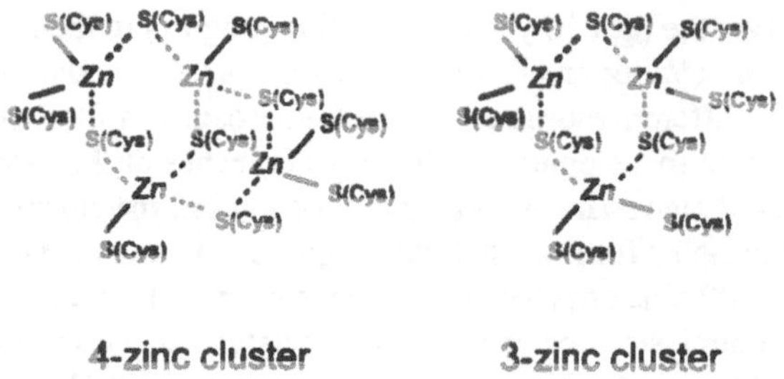

4-zinc cluster   3-zinc cluster

FIG. 1. Zinc coordination in the two types of clusters in metallothionein. End-on coordination of thiol sulfur ligands is shown with solid lines, and bridging thiol sulfur ligands are shown with broken lines. Though every zinc atom resides in a tetracoordinated ligand field, the coordination involves only 20 cysteines and therefore requires sharing of ligands. For further details see text.

crystal and solution structures of MTs (Robbins et al. 1991; Wüthrich 1991) have established that the seven zinc atoms of the molecule are bound exclusively to all sulfur thiolates of the 20 cysteine side chains. Since this coordination is likely to be critical to the function of the protein, it will be briefly described in more detail (Fig. 1). The binding of zinc to sulfur occurs via end-on (terminal) and bridging cysteine thiol ligands. Structural studies also revealed that the protein has two domains, each of which contains a so-called zinc cluster. The cluster of the N-terminal β-domain has three zinc atoms bound by six terminal and three bridging ligands, providing a coordination that is formally identical for each atom in terms of bridges and end-on coordination. The cluster in the C-terminal α-domain with four zinc atoms also has six terminal ligands, but a total of five bridging ligands.

Since its discovery 40 years ago (Margoshes and Vallee 1957), the function of MT has remained completely conjectural. MT is unusual in that its metal composition varies with the type of metals prevalent in the organism, while other metalloproteins tend to be specific for one particular metal ion. Hence, MT can be present as a zinc protein or a mixed zinc–copper protein, or indeed contain other heavy-metal ions (e.g., cadmium). However, there is no iron in MT, a remarkable fact considering the high affinity of iron for sulfur and the iron–sulfur coordination of ferredoxins, but perhaps not all that surprising on allowing for the fact that the distribution of zinc must be under a control as careful as that of iron. The binding of heavy metals and the induction of MT by essential and toxic metals through metal-dependent transcription factors that act on metal-responsive elements of the MT gene promoter has stimulated research on the possible function of MT in metal detoxification. Such a potentially protective role was then extended to include other presumed functions, since MT acts as a scavenger of radicals and reactive oxygen species, and protects against high-energy radiation. A potential physiological function, however—for example, in zinc metabolism (Vallee 1991; Vallee and Maret 1993; Vallee 1995)—has received much less attention.

Before returning to that subject, recent experiments that have attempted to address the possible function of MT by generating transgenic mice lacking functional

MT-I and MT-II genes require discussion. These animals seemed normal, grew and reproduced normally under laboratory conditions (Michalska and Choo 1993; Masters et al. 1994), but exhibited increased sensitivity to cadmium, which was toxic to them, and to oxidative stress (Lazo et al. 1995). Hence, the authors concluded that MT does not play an essential role in cellular zinc metabolism. This interpretation neglected the fact that in the mouse there are at least two other MT genes and that decisive tests for an altered phenotype in the transgenic animals with regard to either zinc or copper metabolism were not applied rigorously. Indeed, studies with these transgenic mice have now led to the suggestion that MT plays a role in glucose homeostasis (Rofe, Philcox, and Coyle 1996) and protects against zinc deficiency and cellular zinc or copper toxicity (Kelly et al. 1996; Kelly and Palmiter 1996). Also, using antisense-RNA–DNA technology, the significance of MT in the cellular physiology of monocytes (Leibbrandt, Khokha, and Koropatnick 1994) or in embryonic development (Peters et al. 1995) has been shown clearly.

Although MT undoubtedly fulfills protective functions, they are not necessarily biological. Thus, it has not been stated or resolved what the teleology of a 'function as a scavenger of either metals or radicals might be. It can be argued that interaction of MT with reactive oxygen species might destroy the clusters and mobilize zinc, with potentially deleterious effects. In a similar vein, the extraordinary reactivity of metals in MT (Otvos, Petering, and Shaw 1989; Maret, Larsen, and Vallee 1997) might mobilize cadmium or transfer it to other zinc-dependent proteins (Misra et al. 1996)—also with undesirable consequences, which might not be compatible with a role for MT as a long-term device to sequester toxic metals. Without delineation of the fate of metals in MT or elucidation of specific export/excretion pathways, a role of MT in detoxification remains hypothetical. Overall, these "functions" fail to include the obvious, namely, a role of MT in zinc and/or copper metabolism. The transcriptional regulation of MT by dietary zinc and its control by several zinc-dependent proteins, its regulation during the cell cycle (Nagel and Vallee 1995) and development by multiple signaling pathways and factors, its role in the acute phase response (Maret 1995a), and the existence of a large gene family with tissue- and isoprotein-specific expression all seem to strongly support an important role in zinc metabolism. What precisely might this role be?

In vivo, zinc in MT seems to be metabolically active in metal transfer/redistribution (Cherian 1977). Proteolysis of MT apparently is not the rate-limiting step of metal release (Krezoski, Shaw, and Petering 1991), though the molecular details of metal transfer remain unknown. In vitro, MT can act either as a metal donor (Brady 1982) or a metal acceptor. Several reports describe a role of MT in metal transfer to and from zinc proteins. Thus, zinc transfer from MT to the apoforms of some zinc enzymes has been demonstrated (Udom and Brady 1980; Li et al. 1980; Ou and Ebadi 1992). Moreover, thionein (apometallothionein) can interfere with and remove zinc from the transcription factors Sp1 (Zeng et al. 1991), IIIA (Zeng, Vallee, and Kägi 1991), and Bicpo (Fraefel et al. 1994). We have used ^{65}Zn and fast chromatographic separation of reaction products to demonstrate direct zinc transfer from MT to the apoforms of zinc finger proteins (Maret 1994a), which has also been inferred from the effect of MT on the DNA-binding activity of the full-length estrogen receptor in metal-depleted cellular extracts (Cano-Gauci and Sarkar 1996). Moreover, we have demonstrated relatively rapid intermolecular self-exchange of zinc between MT isoforms (Maret, Larsen, and Vallee 1997), which is faster between the β-domain than between the α-domain clusters. Such zinc exchange is not limited to MT, since we have observed it between MT and zinc in the cluster structure of the transcription factor Gal4 (Maret Larsen, and Vallee 1997). Thus, the dual functions of MT in serving as a reservoir, in which zinc is tightly sequestered due to high thermodynamic stability and that provides zinc to other zinc-dependent proteins due to kinetic lability, are realized in the unusual zinc cluster structure of this molecule.

3. GLUTATHIONE, GLUTATHIONE DISULFIDE, AND THE THIOL–DISULFIDE REDOX STATE

The functions of the tripeptide glutathione, γ-glutamylcysteinylglycine (GSH), are pleiotropic (Meister, 1988) and resemble those of the MT system in many ways. Both systems are antioxidants and are induced by metals and redox stress. Moreover, owing to their metal-binding properties, glutathionelike polymers (phytochelatins) have been categorized as class III MTs. Synthesis and degradation of GSH are tightly controlled by the γ-glutamyl cycle (Meister 1988). Glutathione represents about 90% of nonprotein sulfur in the cell (Meister 1988).

One major function of GSH is its reaction with electrophiles that constitute a threat to the cell (Mannervik 1996). This protective action can either be enzyme-catalyzed (e.g., by the glutathione transferase family) or occur nonenzymatically. Another major function is that of a reducing agent. In this reaction, GSH is oxidized to glutathione disulfide (GSSG), also either enzymatically or nonenzymatically. The GSH–GSSG couple is a major determinant of the cellular thiol–disulfide redox state and is controlled by at least two enzymes. Glutathione peroxidases form GSSG from GSH in the presence of peroxides, whereas glutathione reductases regenerate GSH from GSSG in the presence of NADPH. The coupling between the GSH–GSSG and the $NADP^+$–NADPH redox states via glutathione reductase ensures metabolic control through the pentose phosphate pathway and ultimately glucose, since in this pathway glucose 6-phosphate dehydrogenase determines the cellular $NADP^+$/NADPH ratio (Fig. 2). This scheme also illustrates the central role of the GSH–GSSG redox pair in coupling the $NADP^+$–NADPH redox state with other redox pairs, e.g., ascorbate–dehydroascorbate, hydroperoxide–alcohol, and thiol–disulfide. Sufficient reducing power must be available to provide thiolate sulfur for the binding of zinc.

As GSSG is biosynthesized from GSH, whose concentration varies in the range 0.1–10 mM (Meister 1988), its concentration is regulated by that of GSH. GSSG is also the major end product of the decomposition of S-nitrosoglutathione (Hogg, Singh, and Kalyanaraman 1996; Singh et al. 1996), a compound that is now considered a biologically significant donor of nitric oxide. Alternatively, the concentration of GSSG can change independently if the redox state changes. In the cytosol, where

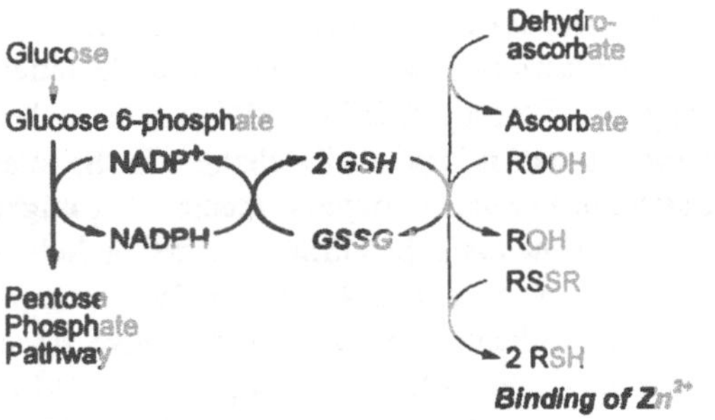

Glucose
↓
Glucose 6-phosphate

NADP⁺ → 2 GSH

NADPH → GSSG

Pentose
Phosphate
Pathway

Dehydro-
ascorbate

Ascorbate

ROOH

ROH

RSSR

2 RSH

Binding of Zn²⁺

FIG. 2. Coupling between the cellular thiol–disulfide and the NADP⁺–NADPH redox states and multiple functions of the GSH–GSSG redox pair. Reducing equivalents in the form of NADPH are generated and controlled by glucose 6-phosphate dehydrogenase in the pentose phosphate pathway. NADPH is the cofactor required for the reduction of GSSG by glutathione reductase. Glutathione peroxidase reacts with GSH and peroxides to generate GSSG. The GSH–GSSG redox pair is coupled enzymatically or nonenzymatically with vitamin C, with vitamin E (not shown), with the thiol–disulfide state of proteins, and, accordingly, with the binding of zinc in proteins with cysteine sulfur coordination sites. The metabolic flux in this scheme reflects an ample supply of reducing power.

the GSH/GSSG ratio typically is between 30 : 1 and 100 : 1, reducing conditions prevail. However, this does not apply to all cellular compartments. Examination of the GSH/GSSG ratio in the endoplasmic reticulum reveals that it is more oxidizing by a factor between 10 and 100 than that of the cytosol, resulting in GSH/GSSG ratios between 3 : 1 and 1 : 1 (Hwang, Sinskey, and Lodish 1992). These are thought to be necessary for disulfide formation in protein folding. Even small changes of the GSH/GSSG redox ratio modify physiological responses (Ziegler 1985; Gilbert 1990), but how such changes (including redox compartmentation) affect mineral metabolism is unknown.

Most glutathione research has focused on the functions of GSH, and very little is known about possible functions of GSSG. In fact, it would be rather surprising if the only purpose of glutathione reductase and peroxidase were the control of the GSH–GSSG redox state and if GSSG were not utilized in any other way. Chemically, the possibility that GSSG has special functions is an attractive hypothesis, because GSSG is an electrophile and a mild oxidizing agent, in contrast to GSH, which is a nucleophile and a reducing agent. Aside from an interaction between MT and GSSG (see section 4) and glutathionylation (S-thiolation) of proteins under conditions of oxidative stress (Thomas, Poland, and Honzatko 1995), specific molecular actions involving GSSG have apparently not been reported. The remaining discussion, therefore, addresses changes of the GSH–GSSG redox state without specifying the precise nature of the underlying chemistry.

Redox modulation is now being recognized widely in many areas of biology. For example, GSSG generated in vivo modulates the Ca^{2+} content of the Ca^{2+} store of endothelial cells (Henschke and Elliott 1995); GSSG in the presence of GSH activates protein kinase C (Kass, Duddy, and Orrenius 1989); the redox state controls receptor-channel proteins (Weir and Archer 1995; Gozlan and Ben-Ari 1995); it has been argued that the redox regulation of T-cell transcription factors as a result of depletion of GSH is actually due to an accompanying deficiency of GSSG (Dröge et al. 1994). The paucity of data precludes a decision whether it is primarily changes in the total amount of glutathione, in the redox state, or in reactive oxygen species that might be responsible for these effects. Some effects mediated by GSSG in vitro, including protein S-thiolation (Thomas et al. 1995; Dafré, Sies, and Akerboom 1996), might in fact be mediated in vivo by glutathionyl radicals generated by oxidative stress (Kwak et al. 1995). Specifically, glutathionyl radicals have now been implicated in the disassembly of the [2Fe-2S] cluster of the *E. coli* regulatory protein SoxR (Ding and Demple 1996).

4. ZINC–SULFUR COORDINATION BONDS AND THEIR REACTIVITY IN BIOLOGY

The control of biological function by thiolate groups of cysteines in proteins is well established. A more recent development is the realization that the reactivity of cysteine thiolate is modulated through zinc coordination and that the special reactivity of the zinc–sulfur bond is also employed for biological control. An early example of this reactivity is the carboxymethylation of horse liver alcohol dehydrogenase with either iodoacetamide or iodoacetate, which specifically modifies just one of the two zinc–sulfur (cysteine) bonds of the catalytic zinc atom without loss of zinc. This specificity is achieved through an arginine side chain, which orients and directs the reagent toward the active site (Li and Vallee 1963; Zeppezauer, Jörnvall, and Ohlsson 1975).

Of potential significance is the marked sensitivity of the zinc–sulfur bond to oxidants. Zinc centers in MT (Fliss and Ménard 1992; Kröncke et al. 1994), zinc complexed with diethyldithiocarbamate (Lapenna et al. 1994), zinc in the active site of alcohol dehydrogenase (Crow, Beckman, and McCord 1995; Gergel and Cederbaum 1996), zinc in the cluster structure of Lac9 (Kröncke et al. 1994), and zinc in a DNA repair enzyme with a zinc finger structure (Wink and Laval 1994) all react either with hypochlorite, peroxide, superoxide, or peroxynitrite. Perhaps the effects of nitric oxide are especially revealing in that they may signify a control of zinc–sulfur centers by nitric oxide in vivo, a possibility that has already been raised (Stamler 1994). Cysteine ligands of the zinc center of the HIV nucleocapsid protein NCp7 are oxidized to disulfides by nitrosobenzamide (Yu et al. 1995). These reactions suggest that in vivo the zinc–sulfur bond is a target for potential damage under conditions of oxidative stress and must be protected from oxidant-induced damage by a proper antioxidant status.

Given such reactivity of the zinc–sulfur bond, why has cysteine then been chosen as a zinc ligand? It might well be that its reactivity serves biological functions. This has been demonstrated in at least one case and postulated in others, and will be elaborated upon further under the next heading. For example, the *E. coli* Ada protein is a DNA repair enzyme (Myers et al. 1993) and a zinc finger protein, in which the zinc-bound sulfhydryl group attacks the methyl group of a phosphotriester as an activated nucleophile. A subsequent study of zinc complexes has now confirmed that zinc coordination with four thiolate sulfurs can indeed activate a thiol as a nucleophile (Wilker and Lippard, 1995). Based on these experiments, it has been postulated that zinc activates homocysteine in *E. coli* methionine synthase (Gonzáles et al. 1996) and a cysteine thiol in farnesyltransferase (Huang, Casey, and Fierke 1997).

It is now established that DNA binding of many transcription factors is controlled by reversible redox reactions of cysteine thiolates and that this is one basis for redox regulation of gene expression (Sun and Oberley 1996). Further, there is growing evidence that zinc-dependent transcription factors also exhibit redox-sensitive DNA binding activity in vivo and in vitro. These include the tumor suppressor p53 (ZnS_3N coordination; S = cysteine sulfur, N = histidine nitrogen), the transcription factors Sp1 and Egr-1 (both ZnS_2N_2 coordination), and members of the nuclear hormone receptor superfamily such as the glucocorticoid and thyroid hormone receptors (ZnS_4 coordination). Since in all these factors zinc participates in DNA recognition, it is not surprising that loss of zinc and oxidation of thiolate ligands certainly is one mechanism of redox regulation. It remains to be shown, however, whether cysteine ligands of zinc or other cysteine side chains are the ones that are critically involved in redox sensitivity of these transcription factors.

We have shown a different type of reactivity of the zinc–sulfur bond, namely, a reaction of the cysteines in MT with biological disulfides (Maret 1994b, 1995b). Glutathione disulfide interacts with MT and releases zinc in a monophasic reaction, indicating similar or equal reactivity of the two clusters in MT and a process in which an initial *S*-thiolation of critical cysteines leads to collapse of each cluster. This reaction provides a new mechanism for release of zinc from MT and could proceed by direct attack of the disulfide on the intact zinc–sulfur bond or in a reaction with the free thiol that dissociates from zinc. Because of the aforementioned activation of cysteine thiolate by zinc, we favor the first possibility and have termed this reaction zinc-thiol–disulfide interchange. Even though this interchange (glutathionylation) itself is not an oxidation, the formation of intra- and intermolecular disulfides in MT in the presence of excess disulfide reagent constitutes an oxidation of MT. It is also an oxidative process in the sense that a disulfide has to be generated for this type of metal release to occur.

The rate of thiol–disulfide interchange in MT could conceivably be enhanced through multiple, conserved lysine side chains that provide positive charges, since it has been observed that rates of thiol–disulfide interchange in proteins can vary by a factor of 10^6 due to the local environment of cysteines (Snyder et al. 1981). For this type of reaction, there is precedent from two other biological systems. Glutathione disulfide activates matrix metalloproteases (Macartney and Tschesche 1983; Tyagi, Kumar, and Borders 1996). In the zymogen form of these enzymes, the catalytic zinc atom is thought to be blocked by a thiol group contributed by a cysteine side chain (Springman et al. 1990). Reaction of this zinc–sulfur bond with glutathione disulfide leads to activation of the protease.

Disulfide benzamides that have been developed as potential drugs for the treatment of AIDS are yet another example of this type of reactivity. Several derivatives react with the zinc finger structure of the HIV nucleocapsid protein NCp7 (ZnS_3N coordination), in which zinc is bound with high thermodynamic stability [$K_D = 3.6 \times 10^{-15}$ M at pH 7.5 (Mély et al. 1996)], and release zinc (Rice et al. 1995; Tummino et al. 1996). To assess the mechanism through which zinc is ejected from the zinc finger structure, the investigators analyzed the reaction products by high-performance liquid chromatography and identified a mixed zinc-finger–glutathione disulfide (Rice et al. 1995). These studies suggest that the cysteines in MT and in other zinc–sulfur coordination sites are important targets for control of their zinc content.

5. A LINK BETWEEN THE CELLULAR THIOL–DISULFIDE REDOX STATE AND MOBILIZATION OF ZINC FROM ZINC–SULFUR COORDINATION SITES

Zinc is redox-inert in biological systems, and hence, unlike iron or copper, it cannot utilize different oxidation states to control its mobility. From the foregoing discussion, it appears that biological chemistry avails itself of the redox chemistry of the cysteine ligand instead of the central zinc atom to ensure reactivity. Thus, the widespread use of cysteine as a ligand in cellular proteins confers chemical properties that are crucial to the mobility of this essential element (Fig. 3).

The reactivity of zinc–sulfur bonds links zinc and the thiol–disulfide redox state of the cell. In the case of MT, this link exists in at least two ways—at the gene and at the protein level. At the gene level, oxidative stress (Baumann et al. 1991) and signaling pathways that function via reactive oxygen intermediates induce MT, a response that is mediated through metal and antioxidant regulatory elements in the MT gene promoter (Ren and Smith 1995; Dalton et al. 1996). At the protein level, MT is sensitive to oxidative reactions and is thought to be an antioxidant and to participate in regulating the intracellular redox state (Lazo et al. 1995; Ebadi et al. 1996). In terms of its protective effects on the cell, the interaction between MT and disulfides (Maret 1994b) implies that (i) stress-elevated levels of MT react with disulfides in order to lower their concentration, and (ii) disulfides generated under chemical or physical stress release zinc for the purpose of using it in cellular repair systems. The observation that derangements in zinc metabolism occur in pathological situations in which oxidative

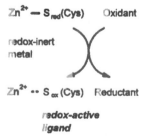

$Zn^{2+} - S_{red}(Cys)$ Oxidant

redox-inert
metal

$Zn^{2+} \cdots S_{ox}(Cys)$ Reductant

*redox-active
ligand*

FIG. 3. Binding of the biologically redox-inert zinc ion can be modulated by reversible or irreversible reactions of the redox-active cysteine ligand.

stress is generated, such as diabetes and neurodegenerative diseases (Maret 1995b), seems to further support the supposition. Looking beyond the mobilization of zinc from proteins by oxidative stress, we envision the intriguing possibility that changes of the thiol–disulfide redox state are used as a general physiological mechanism to regulate zinc distribution.

The relation between the cellular glutathione thiol–disulfide state and MT in vitro is supported by the demonstration of direct molecular interactions with MT that have been demonstrated for both GSH (Brouwer, Hoexum-Brouwer, and Cashon 1993) and GSSG (Maret 1994b). The metabolic relation between MT and glutathione in vivo remains a controversial issue, however. Almost all earlier experiments were performed with a focus on the question how these pools relate regarding cysteine, not zinc; it was concluded either that there is little interaction (Gallant and Cherian 1989) or that the pools are related somehow (Giralt et al. 1993). Further studies are obviously needed in order to define conditions in vivo under which the availability of zinc is controlled by interactions between MT and glutathione. These conditions could include specific or general functions in providing zinc for critical physiological events such as cellular proliferation, in which zinc must be distributed.

6. SPECIFIC ASPECTS PERTAINING TO THE CENTRAL NERVOUS SYSTEM AND TO NEURODEGENERATIVE DISEASES

Another biological function of MT was discovered recently. A growth inhibitory factor (GIF) that is deficient in brain samples of Alzheimer's disease victims turned out to be a new, brain-specific MT, now termed MT-III (Uchida et al. 1991). The protein has the same number of cysteine residues and characteristic pattern of metal ligands as other MTs, but due to a C-terminal extension is composed of 68 amino acids. Its exact molecular function and the medical significance that underlie its inhibition of nerve growth are unknown, and it is unclear whether or not the process involves metal transfer. Yet, it is a function that is specific for MT-III; other MT isoforms are not active in this biological assay of GIF activity. It was claimed subsequently that in Alzheimer's disease patients MT-III is not downregulated, although its GIF activity was confirmed (Erickson et al. 1994). Mutagenesis studies showed the importance of residues in the N-terminal β-domain; neither the unusual α-domain, in which the extension occurs, nor the metal composition is critical for the role of the protein in the bioassay (Sewell et al. 1995). Thus, it was not deemed to be of significance that MT-III (GIF) when isolated from different species invariably contains copper (Uchida et al. 1991; Pountney et al. 1994).

MT-III is present in zinc-containing neurons. MT-I and MT-II are absent in neurons but expressed in astrocytes (Aschner 1996). Zinc is stored in the hippocampus in synaptic secretory vesicles of neurons and acts as a neuromodulator of many ligand- and voltage-gated ion channels (Harrison and Gibbons, 1994). GSH is also considered to be a neuromodulator in the central nervous system (Lanius et al. 1994; see also Pasqualotto et al., chapter 9, this volume).

There are several reasons why imbalances of the cellular redox state are of particular importance for diseases of the central nervous system. First, the most active aerobic metabolism is in the brain. Second, the brain apparently does not have any special protective systems to cope effectively with the resulting oxidative stress. Third, reactive oxygen species (i.e., physiological oxidative stress) are now implicated in cellular proliferation and programmed cell death (apoptosis) of neurons. Many neurological disorders are proliferative diseases and/or are the result of cell death. Fourth, deterioration of neural function is related to oxidative damage. Fifth, decline of antioxidant defense with age is accompanied by an increase of age-related neurodegenerative disorders. Under conditions where redox stress exceeds the regulatory capacity of the system, the abnormal oxidative release of zinc from proteins leads to nonphysiological levels of intracellular zinc that could be an important step in the cascade of events in pathogenesis (Maret 1995b). Oxidative stress is a notable feature of Alzheimer's and Parkinson's disease (Behl et al. 1994; Ebadi, Srinivasan, and Baxi 1996) and Down's syndrome (Kedziora and Bartosz 1988). The role of reactive oxygen species (which are also formed in ischemia and reperfusion), in the pathogenesis of these diseases and in amyotrophic lateral sclerosis and multiple sclerosis has been discussed in a recent review (Knight 1997). In Alzheimer's disease, zinc blocks amyloid β-protein calcium channels (Arispe, Pollard, and Rojas 1996) and induces Aβ-amyloid formation (Bush et al. 1994). Zinc also seems to play a critical role in neural cell death in stroke (Koh et al. 1996).

The antioxidant hypothesis holds that suboptimal levels of antioxidant nutrients compromise the defense against reactive oxygen species and thus constitute a risk factor for many other degenerative diseases of aging, such as cardiovascular disease, cancer, decline of the immune system, and cataracts. Zinc is thought to be an antioxidant (Bray and Bettger 1990), and zinc deficiency elicits oxidative damage in laboratory animals (Oteiza et al. 1995). Hence, the observation of oxidative release of zinc from proteins has important consequences for the progression of zinc deficiency, since it may contribute to a further deterioration of the zinc status. Whether or not successful antioxidant therapy in these diseases is related to amelioration of zinc status and whether or not any intervention in zinc metabolism could be exploited therapeutically is unknown.

7. CONCLUSIONS

This chapter has discussed a novel link between the cellular redox state and zinc metabolism, in which a change of the glutathione–glutathione disulfide redox balance to a more oxidizing state (oxidative stress) mobilizes zinc from proteins in coordination environments containing cysteine sulfur. This suggests a pathophysiological mechanism for disorders of zinc metabolism in various cases of compromised antioxidant status. Thus, both zinc and glutathione have been implicated in neurodegenerative diseases and in the biological aging process. Furthermore, redox regulation may also be an important physiological mechanism in cellular zinc distribution, in the regulation of zinc-dependent signaling pathways, and in gene expression of zinc finger-containing transcription factors. This proposal is based on the concept that zinc itself is redox-inert, and that the thiol sulfur from cysteine, which is a frequent ligand in zinc binding sites, confers redox activity on the zinc–sulfur bond.

ACKNOWLEDGMENTS

I thank Drs. Bert L. Vallee and James F. Riordan for advice and encouragement. This work was supported by the Endowment for Research in Human Biology, Inc.

REFERENCES

Arispe, N., Pollard, H. B., and Rojas, E. 1996. Zn^{2+} interaction with Alzheimer amyloid β protein calcium channels. *Proceedings of the National Academy of Sciences of the U.S.A.* 93:1710–15.

Aschner, M. 1996. The functional significance of brain metallothionein. *FASEB Journal* 10:1129–36.

Atar, D., Backx, P. H., Appel, M. M., Gao, W. D., and Marban, E. 1995. Excitation–transcription coupling mediated by zinc influx through voltage-dependent calcium channels. *Journal of Biological Chemistry* 270:2473–7.

Baumann, J. W., Liu, J., Liu, Y. P., and Klaassen, C. D. 1991. Increase in metallothionein production by chemicals that induce oxidative stress. *Toxicology and Applied Pharmacology* 110:347–54.

Becker, K. G., Nagle, J. W., Canning, R. D., Biddison, W. E., Ozato, K., and Drew, P. D. 1995. Rapid isolation and characterization of 118 novel C2H2-type zinc-finger cDNAs expressed in human brain. *Human Molecular Genetics* 4:685–91.

Behl, C., Davis, J. B., Lesley, R., and Schubert, D. 1994. Hydrogen peroxide mediates amyloid β protein toxicity. *Cell* 77:817–27.

Brady, F. O. 1982. The physiological function of metallothionein. *Trends in Biochemical Sciences* 7:143–5.

Bray, T. M., and Bettger, W. J. 1990. The physiological role of zinc as an antioxidant. *Free Radical Biology & Medicine* 8:281–91.

Brouwer, M., Hoexum-Brouwer, T., and Cashon, R. E. 1993. A putative glutathione-binding site in Cd,Zn-metallothionein identified by equilibrium binding and molecular-modeling studies. *Biochemical Journal* 294:219–25.

Burdon, R. H. 1995. Superoxide and hydrogen peroxide in relation to mammalian cell proliferation. *Free Radical Biology & Medicine* 18:775–94.

Bush, A. I., Pettingell, W. H., Multhaup, G., d. Paradis, M., Vonsattel, J.-P., Gusella, J. F., Beyreuther, K., Masters, C. L., and Tanzi, R. E. 1994. Rapid induction of Alzheimer Aβ-amyloid formation by zinc. *Science* 265:1464–7.

Buttke, T. M., and Sandstrom, P. A. 1995. Redox regulation of programmed cell death in lymphocytes. *Free Radical Research* 22:389–97.

Cano-Gauci, D. F., and Sarkar, B. 1996. Reversible zinc exchange between metallothionein and the estrogen receptor zinc finger. *FEBS Letters* 386:1–4.

Cherian, M. G. 1977. Studies on the synthesis and metabolism of zinc-thionein in rats. *Journal of Nutrition* 107:965–72.

Crow, J. P., Beckman, J. S., and McCord, J. M. 1995. Sensitivity of the essential zinc-thiolate moiety of yeast alcohol dehydrogenase to hypochlorite and peroxynitrite. *Biochemistry* 34:3544–52.

Csermely, P., Sandor, P., Radics, L., and Somogyi, J. 1989. Zinc forms complexes with higher kinetical stability than calcium, 5-f-bapta as a good example. *Biochemical and Biophysical Research Communications* 165:838–44.

Dafré, A. L., Sies, H., and Akerboom, T. 1996. Protein S-thiolation and regulation of microsomal glutathione transferase activity by the glutathione redox couple. *Archives of Biochemistry and Biophysics* 332:288–94.

Dalton, T. P., Li, Q., Bittel, D., Liang, L., and Andrews, G. K. 1996. Oxidative stress activates metal-responsive transcription factor-1 binding activity. *Journal of Biological Chemistry* 271:26233–41.

Ding, H., and Demple, B. 1996. Glutathione-mediated destabilization in vitro of [2Fe-2S] centers in the SoxR regulatory protein. *Proceedings of the National Academy of Sciences of the U.S.A.* 93:9449–53.

Dröge, W., Schulze-Osthoff, K., Mihm, S., Galter, D., Schenk, H., Eck, H.-P., Roth, S., and Gmünder, H. 1994. Functions of glutathione and glutathione disulfide in immunology and immunopathology. *FASEB Journal* 8:1131–8.

Ebadi, M., Leuschen, M. P., El Refaey, H., Hamada, F. M., and Rojas, P. 1996. The antioxidant properties of zinc and metallothionein. *Neurochemistry International* 29:159–66.

Ebadi, M., Srinivasan, S. K., and Baxi, M. D. 1996. Oxidative stress and antioxidant therapy in Parkinson's disease. *Progress in Neurobiology* 48:1–19.

Erickson, J. C., Sewell, A. K., Jensen, L. T., Winge, D. R., and Palmiter, R. D. 1994. Enhanced neurotrophic activity in Alzheimer's disease cortex is not associated with down-regulation of metallothionein-III (GIF). *Brain Research* 649:297–304.

Fliss, H., and Ménard, M. 1992. Oxidant-induced mobilization of zinc from metallothionein. *Archives of Biochemistry and Biophysics* 293:195–9.

Fraefel, C., Zeng, J., Choffat, Y., Engels, M., Schwyzer, M., and Ackermann, M. 1994. Identification and zinc dependence of the bovine herpes-virus I transactivation protein BICPO. *Journal of Virology* 68:3154–62.

Gallant, K. R., and Cherian, M. G. 1989. Metabolic changes in glutathione and metallothionein in newborn rat liver. *Journal of Pharmacology and Experimental Therapeutics* 249:631–7.

Gergel, D., and Cederbaum, A. I. 1996. Inhibition of the catalytic activity of alcohol dehydrogenase by nitric oxide is associated with S nitrosylation and the release of zinc. *Biochemistry* 35:16186–94.

Gilbert, H. F. 1990. Molecular and cellular aspects of thiol–disulfide exchange. *Advances in Enzymology and Related Areas of Molecular Biology* 63:69–172.

Giralt, M., Gasull, T., Hernandez, J., Garcia, A., and Hidalgo, J. 1993. Adrenalectomy and changes in glutathione metabolism on rat kidney metallothionein content: Comparison with liver metallothionein. *Biometals* 6:171–8.

Gonzáles, J. C., Peariso, K., Penner-Hahn, J. E., and Matthews, R. G. 1996. Cobalamin-independent methionine synthase from *Escherichia coli*: A zinc metalloenzyme. *Biochemistry* 35:12228–34.

Gozlan, H., and Ben-Ari, Y. 1995. NDMA receptor redox sites: Are they targets for selective neuronal protection? *Trends in Pharmacological Sciences* 16:368–774.

Harrison, N. L., and Gibbons, S. J. 1994. Zn^{2+}: An endogenous modulator of ligand- and voltage-gated ion channels. *Neuropharmacology* 33:935–52.

Henschke, P. N., and Elliott, S. J. 1995. Oxidized glutathione decreases luminal Ca^{2+} content of the endothelial cell Ins(1,4,5)P3-sensitive Ca^{2+} store. *Biochemical Journal* 312:485–9.

Hogg, N., Singh, R. J., and Kalyanaraman, B. 1996. The role of glutathione in the transport and catabolism of nitric oxide. *FEBS Letters* 382:223–8.

Huang, C.-C., Casey, P. J., and Fierke, C. A. 1997. Evidence for a catalytic role of zinc in protein farnesyltransferase. *Journal of Biological Chemistry* 272:20–3.

Hwang, C., Sinskey, A. J., and Lodish, H. F. 1992. Oxidized redox state of glutathione in the endoplasmic reticulum. *Science* 257:1496–1502.

Kägi, J. H. R. 1993. Evolution, structure and chemical activity of class I metallothioneins: An overview. In *Metallothionein III*, ed. K. T. Suzuki, N. Imura, and M. Kimura, 29–55. Basel, Switzerland: Birkhäuser.

Kass, G. E. N., Duddy, S. K., and Orrenius, S. 1989. Activation of protein kinase C by redox-cycling quinones. *Biochemical Journal* 260:499–507.

Kedziora, J., and Bartosz, G. 1988. Down's syndrome: A pathology involving the lack of balance of reactive oxygen species. *Free Radical Biology & Medicine* 4:317–30.

Kelly, E. J., and Palmiter, R. D. 1996. A murine model of Menkes disease reveals a physiological function of metallothionein. *Nature Genetics* 13:219–22.

Kelly, E. J., Quaife, C. F., Froelick, G. J., and Palmiter, R. D. 1996. Metallothionein I and II protect against zinc deficiency and zinc toxicity in mice. *Journal of Nutrition* 126:1782–90.

Knight, J. A. 1997. Reactive oxygen species and the neurodegenerative disorders. *Annals of Clinical and Laboratory Science* 27:11–25.

Koh, J.-Y., Suh, S. W., Gwag, B. J., He, Y. Y., Hsu, C. Y., and Choi, D. W. 1996. The role of zinc in selective neuronal death after transient global ischemia. *Science* 272:1013–16.

Krezoski, S. K., Shaw III, C. F., and Petering, D. H. 1991. Role of metallothionein in essential, toxic and therapeutic metal metabolism in Ehrlich cells. *Methods in Enzymology* 205:302–11.

Kröncke, K.-D., Fehsel, K., Schmid, T., Zenke, F. T., Dasting, I., Wesener, J. R., Bettermann, H., Breunig, K. D., and Kolb-Bachofen, V. 1994. Nitric oxide destroys zinc–sulfur clusters inducing zinc release from metallothionein and inhibition of the zinc finger-type yeast transcription activator LAC9. *Biochemical and Biophysical Research Communications* 200:1105–10.

Kwak, H.-S., Yim, H.-S., Chock, P. B., and Yim, M. B. 1995. Endogenous intracellular glutathionyl radicals are generated in neuroblastoma cells under hydrogen peroxide stress. *Proceedings of the National Academy of Sciences of the U.S.A.* 92:4582–6.

Lanius, R. A., Shaw, C. A., Wagey, R., and Krieger, C. 1994. Characterization, distribution, and protein kinase C-mediated regulation of [^{35}S]glutathione binding sites in mouse and human spinal cord. *Journal of Neurochemistry* 63:155–60.

Lapenna, D., De Gioia, S., Ciofani, G., Mezzetti, A., Consoli, A., di Ilio, C., and Cuccurullo, F. 1994. Hypochlorous acid–induced zinc release from thiolate bonds: A potential protective mechanism towards biomolecules oxidant damage during inflammation. *Free Radical Research* 20:165–70.

Lazo, J. S., Kondo, Y., Dellapiazza, D., Michalska, A. E., Choo, K. H. A., and Pitt, B. R. 1995. Enhanced sensitivity to oxidative stress in cultured embryonic cells from transgenic mice deficient in metallothionein I and II genes. *Journal of Biological Chemistry* 270:5506–10.

Leibbrandt, M. E. I., Khokha, R., and Koropatnick, J. 1994. Antisense downregulation of metallothionein in a human monocytic cell line alters adherence, invasion, and respiratory burst. *Cell Growth and Differentiation* 5:17–25.

Li, T.-K., and Vallee, B. L. 1963. Selective carboxymethylation of functional sulfhydryl groups at the active center of horse liver alcohol dehydrogenase. *Biochemical and Biophysical Research Communications* 12:44–9.

Li, T.-Y., Kraker, A. J., Shaw III, C. F., and Petering, D. H. 1980. Ligand substitution reactions of metallothioneins with EDTA and apo-carbonic anhydrase. *Proceedings of the National Academy of Sciences of the U.S.A.* 77:6334–8.

Macartney, H. W., and Tschesche, H. 1983. Latent and active human polymorphonuclear leukocyte collagenases. *European Journal of Biochemistry* 130:71–8.

Mannervik, B. 1996. Evolution of glutathione transferases and related enzymes for the protection of cells against electrophiles. *Biochemical Society Transactions* 1996:878–80.

Maret, W. 1994a. Molecular mechanisms in cellular zinc distribution. P7-334. In *Proceedings of the 16th International Congress of Biochemistry*, New Delhi, India, abstracts, 2:301.

Maret, W. 1994b. Oxidative metal release from metallothionein via zinc-thiol/disulfide interchange. *Proceedings of the National Academy of Sciences of the U.S.A.* 91:237–41.

Maret, W. 1995a. Metallothionein and the acute phase response. *Journal of Laboratory and Clinical Medicine* 126:106–7.

Maret, W. 1995b. Metallothionein/disulfide interactions, oxidative stress, and the mobilization of cellular zinc. *Neurochemistry International* 27:111–7.

Maret, W., Larsen, K. S., and Vallee, B. L. 1997. Coordination dynamics of biological zinc "clusters" in metallothionein and in the DNA-binding domain of the transcription factor Gal4. *Proceedings of the National Academy of Sciences of the U.S.A.* 94:2233–7.

Margoshes, M., and Vallee, B. L. 1957. A cadmium protein from equine kidney cortex. *Journal of the American Chemical Society* 79:4813–4.

Masters, B. A., Kelly, E. J., Quaife, C. J., Brinster, R. L., and Palmiter, R. D. 1994. Targeted disruption of metallothionein I and II genes increases sensitivity to cadmium. *Proceedings of the National Academy of Sciences of the U.S.A.* 91:584–8.

Meister, A. 1988. Glutathione metabolism and its selective modification. *Journal of Biological Chemistry* 263:17205–8.

Mély, Y., de Rocquigny, H., Morellet, N., Roques, B. P., and Gérard, D. 1996. Zinc binding of the HIV-1 nucleocapsid protein: a thermodynamic investigation by fluorescence spectroscopy. *Biochemistry* 35:5175–82.

Michalska, A. E., and Choo, K. H. A. 1993. Targeting and germ-line transmission of a null mutation at the metallothionein I and II loci in mouse. *Proceedings of the National Academy of Sciences of the U.S.A.* 90:8088–92.

Misra, R. R., Hochadel, J. F., Smith, G. T., Cook, J. C., Waalkes, M. P., and Wink, D. A. 1996. Evidence that nitric oxide enhances cadmium toxicity by displacing the metal from metallothionein. *Chemical Research in Toxicology* 9:326–32.

Monteiro, H. P., and Stern, A. 1996. Redox modulation of tyrosine phosphorylation-dependent signal transduction pathways. *Free Radical Biology & Medicine* 21:323–33.

Myers, L. C., Terranova, M. P., Ferentz, A. E., Wagner, G., and Verdine, G. L. 1993. Repair of DNA methylphosphotriesters through a metalloactivated cysteine nucleophile. *Science* 261:1164–7.

Nagel, W. W., and Vallee, B. L. 1995. Cell cycle regulation of metallothionein in human colonic cancer cells. *Proceedings of the National Academy of Sciences of the U.S.A.* 92:579–83.

Oteiza, P. I., Olin, K. L., Fraga, C. G., and Keen, C. L. 1995. Zinc deficiency causes oxidation damage to protein, lipids and DNA in rat testes. *Journal of Nutrition* 125:823–9.

Otvos, J. D., Petering, D. H., and Shaw, C. F. 1989. Structure–reactivity relationships of metallothionein, a unique metal-binding protein. *Comments on Inorganic Chemistry* 9:1–35.

Ou, C. Z., and Ebadi, M. 1992. Pineal and retinal protein kinase C isoenzymes: Cooperative activation by calcium and zinc metallothionein. *Journal of Pineal Research* 12:17–26.

Peck, E. J., and Ray, W. J. 1971. Metal complexes of phosphoglucomutase in vivo. *Journal of Biological Chemistry* 246:1160–7.

Peters, J. M., Duncan, J. R., Wiley, L. M., Rucker, R. B., and Keen, C. L. 1995. Effect of a metallothionein antisense oligonucleotide on embryo development. *Reproductive Toxicology* 9:123–30.

Pountney, D. L., Fundel, S. M., Faller, P., Birchler, N. E., Hunziker, P., and Vašák, M. 1994. Isolation, primary structures and metal binding properties of neuronal growth inhibitory factor (GIF) from bovine and equine brain. *FEBS Letters* 345:193–7.

Ren, Y., and Smith, A. 1995. Mechanism of metallothionein gene regulation by heme-hemopexin. *Journal of Biological Chemistry* 270:23988–95.

Rice, W. G., Supko, J. G., Malspeis, L., Buckheit, Jr., R. W., Clanton, D., Bu, M., Graham, L., Schaeffer, C. A., Turpin, J. A., Domagala, J., Gogliotti, R., Bader, J. P., Halliday, S. M., Coren, L., Sowder II, R. C., Arthur, L. A., and Henderson, L. E. 1995. Inhibitors of the HIV nucleocapsid protein zinc fingers as candidates for the treatment of AIDS. *Science* 270:1194–7.

Robbins, A. H., McRee, D. E., Williamson, M., Collett, S. A., Xuong, N. H., Furey, W. F., Wang, B. C., and Stout, C. D. 1991. Refined crystal structure of Cd,Zn metallothionein at 2.0 Å resolution. *Journal of Molecular Biology* 221:1269–93.

Rofe, A. M., Philcox, J. C., and Coyle, P. 1996. Trace metal, acute phase and metabolic response to endotoxin in metallothionein-null mice. *Biochemical Journal* 314:739–97.

Sen, C. K., and Packer, L. 1996. Antioxidant and redox regulation of gene transcription. *FASEB Journal* 10:709–20.

Sewell, A. K., Jensen, L. T., Erickson, J. C., Palmiter, R. D., and Winge, D. R. 1995. Bioactivity of metallothionein-3 correlates with its novel β domain sequence rather than metal binding properties. *Biochemistry* 34:4740–7.

Simons, T. J. B. 1991. Intracellular free zinc and zinc buffering in human red blood cells. *Journal of Membrane Biology* 123:63–71.

Singh, S. P., Wishnok, J. S., Keshive, M., Deen, W. M., and Tannenbaum, S. R. 1996. The chemistry of the *S*-nitrosoglutathione/glutathione system. *Proceedings of the National Academy of Sciences of the U.S.A.* 93:14428–33.

Snyder, G. H., Cennerazzo, M. J., Karalis, A. J., and Field, D. 1981. Electrostatic influence of local cysteine environments on disulfide exchange kinetics. *Biochemistry* 20:6509–19.

Sohal, R. S., and Weindruch, R. 1996. Oxidative stress, caloric restriction, and aging. *Science* 273:59–63.

Springman, E. B., Angleton, E. L., Birkedal-Hansen, H., and Van Wart, H. E. 1990. Multiple modes of activation of latent human fibroblast collagenase: evidence for the role of a Cys[73] active-site zinc complex in latency and a "cysteine switch" mechanism for activation. *Proceedings of the National Academy of Sciences of the U.S.A.* 87:364–8.

Stamler, J. S. 1994. Redox signaling: nitrosylation and related target interactions of nitric oxide. *Cell* 78:931–6.

Sun, Y., and Oberley, L. W. 1996. Redox regulation of transcriptional activators. *Free Radical Biology & Medicine* 21:335–48.

Thomas, J. A., Poland, B., and Honzatko, R. 1995. Protein sulfhydryls and their role in the antioxidant function of protein *S*-thiolation. *Archives of Biochemistry and Biophysics* 319:1–9.

Tummino, P. J., Scholten, J. D., Harvey, P. J., Holler, T. P., Maloney, L., Gogliotti, R., Domagala, J., and Hupe, D. 1996. The in vitro ejection of zinc from human immunodeficiency virus (HIV) type 1 nucleocapsid protein by disulfide benzamides with cellular anti-HIV activity. *Proceedings of the National Academy of Sciences of the U.S.A.* 93:969–73.

Tyagi, S. C., Kumar, S. G., and Borders, S. 1996. Reduction–oxidation (redox) state regulation of extracellular matrix metalloproteinases and tissue inhibitors in cardiac normal and transformed fibroblast cells. *Journal of Cellular Biochemistry* 61:139–51.

Uchida, Y., Takio, K., Titani, K., Ihara, Y., and Tomonaga, Y. M. 1991. The growth inhibitory factor that is deficient in the Alzheimer's disease brain is a 68 amino acid metallothionein-like protein. *Neuron* 7:337–47.

Udom, A. O., and Brady, F. O. 1980. Reactivation in vitro of zinc-requiring apoenzymes by rat liver zinc-thionein. *Biochemical Journal* 187:329–35.

Vallee, B. L. 1991. Introduction to metallothionein. *Methods in Enzymology* 205:3–7.

Vallee, B. L. 1995. The function of metallothionein. *Neurochemistry International* 27:23–33.

Vallee, B. L., and Auld, D. S. 1993. Zinc: Biological functions and coordination motifs. *Accounts of Chemical Research* 26:543–51.

Vallee, B. L., and Falchuk, K. H. 1993. The biochemical basis of zinc physiology. *Physiological Reviews* 73:79–118.

Vallee, B. L., and Maret, W. 1993. The functional potential and potential functions of metallothionein: A personal perspective. In *Metallothionein III*, ed. K. T. Suzuki, N. Imura, and M. Kimura, 1–27. Basel, Switzerland: Birkhäuser.

Weir, E. K., and Archer, S. L. 1995. The mechanism of acute hypoxic pulmonary vasoconstriction: the tale of two channels. *FASEB Journal* 9:183–9.

Wilker, J. J., and Lippard, S. J. 1995. Modeling the DNA methylphosphotriester repair site in *Escherichia coli* Ada. Why zinc and four cysteines? *Journal of the American Chemical Society* 117:8682–3.

Williamson, J. R., Chang, K., Frangos, M., Hasan, K. S., Ido, Y., Kawamura, T., Nyengaard, J. R., van den Enden, M., Kilo, C., and Tilton, R. G. 1993. Hyperglycemic pseudohypoxia and diabetic complications. *Diabetes* 42:801–13.

Wink, D. A., and Laval, J. 1994. Inhibition by nitric oxide of the repair protein O^6, methylguanine-DNA-methyltransferase. *Carcinogenesis* 15:2125–9.

Wüthrich, K. 1991. Determination of the three-dimensional structure of metallothioneins by nuclear magnetic resonance spectroscopy in solution. *Methods in Enzymology* 205:502–20.

Yu, X., Hathout, Y., Fenselau, C., Sowder II, R. C., Henderson, L. E., Rice, W. G., Mendeleyev, J., and Kun, E. 1995. Specific disulfide formation in the oxidation of HIV-1 zinc finger protein nucleocapsid p7. *Chemical Research in Toxicology* 8:586–90.

Zeng, J., Heuchel, R., Schaffner, W., and Kägi, J. H. R. 1991. Thionein (apometallothionein) can modulate DNA binding and transcription activation by zinc finger containing factor Sp1. *FEBS Letters* 279:310–2.

Zeng, J., Vallee, B. L., and Kägi, J. H. R. 1991. Zinc transfer from transcription factor IIIA fingers to thionein clusters. *Proceedings of the National Academy of Sciences of the U.S.A.* 88:9984–8.

Zeppezauer, E., Jörnvall, H., and Ohlsson, I. 1975. Carboxymethylation of horse liver alcohol dehydrogenase in the crystalline state. *European Journal of Biochemistry* 58:95–104.

Ziegler, D. M. 1985. Role of reversible oxidation–reduction of enzyme thiols–disulfides in metabolic regulation. *Annual Review of Biochemistry* 54:305–29.

Glutathione in the Nervous System
Edited by Christopher A. Shaw
Copyright © 1998 Taylor & Francis

13

Glutathione Release and Nitrosoglutathione Presence in the CNS: Implications for Schizophrenia

Michel Cuénod and Kim Quang Do

Brain Research Institute, University of Zürich, Zürich, Switzerland

ABBREVIATIONS

CSF cerebrospinal fluid
Cys cysteine
GSH glutathione
GSNO S-nitrosoglutathione
γ-Glu-Gln γ-glutamylglutamine
HPLC high-performance liquid chromatography
NMDA N-methyl-D-aspartate
NO nitric oxide

1. INTRODUCTION

Glutathione (GSH) plays an important role in the metabolism of the cell, and its specific functions in the central nervous system are not entirely elucidated. The present chapter reviews results from our laboratory on the release of GSH, the presence of S-nitrosoglutathione (GSNO), and a deficit of GSH in schizophrenia, as well as an hypothesis on the possible implication of GSH in the pathophysiology of this disease.

2. RELEASE OF GLUTATHIONE AND CYSTEINE FROM BRAIN TISSUE

Intercellular communication by way of specific molecules is central to the functions of the nervous system. Neurotransmitters and neuromodulators, or more generally neuroactive compounds, are released from their intracellular compartment to the intercellular space under controlled conditions in order to influence other cells. When rat brain slices are exposed to 50 mM $[K^+]$, the cell membranes depolarize, and compounds are released in a Ca^{2+}-dependent manner. Such a release can be either direct, due to the depolarization-induced release of compounds, or indirect, triggered by neuroactive substances that are released through depolarization and that in turn act on other cells or compartments, where they activate the efflux of compounds.

Using this approach, cysteine (Cys), the precursor of GSH, was shown to be released differentially from various regions of the rat brain (Keller et al. 1989), in addition to some amino acids, including glutamate and glycine. Similarly, GSH was shown to be released from rat brain slices by depolarization induced by high $[K^+]$ in a Ca^{2+}-dependent manner (Zängerle et al. 1992). For both GSH and Cys, this release was most prominent in the telencephalic parts of the rat brain and reached 2–3 times the resting efflux, amounting to a few picomoles per milligram of protein per minute. Over 85 percent of the GSH was in the reduced form. Whereas a resting efflux of GSH has been described by Yudkoff et al. (1990) and quantified by Sagara et al. (1996), the depolarization-induced release of GSH indicates that its extracellular level is influenced by cellular activity. The mechanism and the role of this depolarization-induced GSH release are unknown at present. The fact that it is Ca^{2+}-dependent suggest that either it is released by a vesicular mechanism similar to that of classical transmitters, or its efflux is under the control of a released transmitter. One can only speculate about its extracellular functions, if any: it could modulate receptors (the NMDA receptor or others), it could be involved in the detoxification of free radicals (such as the metabolites of dopamine), or it could be involved in the transport of amino acids.

Cys displays a strong cytotoxic effect (Olney et al. 1972; Karlsen et al. 1981), a high affinity for the L-APB binding site, which is one of the Na^+-independent, presynaptic glutamate binding sites (Pullan et al. 1987), and an ability to activate the glutamate metabotropic receptor (Charpak et al. 1990). GSH has been shown to selectively inhibit the binding of glutamate to the NMDA receptor with concomitant potentiation of the MK-801 binding to the NMDA channel (Ogita and Yoneda 1987, 1989; Ogita et al. 1995). Moreover, GSH can act at the redox modulatory site(s) of the NMDA receptor (Aizenman, Lipton, and Loring 1989; for a review see Gozlan and Ben-Ari 1995; Sucher et al. 1996), where it potentiates the L-glutamate-activated currents in cells, expressing the NR_1–NR_{2A} receptor channels (Köhr et al. 1994). GSH binding sites have been localized predominantly on CNS neuroglial cells and on cultured astrocytes (Guo, McIntosh, and Shaw 1992). However, with antisera against GSH, Hjelle et al. (1994; see also Hjelle et al., chapter 4, this volume) reported significant levels of GSH-like immunoreactivity at the electron-microscopic level in neuronal as well as in glial compartments. Moreover, GSH elicited increased levels of intracellular inositol triphosphate when applied to cultured astrocytes (Guo et al. 1992). Recently Shaw et al. (1996) reported that GSH elicits a large, fast depolarizing potential when applied to cortical slices. These sodium currents are not blocked by antagonists of the excitatory receptors. On the other hand, GSH and GSSG applied by microiontophoresis in the rat ventrobasal thalamus inhibit the synaptic responses (Shaw and Salt 1997).

3. PRESENCE OF S-NITROSOGLUTATHIONE IN RAT BRAIN

Nitric oxide (NO) plays a role in many physiological responses, including the regulation of blood pressure, the inhibition of platelet aggregation, and the modulation of the immunity and inflammation (for reviews, see Moncada 1994; Loscalzo and Welch 1995). In the CNS, NO acts as a neuronal messenger (Garthwaite, Charles, and Russell 1988; Bredt and Snyder 1989) and is involved in various processes such as synaptic plasticity and neurotoxicity (for reviews, see Jaffrey and Snyder 1995; Garthwaite and Boulton 1995). NO is a gas that readily diffuses across membranes and thus does not act as a conventional neurotransmitter. It is not stored in synaptic vesicles or released by exocytosis. Further, it does not act at conventional membrane-associated receptors.

Although the evidence is that NMDA receptor activation induces the release of NO from cerebellar cells (Garthwaite et al. 1988), the form in which NO is stored, delivered, and transported in CNS has not been directly investigated. Based on the strong reactivity of NO for thiols and on the presence of Cys and GSH at the millimolar level intracellularly and the micromolar level extracellularly (Zängerle et al. 1992), we have investigated whether S-nitrosothiols, that is, S-nitrosocysteine and/or S-nitrosoglutathione (GSNO), may be the package form in which NO could be stored. The extraction was optimized in order to avoid partial degradation of S-nitrosothiols, and a sensitive and selective analytical method was developed, based on reversed-phase high-performance liquid chromatography (HPLC) combined with multiwavelength detection and an on-line absorption spectrum, which allows us to quantify S-nitrosothiols at a level of 150–300 pmol.

In extracts of cerebellar slices from 8–10-day-old rats incubated with radioactive [35S]Cys, a radiolabeled peak corresponding to GSNO has been detected. Its identification was confirmed by spiking with a reference compound. Indeed, labeled Cys was taken up and incorporated into GSH and GSNO. Moreover, the endogenous compound eluting at the retention time of GSNO was chemically characterized by micro HPLC coupled to continuous-flow fast-atom-bombardment mass spectrometry: Its fluorenylmethyloxycarbonyl derivative has a mass spectrum identical to that of authentic GSNO. To quantify the level of endogenous GSNO, [15N]GSNO was added to brain extracts as an internal standard. Micro HPLC mass-spectrometry analysis of the fluorenylmethyloxycarbonyl derivatives of GSNO and [^{15}N]GSNO, which were detected by selected ion monitoring of their [M–H]$^-$ ions at $m/z = 557$ and 558, allowed us to determine the absolute amounts of GSNO in the extract. In the adult rat cerebellum it reached 15.3 ± 1.4 pmol per milligram of protein. Using a protein-to-wet-weight ratio of 0.35–0.37 g of protein per gram of wet weight (Kudo et al. 1990), the tissue concentration of GSNO was estimated to 6–7 μM. This GSNO concentration is comparable to the level of S-nitrosothiols in plasma (Stamler et al. 1992) or in human airways (Gaston et al. 1993). This GSNO concentration is about 0.3–0.6 percent of the tissue GSH level in rat cerebellum [1 to 2 mM (Kudo et al. 1990)] and comparable to the cerebellar extracellular GSH level (Zängerle et al. 1992). Moreover, incubation of cerebellar slices with NMDA increased the level of GSNO level by four times. This study demonstrated unequivocally the presence of an endogenous nitrosothiol, GSNO, in rat brain tissue (Do et al. 1996).

The physiological significance of this nitrosothiol in the CNS is yet unknown. GSNO has been used widely as a NO donor, and it is the most potent activator of guanylate cyclase (Garthwaite 1993). Whether endogenous GSNO acts through release of the free radical NO* or independently of the latter has yet to be determined. In support of the first possibility, GSNO may play a role as a more stable carrier for the radical NO* that can freely cross the cell membrane and can act on soluble guanylate cyclase in the target cell. However, the presence in rat brain of a specific [^{3}H]GSNO binding site (Taguchi, Ohta, and Talman 1995) favors the second possibility, where endogenous GSNO may bind to the target cell membrane and elicit its activity per se. Another possibility is that GSNO, through the transfer or reaction of NO$^+$, may nitrosylate sulfhydryl centers of proteins such as key enzymes or receptors (Meffert et al. 1996) in the target cell. Moreover, a neuroprotective antioxidative effect has been found for GSNO when injected in vivo in rat striatum (Rauhala et al. 1996).

One potential role of the endogenous intra- and extracellular GSH is the packaging of NO in the form of nitrosothiols. This might serve to prolong the life of this short-lived messenger and to target its delivery to specific effectors. That could confer a specificity of action of this widely diffusable messenger, determine its range of effectiveness, and mitigate its adverse cytotoxic effects.

4. PROPOSAL FOR A GLUTATHIONE MECHANISM IN SCHIZOPHRENIA: GLUTATHIONE AND γ-GLUTAMYLGLUTAMINE DEFICIT IN CEREBROSPINAL FLUID OF SCHIZOPHRENIC PATIENTS

Schizophrenia is an endogenous psychosis characterized by positive symptoms, such as delusions, hallucinations, thought disorder, and incoherence, and negative ones, such as deficits in cognitive and social abilities and motivation, and poverty of speech accompanied by a loss of emotional content. It is a most devastating disease for the individual patient.

Recent neuroimaging studies have shown structural abnormalities in cerebral cortical regions (temporal, entorhinal, cingulate, and prefrontal) (Selemon, Rajkowska, and Goldman-Rakic 1995). These cortical areas are richly innervated with dopamine terminals and play an essential role in working memory in primates (Goldman-Rakic 1994). The subcortical regions of the temporal lobe, the basal ganglia, and the thalamus are affected as well. Functional imaging studies also suggest prefrontal cortical hypofunction and a functional disorganization of the prefrontal–temporal–limbic cortical connectivity.

Presently, two main biological hypothesis have been proposed to explain the pathophysiology of schizophrenia.

1. The dopamine theory proposes that schizophrenic illness is a manifestation of increased dopaminergic activity in certain brain areas (Matthysse 1973). This interpretation gains support from the observation that most neuroleptics are antagonists of the dopamine receptors, whereas psychotic symptoms can be induced by drugs that enhance dopaminergic activity, such as amphetamine (Seeman et al. 1976). However, this initial view was not unequivocally supported by subsequent studies, because conflicting observations were reported regarding dopamine and its metabolites' levels in CSF, plasma, and post mortem brain tissue of individuals with schizophrenia (for an overview, see Davis et al. 1991). During the past few years, however, evidence has been accumulating that schizophrenic disorders may be characterized by both hypodopaminergia in the mesocortical dopamine neurons and hyperdopaminergia in the mesolimbic dopamine neurons (Davis et al. 1991). D2-like receptors were reported to be elevated in post mortem brain tissue of schizophrenic patients (Seeman, Nizinik, and Guan 1990), which supports the hypothesis that dopaminergic neurotransmission in schizophrenia is enhanced through overactive response by D2 receptors. This idea has been criticized because clozapine, which has only low affinity for D2 receptors, is clinically a highly effective neuroleptic. Recently, however, the group led by Seeman demonstrated the existence of a polymorphic D4 receptor to which clozapine binds with an affinity 10 times higher than to any other yet characterized dopamine receptors (van Tol et al. 1992). Interestingly, the same group found D4 receptors to be elevated in the brain of patients with schizophrenia (Seeman, Guan, and van Tol 1993). Moreover, although the dopaminergic system is likely to be involved in the pathophysiology of the disease, there is little evidence that it is the primary factor in the development of schizophrenia. A further line of research suggested that lesions possibly occurring during the fetal period or early in the development, due to hypoxia, nutritional or toxic factors, or viral infection, might play a central role in the developmental cortical defect, which has a delayed effect on cortical function and dopamine regulation (Lillrank, Lipska, and Weinberger 1995).

2. The second major hypothesis, as originally formulated by Kim et al. (1980), claimed a role for excitatory amino acids, i.e., glutamate, in the pathophysiology of schizophrenia. On the basis of decreased levels of glutamate in the CSF of individuals with schizophrenia, these authors postulated that either a hyperactivity of dopaminergic neurons leads to an enhanced inhibition of glutamate release, or a hypofunction

of the glutamate receptor causes the decreased glutamate release. Indeed, in rodents, the release of glutamate from the striatal terminals is inhibited by dopamine (see Carlsson and Carlsson 1990). The glutamatergic hypothesis has been reformulated, based on the observation that phencyclidine (PCP) elicits both productive and deficit symptoms of schizophrenia (Javitt and Zukin 1991). Although PCP exerts its effects in part via central dopamine systems, it is a potent and noncompetitive antagonist at the N-methyl-D-aspartate (NMDA) receptor. Thus, the current hypoglutamatergic theory of schizophrenia proposes diminished glutamatergic neurotransmission at the level of the NMDA receptor site (Deutsch et al. 1989; Squires and Saederup 1991). The glutamate hypothesis also gains support from the increased binding of glutamate receptors in post mortem brain tissue, which might reflect upregulation due to decreased glutamatergic activity (Nishikawa et al. 1983; Deakin et al. 1989; Kornhuber et al. 1989; Eastwood et al. 1995). However, concerning abnormalities in glutamate levels in the brain and CSF of schizophrenic patients, contradictory results have been published (Kim et al. 1980; Gattaz, Gasser, and Beckmann 1985; Perry 1982; Korpi et al. 1987; Swahn 1990; Do et al. 1995). In most of these studies, metabolic side effects due to neuroleptic treatment could not be excluded.

Although both the dopaminergic and glutamatergic hypotheses of schizophrenia have stimulated intense work in the field, there has been neither unequivocal support nor clear rejection of either theory.

Making the working hypothesis that changes in the cerebrospinal fluid (CSF) composition are a direct or indirect reflection of changes in neuroactive substances involved in the pathology of schizophrenia, a high-powered bioanalytical method (HPLC, gas chromatography, and mass spectrometry; or HPLC and mass spectrometry) was used to analyze the CSF of patients. This enables determination of amino acids, monoamine metabolites, N-acetylaspartate, and N-acetylaspartylglutamate levels and identification of heretofore unknown compounds. As a study sample, a group of 26 patients with schizophrenia was investigated, in whom long-term changes secondary to previous antipsychotic treatment could be excluded. Fifteen age- and sex-matched subjects in whom internal, neurological, and psychiatric disorders were excluded served as the control group.

Among the 26 compounds (amino acids, small peptides, and monoamine metabolites) analyzed in the CSF of drug-naïve schizophrenics, a significantly decreased level was observed for GSH (controls: 565 ± 508 [SD] pmol/ml; patients: 327 ± 127 pmol/ml; -42 percent, $p < 0.05$) (Do, Lauer, Schreiber, Holsboer, Cuenod, unpublished observations), for its metabolite γ-glutamylglutamine (γ-Glu-Gln) (controls: $2,469 \pm 473$ [SD] pmol/ml; patients: $2,077 \pm 522$; -16 percent, $p < 0.01$), and for taurine (controls: $5,871 \pm 630$ [SD] pmol/ml; patients: $4,988 \pm 865$; -15 percent, $p < 0.001$) (Do et al. 1995) as compared to controls. In contrast, dopamine and serotonine metabolites were not different from controls, an observation that does not support the dopamine hypothesis. The concentration of glutamate, on average, was lower in the patients, but this decrease did not reach significance. However, the use of GSH, γ-Glu-Gln, glutamate, aspartate, taurine, and isoleucine as discriminating variables allowed us to classify 85 percent of the subjects correctly, with a specificity of 96 percent (i.e., 96 percent of the schizophrenic patients were correctly classified). These results suggest the implication of at least some of those compounds in the physiopathology of schizophrenia. Up to now, no consideration has been given to a possible involvement of GSH and/or γ-Glu-Gln in the pathophysiology of schizophrenia. A hypothesis including the role of GSH in schizophrenia is proposed below.

Assuming that the GSH deficit in CSF reported above is a direct or indirect reflection of an important GSH deficit in brain areas involved in the pathology of schizophrenia, the following hypothesis can be formulated. This deficit could be due to either a genetic deficit of the enzymes involved in the GSH metabolism or a deficit of their activity. Indeed, a decrease in the activity of the enzyme GSH peroxidase in platelets and in erythrocytes was found to be associated with schizophrenia (Buckman et al. 1987, 1990). This GSH deficit would have many consequences:

1. GSH is considered to be a major antioxidant in the brain. Because of its sulfhydryl group, it plays an important role in reducing free radicals. Thus GSH has been shown to detoxify (organic and inorganic) activated oxygen radicals resulting from dopamine catabolism (Maker et al. 1981; Spina and Cohen 1989). Toxic oxidation products, among them metabolites of dopamine, will not be adequately reduced, leading to an accumulation of free radicals and thus to (sub)cellular degeneration. Although there are no data to our knowledge on this issue, it is possible that a GSH deficit during development has drastic consequence leading to abnormal connectivity. Such a process is compatible with the fact that neuroleptic drugs effective in schizophrenia not only block the dopamine receptors, but also reduce the dopamine level, an effect that takes a few days to be achieved, paralleling the delay needed for the treatment to be effective. Thus the deficit in GSH might in part explain the phenomena that led to the proposal of the dopamine hypothesis. The toxicity of the dopamine metabolites might be restricted to the microenvironment of the terminals of the dopamine fibers innervating the cortex (Goldman-Rakic et al. 1989), leading to the degeneration of spines and dendrites rather than of the entire cell bodies. This would be consistent with the findings that post mortem histological analysis of the frontal cortex of schizophrenic patients reveals an increased density of neurons, indicating a decrease in neuropil (Selemon et al. 1995). Furthermore, it is interesting to note that the habenula, which has been implicated in the dopamine theory of schizophrenia (Ellison 1994), shows a high level of [3][H] GSH binding (C. A. Shaw, personal communication). Moreover, depletion in rats of GSH with buthionine sulfoximine followed by dopamine treatment produced deficits in psychomotor behavior (Shukitt-Hale et al. 1997).

2. A deficit in GSH would reduce the formation of the nitrosothiol GSNO (see above), thus decreasing the control of the toxicity of NO. In this connection it is interesting that a reduced density of cells histochemically staining for nicotinamide adenine dinucleotide phosphate diaphorase (NO synthase staining) has been reported in frontal lobe of schizophrenics (Akbarian et al. 1993).

3. GSH was effective in displacing Glu and NMDA agonist binding (Ogita et al. 1995; see also Ogita et al., chapter 7, this volume). It can act at the redox modulatory

site(s) of the NMDA receptor (Aizenman et al. 1989; for a review see Gozlan and Ben-Ari 1995; Sucher et al. 1996), where it potentiates the L-glutamate-activated currents in cells expressing the NR_1–NR_{2A} receptor channels (Köhr et al. 1994). Although the concentration of GSH required is in the millimolar range, that could well be the local concentration available in the synaptic microenvironment following release of GSH, as described by our group (Zängerle et al. 1992). Indeed, the intracellular concentration of GSH is in the millimolar range. At a low level of GSH, the potentiation of the NMDA receptor might be deficient. Such an inadequate activation of the NMDA receptors could be related to some of the symptoms of schizophrenia, as the phencyclidines, the NMDA receptors' antagonists, induce psychopathological symptoms reminiscent of those observed in schizophrenic patients.

4. Meister and his collaborators have proposed that the GSH cycle is involved in the transmembrane transport of amino acids. The enzyme γ-glutamyltranspeptidase, which transfers the γ-glutamyl moiety of GSH to an amino acid acceptor, has a particularly high affinity for glutamine and methionine (Tate and Meister 1974). The transport of glutamine is very important in neurons, in order to reload the precursor pool of the major excitatory transmitter glutamate. Moreover, Yudkoff et al. (1990) reported that a partial depletion of intracellular GSH in astrocytes was associated with a significant reduction of intracellular glutamate. The transmembrane transport of amino acids, particularly that of glutamine and methionine, will be deficient when the GSH availability is below a critical level. This will affect the glutamatergic presynaptic boutons, which rely on a supply of glutamine as precursor of glutamate. This view is supported by the reported observation that the γ-glutamylglutamine level was decreased in CSF of schizophrenic patients (Do et al. 1995), suggesting a decrease in glutamine transport. These processes, together with the deficient efficacy of the NMDA receptors mentioned above, would lead to a deficit of glutamatergic transmission, which has been the basis for the glutamate hypothesis. In this context, it is worth noting that methionine given per os has been shown to be the only amino acid that exacerbates the psychotic symptoms in schizophrenic patients (Park, Baldessarini, and Kety 1965). This observation might be related to a deficit in the γ-glutamyltransferase system in which methionine competes with glutamine. An excess of methionine would impede the already deficient transport of Gln. Indeed, methionine induced decreases in glutamate, glutamine, and aspartate levels in synaptosomes, probably due to some impairment in the uptake of glutamine into the nerve terminals. The primary cause of this glutamine insufficiency was not linked to the enzymes of glutamate and glutamine metabolism (Wood, Kurylo, and Geddes 1985). The role of γ-glutamyltransferase in this context has not been studied. Alternatively, GSH could be an important reservoir for glutamate (Yudkoff et al. 1990), and a depletion in GSH level would lead to a deficit of glutamate.

5. CONCLUSIONS

In conclusion, a deficit in GSH and GSH-related enzymes might play an essential role in the pathophysiology of schizophrenia, or of some forms of schizophrenia.

If the hypothesis proposed is correct, it would have the advantage of integrating a majority of elements known about the disease and could lead to new approaches to its treatment.

ACKNOWLEDGMENTS

This work was supported by grants from the Swiss National Foundation (NF31-39681.93), the Sandoz Research Institute, the Human Frontier Science Program (HFSP RG-2/95), and the E. Slack-Gyr Foundation.

REFERENCES

Aizenman, E., Lipton, S. A., and Loring, R. H. 1989. Selective modulation of NMDA responses by reduction and oxidation. *Neuron* 2:1257–63.

Akbarian, S., Vinuela, A., Kim, J. J., Potkin, S. G., Bunney, W. E., Jr., and Jones, E. G. 1993. Distorted distribution of nicotinamide–adenine dinucleotide phosphate–diaphorase neurons in temporal lobe of schizophrenics implies anomalous cortical development. *Archives of General Psychiatry* 50:178–87.

Bredt, D. S., and Snyder, S. H. 1989. Nitric oxide mediates glutamate-linked enhancement of cGMP levels in the cerebellum. *Proceedings of the National Academy of Sciences of the U.S.A.* 86:9030–3.

Buckman, T. D., Kling, A. S., Eiduson, S., Sutphin, M. S., and Steinberg, A. 1987. Glutathione peroxidase and CT scan abnormalities in schizophrenia. *Biological Psychiatry* 22:1349–56.

Buckman, T. D., Kling, A., Sutphin, M. S., Steinberg, A., and Eiduson, S. 1990. Platelet glutathione peroxidase and monoamine oxidase activity in schizophrenics with CT scan abnormalities: relation to psychosocial variables. *Psychiatry Research* 31:1–14.

Carlsson, M., and Carlsson, A. 1990. Interactions between glutamatergic and monoaminergic systems within the basal ganglia—implications for schizophrenia and Parkinson's disease. *Trends in Neuroscience* 13:272–6.

Charpak, S., Gahwiler, B. H., Do, K. Q., and Knopfel, T. 1990. Potassium conductances in hippocampal neurons blocked by excitatory amino-acid transmitters. *Nature* 347:765–7.

Davis, K. L., Kahn, R. S., Ko, G., and Davidson, M. 1991. Dopamine in schizophrenia: a review and reconceptutualization. *American Journal of Psychiatry* 148:1474–86.

Deakin, J. F., Slater, P., Simpson, M. D., Gilchrist, A. C., Skan, W. J., Royston, M. C., Reynolds, G. P., and Cross, A. J. 1989. Frontal cortical and left temporal glutamatergic dysfunction in schizophrenia. *Journal of Neurochemistry* 52:1781–6.

Deutsch, S. I., Mastropaolo, J., Schwartz, B. L., Rosse, R. B., and Morihisa, J. M. 1989. A "glutamatergic hypothesis" of schizophrenia. Rationale for pharmacotherapy with glycine. *Clinical Neuropharmacology* 12:1–13.

Do, K. Q., Lauer, C. J., Schreiber, W., Zollinger, M., Gutteck Amsler, U., Cuenod, M., and Holsboer, F. 1995. Gamma-glutamylglutamine and taurine concentrations are decreased in the cerebrospinal fluid of drug-naive patients with schizophrenic disorders. *Journal of Neurochemistry* 65:2652–62.

Do, K. Q., Benz, B., Grima, G., Gutteck-Amsler, U., Kluge, I., and Salt, T. E. 1996. Commentary—nitric oxide precursor arginine and S-nitrosoglutathione in synaptic and glial function. *Neurochemistry International* 29(3):213–24.

Eastwood, S. L., McDonald, B., Burnet, P. W., Beckwith, J. P., Kerwin, R. W., and Harrison, P. J. 1995. Decreased expression of mRNAs encoding non-NMDA glutamate receptors GluR1 and GluR2 in medial temporal lobe neurons in schizophrenia. *Molecular Brain Research* 29:211–23.

Ellison, G. 1994. Stimulant-induced psychosis, the dopamine theory of schizophrenia, and the habenula. *Brain Research Review* 19:223–39.

Garthwaite, J. 1993. Nitric oxide signalling in the nervous system. *Seminars in the Neurosciences* 5:171–80.

Garthwaite, J., and Boulton, C. L. 1995. Nitric oxide signaling in the central nervous system. *Annual Review of Physiology* 57:683–706.

Garthwaite, J., Charles, S. L., and Russell, C. W. 1988. Endothelium-derived relaxing factor release on activation of NMDA receptors suggests role as intercellular messenger in the brain. *Nature* 336:385–8.

Gaston, B., Reilly, J., Drazen, J. M., Fackler, J., Ramdev, P., Arnelle, D., Mullins, M. E., Sugarbaker, D. J., Chee, C., Singel, D. J., Loscalzo, J., and Stamler, J. S. 1993. Endogenous nitrogen oxides and bronchodilator S-nitrosothiols in human airways. *Proceeding of the National Academy of Sciences of the U.S.A.* 90:10957–61.

Gattaz, W. F., Gasser, T., and Beckmann, H. 1985. Multidimensional analysis of the concentrations of 17 substances in the CSF of schizophrenics and controls. *Biological Psychiatry* 20:360–66.

Goldman-Rakic, P. S. 1994. Working memory dysfunction in schizophrenia. *Journal of Neuropsychiatry and Clinical Neuroscience* 6:348–57.

Goldman-Rakic, P. S., Leranth, C., Williams, S. M., Mons, N., and Geffard, M. 1989. Dopamine synaptic complex with pyramidal neurons in primate cerebral cortex. *Proceedings of the National Academy of Sciences of the U.S.A.* 86:9015–19.

Gozlan, H., and Ben-Ari, Y. 1995. NMDA receptor redox sites: are they targets for selective neuronal protection? *Trends in Pharmacological Science* 16:368–74.

Guo, N., McIntosh, C., and Shaw, C. 1992. Glutathione: new candidate neuropeptide in the central nervous system. *Neuroscience* 51:835–42.

Hjelle, O. P., Chaudhry, F. A., and Ottersen, O. P. 1994. Antisera to glutathione: characterization and immunocytochemical application to the rat cerebellum. *European Journal of Neuroscience* 6:793–804.

Jaffrey, S. R., and Snyder, S. H. 1995. Carbon monoxide, cyclic gmp, neurotoxicity, and glutamate. Nitric oxide—a neural messenger. *Annual Review of Cell & Developmental Biology* 11:417–40.

Javitt, D. C., and Zukin, S. R. 1991. Recent advances in the phencyclidine model of schizophrenia. *American Journal of Psychiatry* 148:1301–8.

Karlsen, R. L., Grofova, I., Malthe-Sørenssen, D., and Fonnum, F. 1981. Morphological changes in rat brain induced by L-Cys injection in newborn animals. *Brain Research* 208:167–80.

Keller, H. J., Do, K. Q., Zollinger, M., Winterhalter, K. H., and Cuénod, M. 1989. Cysteine: depolarization-induced release from rat brain in vitro. *Journal of Neurochemistry* 52:1801–6.

Kim, J. S., Kornhuber, H. H., Schmid-Burgk, W., and Holzmüller, B. 1980. Low cerebrospinal fluid glutamate in schizophrenic patients and a new hypnosis of schizophrenia. *Neuroscience Letters* 20:379–82.

Köhr, G., Eckardt, S., Luddens, H., Monyer, H., and Seeburg, P. H. 1994. NMDA receptor channels: subunit-specific potentiation by reducing. *Neuron* 12:1031–40.

Kornhuber, J., Mack Burkhardt, F., Riederer, P., Hebenstreit, G. F., Andrews, H. B., and Beckmann, H. 1989. (^{3}H)MK-801 binding sites in postmortem brain regions of schizophrenic patients. *Journal of Neural Transmission* 77:231–6.

Korpi, E. R., Kaufmann, C. A., Marnela, K. M., and Weinberger, D. R. 1987. Cerebrospinal fluid amino acid concentrations in chronic schizophrenia. *Psychiatry Research* 20:337–45.

Kudo, H., Kokunai, T., Kondoh, T., Tamaki, N., and Matsumoto, S. 1990. Quantitative analysis of glutathione in rat central nervous system: comparison of GSH in infant brain with that in adult brain. *Brain Research* 511:326–8.

Lillrank, S. M., Lipska, B. K., and Weinberger, D. R. 1995. Neurodevelopmental animal models of schizophrenia. *Clinical Neuroscience* 3:98–104.

Loscalzo, J., and Welch, G. 1995. Nitric oxide and its role in the cardiovascular system. *Progress in Cardiovascular Disease* 38:87–104.

Maker, H. S., Weiss, C., Silides, D. J., and Cohen, G. 1981. Coupling of dopamine oxidation (monoamine oxidase activity) to glutathione oxidation via the generation of hydrogen peroxide in rat brain homogenates. *Journal of Neurochemistry* 36:589–93.

Matthysse, S. 1973. Antipsychotic drug actions: a clue to the neuropathology of schizophrenia? *Federation Proceedings* 32:200–5.

Meffert, M. K., Calakos, N. C., Scheller, R. H., and Schulman, H. 1996. Nitric oxide modulates synaptic vesicle docking/fusion reactions. *Neuron* 16:1229–36.

Moncada, S. 1994. Nitric oxide. [Review]. *Journal of Hypertension Supplement* 12:S35–9.

Nishikawa, T., Takashima, M., and Toru, M. 1983. Increased [^{3}H]kainic acid binding in the prefrontal cortex in schizophrenia. *Neuroscience Letters* 40:245–50.

Ogita, K., and Yoneda, Y. 1987. Possible presence of [^{3}H]glutathione (GSH) binding sites in synaptic membranes from rat brain. *Neuroscience Research* 4:486–96.

Ogita, K., and Yoneda, Y. 1989. Selective potentiation by L-cysteine of apparent binding activity of [^{3}H]glutathione in synaptic membranes of rat brain. *Biochemical Pharmacology* 38:1499–505.

Ogita, K., Enomoto, R., Nakahara, F., Ishitsubo, N., and Yoneda, Y. 1995. A possible role of glutathione as an endogenous agonist at the N-methyl-D-aspartate recognition domain in rat brain. *Journal of Neurochemistry* 64:1088–96.

Olney, J. W., Ho, O.-L., Rhee, V., and Schainker, B. 1972. Cysteine-induced brain damage in infant and fetal rodents. *Brain Research* 45:309–13.

Park, L. C., Baldessarini, R. J., and Kety, S. S. 1965. Methionine effects on chronic schizophrenics. *Archives of General Psychiatry* 12:346–51.

Perry, T. L. 1982. Normal cerebrospinal fluid and brain glutamate levels in schizophrenia do not support the hypothesis of glutamatergic neuronal dysfunction. *Neuroscience Letters* 28:81–5.

Pullan, L. M., Olney, J. W., Price, M. T., Compton, R. P., Hood, W. F., Michel, J., and Monahan, J. B. 1987. Excitatory amino acid receptor potency and subclass specificity of sulfur-containing amino acids. *Journal of Neurochemistry* 49:1301–7.

Rauhala, P., Mohanakumar, K. P., Sziraki, I., Lin, A. M. Y., and Chiueh, C. C. 1996. Nitric oxide, lipid peroxidation, ferrous citrate, free radical, and S-nitrosothiols. S-nitrosothiols and nitric oxide, but not sodium nitroprusside, protect nigrostriatal dopamine neurons against iron-induced oxidative stress in vivo. *Synapse* 23:58–60.

Sagara, J. I., Makino, N., and Bannai, S. 1996. Glutathione, cystine, transport, astrocytes, and oxidative stress. Glutathione efflux from cultured astrocytes. *Journal of Neurochemistry* 66:1876–81.

Seeman, P., Lee, T., Chau-Wong, M., and Wong, K. 1976. Antipsychotic drug doses and neuroleptic/dopamine receptors. *Nature* 261:717–19.

Seeman, P., Niznik, H. B., and Guan, H. C. 1990. Elevation of dopamine D2 receptors in schizophrenia is underestimated by radioactive raclopride. *Archives of General Psychiatry* 47:1170–2.

Seeman, P., Guan, H. C., and van Tol. H. H. M. 1993. Dopamine D4 receptors elevated in schizophrenia. *Nature* 365:441–5.

Selemon, L. D., Rajkowska, G., and Goldman-Rakic, P. S. 1995. Abnormally high neuronal density in the schizophrenic cortex. A morphometric analysis of prefrontal area 9 and occipital area 17. *Archives of General Psychiatry* 52:805–18.

Shaw, P. J. and Salt, T. E. 1997. Modulation of sensory and excitatory amino acid responses by nitric oxide donors and glutathione in the ventrobasal thalamus of the rat. *European Journal of Neuroscience* 9:1507–13.

Shaw, C. A., Pasqualotto, B. A., and Curry, K. 1996. Glutathione, cortex, sodium currents, excitatory amino acid receptors, cortical wedge, and ionotropic receptors. Glutathione-induced sodium currents in neocortex. *Neuroreport* 7(6):1149–52.

Shukitt-Hale, B., Denisova, N. A., Strain, J. G. and Joseph, J. A. 1997. Psychomotor effects of dopamine infusion under decreased glutathione conditions. *Free Radical Biology and Medicine* 23:412–18.

Spina, M. B., and Cohen, G. 1989. Dopamine turnover and glutathione oxidation: implications for Parkinson disease. *Proceedings of the National Academy of Sciences of the U.S.A.* 86:1398–400.

Squires, R. F., and Saederup, E. 1991. A review of evidence for GABergic predominance/glutamatergic deficit as a common etiological factor in both schizophrenia and affective psychoses: more support for a continuum hypothesis of "functional" psychosis. *Neurochemical Research* 16:1099–111.

Stamler, J. S., Jaraki, O., Osborne, J., Simon, D. I., Keaney, J., Vita, J., Singel, D., Valeri, C. R., and Loscalzo, J. 1992. Nitric oxide circulates in mammalian plasma primarily as an S-nitroso adduct of serum albumin. *Proceedings of the National Academy of Sciences of the U.S.A.* 89:7674–7.

Sucher, N. J., Awobuluyi, M., Choi, Y. B., and Lipton, S. A. 1996. NMDA receptors: from genes to channels. *Trends in Pharmacological Sciences* 17:348–55.

Swahn, C.-G. 1990. Determination of N-acetylaspartic acid in human cerebrospinal fluid by gas chromatography–mass spectrometry. *Journal of Neurochemistry* 54:1584–8.

Taguchi, J., Ohta, H., and Talman, W. T. 1995. Identification and pharmacological characterization of an S-nitroglutathione binding site in rat brain (abstract). *Society for Neuroscience Abstracts* 21:626.

Tate, S. S., and Meister, A. 1974. Interaction of gamma-glutamyl transpeptidase with amino acids, dipeptides, and derivatives and analogs of glutathione. *Journal of Biological Chemistry* 249:7593–602.

van Tol, H. H., Wu, C. M., Guan, H. C., Ohara, K., Bunzow, J. R., Civelli, O., Kennedy, J., Seeman, P., Niznik, H. B., and Jovanovic, V. 1992. Multiple dopamine D4 receptor variants in the human population. *Nature* 358:149–52.

Wood, J. D., Kurylo, E., and Geddes, J. W. 1985. Methionine-induced changes in glutamate, aspartate, glutamine, and gamma-aminobutyrate levels in brain tissue. *Journal of Neurochemistry* 45:777–83.

Yudkoff, M., Pleasure, D., Cregar, L., Lin, Z. P., Nissim, I., and Stern, J. 1990. Glutathione turnover in cultured astrocytes: studies with. *Journal of Neurochemistry* 55:137–45.

Zängerle, L., Cuénod, M., Winterhalter, K. H., and Do, K. Q. 1992. Screening of thiol compounds: depolarization-induced release of glutathione and cysteine from rat brain slices. *Journal of Neurochemistry* 59:181–9.

Glutathione in the Nervous System
Edited by Christopher A. Shaw
Copyright © 1998 Taylor & Francis

14

The Role of Altered Glutathione Status in the Development of Parkinson's Disease

J. Sian

Division of Clinical Neurochemistry, Department of Psychiatry, University of Würzburg, Würzburg, Germany

M. Gerlach

Division of Clinical Neurochemistry, Department of Psychiatry, University of Würzburg, Würzburg, Germany and Division of Clinical Neurochemistry, Department of Neurology, St. Josef's Hospital, University of Bochum, Bochum, Germany

P. Riederer

Division of Clinical Neurochemistry, Department of Psychiatry, University of Würzburg, Würzburg, Germany

---

ABBREVIATIONS

ATP adenosine triphosphate
GSH reduced glutathione
GSSG oxidized glutathione
L-BSO L-buthionine(S,R)sulphoximine
L-dopa 3,4-dihydroxyphenylalanine
MPP$^+$ 1-methyl-4-phenylpyridinium
MPTP 1-methyl-4-phenyl-1,2,3,6-tetrahydropyridine
PD Parkinson's disease
SN substantia nigra
SOD superoxide dismutase

1. INTRODUCTION: THE ANTIOXIDANT GLUTATHIONE

The ubiquitous thiol-containing tripeptide, glutathione, serves a host of functions in mammalian cells (Booth, Boyland, and Sims 1961; Meister 1983, 1991), as discussed in previous chapters. The antioxidant properties of this thiol appears to be its main interesting feature, and this has been extensively investigated in the past and the present.

Glutathione plays its role as a free-radical scavenger by employing three different pathways (Siesjö and Rehncrona 1980). Firstly, it can serve as hydrogen donor and thus directly scavenge toxic free-radical species produced during aberrant occurrence of various metabolic pathways (Chance, Sies, and Boveris 1979; Cohen 1983). Secondly, it can also regenerate other cellular antioxidants, such as α-tocopherol, ascorbate, and ubiquinones (Siesjö 1981). Finally, it can induce the reduction of organic peroxides, which are believed to be produced during lipid peroxidation (Halliwell 1994).

A state of oxygen tension, during the occurrence of normal metabolic reactions, results in the formation of peroxides (Meister 1983, 1991). Reactions such as dismutation of superoxide and dopamine deamination may yield the production of the cellular oxidant hydrogen peroxide (Cohen 1983). Although hydrogen peroxide itself does not qualify as a free radical, nevertheless it can induce cell death at high concentrations (Eaton 1991). In addition, in the presence of reduced transition metals such as iron(II) and copper(I), it can produce cellular toxic hydroxyl radicals (Halliwell and Gutteridge, 1984; Halliwell 1994). Glutathione represents one of the two major cellular defense systems involved in curtailing the production of toxic species from hydrogen peroxide, organic peroxides, and free radicals (Aebi and Suter 1974; Wellner et al. 1984; Umemura et al. 1991). It performs this task by employing the enzyme glutathione peroxidase for decomposition of the peroxide (Fig. 1) to produce water and oxygen (Meister 1983). This reaction is coupled with the oxidation of reduced glutathione (GSH) to the oxidised form (GSSG).

It has been postulated that during the state of oxidative stress, there is a large influx of cellular hydrogen peroxide due to either an increase in its production or an inability to metabolize the high concentrations (Halliwell and Gutteridge 1984; Spina and Cohen 1989). Subsequently, this may lead to a stage when the glutathione system

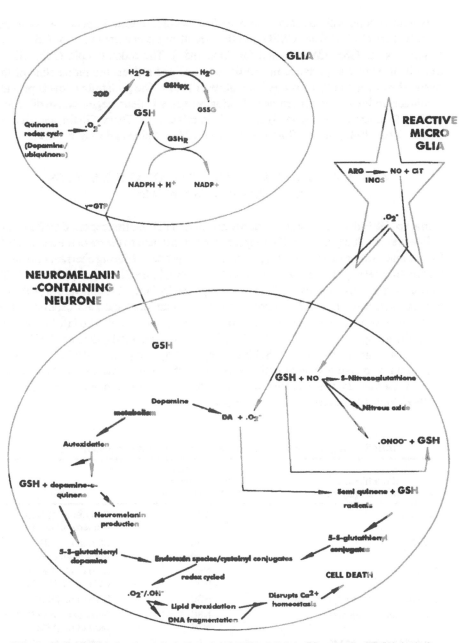

FIG. 1. Proposed mechanism(s) eliciting nigral GSH deficit in PD: neuronal GSH depletion, γ-GTP-mediated translocation of GSH from glial cells to neuromelanin containing cells, irreversible consumption of glial GSH, and resultant overall GSH deficit in glial and neuronal cells in SN in PD.

is unable to cope with the large turnover. This would, in turn, impede the recycling or reduction of GSH from GSSG and thus result in the accumulation of GSSG and a depletion of GSH (Meister and Griffiths 1985). The redox-couple GSH : GSSG ratio is of pivotal importance in cellular survival and reflects the redox state of the tissue (Kosower and Kosower 1978; Halliwell 1992). The glutathione levels provide a valuable index for the occurrence of oxidative stress in neurodegenerative disorders such as Parkinson's disease (PD) and Alzheimer's disease (Perry, Godin, and Hansen 1982; Cohen 1983; Slivka, Spina, and Cohen 1987a; Spina and Cohen 1989).

2. GLUTATHIONE LEVELS AND OXIDATIVE STRESS IN PARKINSON'S DISEASE

A marked GSH deficit in the substantia nigra (SN) in PD was first reported by Perry and colleagues in 1982 (Table 1). This depletion was attributed to the involvement of GSH for the inactivation of some toxic species, perhaps produced during aberrant dopamine metabolism (Perry et al. 1982; Perry and Yong 1986; Perry, Hansen, and Jones 1988). However, the unusually high GSSG levels reported by Perry and colleagues (1982) in normal subject brains (89.3 percent of the total glutathione) and Parkinsonian brains (100 percent) have been questioned and criticized (Slivka et al. 1987a). In contrast, normally a high intracellular GSH : GSSG ratio is maintained (Meister 1983). Thus it would appear that the high GSSG levels reported by Perry et al. (1982) are unlikely to reflect true physiological glutathione levels, both in the normal and in the diseased state. Subsequently, these unusual brain glutathione concentrations were argued to be methodologically related (Sian et al. 1991a).

TABLE 1. *Nigral changes in the glutathione pathway in Parkinson's disease*

Glutathione or related enzyme	Change as a fraction of normal control value (%)[a]	Authority
Glutathione	↓ 28	Perry et al. (1982)
	↓	Riederer et al. (1989)
GSH	↓ 40	Sian et al. (1991b)
	↓ 46	Sofic et al. (1992)
GSSG	NSC	Sian et al. (1991b)
	NSC	Sofic et al. (1992)
γ-Glutamylcysteine synthetase	NSC	Sian et al. (1992b)
Glutathione transferase	NSC	Perry and Yong (1986)
	NSC	Sian et al. (1992b)
Glutathione peroxidase	↓ 61	Ambani et al. (1975)
	↓ 81	Kish et al. (1985)
	NSC	Martilla et al. (1988)
	NSC	Sian et al. (1992b)
	↑ 83	Damier et al. (1993)
γ-Glutamyl transpeptidase	↑ 76	Sian et al. (1992b)

[a]NSC: no significant changes.

Nevertheless, the nigral GSH deficit in PD found by Perry and colleagues (1982) appears to be valid, inasmuch as it was subsequently confirmed by others (Table 1). Indeed, Riederer et al. (1989) reported a reduction of total glutathione, and Sian et al. (1991b) and Sofic et al. (1992) found a depletion of GSH but no changes in GSSG levels (Table 1).

The nigral GSH deficit mirrors the involvement of the glutathione system in the disease; this may, in turn, have important implications with reference to the pathogenic processes. It has been postulated that the catecholaminergic neurons in the SN are particularly vulnerable to cytotoxic-species-induced oxidative stress as a result of dopamine metabolism via either autooxidation or monoamine oxidase–catalyzed deamination (Cohen 1983, 1987; Youdim and Ben-Shachar 1987). Therefore, perhaps this major impairment in a cellular protective system may compromise the ability of the brain to metabolize hydrogen peroxide and reactive oxygen species, resulting in the imposition of oxidative stress (Spina and Cohen 1989; Sian et al. 1991b, 1994a). Thus, the alteration in the GSH content in the SN provides support for the occurrence of oxidative stress in PD. In view of the finding that GSH is up-regulated during mild oxidative stress (Han, Mytilineou, and Cohen 1996), perhaps the degree of oxidative stress is relatively severe in PD and thus contributes to GSH depletion.

Suprisingly, the significant reduction of nigral GSH and total glutathione levels in PD occurs without a corresponding elevation in its oxidized form, GSSG. This suggests an alternative route for GSH depletion as opposed to the oxidation pathway.

The importance and relevance of the GSH deficit to the pathogenesis of PD are clearly demonstrated by two important findings. The depletion appears to be area-selective, being confined to the SN (Perry et al. 1982; Riederer et al. 1989; Sian et al. 1991b; Sofic et al. 1992), an area where marked pathology of the disorder is observed (Gibb and Lees 1988; Jellinger 1989). Secondly, this change is disease-specific: it is not found in other basal-ganglia disorders, including, striatonigral degeneration, Steele–Richardson–Olszewski disease, and Huntington's disease (Sian et al. 1992a, 1994b).

The importance of the antioxidant in the etiology of the illness is further highlighted by a similar nigral reduction of GSH content without changes in GSSG in incidental Lewy-body disease (Sian et al. 1991b). In addition, no other nigral changes found in PD (such as iron and ferritin levels or mitochondrial dysfunction) were reported in these subjects (Dexter et al. 1994). Incidental Lewy-body disease represents an incidental finding of nigral pathology in 7–10 percent of the normal population, usually above the age of 60 years (Gibb 1986). Lewy bodies and early neuronal cell loss [20–50 percent (Fearnley and Lees 1991)] in the ventrolateral area of the SN pars compacta (a region severely affected in PD) are found in these individuals. These pathological observations furnish support for the hypothesis that incidental Lewy-body disease represents an early asymptomatic stage in the progression of PD (Gibb and Lees 1988). Thus, the nigral GSH deficit in PD represents a primary event. Also, the almost equal GSH reduction in PD and in incidental Lewy-body disease suggests that, surprisingly, the antioxidant appears to be involved only in the early presymptomatic phase of the disorder (Sian et al. 1991b). Consequently, it is imperative to elucidate the mechanism(s) involved in the consumption of GSH in PD, as this may provide an insight into the pathogenic processes occurring during the illness (Fig. 1). Unfortunately, this vital information largely remains obscure, although a number of hypothesis have been proposed (Perry et al. 1982; Riederer et al. 1989; Sian et al. 1991b; Sofic et al. 1992; Sian et al. 1994a; Shen et al. 1996).

2.1 Relevance of GSH Deficit to Other Nigral Changes in Parkinson's Disease

The GSH changes may be related to either or both the nigral cell loss and the Lewy-body occurrence found in incidental Lewy-body disease (Gibb and Lees 1988). However, despite the significantly different degrees of nigral neuronal cell loss in this area in incidental Lewy-body disease [20–50 percent (Fearnley and Lees 1991)] and PD [70–80 percent (Gibb and Lees 1988)], there appeared to be little or no further reduction of GSH in SN in these two disease states [39 percent in incidental Lewy-body disease and 40 percent in PD (Sian et al. 1991b); see Table 1]. This suggests that the GSH change is unlikely to be a consequence of dopaminergic neuronal cell loss. This contention is supported by results obtained from experimental studies. Indeed, rats acutely treated with L-buthionine(S,R)sulphoximine (L-BSO, an irreversible inhibitor of γ-glutamylcysteine synthetase) exhibited striatal GSH loss with no changes in tyrosine hydroxylase–positive cells in the SN or [^{3}H]mazindol binding to dopamine uptake sites in the striatum (Sian 1993). Subsequently, Toffa et al. (1997) reported similarly findings in rats chronically treated with L-BSO. In contrast, Riederer et al. (1989) reported a positive correlation between cellular deficit and GSH content in pooled brain samples in PD.

The nigral GSH deficit is unlikely to be associated to the occurrence or formation of Lewy bodies, for no glutathione changes were found in the substantia innominata [which is mainly cholinergic with some catecholaminergic innervations, and also exhibits Lewy-body pathology (Sian 1993)]. Although Lewy-body inclusions are not exclusive to PD, their presence in the SN is as essential for the neuropathological diagnosis of the illness as the specific distribution of SN neuronal loss (Gibb and Lees 1988; Forno 1996). They are believed to be the hallmark of neurodegeneration occurring in PD, and therefore their occurrence may be a manifestation of the cell loss cascade, or alternatively they may be toxic and thus elicit or contribute to neuronal death. Therefore, they may represent a consequence of oxidative stress and/or increased oxidation of catecholamines, whereas in noncatecholaminergic neurons they may be formed by different mechanism(s) but share the same staining properties (Forno 1996).

The possibility of the antioxidant depletion being a consequence of levodopa (3,4-dihydroxphenylalanine, L-dopa) therapy is also unlikely, since the incidental-Lewy-body subjects did not receive any anti-Parkinsonian treatment. This notion is further confirmed by the unaltered caudate–putamen dopamine levels in these subjects (Dexter et al. 1994). This is in contrast with the view that L-dopa treatment may exacerbate the events leading to oxidative stress and eventually to nigral neuronal death (Cohen 1985).

The unaltered activity of the GSH synthetic enzyme γ-glutamylcysteine synthetase in PD (Sian et al. 1992b, 1994b) negates the possibility of an alteration in the glutathione synthetic mechanisms or a reduction in the content of the precursor L-cysteine. These findings also indicate that the SN iron overload in PD does not affect the glutathione generating pathway(s). In contrast, in mice, iron induced an increase in the activity of the glutathione synthetic enzyme and a consequent elevation of GSH content (Ogino, Kawabat, and Awai 1989).

The GSH deficit may represent a secondary event to the maintenance of other cellular antioxidants. Indeed, it has been postulated that the GSH depletion may be a consequence of the maintanence of high α-tocopherol in the cell membrane as a cellular protective mechanism against the disease process (Adams, Klaidman, and Odunze 1991). It has been suggested that this phenomenon operates almost exclusively in the midbrain. Similarly, the utilization of GSH is favored as a cytosolic antioxidant as opposed to ascorbate (McCay et al. 1989). Therefore the GSH deficit in PD may be a result of the operation of some cellular protective mechanism involved in the maintenance and regulation of other antioxidants.

Alternatively, the alterations in the glutathione content may be associated with or mirror the defect in the mitochondrial respiratory enzyme for complex I (Reichmann and Riederer 1989; Schapira et al. 1989). Indeed, like the GSH deficit, the nigral mitochondrial dysfunction reported in PD (Reichmann and Riederer 1989; Schapira et al. 1989) also appears to be brain-region-specific, independent of nigral cell loss or L-dopa therapy [no change in mitochondrial activity was found in multiple-system-atrophy patients treated with L-dopa (Dexter et al. 1991; Reichmann et al., personal communication)]. Furthermore, in vitro studies using rat hepatocytes incubated with mitochondrial toxins such as MPP$^+$, antimycin A, and potassium cyanide induced a depletion of ATP and a subsequent loss of intracellular GSH mediated through some efflux mechanism (Mithöfer et al. 1992). However, in the primate model, MPTP treatment did not alter the glutathione content in the putamen (Gerlach, Riederer, and Youdim 1996) or in the SN (Marzatico et al. 1993). In contrast, a transient and reversible GSH depletion was reported in the striatum from MPTP-treated mice (Riederer et al. 1988). This suggests that the mode of MPTP toxicity is unlikely to be directly related to the oxidative-stress-mediated GSH depletion pathway(s). Furthermore, it argues against the possible relation between mitchondrial dysfunction and GSH depletion found in SN in PD. Indeed, this conclusion is supported by the finding that L-BSO induced GSH depletion in the rat model, whereas it did not appear to alter the mitochondrial function or superoxide dismutase activity in the SN (Sian 1993; Seaton, Jenner, and Marsden 1996).

2.2 Putative Mechanism(s) Eliciting the Alterations in the Glutathione Pathway in Parkinson's Disease

The marked elevation of the activity of the glutathione translocation enzyme, γ-glutamyltranspeptidase in the SN in PD (Sian et al. 1992b, 1994b; see Table 1) may be related to the translocation of GSH (Fig. 1) out of the glial cell, resulting in the GSH deficit reported in PD (Sian 1993). The GSH transport out of the cell may serve as an additional protective role in that there are no extracellular mechanisms for GSSG reduction.

It has been postulated that γ-glutamyltranspeptidase mediates GSH or cysteine translocation out of glial cells and into neuromelanin containing nigral neurons, where those substances would interrupt the normal pathway for neuromelanin production (Fig. 1; Cheng et al. 1996). This would account for the characteristic neuromelanin depletion observed in PD (Kastner, Hirsch, and Lejeune 1992). Subsequently, GSH or cysteine would inactivate the dopamine-o-quinones (liberated from dopamine autooxidation) to 5-S-glutathionyldopamine (Zhang and Dryhurst 1994) or 5-S-cysteinyldopamine (Cheng et al. 1996). Indeed, the 5-S-cysteinyldopamine/dopamine ratio was reported to be increased in a de novo PD patient (Carlsson and Fornstedt 1991). This accords with the observation that the precursor, cysteine, is transported to neurons from glial cells for the synthesis of glutathione (Sagara, Miura, and Bannai 1993). The surviving dopaminergic neurons (circa 20 percent) may elicit a compensatory response resulting in an increase in dopamine turnover (Hefti, Melamed, and Wurtman 1980). Subsequently, superoxide radicals generated from reactive microglia (Meldrum and Garthwaite 1991) may combine with dopamine to yield semiquinone radicals (Fig. 1). The semiquinone radicals may then be scavenged by GSH to produce 5-S-glutathionyl conjugates (Spencer, Jenner, and Halliwell 1995).

Finally, the 5-S-glutathionyl dopamine and 5-S-glutathionyl conjugates may induce the production of some endotoxic species (Shen and Dryhurst 1996). These endotoxin(s) are likely to be cysteinyl-derived conjugates such as benzothiazines and dihydrobenzothiazines (Fig. 1). The subsequent redox cycling of these conjugates would result in the formation of superoxide radicals, which would lead to the production of hydrogen peroxide–derived hydroxyl radicals. The generation of these cytotoxic free-radical species may further propagate nigral neuronal loss via mechanisms such as lipid peroxidation and DNA strand breakage (Zhang and Dryhurst 1994). The operation of this pathway may be attributed to the irreversible consumption of GSH, which appears to be highly likely in that none of the other glutathione enzyme activities besides that of γ-glutamyltranspeptidase is altered in SN in PD (Sian et al. 1992b). The failure to detect sufficient amounts of either of these products (5-S-glutathionyl or 5-S-glutathionyl dopamine conjugates) in the SN in PD may be related to their intraneuronal oxidation and extraneuronal metabolism (Shen et al. 1996). Also, the nonenzymatic inactivation of both singlet oxygen and hydroxyl radicals by GSH can contribute further to the irreversible nigral GSH depletion in PD (Coyle and Puttfarcken 1993).

This prompts the suggestion of the involvement of the superoxide radical as opposed to the previously suspected hydroxyl radical and thereby indicates an alternative route of free-radical generation to the one formerly hypothesized (Halliwell 1992, 1994).

The proposal of an alternative route for GSH decrease is supported by studies on ischemia. It has been reported that during postischemia reperfusion a marked depletion of GSH occurs without a corresponding increase in GSSG content in rat brain (Cooper, Pusinelli, and Duff 1980), similarly to the changes in PD (Sian et al. 1991b; Sofic et al. 1992). Thus, the Parkinsonian GSH deficit may be elicited by some other mechanism than the usual GSH-mediated scavenging pathways (Shen et al. 1996). However, the sequence of events manifesting the depletion of GSH in the in vitro system or in the animal model may not necessarily reflect the series of cellular event(s) occurring in a diseased state.

Although many attractive hypotheses have been proposed suggesting the involvement of an endogenous/exogenous toxin (Perry et al. 1982; Jenner 1991) in the pathogenesis of the disease, unfortunately no such toxin has been isolated in PD. Indeed, Ikeda, Markey, and Markey (1992) found no MPP^+ or endogenous MPTP-like compounds (β-carbolines, tetrahydroisoquinolines) in the Parkinsonian brain. Nevertheless, this does not negate the involvement of these endogenous toxins completely. β-carbolines are rapidly metabolized, and this may account for their absence in the PD brain. Also, the findings of a recent study suggest the involvement of a putative endotoxin in the neuronal-cell-loss cascade in PD. This phenomenon is based on the increased level of N-methyl(R)salsolinol found in the cerebrospinal fluid (CSF) of PD patients compared to normal control subjects (Naoi et al. 1997). This isoquinoline is believed to be generated by a specific biochemical mechanism related to dopamine metabolism. More importantly, it has been proposed that these high levels in the CSF may also mirror increased levels of the compound in the brain. Although N-methyl(R)salsolinol is not acutely toxic, nevertheless it may manifest cellular cytotoxic effects over a chronic period. Therefore, it represents a highly suitable endotoxin candidate in the progressive degeneration process that occurs in PD (Gibb and Lees 1988). Thus the possibility of some toxin in the pathogenesis of PD must not be completely rejected and clearly warrants further investigation.

Alternatively, the toxin(s) in question may not be a MPP^+-like mitochondrial toxin, but a compound that interacts exclusively and irreversibly with the glutathione system, consumes neuronal GSH, and consequently increases the susceptibility to some toxin. This may, in turn, elicit the mitochondrial dysfunction and propagate the cell degeneration cycle.

This hypothesized toxin may be generated by the cytotoxic nitric oxide pathway (Fig. 1). Indeed, the reactive microglia in PD may generate nitric oxide (McGeer, McGeer, and Suzuki 1988; Meldrum and Garthwaite 1991; McCall and Vallance 1991). Thus, the deficit of GSH may be ascribed to its reactions (Fig. 1) with nitric oxide to produce S-nitrosoglutathione or nitrous oxide or peroxynitrite radical (Hogg, Singh, and Kalyanaraman 1996; Gergel and Cederbaum 1997). Subsequently, GSH may be involved in the inactivation of peroxynitrite radical, thus precipitating a further deficit (Gerlach et al. 1994; Barker, Heales, and Cassidy 1996). A pertinent interaction has been suggested between the glutathione and nitric oxide pathways. This has evolved from experimental studies in cultured neurones and in vivo, which have shown that GSH depletion resulted in an increase in the activity of the synthetic enzyme nitric oxide synthase (Heales, Bolanos, and Clark 1996). In the primate, nitric oxidase appears to be highly concentrated in the putamen and striatum area; extrapolation of the finding (although the distribution exhibits species differences) would support its possible involvement in the basal ganglia degeneration in PD (Gerlach et al. 1995). It has been reported that high levels of nitric oxide elicit DNA damage by inhibiting the synthetic enzyme ribonucleoside diphosphate reductase. Indeed, 8-hydroxy-2′-deoxyguanosine (a product of DNA free-radical damage) has been found to be elevated in the SN in PD (Sanchez-Ramos, Övervik, and Ames 1994). Nitric oxide can mediate cell death via several mechanisms, including inhibition of mitochondrial enzymes for complexes I, II, and IV; perhaps this results in disruption of iron homeostasis, generation of free radicals, lipid peroxidation, and disturbance in calcium homeostasis (Reif and Simmons 1990; Lancaster and Hibbs 1990; Gerlach et al. 1996). However, the involvement of nitric oxide in the neurodegeneration in PD still merits further evaluation. Indeed, a recent study has suggested that under both aerobic and anaerobic conditions, nitric oxide and GSH react to produce GSSG (Hogg et al. 1996). In contrast, the nigral GSH deficit in PD is not coupled with any corresponding changes in GSSG content. Perhaps there is a marginal elevation in GSSG content corresponding to the GSH depletion; due to the very small content of GSSG, this change might not be detectable by the methods employed.

The reduction of GSH in early asymptomatic PD [incidental Lewy-body disease (Sian et al. 1991b)] indicates a state of chronic deficiency in a principal cellular protective system. This chronic state of GSH deficit in SN in PD may have important implications and consequences in relation to the progression of the disorder and manifestation of clinical symptoms.

3. CONSEQUENCES OF NIGRAL GSH DEFICIT IN PARKINSON'S DISEASE

The effects of a GSH-deficient state has been extensively investigated in an animal model, both in the whole system and in in vitro studies, by employing various GSH depleting drugs. A GSH deficit induced by the administration of L-BSO to neonatal rats led to mitochondrial damage and cell death in skeletal muscle (Jain et al. 1991). Similarly, perhaps in PD the reduction of GSH exacerbates nigral neuronal cell loss by rendering the remaining nigral cells more vulnerable to the effects of some endogenously produced or an exogenous cellular toxin (Perry et al. 1982; Spina and Cohen 1989; Jenner, Schapira, and Marsden 1992). Therefore, a GSH deficit per se may not be as important to the neuronal cell loss process as the secondary consequences that it induces.

It has been suggested that alterations in the GSH levels may increase the sensitivity of dopaminergic cells to cytotoxins by interferrence with energy-producing mechanism(s) (Zeevalk et al. 1997). Indeed, it has been shown that cells depleted of GSH (L-BSO treatment) are more susceptible to the toxic effects of morphine (McCartney 1989), mercury chloride (Nagamura, Anderson, and Meister 1990), tetrabutylhydroperoxide (Shan et al. 1993), MPTP and MPP$^+$ (Wüllner et al. 1996), and malonate (Zeevalk et al. 1997). However, caution must be used in extrapolating the findings from animal experiments using L-BSO-induced GSH depletion and relating them to the nigral GSH-depleted state in PD. For instance, in PD the GSH deficit appears to be localized in the SN only. In contrast, L-BSO depletes GSH content both centrally and periperally, which may consequently manifest itself in other cellular events. Indeed, rats acutely treated with L-BSO showed a depletion of striatal GSH and an elevation of iron, but no alteration in mitochondrial function (complex I activity) or in their number (citrate synthase activity) (Sian 1993). The absence of mitochondrial changes has also reported in the cortex from rats administered with L-BSO intracerebroventric-ularly (Seaton et al. 1996). This may suggest that L-BSO depletes only the cytosolic GSH pool in the brain and therefore does not affect the mitochondrial GSH content and thus induces no changes in mitochondrial activity or number. However, this pos-sibilty is unlikely in view of the fact that GSH is synthesized in the cytoplasm and transported into the mitochondria by a high-affinity transport system (Märtensson, Lal, and Meister 1990). Also, the mechanism(s) involved in eliciting GSH depletion by L-BSO and in PD appear to differ. L-BSO inhibits the glutathione synthetic enzyme γ-glutamylcysteine synthetase (Meister 1983), whereas no changes in the activity of this enzyme were found in SN in PD (Sian et al. 1992b, 1994b).

Ideally, for investigating the effects of chronic GSH deficiency that occurs in PD, it would be more relevant to employ drugs that specifically alter dopamine handling, such as haloperidol or reserpine. For instance, in mice acutely treated with reserpine, selectively reduced GSH levels were found in the striatum but not the frontal cortex (Spina and Cohen 1989). Similarly, 6-hydroxydopamine depletes GSH in the caudate–putamen, perhaps mediated via some free-radical pathway without affecting the levels in the hippocampus (Perumal et al. 1989). Studies using rat hepatocytes have shown that when GSH depletion reaches a critical threshold (approximately 90 percent) lipid peroxidation occurs (Maellaro et al. 1990). In view of the rapid glutathione turnover in the liver compared to the brain (Meister 1991), perhaps in the brain such a marked GSH deficit is not required to precipitate lipid peroxidation. This offers some fascinating implications for the nigral GSH deficit and a possible association with lipid peroxidation found in PD (Dexter et al. 1989).

4. SITE OF OCCURRENCE OF GSH DEPLETION: IS PARKINSON'S DISEASE A NEURONAL OR A GLIAL DISORDER?

The localization of the cell type (e.g., neuronal or glial) in which this alteration of GSH occurs in the disease would be important to elucidate, as this site may provide a better understanding of the sequence of events occurring during the degenerative processes. Up to the present time, due to the marked nigral neuronal loss in PD, neurons have been the major focus in the illness. However, accumulating evidence provided by post mortem studies has shifted the focus to glial cell types (McGeer et al. 1988). Indeed, histochemical studies suggest that GSH is primarily found in glia and nonneuronal components in both animals (Slivka, Mytilineou, and Cohen 1987b; Raps et al. 1989; Philbert et al. 1991) and the human brain (Sian 1993). Furthermore, a marked depletion of GSH histofluorescence has been reported in the neuropil of the SN in PD (Sian

1993). These findings are in accordance with other studies that have demonstrated the GSH-dependent enzyme glutathione peroxidase to be present almost exclusively in glial cells, both in control and in PD brains (Ushijima, Miyazaki, and Moriokaz 1986; Damier et al. 1993). Glia have also been implicated in the induction of the glutathione translocation–transport enzyme γ-glutamyltranspeptidase (Bauer et al. 1990).

The importance of glial cells in PD is further emphasized by the localization of other PD biochemical changes in this cell type. The biochemical increase in the iron content in the SN pars compacta in PD (Riederer et al. 1989; Dexter et al. 1991) has been demonstrated in glial cells (Jellinger et al. 1990). Also, the dysfunction in mitochondrial respiratory enzyme complex I activity in PD (Reichmann and Riederer 1989; Schapira et al. 1989) is believed to reflect the activity in glia as opposed to "sick neurons" (Benzi et al. 1991). Collectively these observations suggest that perhaps PD is primarily a glial disorder.

An early glial GSH deficit would reduce neuronal GSH supplies (Yudkoff et al. 1990; Sagara et al. 1993), consequently increasing the susceptibility to cytotoxins [including those derived from dopamine metabolism (Cohen 1985)] and finally precipitating neuronal cell loss. Indeed, the occurrence of some active process in the cascade of cell loss is exhibited by the reactive microgliosis found in the SN in PD (McGeer et al. 1988). In addition, microgliosis can further propagate secondary neuronal cell loss by releasing cytotoxins such as nitric oxide, superoxide radicals, glutamate, proteolytic enzymes, and hydrogen peroxide (Fig. 1; Meldrum and Garthwaite 1991).

Alternatively, the GSH deficit in PD may not be a consequence of its protective role, but rather due to the induction of its cytotoxic action. Indeed, glutathione itself can also generate thiyl radicals (GS^0) and is involved in toxicity of certain quinones (Butler and Hoey 1992). In vitro studies have shown GSH reacts with quinones to yield quinone–SG glutathione conjugates, GSSG, and semiquinone free radicals (Takahashi, Schreiber, and Fisher 1987; Gant et al. 1988). These products are believed to be formed from the addition of glutathione without any electron transfer (Wardman 1990). However, the thiyl radical is fairly inactive compared to the superoxide and hydroxyl radicals. More importantly, superoxidase dismutase (SOD) is believed to inhibit radical-mediated oxidation of GSH and thereby ameliorate its actions as a physiological antioxidant (Munday and Winterbourn 1989). This notion is supported by the elevation of SOD activity reported in the SN in PD (Martilla, Lorenzo, and Rinne 1988; Saggu et al. 1989).

5. GSH, A POTENTIAL CANDIDATE FOR NEUROPROTECTIVE TREATMENT IN PARKINSON'S DISEASE?

Excitotoxic mechanisms have also been implicated in the pathogenesis of PD (Olney 1978). Indeed, findings from an in vitro study using neonatal cortical slices suggest that GSH can exert excitatory neurotransmission not only through NMDA receptors (Ogita et al. 1995), but also via its own Na^+-permeable channels (Shaw, Pasqualotto, and Curry 1996). The implications of these results need to be carefully considered, particularly when considering GSH treatment either for monotherapy or as an adjunct in the treatment of PD. A recent open study using GSH therapy in de novo PD patients for a 30-day period showed a marked improvement in motor disability for the next four months after treatment (Sechi et al. 1996). However, it is difficult to comprehend on what basis the authors concluded, from this acute study, that GSH treatment may slow the rate of progression of the disease. In addition, the progression of the disease and thus the manifestation of clinical symptoms appear to vary between PD patients (Agid et al. 1993). This is fundamentally because PD is a spectrum disease that encompasses various forms of illness, as demonstrated by the dominance of certain PD clinical features. A longer GSH therapy study with de novo PD patients presenting different severities of clinical disability needs to be conducted to truly support this hypothesis suggested by Sechi et al. (1996).

γ-Glutamylcysteine administration may be considered rather than glutathione, because the precursor appears to be more effective at increasing brain GSH levels (Pileblad and Magnusson 1992). Alternatively, GSH replenishment may also be achieved indirectly by employing agents such as α-lipoic acid (thioctic acid). This antioxidant effectively reduces GSSG to GSH and thus regenerates GSH (Kagan et al. 1992). However, there are conflicting reports regarding its ability to increase intracellular GSH. Indeed, Busse et al. (1992) reported it to increase cellular GSH content both in vivo and in vitro in mice. In contrast, in rat cortical cells depleted of GSH (by L-BSO treatment), α-lipoic acid did not reverse the GSH deficit (Seaton et al. 1996). This discrepancy should be resolved before considering the clinical use of α-thioctic acid.

Finally, the primary involvement of the glutathione system in the progression of the disease suggests that it may perhaps hold the key to disclosing the cause of the disease or furnishing a better understanding about the underlying mechanisms operating during neurodegeneration. Therefore, the very early change in GSH in the disease process offers an attractive possibility of employing it or a related enzyme as a marker for screening potential subjects (those with incidental Lewy-body disease) for PD. The detection of these subjects in the asymptomatic phase would allow the early application of neuroprotective therapy, which may preserve the few remaining nigral neurones (circa 20 percent) and thus prevent or delay the clinical manifestation of the disease.

6. CONCLUSIONS

In summary, the specificity of the alterations found in the glutathione pathway in the SN in PD are clearly indicated by the evidence reviewed. This in turn lends support for the fundamental role played by this antioxidant and thus its relevance to a selective neurodegenerative process operating in PD. The absence of a corresponding increase of GSSG content in addition to the GSH depletion in SN suggests the irreversible consumption of the reduced form in the cascade of neuronal cell loss. The γ-glutamyl-transpeptidase-mediated translocation of GSH from the glial cells into the neuromelanin containing dopaminergic neurones may represent an attempt by the glial cells to rescue neurons by providing GSH.

REFERENCES

Adams, J. D., Klaidman, L. K., and Odunze, I. N. 1991. Alzheimer's and Parkinson's disease, brain levels of glutathione, glutathione disulphide and vitamin E. *Molecular and Chemical Neuropathology* 14: 213–26.

Aebi, H., and Suter, H. 1974. Protective function of reduced glutathione against the effect of pro-oxidant substances and of irradiation in the red cell. In *Glutathione*, ed. L. Flohe, H. C. Benhor, H. Sies, H. D. Waller, and A. Wendel, 199–299. Stuttgart: George Thieme.

Agid, Y., Ruberg, M., Javoy-Agid, F., Hirsch, E., Raisman, V. R., Vyas, S., Faucheux, B., Michel, P., and Kestner, A. 1993. Are dopaminergic neurones selectively vulnerable to Parkinson's disease? In *Advances in Neurology: Parkinson's disease, from basic research to treatment*, eds. H. Narabayashi, T. Nagatsu, N. Yanigasawa, et al., 60:148–64, Lippincott-Raven.

Ambani, L. M., Van Woert, M. H., and Murphy, S. 1975. Brain peroxidase and catalase activity in Parkinson's disease brain. *Archives of Neurology* 32:114–6.

Barker, J. E., Heales, S. J. R., and Cassidy, A. 1996. Depletion of brain glutathione results in a decrease of glutathione reductase activity; an enzyme susceptible to oxidative damage. *Brain Research* 716(1–2):118–22.

Bauer, H. C., Tontsch, U., Amberger, A., and Bauer, H. 1990. γ-glutamyltranspeptidase and Na^+, K^+-ATPase activities in different subpopulations of cloned cerebral endothelial cells: responses to glial stimulation. *Biochemical and Biophysical Research Communications* 18:358–63.

Benzi, G., Curti, D., Pastoris, O., Marzatico, F., Villa, R. F., and Dagani, F. 1991. Sequential damage in mitochondrial complexes by peroxidative stress. *Neurochemical Research* 16(12):1295–302.

Booth, J., Boyland, E., and Sims, P. 1961. An enzyme from rat liver catalysing the conjugation with glutathione. *Biochemical Journal* 79:516–24.

Busse, E., Zimmer, G., Schophol, B., and Kornhuber, B. 1992. Influence of *a*-lipoic acid on intracellular glutathione in vitro and in vivo. *Arzneimittel-Forschung/Drug Research* 42(1):829–31.

Butler, J., and Hoey, B. M. 1992. Reactions of glutathione and gluathione radicals with benzoquinones. *Free Radicals in Biology and Medicine* 12:337–45.

Carlsson, A., and Fornstedt, B. 1991. Possible mechanisms underlying the special vulnerability of dopaminergic neurones. *Acta Neurologica Scandinavica* 84(1336):16–18.

Chance, B., Sies, H., and Boveris, A. 1979. Hydrogen peroxide metabolism in mammalian organs. *Physiolical Review* 59:527–605.

Cheng, F. C., Kuo, J. S., Chia, L. G., and Dryhurst, G. 1996. Elevated 5-*S*-cysteinyldopamine/ homovanillic acid ratio and reduced homovanillic acid in cerebrospinal fluid: possible markers for and potential insights into the pathoetiology of Parkinson's disease. *Journal of Neural Transmission* 130:433–46.

Cohen, G. 1983. The pathobiology of Parkinson's disease: biochemical aspects of dopamine neurone. *Journal of Neural Transmission* 19:89–103.

Cohen, G. 1985. Oxidative stress in the nervous system. In *Oxidative stress*, ed. H. Sies, 383–402. London Press.

Cohen, G. 1987. Oxygen radicals and Parkinson's disease. In *Oxygen Radicals and Tissue Injury*, ed. B. Halliwell, 130–5. Bethesda, MD: Federation of American Societies for Experimental Biology.

Cooper, A. J. L., Pusinelli, W. A., and Duff, T. E. 1980. Glutathione and ascorbate during ischaemia and post ischaemia reperfusion in rat brain. *Journal of Neurochemistry* 35:1242–5.

Coyle, J. T., and Puttfarcken, P. 1993. Oxidative stress, glutamate and neurodegenerative disorders. *Sciences* 262:689–95.

Damier, P., Hirsch, E. C., Zhang, P., Agid, Y., and Javoy-Agid, F. 1993. Glutathione peroxidase, glial cells, and Parkinson's disease. *Neuroscience Letters* 32(1):1–6.

Dexter, D. T., Carter, C. J., Wells, F. R., Lees, A., Jenner, P., and Marsden, C. D. 1989. Basal lipid peroxidation in substantia nigra is increased in Parkinson's disease. *Journal of Neurochemistry* 52: 381–9.

Dexter, D. T., Carayon, A., Javoy-Agid, F., Wells, F. R., Daniel, S. E., Lees, A. J., Jenner, P., and Marsden, C. D. 1991. Alterations in the levels of iron, ferritin and other trace metals in Parkinson's disease and other neurodegenerative diseases affecting the basal ganglia. *Brain* 114:1953–75.

Dexter, D. T., Sian, J., Rose, S., Hindmarsh, J. G., Mann, V. M., Cooper, J. M., Wells, F. R., Daniel, S. E., Lees, A. J., Schapira, A. H. V., Jenner, P., and Marsden, C. D. 1994. Indices of oxidative stress and mitochondrial function in individuals with incidental Lewy body disease. *Annals of Neurology* 35(1):38–44.

Eaton, J. W. 1991. Is the lens canned? *Free Radical Biology and Medicine* 11:207–13.

Fearnley, J. M., and Lees, A. J. 1991. Ageing and Parkinson's disease: substantia nigra regional selectivity. *Brain* 114:2283–301.

Forno, L. S. 1996. Neuropathology of Parkinson's disease. *Journal of Neuropathology and Experimental Neurology* 55(3):259–72.

Gant, T. W., Rao, D. N. R., Mason, R. P., and Cohen, G. 1988. Redox cycling and sulphydryl arylation; their relative importance in the mechanism of quinone cytotoxicity to isolated hepatocytes. *Chemico-Biological Interactions* 65:157–73.

Gergel, D., and Cederbaum, A. I. 1997. Inhibition of the catalytic activity of alcohol dehydrogenase by nitric oxide is associated with *S*-nitrosylation and the release of zinc. *Biochemistry* 35(50):16186–94.

Gerlach, M., Ben-Shachar, D., Riederer, P., and Youdim, M. B. H. 1994. Altered brain metabolism of iron as a cause of neurodegenerative diseases? *Journal of Neurochemistry* 63:793–807.

Gerlach, M., Oehler, D., Blum-Degen, D., Lange, K. W., Mayer, B., Reichmann, H., and Riederer, P. 1995. Regional distribution and characterisation of nitric oxide synthase activity in the brain of the common marmoset. *Neuroreport* 6:1141–5.

Gerlach, M., Riederer, P., and Youdim, M. B. H. 1996. Molecular mechanisms for neurodegeneration: synergism between reactive oxygen species, calcium and excitotoxic amino acids. In *Advances in neurology*, ed. L. Battistin, G. Scarlato, T. Caraceni, and S. Ruggieri. 69:177–94. Philadelphia: Lippincott-Raven.

Gibb, W. R. G. 1986. Idiopathic Parkinson's disease and Lewy body disorders. *Neuropathology and Applied Neurobiology* 12:223–34.

Gibb, W. R. G., and Lees, A. J. 1988. The relevance of Lewy body to the pathogenesis of idiopathic Parkinson's disease. *Journal of Neurological and Neurosurgical Psychiatry* 51:745–52.

Halliwell, B. 1992. Reactive oxygen species and the central nervous system. *Journal of Neurochemistry* 59:1609–23.

Halliwell, B. 1994. Free radicals, antioxidants, and human disease: curiosity, cause, or consequence? *Lancet* 344:721–4.

Halliwell, B., and Gutteridge, J. M. C. 1984. Lipid peroxidation, oxygen radicals cell damage and antioxidant therapy. *Lancet* i:1396–8.

Han, S.-K., Mytilineou, C., and Cohen, G. 1996. L-Dopa upregulates glutathione and protects against oxidative stress. *Journal of Neurochemistry* 66:501–10.

Heales, S. J. R., Bolanos, J. P, and Clark, J. B. 1996. Glutathione depletion is accompanied by increased neuronal nitric oxide synthase activity. *Neurochemical Research* 21(1):35–9.

Hefti, F., Melamed, E., and Wurtman, R. J. 1980. Partial lesions of the dopaminergic nigrostriatal system in rat brain. Biochemical characterisation. *Brain Research* 79:95–101.

Hogg, N., Singh, R., and Kalyanaraman, B. 1996. The role of glutathione in the transport and catabolism of nitric oxide. *FEBS Letters* 382:223–8.

Ikeda, H., Markey, C. J., and Markey, S. P. 1992. Search for neurotoxins structurally related to MPP+ in the pathogenesis of Parkinson's disease. *Brain Research* 575:285–98.

Jain, A., Mårtennson, J., Store, E., Auld, P. A., and Meister, A. 1991. Glutathione deficiency leads to mitochondrial damage in brain. *Proceedings of the National Academy of Sciences of the U.S.A.* 88:1913–17.

Jellinger, K. 1989. Pathology of Parkinson's syndrome. In *Handbook of experimental pharmacology*, ed. D. B. Calne, 88:47–112. Berlin Springer.

Jellinger, K., Paulus, W., Grundke-Iqbal, I., Riederer, P., and Youdim, M. B. 1990. Brain iron and ferritin in Parkinson's and Alzheimer's disease. *Journal of Neural Transmission* 2:327–40.

Jenner, P. 1991. Oxidative stress as a cause of Parkinson's disease. *Acta Neurologica Scandinavica.* 89(suppl. 136):6–15.

Jenner, P., Schapira, A. H. V., and Marsden, C. D. 1992. New insights into the cause of Parkinson's disease. *Neurology* 42:2241–50.

Kagan, V. E., Shveddova, A., Serbinova, A., and Packer, L. 1992. Dihydrolipoic acid—a universal antioxidant both in the membrane and the aqueous phase. *Biochemical Pharmacology* 41(1):1–13.

Kastner, A., Hirsch, E. C., Lejeune, O., et al. 1992. Is the vulnerbility of neurones in the substantia nigra of patients with Parkinson's disease related to their neuromelanin content? *Journal of Neurochemistry* 59:1080–9.

Kish, S. J., Morito, C., and Hornykiewicz, O. 1985. Gluthione peroxidase activity in Parkinson's disease brain. *Neuroscience Letters* 58:343–6.

Kosower, N. S., and Kosower, E. M. 1978. The glutathione status of cells. *International Review of Cytology* 78:516–24.

Lancaster, J. R., and Hibbs, J. B. 1990. Demonstration of iron–nitrosyl complex by cytotoxic activated macrophages. *Proceedings of the National Academy of Sciences of the U.S.A.* 87:1223–7.

Maellero, E., Casini, A. F., Del Bello, B., et al. 1990. Lipid peroxidation and autoxidant systems in the liver injury produced by glutathione depleting agents. *Biochemical Pharmacology* 39:1513–21.

Mårtensson, J., Lal, J. C. K., and Meister, A. 1990. High affinity transport of glutathione is part of a multicom system, essential for mitochondrial function. *Proceedings of the National Academy of Sciences of the U.S.A.* 87:7185–7.

Martilla, R. J., Lorenzo, H., and Rinne, U.K. 1988. Oxygen toxicity protecting enzymes in Parkinson's disease; increase of superoxide dismutase-like activity in substantia nigra and basal nucleus. *Journal of Neurological Science* 86:321–31.

Marzatico, F., Cafe, C., Taborelli, M., and Benzi, G. 1993. Experimental Parkinson's disease in monkeys. Effect of ergot alkaloid derivative on lipid peroxidation in different brain areas. *Neurochemical Research* 18:1101–6.

McCall, T., and Vallance, P. 1991. Nitric oxide takes centre stage with newly defined roles. *Trends in Pharmacological Science* 13:1–6.

McCartney, M. A. 1989. Effect of glutathione depetion on morphine toxicity in mice. *Biochemical Pharmacology* 38:207–9.

McCay, P. B., Brueggemann, G., Lai, E. K., and Powell, S. R. 1989. Evidence that a-tocopherol functions cyclically to quench free radicals in hepatic microsomes. *Annals of the New York Academy of Science* 570:32–45.

McGeer, P. L., McGeer, E. G., and Suzuki, J. S. 1988. Rate of cell death in Parkinsonism. *Annals of Neurology* 24:574–6.

Meister, A. 1983. Selective modification of glutathione metabolism. *Science* 220:472–7.

Meister, A. 1991. Glutathione deficiency produced by inhibition of its synthesis and its reversal; applications in research and therapy. *Pharmacology and Therapeutics* 51:155–94.

Meister, A., and Griffiths, G. 1985. Trend in pharmacological science. In *Methods in enzymology*, ed. A. Meister, 113:500–10. Alton: Academic Press.

Meldrum, B., and Garthwaite, J. 1991. Excitatory amino acids and neurotoxicity and neurodegenerative diseases. In *The pharmacology of excitatory amino acids*, A Tips special report, ed. D. Lodge, 54–62. Cambridge: Elsevier.

Mithöfer, K., Sandy, M. S., Smith, M. T., and Di Monte, D. 1992. Mitochondrial poisons cause depletion of reduced glutathione in isolated hepatocytes. *Archives of Biochemistry and Biophysics* 295:132–6.

Munday, R., and Winterbourn, C. C. 1989. Reduced glutathione in combination with superoxide dismutase as an important biological antioxidant defence mechanism. *Biochemical Pharmacology* 38:4349–52.

Nagamura, A., Anderson, M. E., and Meister, A. 1990. Cellular glutathione is a determiner of sensitivity to mercury chloride toxicity: prevention of toxicity by giving glutathione monoester. *Biochemical Pharmacology* 40:693–7.

Naoi, M., Maruyama, W., Dostert, P., and Hashizume, Y. 1997. N-Methyl-(R)salsolinol as a dopaminergic neurotoxin: from an animal model to an early marker of Parkinson's disease. In *Advances in research on neurodegeneration*, ed. P. Riederer, D. B. Calne, R. Horowski, et al. 5:89–105. New York: Springer, Wien.

Ogino, T., Kawabat, T., and Awai, M. 1989. Simulation of glutathione synthesis in iron-loaded mice. *Biochemistry and Biophysics* 1006:131–5.

Ogita, K., Enomoto, R., Nakahara, F., and Yoneda, Y. 1995. A possible role of glutathione as an endogenous agonist at the NMDA recognition domain in rat brain. *Journal of Neurochemistry* 64(3):1088–96.

Olney, J. W. 1978. Neurotoxicity of excitatory amino acids. In *Kainic acid as a tool in neurobiology*, ed. E. G. McGeer, J. W. Olney, 95–121. New York: Raven Press.

Perry, T. L., and Yong, V. W. 1986. Idiopathic Parkinson's disease, progressive supranuclear palsy and glutathione metabolism in the substantia nigra of patients. *Neuroscience Letters* 67:269–74.

Perry, T. L., Godin, D. V., and Hansen, S. 1982. Parkinson's disease: a disorder due to nigral glutathione deficiency? *Neuroscience Letters* 33:305–10.

Perry, T. L., Hansen, S., and Jones, K. 1988. Brain amino acids and glutathione in progressive supranuclear palsy. *Neurology* 38:943–6.

Perumal, A. S., Tordzro, W. K., Katz, M., Jackson-Lewis, V., Thomas, B. C., Fahn, S., and Cadet, J. L.

1989. Regional effects of 6-hydroxydopamine (6-OHDA) on free radical scavengers in rat brain. *Brain Research* 504:139–41.

Philbert, M. A., Beiswanger, C. M., Waters, D. L., Beiswanger, C. M., Waters, S. D. K., Reuhl, K. R., and Lowndes, H. E. 1991. Cellular and regional distribution of reduced glutathione in the nervous system of the rat: histochemical localisation by mercury orange and o-pthaldialdehyde induced histofluoresence. *Toxicology and Applied Pharmacology* 107:215–27.

Pileblad, E., and Magnusson, T. 1992. Increase in rat brain following intracerebroventricular administration of γ-glutamyl cysteine. *Biochemical Pharmacology* 44(5):895–903.

Raps, S. P., Lai, J. C. K., Hertz, L., and Cooper, A. J. 1989. Glutathione is present in high concentrations in cultured astrocytes but not in cultured neurones. *Brain Research* 443:398–401.

Reichmann, H., and Riederer, P. 1989. Biochemische Analyse der Atmungskettenkomplexe verschiedener Hirnregionen von Patienten mit M. Parkinson. In *Morbus Parkinson und andere Basalganglienerkrankungen*, Symposium zu einem Föderschwerpunkt des Bundesministerium für forschung und Technik (BMFT) 44.

Reif, D. W., and Simmons, R. D. 1990. Nitric oxide mediates iron release from ferritin. *Archives of Biochemistry and Biophysics* 283:537–41.

Riederer, P., Sofic, E., Heuschneider, G., Strolin Benedetti, M., and Dostert, P. 1988. Secondary (toxic) Parkinsonism as model of Parkinson's disease. *Functional Neurology* 3:449–57.

Riederer, P., Sofic, E., Rausch, W.-D., Schmidt, B., Reynolds, G. P., Jellinger, K., and Youdim, M. B. 1989. Transition metals, ferritin, glutathione, and ascorbic acid in Parkinsonian brains. *Journal of Neurochemistry* 52:515–20.

Sagara, J.-I., Miura, K., and Bannai, S. 1993. Maintenance of neuronal glutathione by glial cells. *Journal of Neurochemistry* 61:1672–6.

Saggu, H., Cooksey, J., Dexter, D., Wells, F. R., Lees, A., Jenner, P., and Marsden, C. D. 1989. A selective increase in particulate superoxide dismutase activity in Parkinsonian substantia nigra. *Journal of Neurochemistry* 53:692–7.

Sanchez-Ramos, J. R., Övervik, E., and Ames, B. N. 1994. A marker of oxyradical-mediated DNA damage (8-hydroxy-2'-deoxyguanosine) is increased in nigro–striatum of Parkinson's disease brain. *Neurodegeneration* 3:197–204.

Schapira, A. H. V., Cooper, J. M., Dexter, D. T., Jenner, J., Clark, J. B., and Marsden, C. D. 1989. Mitochondrial complex I deficiency in Parkinson's disease. *Lancet* 1:1269–72.

Seaton, T. A., Jenner, P., and Marsden, C. D. 1996. Mitochondrial respiratory enzyme function and superoxide dismutase activity following brain glutathione depletion in the rat. *Biochemical Pharmacology* 52(11):1657–63.

Sechi, G., Deledda, M. G., Bua, G., Satta, W. M., Deinana, G. A., Pes, G. M., and Rosati, G. 1996. Reduced intravenous glutathione in the treatment of early Parkinson's disease. *Progress in Neuro. Psycho. & Biological Psychiatry* 20(7):1159–70.

Shan, X., Jones, D. P., Hashmi, M., and Anders, M. W. 1993. Selective depletion of mitochondrial glutathione concentrations by (R,S)-3-hydroxyl-4-petenoate potentiates oxidative cell death. *Chemical Research in Toxicology* 6:75–81.

Shaw, C. A., Pasqualotto, B. A., and Curry, K. 1996. Glutathione-induced sodium currents in neocortex. *Neurochemistry* 7(6):1149–52.

Shen, X. M., and Dryhurst, G. 1996. Further insights into the influence of L-cysteine on oxidation chemistry of dopamine: reaction pathways of potential relevance of Parkinson's disease. *Chemical Research in Toxicology* 9(7):751–63.

Shen, X. M., Xia, B., Wrona, M. Z., and Dryhurst, G. 1996. Synthesis, redox properties, vivo formation and neurobehavioral effects of N-acetylcysteinyl conjugates of dopamine: possible metabolites of relevance to Parkinson's disease. *Chemical Research Toxicology* 9(7):1117–26.

Sian, J. 1993. Brain glutathione in neurodegenerative disorders affecting the basal ganglia in man. Ph.D. Thesis, University of London.

Sian, J., Dexter, D. T., Cohen, G., Jenner, P., and Marsden, C. D. 1991a. Comparison of HPLC and enzymatic recycling assay for the measurement of oxidised glutathione levels in brain. *British Journal of Pharmacology*, December.

Sian, J., Dexter, D. T., Jenner, P., and Marsden, C. D. 1991b. Decrease in nigral glutathione in Parkinson's disease. *British Journal of Pharmacology* 104:281P.

Sian, J., Dexter, D. T., Jenner, P., and Marsden, C. D. 1992a. Glutathione levels in brain neurodegenerative disease affecting basal ganglia. *British Journal of Pharmacology*, April.

Sian, J., Dexter, D. T., Jenner, P., and Marsden, C. D. 1992b. Glutathione related enzymes in brain in basal ganglia degenerative disorders. *British Journal of Pharmacology*, September.

Sian, J., Dexter, D. T., Lees, A. Daniel, S. E., Agid, Y., Javoy-Agid, F., Jenner, P., and Marsden, C. D. 1994a. Alterations in glutathione levels in Parkinson's disease and other neurodegenerative disorders affecting basal ganglia. *Annals of Neurology* 36:348–55.

Sian, J., Dexter, D. T., Lees, A., Daniel, S. E., Agid, Y., Javoy-Agid, F., Jenner, P., and Marsden, C. D. 1994b. Glutathione-related enzymes in brain in Parkinson's disease. *Annals of Neurology* 36:356–61.

Siesjö, B. K. 1981. Cell damage in brain, a speculative synthesis. *J. Cere and Metab.*1:135–85.

Siesjö, B. K., and Rehncrona, S. 1980. Adverse factors affecting neuronal metabolism: relevance to the dementias. In *Biochemistry of dementia,* ed. P. J. Roberts, 91–120 New York: Wiley.

Slivka, A., Spina, M. B., and Cohen, G. 1987a. Reduced and oxidised glutathione in human and monkey brain. *Neuroscience Letters* 74:112–8.

Slivka, A., Mytilineou, C., and Cohen, G. 1987b. Histochemical evaluation of glutathione in brain. *Brain Research* 409:275–84.

Sofic, E., Lange, K. W., Jellinger, K., and Riederer, P. 1992. Reduced and oxidized glutathione in the substantia nigra of patients with Parkinson's disease. *Neuroscience Letters* 142:128–30.

Spencer, J. P. E., Jenner, P., and Halliwell, B. 1995. Superoxide-dependent depletion of reduced glutathione by L-dopa and dopamine. Relevance to Parkinson's disease. *NeuroReport* 6:1480–4.

Spina, M. B., and Cohen, G. 1989. Deprenyl suppresses the oxidative stress associated with increased dopamine turnover. *Annals of Neurology* 26:689–90.

Takahashi, N., Schreiber, J., and Fisher, V. 1987. Formation of glutathione-conjugated semiquinones by the reaction of quinones with glutathione: an ESR study. *Archives of Biochemistry and Biophysics* 252:41–8.

Toffa, S., Kunkowska, G. M., Zeng, B. Y., Jenner, P., and Marsden. C. D. 1997. Glutathione depletion in rat brain does not cause nigrostriatal pathway degeneration. *Journal of Neural Transmission* 104(1):67–75.

Umemura, T., Sai, K., Takagi, A., Hasegawa, R., and Kurokawa, Y. 1991. The effect of exogenous glutathione and cysteine on oxidative stress induced by ferric nitroltriacetate. *Cancer Letters* 58:49–56.

Ushijima, K., Miyazaki, H., and Moriokaz, T. 1986. Immunohistochemical localisations of glutathione peroxidase in the brain of the rat. *Resuscitation* 13:97–105.

Wardman, P. 1990. Bioreductive activation of quinones: redox properties and thiol reactivity. *Free Radical Research Communications* 8:219–29.

Wellner, V. P., Anderson, M. E., Puri, R. N., Jensen, G. L., and Meister, A. 1984. Radio protection by glutathione ester into human lymphoid cells and fibroids. *Proceedings of the National Academy of Sciences of the U.S.A.* 81:4732–5.

Wüllner, U., Löschmann, P.-A., Schulz, J. B., Schmid, A., Dringen, R., Eblen, F., Turski, L., and Klockgether, T. 1996. Glutathione depletion potentiates MPTP and MPP$^+$ toxicity in nigral dopaminergic neurones. *Neuroreportoe* 7:921–3.

Youdim, M. B. H., and Ben-Shachar, D. 1987. Minimal brain damage induced by early iron deficiency: modified dopaminergic neurotransmission. *Israel Journal of Medical Science* 23:19–25.

Yudkoff, M., Pleasure, G., Cregar, L., Lin, Z. P., Nissim, J., Stern, J., and Nissim, I. 1990. Glutathione turnover in cultural astrocytes: studies with [^{15}N] glutamate. *Journal of Neurochemistry* 55:137–45.

Zeevalk, G. D., Bernard, L. P., Albers, D. S., Mirochnitchenko, O., Nicklas, W. S., and Sonasaka, P. K. 1997. Energy stress–induced dopamine loss in glutathione peroxidase overexpressing transgenic mice and in glutathione depleted mesencephalic cultures. *Journal of Neurochemistry* 68(1):426–9.

Zhang, F., and Dryhurst, G. 1994. Effects of L-cysteine on the oxidation chemistry of dopamine: new reaction pathways of potential relevance to idiopathic Parkinson's disease. *Journal of Medicinal Chemistry* 37:1084–98.

Glutathione in the Nervous System
Edited by Christopher A. Shaw
Copyright © 1998 Taylor & Francis

15

Glutathione, Cysteine, and the Neuromelanin Pathway: Potential Roles in the Pathogenesis of Parkinson's Disease—a New Hypothesis

Glenn Dryhurst

Department of Chemistry and Biochemistry, University of Oklahoma, Norman, Oklahoma 73019

ABBREVIATIONS

ADP adenosine diphosphate
ATP adenosine triphosphate
BBB blood–brain barrier
2,5-bis-*S*-CyS-DA 2,5-bis-*S*-cysteinyldopamine
BT benzothiazine
complex I NADH–coenzyme Q reductase

CSF cerebrospinal fluid
CySH L-cysteine
DA dopamine
DHBT dihydrobenzothiazine
DHBT-1 7-(2-aminoethyl)-3,4-dihydro-5-hydroxy-2*H*-1,4-benzothiazine-
   3-carboxylic acid
5,6-DHI 5,6-dihydroxyindole
L-Dopa L-3,4-dihydroxyphenylalanine
DOPAC 3,4-dihydroxyphenylacetic acid
GABA gamma aminobutyric acid
GAD glutamic acid decarboxylase
Glu glutamic acid
GSH glutathione
GSSG glutathione disulfide
γ-GT γ-glutamyl transpeptidase
5-HT 5-hydroxytryptamine, or serotonin
HVA homovanillic acid
HO• hydroxyl radical
KCN potassium cyanide
LC locus ceruleus
MAO monoamine oxidase
MPP^+ 1-methyl-4-pyridinium
MPTP 1-methyl-4-phenyl-1,2,3,6-tetrahydropyridine
NADH reduced nicotine adenine dinucleotide
NE norepinephrine
$O_2^{-\bullet}$ superoxide radical anion
PD Parkinson's disease
ROS reactive oxygen species
2-*S*-CyS-DA 2-*S*-cysteinyldopamine
5-*S*-CyS-DA 5-*S*-cysteinyldopamine
5-*S*-Glu-DA 5-*S*-glutathionyldopamine
SN substantia nigra
SOD superoxide dismutase

1. INTRODUCTION

In 1817 James Parkinson published his *Essay on the Shaking Palsy* in which he described a disorder characterized by three cardinal symptoms: resting tremor, akinesia,
and rigidity. This disorder is now known as Parkinson's disease (PD) and is the second most common neurodegenerative disease of the brain, after Alzheimer's disease.
Exact figures on the incidence of PD are not available, but it has been estimated that
more than one million Americans are afflicted with the disorder along with many
more millions of people worldwide. The prevalence of PD increases dramatically

with age and becomes highest in the seventh and eighth decade of life (Schoenberg 1986). Although PD is primarily a movement disorder, neuropsychiatric deficits are also common and include depression and dementia along with one or more vegetative disturbances (Birkmayer and Riederer 1985; Gerlach and Riederer 1996).

Although the Parkinsonian syndrome has now been recognized for more than 180 years, the fundamental causes of the disorder are unknown. The degeneration of pigmented neuronal cell bodies located in the brain stem, particularly in the substantia nigra (SN) pars compacta, is the primary neuropathological feature of PD. Much is now known about changes that occur in the Parkinsonian SN, and some neurochemical and neurobiological processes have been identified that may contribute to the degeneration of pigmented neurons in this structure. However, the fundamental mechanisms that both trigger and underlie these selective neuropathological processes are unknown. An understanding of both the trigger mechanisms and the fundamental neuropathological mechanisms is essential in order to develop methods to detect the disorder in its early presymptomatic stage and to develop therapeutic strategies that can be applied to halt or reverse the normally inexorable progression of the neurodegenerative processes.

One of the earliest and most profound changes that occur in the Parkinsonian SN is a massive and irreversible loss of glutathione (GSH). In principle, there are many reasons that might account for this loss of GSH. These include consumption of GSH, to form glutathione disulfide (GSSG), as a consequence of elevated production of oxygen radicals and H_2O_2. Alternatively, enzymes associated with the biosynthesis of GSH might be defective, or ATP supplies, necessary for the biosynthesis of the tripeptide, might be inadequate. However, it is also conceivable that the loss of GSH might reflect a more ominous metabolic alteration in the Parkinsonian SN. In other words, this earliest known change in the SN in PD might point to an important role for GSH or its biosynthetic precursor L-cysteine (CySH) in the fundamental mechanisms underlying the pathogenesis of the disorder. Indeed, this laboratory is exploring the hypothesis that GSH or, more probably, CySH plays a key role in mediating formation of toxic endogenous metabolites that contribute to the degeneration of pigmented SN neurons in PD. Work in support of this hypothesis is at a very early stage. Nevertheless, the evidence bearing on possible roles for GSH and CySH in the neurodegenerative mechanisms that occur in PD will be reviewed. The ideas and interpretations that will be presented are speculative and are likely to be controversial. Nevertheless, they may serve to focus attention on novel approaches to understanding the pathoetiology of PD.

2. IDIOPATHIC PARKINSON'S DISEASE

The major neuropathological feature of PD is severe damage to the neuromelanin-pigmented cell bodies of nigrostriatal dopaminergic neurons in the SN pars compacta. This damage is evidenced by the depigmentation of the Parkinsonian SN, loss of neuromelanin-containing tyrosine hydroxylase–immunoreactive neurons, and severe depletion of the catecholaminergic neurotransmitter dopamine (DA) in the striatum

(Hornykiewicz and Kish 1986; Hornykiewicz 1988; Jellinger 1988, 1990, 1991). The diagnostic hallmark of idiopathic PD at autopsy is the presence of Lewy bodies (Gibb and Lees 1988; Gibb 1989; Lowe 1994). Lewy bodies are cytoplasmic inclusions that are present principally in neurons of the SN and of pigmented noradrenergic cell bodies in the locus ceruleus (LC) of patients with PD (Forno 1986). These inclusions are composed of a filamentous mesh of neurofilaments that surrounds a dense core.

Based upon the analysis of neurochemical and clinical correlations, it becomes apparent that the Parkinsonian syndrome appears when levels of DA in the striatum are decreased by more than 70–80 percent compared to the nondiseased brain (Bernheimer et al. 1973; Gerlach and Riederer 1996). Although reduced levels of DA occur in other regions of the Parkinsonian brain, possibly reflecting degeneration of other dopaminergic pathways, these are not as profound as in the nigrostriatal system (Scatton et al. 1982; Uhl, Hedreen, and Price 1985; Torack and Morris 1988; Jellinger 1988; Hedera and Whitehouse 1994; Gerlach and Riederer 1996).

The neurodegeneration in PD, however, is not restricted to dopaminergic neurons. To illustrate, noradrenergic neurons that project from the neuromelanin-pigmented locus ceruleus (LC) to the prefrontal cortex, motor cortex, and limbic regions also degenerate in PD (Forno 1982; Scatton et al. 1983; Chui et al. 1986; D'Amato et al. 1987a; Gaspar et al. 1991). Reduced levels of norepinephrine (NE) in these regions of the brain have been implicated in dementia, depression, and other nonmotor symptoms of PD (Gerlach et al. 1994). Levels of the serotonergic neurotransmitter 5-hydroxytryptamine (5-HT; serotonin) decrease significantly in many regions of the Parkinsonian brain (Cash et al. 1985; Scatton et al. 1982; D'Amato et al. 1987a; Jellinger 1990), including the SN (Hornykiewicz and Kish 1986). However, it is not clear if decreased 5-HT levels reflect the degeneration of serotonergic neurons (Gerlach and Riederer 1996). PD patients suffering from dementia also often exhibit profound degeneration of cholinergic neurons that project from the nucleus basalis of Meynert (NBM) (Whitehouse et al. 1983; Tagliavini et al. 1984; Jellinger, 1988). There is also a decrease of glutamic acid decarboxylase (GAD) activity in the striatum and SN of PD patients that would normally be expected to lead to lowered levels of the inhibitory neurotransmitter γ-aminobutyric acid (GABA) (Lloyd and Hornykiewicz, 1973). However, GABA activity in the Parkinsonian caudate nucleus is unchanged, and increased levels have been measured in the putamen and pallidum (Perry et al. 1983; Kish et al. 1986). Indeed, the role of GABAergic neurotransmission in the pathology of PD is controversial (Hedera and Whitehouse 1994). Levels of various neuropeptides are also reduced in the nigrostriatal and other neuronal systems in PD, although the roles of these neurotransmitters in the Parkinsonian syndrome and pathology are not clear (Gerlach and Riederer 1993, 1996).

Despite the fact that several neurotransmitter systems are affected in PD, the degeneration of nigrostriatal DA neurons is clearly the principal change that occurs in the brain in this disorder. Accordingly, most investigations into the causes and treatment of Parkinsonism have focused on this dopaminergic system. Thus, the striatal DA deficit in PD is currently treated by DA-replacement therapy using L-Dopa (L-3,4-dihydroxyphenylalanine) and a peripheral L-dopa decarboxylase inhibitor, inhibitors of DA metabolism, and certain DA-receptor agonists.

The results of many investigations strongly suggest that the pathological processes that lead to the degeneration of nigrostriatal dopaminergic neurons occur in the cell bodies of these neurons located in the SN pars compacta (Hornykiewicz 1989). One of these pathological processes is oxidative stress, a condition where the production of partially reduced or reactive oxygen species (ROS) such as superoxide radical anion ($O_2^{-\bullet}$), hydrogen peroxide (H_2O_2), and hydroxyl radical ($HO\bullet$) exceeds the protective capacities of the endogenous defense mechanisms of the cell (Simonian and Coyle 1996). To illustrate, compared to age-matched controls, the SN of PD patients exhibits elevated levels of lipid peroxidation (Dexter et al. 1989, 1994a; Jenner et al. 1992) and evidence for oxygen-radical-mediated damage to mitochondrial DNA (Sanches-Ramos and Ames 1994). Increased activity of manganese-dependent superoxide dismutase (SOD) associated with the mitochondrial membrane (Saggu et al. 1989) and, possibly, cytoplasmic Cu/Zn SOD (Marttila, Lorentz, and Rinne 1988) has also been measured in the Parkinsonian SN, perhaps as a compensatory response to increased $O_2^{-\bullet}$ production. Dopaminergic SN cell bodies are probably normally exposed to significant levels of oxidative stress because $O_2^{-\bullet}$ and H_2O_2 are generated in the biosynthesis of DA from tyrosine catalyzed by tyrosine hydroxylase, in the catabolism of DA by monoamine oxidase (MAO), and as a result of the autooxidation of DA.

Normally, $O_2^{-\bullet}$ is rapidly transformed into H_2O_2 and molecular oxygen by SOD, and then H_2O_2 is converted to water and molecular oxygen by glutathione peroxidase or catalase. However, under abnormal conditions these protective mechanisms may be inadequate, with the result that $HO\bullet$ is formed from H_2O_2 by the Fenton reaction (1) or from $O_2^{-\bullet}$ and H_2O_2 in the Haber–Weiss reaction (2):

$$Fe^{2+} + H_2O_2 \rightarrow Fe^{3+} + OH^- + HO\bullet \tag{1}$$

$$Fe^{3+} + O_2^{-\bullet} \rightarrow Fe^{2+} + O_2^{-2} \tag{2}$$

$$HO^- + HO\bullet \quad H_2O_2 \quad 2H^+$$

Both of these reactions require catalytic concentrations of low-molecular-weight iron or certain other transition-metal ion species (Pryor 1986; Halliwell and Gutteridge, 1987). Thus, it is of interest that, compared to age-matched controls, the total concentration of iron in the Parkinsonian SN is significantly increased (Sofic et al. 1988; Riederer et al. 1989), levels of the iron-binding protein ferritin are decreased (Dexter et al. 1990), and a neuromelanin–iron complex has been detected along with a shift of the Fe^{2+}/Fe^{3+} ratio from 2 : 1 in controls to 1 : 2 in the brains of PD patients (Jellinger et al. 1992).

Because of the ability of low-molecular-weight iron species or the neuromelanin–iron complex to catalyze formation of cytotoxic $HO\bullet$ from H_2O_2 and $O_2^{-\bullet}$ (Pilas et al. 1988; Youdim, Ben-Shachar, and Riederer 1989; Ben-Shachar, Riederer, and Youdim 1991), formed as a result of the synthesis, catabolism, and autooxidation of

DA, it has been suggested that iron is the endotoxin that mediates SN cell death in PD (Youdim et al. 1989; Olanow 1990; Sengstock et al. 1992). Indeed, infusions of iron(III) salts into the SN of rats evokes the progressive degeneration of nigrostriatal dopaminergic neurons (Ben-Shachar and Youdim 1991; Sengstock et al. 1992, 1993, 1994; Arendash, Olanow, and Sengstock 1993; Wesemann et al. 1993, 1994). It has been proposed that Fe^{3+} is taken up into dopaminergic SN neurons by transferrin receptors (Arendash et al. 1993), where it is reduced by cytoplasmic reductants to Fe^{2+}, the form of iron that generates cytotoxic HO• from H_2O_2 [Eqs. (1) and (2). Indeed, increased levels of lipid peroxidation have been measured in the rat SN following infusions of Fe^{3+} (Arendash et al. 1993; Wesemann et al. 1993). The facts that the SN of Parkinsonian patients contains increased levels of iron (Sofic et al. 1988; Riederer et al. 1989), decreased ferritin (Dexter et al. 1990), decreased concentrations of polyunsaturated fats and elevated levels of lipid peroxidation (Dexter et al. 1989), greatly increased concentrations of 8-hydroxy-2'-deoxyguanosine (formed by HO• attack on mitochondrial DNA) (Sanches-Ramos and Ames 1994), and decreased glutathione peroxidase (Kish, Morito, and Hornykiewicz 1985) and catalase (Ambani, Van Woert, and Murphy 1975) activities all support a role for iron in the pathogenesis of PD.

The massive (40–50 percent) decrease of nigral GSH levels that occurs in the Parkinsonian brain (Perry, Godin, and Hansen 1982; Riederer et al. 1989; Sofic et al. 1992; Sian et al. 1994a) might also indicate the consumption of this tripeptide by removal of elevated levels of H_2O_2 (in conjunction with glutathione peroxidase) and by scavenging HO•. However, increases of nigral iron levels are not specific to PD and occur in other neurodegenerative disorders of the basal ganglia (Dexter et al. 1993). Furthermore, the SN of patients who died with incidental Lewy-body disease, believed to be an early presymptomatic stage of PD (Gibb and Lees, 1988), have normal levels of both iron and ferritin (Jenner et al. 1992; Dexter et al. 1994b). Similarly, increases in brain iron have not been observed in patients with mild PD (Riederer et al. 1989). Taken together, these observations suggest that increased nigral iron levels are probably not the initiator of idiopathic PD, but rather a secondary response to some form of primary dopaminergic nerve cell damage (Dexter et al. 1993).

However, it has recently been reported that a rare, genetic form of PD results from a mutation of the ceruloplasmin gene located on chromosome 3 (Harris et al. 1995). Patients with aceruloplasminemia have very low levels of ceruloplasmin, a protein that appears to be involved in iron transport out of cells. Thus, it is probably the high levels of iron present in the basal ganglia of these patients that mediates HO• formation and oxidative damage to dopaminergic neurons. It is also of relevance to note that the decreased levels of GSH measured in the SN in PD also occur in incidental Lewy-body disease (Dexter et al. 1993).

These observations might indicate that PD is a disorder related to low levels of nigral GSH with consequent damage to dopaminergic SN neurons by ROS (Perry et al. 1982). However, several lines of evidence argue against a role for reduced nigral levels of GSH, and hence decreased antioxidant capacity, as an *initiator* of PD. To illustrate, drastically reduced brain levels of GSH resulting from a GSH-synthetase deficiency do not lead to SN cell loss or Parkinsonism (Skellerund et al. 1980; Marstein et al. 1981). Similarly, injection of mice with diethylmaleate causes a profound depletion of GSH in the brain stem (of which the SN is part), but no alteration of striatal DA content (Perry et al. 1986). Perhaps the most persuasive argument against a nigral GSH deficiency as the cause of PD is histochemical evidence indicating that neuronal cell bodies throughout the brain normally contain little or none of the tripeptide or, presumably, its biosynthetic precursor CySH (Slivka, Mytilineou, and Cohen 1987; Philbert et al. 1990). (For an alternative view, see Hjelle et al., chapter 4, this volume).

A second contributor to SN cell death in PD appears to be a deficiency in mito-chondrial respiration at the complex I (NADH-coenzyme Q reductase) stage (Mizuno et al. 1989; Schapira et al. 1990a,b). Complex I activity in the SN of patients with in-cidental Lewy-body disease is reduced to levels intermediate between those in control subjects and those with overt PD (Dexter et al. 1994b). The complex I deficiency in PD is of considerable significance in that it is anatomically specific to the SN (at least of the brain regions studied) and is not present in other neurodegenerative disorders involving the SN (Dexter et al. 1994b). Decreased mitochondrial complex I activity in dopaminergic SN cells would be expected to result in reduced ATP production, which in turn would compromise many metabolic processes, eventually leading to neuronal death.

3. NEUROMELANIN

Oxidative stress and a mitochondrial complex I deficiency (which together may con-tribute to the degeneration of nigrostriatal dopaminergic neurons) and a massive loss of GSH all appear to be specific to the SN in PD. This is the region of the brain in which the cell bodies of nigrostriatal DA neurons are located. These cell bodies are unusual in that they are pigmented with dark brown/black neuromelanin. It is also of relevance that noradrenergic neurons that project from neuromelanin-pigmented cell bodies in the LC also degenerate in PD. Indeed, PD has a clear predilection for neurons having pigmented cell bodies (Fearnley and Lees 1994). Although it is not yet known whether oxidative stress, a defect in mitochondrial complex I, and loss of GSH also occur in the Parkinsonian LC, it has nevertheless long been suspected that neuromelanin itself might play a role in the pathoetiology of PD. Alternatively, the conditions that permit formation of neuromelanin pigment may in some way predispose SN and LC neurons to degeneration in PD.

The neuromelanin present in the cytoplasm of dopaminergic SN cells is formed as a result of the autooxidation (Rodgers and Curzon 1975; Graham 1978) or, perhaps, the enzyme-mediated oxidation (Van Woert and Ambani 1974; Hastings 1995) of DA to DA-*o*-quinone that subsequently polymerizes. Studies of the in vitro autooxidation (Rodgers and Curzon 1975; Scheulen et al. 1975; Graham 1978) and particularly the electrochemically driven oxidation (Tse, McCreery, and Adams 1976; Dryhurst et al. 1982; Young and Babbitt 1983; Zhang and Dryhurst 1993) of DA and re-lated catecholamines have permitted considerable insights into the initial stages of

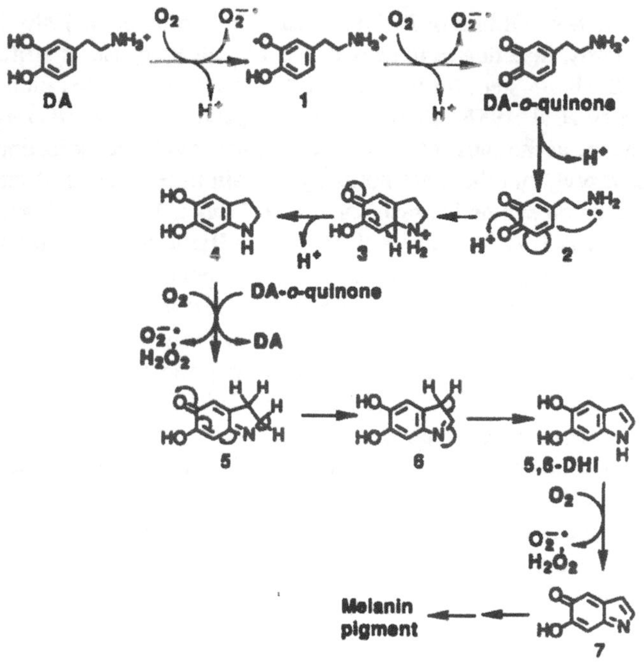

FIG. 1. Reaction pathway from DA to indolic melanin pigment.

this reaction. Thus, DA is initially oxidized by molecular oxygen to DA-o-quinone, probably via radical 1, with concomitant formation of $O_2^{-\bullet}$ (Fig. 1). At pH 7.4 the ethylamino side chain of DA-o-quinone is largely protonated. However, the small fraction that is deprotonated (i.e., 2, Fig. 1) undergoes an intramolecular cyclization reaction to 3 that rearranges to 5,6-dihydroxyindoline 4. Compound 4 is then oxidized by DA-o-quinone (or by molecular oxygen) to p-quinone imine 5, which via 6 aromatizes to 5,6-dihydroxyindole (5,6-DHI). The latter compound is the last known intermediate species in the oxidation pathway that leads from DA to black insoluble melanin polymer in vitro (Zhang and Dryhurst 1993). The polymerization reaction has been proposed to result from the oxidation of 5,6-DHI to p-quinone imine 7, which reacts with precursor species to give an indolic melanin polymer (Young and Babbitt 1983). Neuromelanin has been proposed to be formed by incorporation of melanin into lipofuscin derived from lyosomes (Fearnley and Lees 1994).

However, it is also relevant to note that the reaction pathway shown in Fig. 1 involves a number of electrophilic intermediates that would be expected to react with available nucleophiles. The most powerful and common nucleophiles present in biological systems are the sulfhydryl groups present in the cysteinyl residues of proteins and, particularly, of GSH and CySH. Indeed, evidence has been presented that human neuromelanin pigment from the SN is a mixed melanin consisting of oxidation products derived from both DA and 5-S-cysteinyldopamine (5-S-CyS-DA) (Carstam et al. 1991; Odh et al. 1994). The latter residue was thought to be formed by nucleophilic addition of CySH to DA-o-quinone. Alternatively, addition of GSH to DA-o-quinone would give 5-S-glutathionyldopamine (5-S-Glu-DA), which would then be hydrolyzed by peptidase enzymes to 5-S-CyS-DA (Agrup et al. 1975; Fehling et al. 1981). Based on analysis of the chemical degradation products of human SN neuromelanin, one study has concluded that sulfur-containing units formed by oxidation of 5-S-CyS-DA constitute about one-third to one-half of the polymer, the remainder being derived from oxidation of 5,6-DHI (Odh et al. 1994). However, a similar investigation failed to detect significant amounts of products derived from 5-S-CyS-DA upon chemical degradation of human SN neuromelanin (Wakamatsu, Ito, and Nagatsu 1991). The latter conclusion is in accord with the results of histochemical studies that indicate that neuronal cell bodies normally contain little or no cytoplasmic GSH or CySH (Slivka et al. 1987).

It is also relevant to note that the in vitro oxidation of 5-S-CyS-DA or DA in the presence of relatively large molar excesses of CySH at pH 7.4 ultimately lead to the formation of a yellow, alkali-soluble polymeric material (unpublished results), not a dark brown/black alkali-insoluble pigment such as is found in the human SN (Odh et al. 1994). The large amount of indole-derived residues in human neuromelanin also strongly points to the conclusion that cytoplasmic DA must normally be oxidized either in the absence of GSH or CySH or in the presence of only very low concentrations of these sulfhydryl compounds (Carstam et al. 1991; Odh et al. 1994). Thus, incorporation of residues formed from oxidation of 5-S-CyS-DA into neuromelanin might reflect low fluxes of CySH or perhaps GSH into the cytoplasm of SN neurons or, alternatively, sporadic influxes of much larger concentrations of these sulfhydryl compounds.

The fact that DA is normally oxidized to neuromelanin in dopaminergic SN neurons strongly supports the conclusion that these cell bodies must have a weak endogenous antioxidant defense system. This suggestion is supported by the fact that in the mouse SN the levels of ubiquinols Q_9 and Q_{10} are lower than in any other region of the brain and that the quinol/quinone ratio is heavily skewed in favor of the oxidized forms of these compounds (Fariello et al. 1988a, 1988b). Similarly, concentrations of α-tocopherol in the SN are low compared to many other brain structures (Fariello et al. 1988b). Furthermore, the SN normally has by far the highest basal levels of malondialdehyde, an index of peroxidative damage, of any brain region (Calabrese and Fariello 1988). The normal loss of dopaminergic SN neurons with age (McGeer, McGeer, and Suzuki 1976) could well be, therefore, a consequence of a suboptimal antioxidant defense system that permits autooxidation of DA to cytotoxic o-quinones and ROS (Scheulen et al. 1975; Graham et al. 1978; Wick 1978). Indeed, the dorsal tier of the human SN contains dopaminergic cell bodies with the heaviest neuromelanin pigmentation, and it is these cells that sustain the heaviest losses with normal aging (Gibb, Fearnley, and Lees 1990; Fearnley and Lees 1991).

In PD neuronal loss appears to be greatest in the lateroventral tier, followed by the medioventral tier, and least in the dorsal tier, although in fact all regions of the SN undergo significant degeneration (Fearnley and Lees 1991). A number of studies have demonstrated that the more heavily neuromelanin-pigmented cells in various regions of the SN are preferentially vulnerable to degeneration in PD (Hirsch, Graybiel, and Agid 1988; Kastner et al. 1992). However, one study has reported that the selective neuronal vulnerability in PD shows an inverse correlation with the neuromelanin content of dopaminergic SN cells (Gibb et al. 1990). In this study, it was concluded that Lewy-body degeneration spreads from the lightly to heavily pigmented zones of the SN. It has also been observed that there is a significant decrease in the neuromelanin content of surviving pigmented cells of the Parkinsonian SN (Mann and Yates 1983; Kastner et al. 1992).

The link between neuromelanin content and the vulnerability of nigral dopaminergic neurons to PD might be related to the ability of the pigment to bind exogenous compounds (Salazar, Sokoloski, and Patil 1978; D'Amato, Lipman, and Snyder 1986; D'Amato et al. 1987b) such as 1-methyl-4-phenylpyridinium (MPP^+) that causes a Parkinsonian syndrome (D'Amato et al. 1987b) or metals such as iron (Youdim et al. 1989; Hirsch et al. 1991) that catalyze formation of cytotoxic HO• from H_2O_2. However, there is no evidence that MPP^+ or structurally related compounds are present in the natural environment and play any role in the pathogenesis of idiopathic PD (Ikeda, Markey, and Markey 1992). Furthermore, as noted earlier, increased nigral levels of iron appear only relatively late in the progression of PD (Riederer et al. 1989; Jenner et al. 1992; Dexter et al. 1994b); hence, increased nigral iron is probably not involved in the initiation of PD, although it could well assume an important role in more advanced stages of the disorder.

As discussed previously, intraneuronal formation of neuromelanin would be expected to generate cytotoxic products and byproducts and hence contribute to the pathogenesis of PD. However, the presence of neuromelanin in the cytoplasm of dopaminergic neurons cannot alone account for SN cell death in PD. This is so because very heavily neuromelanin-pigmented cell bodies in certain regions of the SN appear to be somewhat resistant to the pathological processes in PD (Kastner et al. 1992).

4. POTENTIAL ROLE OF GSH AND CySH
IN THE PATHOGENESIS OF PD

Although some controversial issues remain to be resolved, the predominance of information presently available suggests that in certain anatomic regions of the SN dopaminergic cell bodies that contain heavier neuromelanin pigment are preferentially vulnerable to degeneration in PD (Mann and Yates 1983; Hirsch et al. 1988; Kastner et al. 1992). These are neurons that must sustain relatively high basal levels of cytoplasmic DA autooxidation, as a result of particularly weak endogenous antioxidant systems (including low CySH and GSH levels) and/or usually high concentrations of unbound (free) DA. The reports that less pigment is apparently deposited in the cytoplasm of surviving dopaminergic SN neurons in the Parkinsonian brain than in the nondiseased brain (Mann and Yates 1983; Kastner et al. 1992) suggests that the fundamental mechanisms that underlie PD might divert or block the normal neuromelanin pathway. There are several potentially important clues that might have a bearing on such mechanisms. One key clue relates to the massive loss of GSH in the Parkinsonian SN, which is not accompanied by a corresponding increase in levels of the oxidized form of the tripeptide glutathione disulfide (GSSG) (Riederer et al. 1989; Sofic et al. 1992; Sian et al. 1994a). This loss of nigral GSH without an increase of GSSG is specific to the Parkinsonian SN (at least in the brain regions analyzed) (Dexter et al. 1993), occurs at early presymptomatic stages of PD (Jenner et al. 1992; Dexter et al. 1994b), and is not related to decreased activities of biosynthetic enzymes (Sian et al. 1994b). Thus, a key question is raised concerning the fate of the GSH that is lost from the Parkinsonian SN.

In the brain, CySH and thence GSH are believed to be largely manufactured in glial cells (Raps et al. 1989; Sagara et al. 1993, 1996). The GSH present in neuronal axons and terminals (Slivka et al. 1987) appears to be maintained by export of the tripeptide from glia (Sagara et al. 1993, 1996). However, very few cells can directly import GSH. Rather, extracellular GSH is initially hydrolyzed to cysteinylglycine by γ-glutamyl transpeptidase (γ-GT), a membrane-bound enzyme with its active site directed toward the outside of the cell. Cysteinylglycine is then further hydrolyzed by a dipeptidase to CySH. Free CySH and/or cystine (CySSCy) is then transported into cells, including neurons, where it is normally utilized for the intracellular biosynthesis of GSH (Bannai 1984; Meister 1988; Liu et al. 1996). Thus, γ-GT is a key enzyme involved in the translocation of CySH from glia into neurons. It is therefore of considerable interest that the activity of γ-GT is significantly upregulated in the Parkinsonian SN (Sian et al. 1994b).

These observations raise the possibility that in PD elevated γ-GT activity might mediate the translocation of CySH from nigral glia into dopaminergic SN cell bodies that normally contain little or none of this sulfhydryl compound or GSH (Slivka et al. 1987). In vitro studies have established that both CySH and GSH divert the oxidation of DA to black indolic melanin polymer by scavenging DA-*o*-quinone to form, initially, soluble cysteinyl or glutathionyl conjugates of the neurotransmitter, respectively (Carstam et al. 1991; Zhang and Dryhurst 1994, 1995a,b; Shen and Dryhurst 1996). Thus, it seems plausible to suggest that the decreased pigmentation of dying dopaminergic SN neurons in the Parkinsonian brain might be related to an aberrant influx of CySH (or GSH) into these cell bodies, resulting in diversion of the neuromelanin pathway. That increased γ-GT activity might indeed mediate such a translocation of CySH is supported not only by the profound loss of nigral GSH without a corresponding increase in GSSG and decreased pigmentation of dopaminergic SN cells, but also by significant increases in the 5-*S*-CyS-DA/DA and 5-*S*-CyS-DA/homovanillic acid (HVA) concentration ratios in Parkinsonian SN tissue (Fornstedt et al. 1989) and cerebrospinal fluid (CSF) (Cheng et al. 1996), respectively, compared to age-matched controls. Very low concentrations of the 5-*S*-cysteinyl conjugates of DA, L-dopa, and 3,4-dihydroxyphenylacetic acid (DOPAC) are normally present in human and other mammalian brains (Rosengren, Linder-Eliasson, and Carlsson 1985; Fornstedt et al. 1986, 1989, 1990a,b; Fornstedt and Carlsson 1989; Carlsson 1991). The increased 5-*S*-CyS-DA/DA ratio measured in the Parkinsonian SN has been interpreted to reflect increased levels of DA autooxidation in surviving dopaminergic neurons (Fornstedt et al. 1989).

The fact that higher 5-S-CyS-DA/DA ratios were measured in the SN than in dopaminergic terminal fields in the striatum, where DA is largely sheltered in storage vesicles, suggests that only the cytoplasmic pool of the neurotransmitter is available for autooxidation. The results of pharmacological manipulations of cytoplasmic DA levels in animal brains on the formation of 5-S-CyS-DA appear to confirm this conclusion (Fornstedt and Carlsson 1989; Carlsson 1991). An increased rate of autooxidation of cytoplasmic DA, reflected in an increased 5-S-CyS-DA/DA ratio in the Parkinsonian SN, would be expected to be paralleled by elevated fluxes of $O_2^{-\bullet}$, H_2O_2, and, in the presence of low-molecular-weight iron species, HO•, which together might account for the oxidative stress that contributes to dopaminergic cell death. However, several factors tend to argue against this idea. For example, from a purely chemical perspective, there are no obvious mechanisms that could potentiate cytoplasmic DA autooxidation. Furthermore, an increased rate of cytoplasmic DA autooxidation would be expected to result in heavier neuromelanin pigmentation within surviving but dying dopaminergic cell bodies in the Parkinsonian SN. However, exactly the opposite effect occurs, that is, degenerating SN neurons appear to contain less neuromelanin pigment than their counterparts in the nondiseased brain (Mann and Yates 1983; Hirsch et al. 1988; Kastner et al. 1992).

The massive and anatomically selective loss of GSH, upregulation of γ-GT, and increased 5-S-CyS-DA/DA ratio that all occur in the Parkinsonian SN, along with the ability of CySH to divert the oxidation of DA to black indolic melanin polymer, together suggest that a key early step in the pathogenesis of PD might be an elevated translocation of CySH into the cytoplasm of neuromelanin-pigmented dopaminergic cell bodies in the SN (Zhang and Dryhurst 1994; Shen and Dryhurst 1996). Such an influx of CySH would be expected to scavenge the proximate autooxidation product of DA (i.e., DA-o-quinone), thus diverting the indolic neuromelanin pathway by forming, initially, soluble cysteinyl conjugates of the neurotransmitter. A central question, therefore, is whether such a hypothetical CySH-mediated diversion of the normal neuromelanin pathway might have any relevance to the pathogenesis of PD. In order to address this question, the following sections will summarize the influence of CySH (and GSH) on the in vitro oxidation chemistry of DA with particular focus on the products (i.e., putative nigral metabolites) of these reactions. The results of some preliminary experiments aimed at exploring mechanisms that might trigger the hypothetical CySH-mediated diversion of the neuromelanin pathway will be also presented. Finally, some relevant chemical and in vitro and in vivo biochemical properties of the putative metabolites will be described.

5. INFLUENCE OF CYSH AND GSH ON THE OXIDATION CHEMISTRY OF DOPAMINE

The autooxidation (Scheulen et al. 1975; Graham 1978) and electrochemically driven (Tse et al. 1976; Young and Babbitt 1983; Zhang and Dryhurst 1993) oxidation

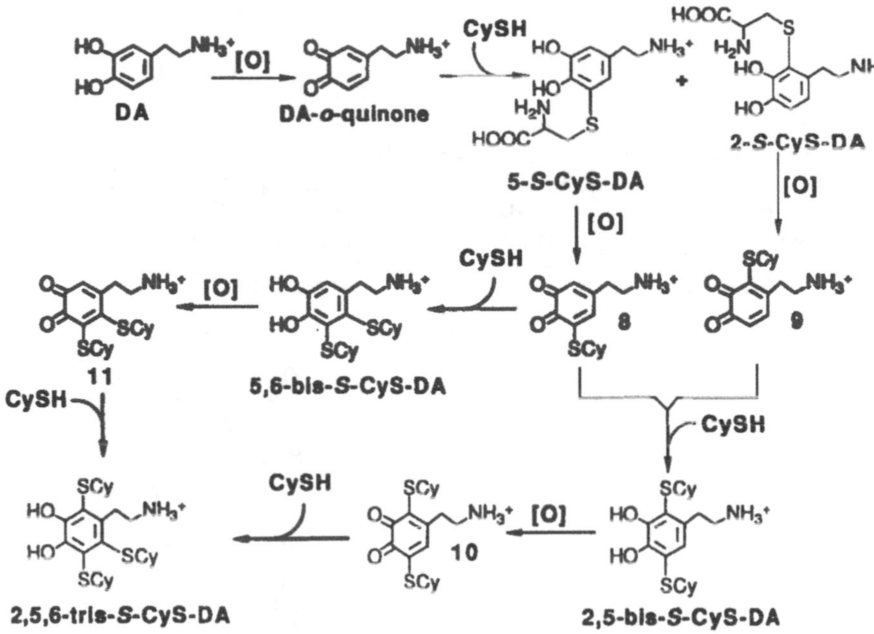

FIG. 2. Cysteinyldopamines formed by oxidation of DA in the presence of CySH at pH 7.4.

of DA at pH 7.4 generate DA-*o*-quinone as the proximate product, probably via transient intermediate radical 1 (Fig. 1). Because the autoxidation of DA at physiological pH is a relatively slow reaction, most studies of the influence of CySH and GSH on this reaction have employed electrochemical methods to generate DA-*o*-quinone. Such oxidation reactions in the absence of CySH or GSH rapidly lead to the formation of a precipitate of dark melanin polymer (Fig. 1). CySH diverts this reaction by scavenging DA-*o*-quinone to give, initially, 5-*S*-CyS-DA (major product) and 2-*S*-cysteinyldopamine (2-*S*-CyS-DA; minor product; Fig. 2) (Zhang and Dryhurst 1994; Shen and Dryhurst 1996). Both 5-*S*-CyS-DA and 2-*S*-CyS-DA are more easily oxidized than DA at pH 7.4 (as discussed in section 7) to give *o*-quinones 8 and 9, respectively.

The subsequent reactions of 8 and 9 are dependent to some extent on the relative concentrations of free CySH. Thus, particularly in the presence of large molar excesses of CySH, a major reaction pathway involves nucleophilic addition of this sulfhydryl compound to give 2,5-bis-*S*-cysteinyldopamine (2,5-bis-*S*-CyS-DA), which is even more easily oxidized to *o*-quinone 10. Nucleophilic addition of CySH to 10 then yields 2,5,6-tris-*S*-cysteinyldopamine (2,5,6-tris-*S*-CyS-DA). Free CySH can also attack *o*-quinone 8 to give 5,6-bis-*S*-cysteinyldopamine (5,6-bis-*S*-CyS-DA)—although this

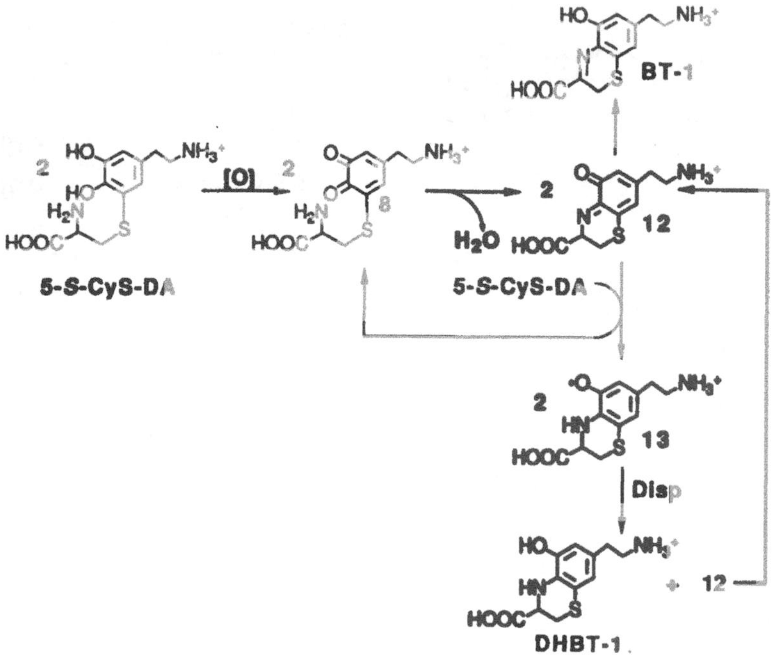

FIG. 3. Oxidation pathway from 5-S-CyS-DA to DHBT-1.

is a very minor pathway—which, after oxidation to **11** and attack by CySH, also yields 2,5,6-tris-S-CyS-DA (Fig. 2). At the point in these reactions when CySH becomes depleted or when DA is oxidized in the presence of low molar excesses of free CySH, alternative reaction pathways begin to predominate, resulting in the formation of dihydrobenzothiazines (DHBTs) and benzothiazines (BTs). Such a reaction pathway is illustrated in Fig. 3 for the oxidation of 5-S-CyS-DA to 7-(2-aminoethyl)-3,4-dihydro-5-hydroxy-2H-1,4-benzothiazine-3-carboxylic acid (DHBT-1) (Zhang and Dryhurst 1994, 1995b; Shen and Dryhurst 1996). Thus, the proximate oxidation product of 5-S-CyS-DA, o-quinone **8**, can undergo an intramolecular cyclization reaction to give the bicyclic o-quinone imine **12**, which is capable of *chemically* oxidizing 5-S-CyS-DA to **8** with concomitant formation of radical **13**. Disproportionation of **13** then yields DHBT-1 and **12**. In part, o-quinone imine **12** can also rearrange (tautomerize) to the 1,4-benzothiazine BT-1 (Fig. 3) (Zhang and Dryhurst 1995b; Shen, Zhang, and Dryhurst 1997).

The reaction pathway shown in Fig. 3 not only occurs when 5-S-CyS-DA is oxidized electrochemically (Zhang and Dryhurst 1995b), but also when DA is electrochemically oxidized in the presence of CySH. Similarly, oxidations of DA by $O_2^{-\bullet}$, HO• or molecular oxygen (autooxidation) and the Fe^{2+}/Fe^{3+}-catalyzed autoxidation reaction in the presence of free CySH also give 5-S-CyS-DA and, under certain conditions,

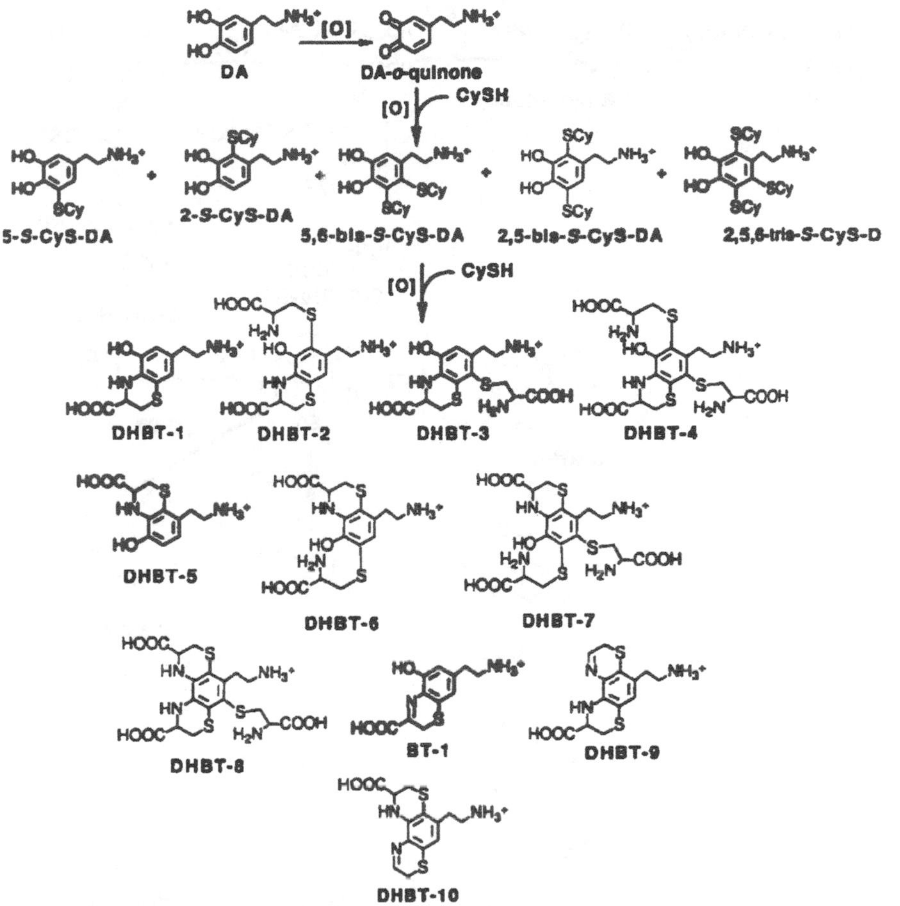

FIG. 4. Cysteinyldopamines, dihydrobenzothiazines, and benzothiazines formed when DA is electrochemically oxidized in the presence of CySH at pH 7.4.

DHBT-1, along with several other products that appear to depend on the oxidant and oxidation reaction conditions (unpublished results).

A summary of the various cysteinyldopamines, DHBTs, and BTs that are formed as a result of the electrochemically driven oxidation of DA in the presence of free CySH at pH 7.4 is presented in Fig. 4 (Zhang and Dryhurst 1994; Shen and Dryhurst 1996; Shen et al. 1997). Preliminary results indicate that 5-*S*-, 2-*S*-, and 2,5-bis-*S*-CyS-DA are also formed in the Mn^{2+}- and Fe^{2+}/Fe^{3+}-catalyzed oxidations of DA by molecular oxygen in the presence of free CySH. Similarly, DHBT-1, DHBT-5, and DHBT-10 also appear to be significant early products in these chemically mediated oxidation reactions. Indeed, the Mn^{2+}-catalyzed autooxidation of DA in the presence of CySH at pH 7.4 exhibits remarkably similar product profiles to those observed in

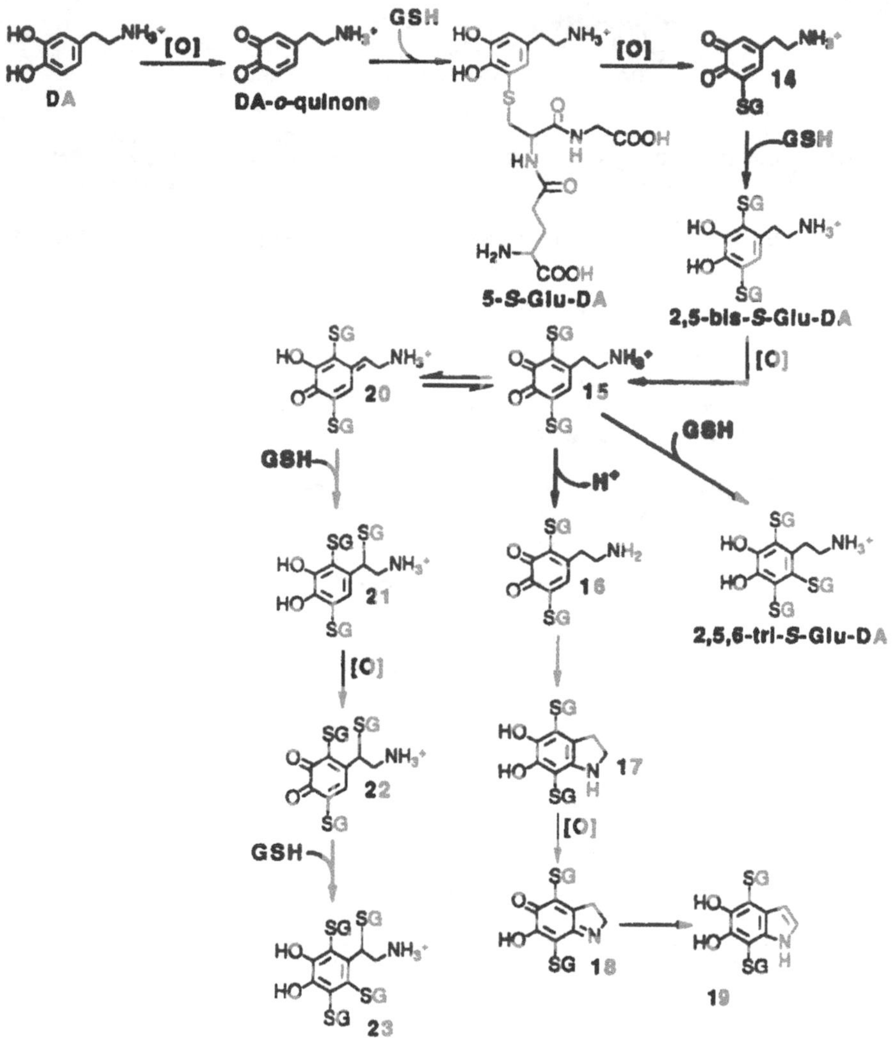

FIG. 5. Influence of GSH on the oxidation chemistry of DA at pH 7.4.

the electrochemically driven reactions. However, the Fe^{2+}/Fe^{3+}-catalyzed oxidation of DA by molecular oxygen in the presence of CySH generates a large number of rather minor additional products that remain to be fully identified.

The electrochemically driven oxidation of DA at pH 7.4 as well as the oxidation by molecular oxygen both with and without transition metal ion catalysis ultimately results in formation of a dark brown/black precipitate of indolic melanin that is insoluble in dilute alkali. The presence of free CySH in the reaction inhibits (delays) the appearance of an insoluble reaction product and, depending on the concentration of this sulfhydryl compound relative to DA, causes the ultimate formation of a yellow (high CySH concentrations) to brown (low CySH concentrations) precipitate that is soluble in dilute NaOH to give a bright yellow solution. The fact that human neuromelanin is not soluble in dilute alkali (Odh et al. 1994) suggests that it is predominantly an indolic melanin polymer.

Glutathione (GSH) also diverts and, at sufficiently high concentrations, blocks the oxidation of DA to insoluble indolic melanin polymer by initially scavenging DA-*o*-quinone to give 5-*S*-Glu-DA (Fig. 5) (Zhang and Dryhurst 1995a). The latter conjugate is more easily oxidized than DA to give *o*-quinone **14**, which reacts with free GSH to give 2,5-bis-*S*-Glu-DA, an even more easily oxidized compound. The resulting *o*-quinone **15** can then react by several pathways. Nucleophilic addition of GSH gives 2,5,6-tri-*S*-Glu-DA. Deprotonation gives **16**, which cyclizes to 2,5-bis-*S*-glutathionyl-5,6-dihydroxyindoline (**17**), which is very easily oxidized to *p*-quinone imine **18**. Rearrangement (aromatization) of **18** then gives 4,7-bis-*S*-glutathionyl-5,6-dihydroxyindole (**19**). *Ortho*-quinone **15** also tautomerizes to *p*-quinone methide **20**, which serves as the precursor of glutathionyl conjugates **21** and **23** containing glutathionyl residues substituted at the *β*-position of the ethylamino side chain of DA.

6. MECHANISMS THAT MIGHT TRIGGER TRANSLOCATION OF CYSTEINE INTO PIGMENTED SUBSTANTIA NIGRA NEURONS

In the event that elevated translocation of CySH (or GSH) into the cytoplasm of pigmented dopaminergic cell bodies in the SN is of relevance to the pathoetiology of PD, then mechanisms must exist to trigger this process. A number of lines of indirect but converging evidence provide some clues to such potential mechanisms. To illustrate, epidemiological investigations suggest a possible link between chronic exposure to high levels of environmental (agricultural, industrial, pharmaceutical) chemicals and an increased incidence of PD (Barbeau 1983; Barbeau et al. 1985; Tanner et al. 1989).

Furthermore, it is becoming increasingly evident that genetic factors might also play important roles in the susceptibility to PD (Golbe et al. 1990; Johnson, Hodge, and Duvoisin 1990). It is conceivable that these genetic factors might be linked to the observations that an extraordinarily high percentage of PD patients have very low activities of cysteine dioxygenase (Steventon et al. 1989) and thiolmethyl transferase (Waring et al. 1989). These are peripheral enzymes that play key roles in the detoxification and elimination of environmental toxicants and xenobiotics. Thus, it has been suggested that chronic exposure of individuals genetically equipped with these defective enzyme systems to certain environmental toxicants might permit some of these substances to cross the blood–brain barrier (BBB), enter the brain, and contribute to the degeneration of nigrostriatal DA neurons and PD (Steventon et al. 1989; Waring et al. 1989). However, no agricultural or industrial toxicant has been identified in the environment that can act as a selective nigrostriatal dopaminergic neurotoxin and cause PD.

Nevertheless, it is conceivable that some, perhaps many, environmental chemicals that are not adequately detoxified and eliminated might enter the brain and trigger translocation of CySH into neuromelanin-pigmented SN cell bodies. Such a response would represent the initial step in the translocation of GSH from glial cells into these and other neurons in order to provide protection against oxidative stress and to conjugate toxic substances (Meister and Anderson 1983). In this respect, it is of particular interest that the SN and LC are two of very few regions of the brain in which the BBB is ineffective or even absent (Ross, Romrell, and Kaye 1995). Thus, the SN and LC, which both contain pigmented neurons that degenerate in PD, are apparently exquisitely vulnerable to exposure to environmental toxicants carried by the blood.

It is also worth noting that chronic exposure to high levels of manganese evokes a syndrome that is remarkably similar to idiopathic PD (Barbeau 1984) along with depletion of striatal DA. Additionally, Mn^{2+} exposure also evokes a marked decrease of brain GSH levels and upregulation of γ-GT (Chandra and Shukla 1981; Parenti et al. 1986, 1988; Liccione and Maines 1988). Similarly, 1-methyl-4-phenyl-1,2,3,6-tetrahydropyridine (MPTP) causes not only the degeneration of the nigrostriatal dopaminergic pathway (Kinemuchi, Fowler, and Tipton 1987; Gerlach et al. 1991; Tipton and Singer 1993), but also a progressive depletion of brain-stem GSH when administered to mice (Yong, Perry, and Krisman 1985; Ferraro et al. 1986). Although Mn^{2+} (Chandra and Shukla 1981; Barbeau 1984; Liccione and Maines 1988) and MPTP (or its neurotoxic metabolite MPP^+) (Nicklas, Vyas, and Heikkila 1985; Sayre 1989; Hasegawa et al. 1990; Cleeter, Cooper, and Schapira 1992) are themselves directly toxic to dopaminergic neurons, they might also trigger upregulation of γ-GT and changes in GSH metabolism and CySH translocation.

In order to explore this idea, in effect using MPP^+ or Mn^{2+} as model environmental toxicants, we have recently carried out experiments in which these compounds were perfused through a microdialysis probe implanted in the striatum or SN of the anesthetized rat and the dialysate analyzed for extracellular GSH and CySH. Some representative results are presented in Fig. 6. Thus, shortly after completion of a 30-min perfusion of MPP^+ (≥ 200 μM in artificial CSF), extracellular levels of GSH begin to increase significantly reaching, almost 300 percent of basal levels after 60 min (Fig. 6 A). Subsequently, GSH declines to concentrations somewhat below basal levels. By contrast, MPP^+ evokes a somewhat slower increase of extracellular CySH concentrations, which again reach approximately 300 percent of basal levels after about 90 min (Fig. 6 B). Interestingly, however, the decline of CySH is much slower than is observed with GSH and does not return to basal levels until several hours have elapsed. Perfusions of MPP^+ into the rat SN evokes similar time-dependent effects on extracellular GSH and CySH concentrations. Perfusion of higher concentrations of MPP^+ appear to give larger changes in extracellular GSH and CySH levels, that is, the in vivo response is dose-dependent. Much higher concentration (≥ 20 mM) of Mn^{2+} ($MnCl_2$ dissolved in artificial CSF) are required to evoke a similar effect as MPP^+ on extracellular CySH and GSH concentrations.

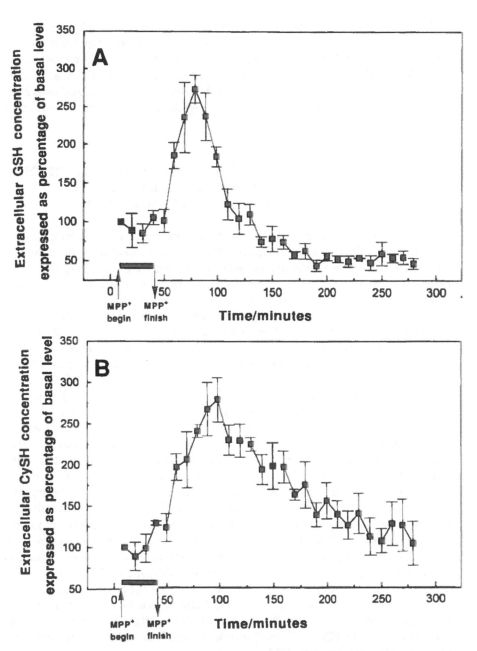

FIG. 6. Influence of MPP+ (500 μM dissolved in artificial CSF) on extracellular levels of (A) GSH and (B) CySH when perfused for 30 min at 1 μL/min through a microdialysis probe (4 mm) implanted in the striatum of an anesthetized male albino Sprague–Dawley rat. GSH and CySH were analyzed at 10-min intervals using an on-line injector and HPLC with electrochemical detection. The times at which MPP+ perfusion were started and stopped are shown in the figure. Error bars show SEM values.

It remains to be determined whether perfusion of other chemical species that are not intrinsic dopaminergic neurotoxins also evoke similar time-dependent changes of extracellular GSH and CySH concentrations. However, the preliminary results using MPP$^+$ and Mn^{2+} perfusions indicate that the appearance of certain environmental chemicals in the SN might indeed evoke release of GSH from glial cells and a rather rapid upregulation of γ-GT, leading to significant and sustained elevations of extracellular CySH.

7. NEUROCHEMICAL AND NEUROBIOLOGICAL PROPERTIES OF CYSTEINYL AND GLUTATHIONYL CONJUGATES OF DOPAMINE AND DIHYDROBENZOTHIAZINES

That MPP$^+$, Mn^{2+}, or other environmental/xenobiotic chemicals result in the elevated translocation of CySH into dopaminergic SN cells with resultant diversion of the neuromelanin pathway and formation of cysteinyldopamines and DHBTs remains to be established. However, if such an intraneuronal metabolic diversion is of relevance to the pathoetiology of PD, then products (metabolites) of this process should have chemical properties and/or biological activities that could contribute to the degeneration of dopaminergic SN neurons. An interesting and potentially significant chemical property of the cysteinyl conjugates of DA is that most are appreciably more easily oxidized than the neurotransmitter itself. Table 1 presents a summary of the voltammetric peak oxidation potentials for these conjugates and various DHBTs formed by oxidation of cysteinyldopamines. Many of the structurally simpler DHBTs such as DHBT-1 are also more easily oxidized than is DA at pH 7.4.

It is also potentially of considerable interest that some cysteinyldopamines and several DHBTs are quite toxic (lethal) when administered into the brains of laboratory mice (Zhang and Dryhurst 1994; Shen and Dryhurst 1996; Shen et al. 1996, 1997). Table 2 shows the LD$_{50}$ values for the toxic compounds thus far discovered as a result of the oxidation of DA in the presence of CySH at pH 7.4. Shortly after intracerebroventricular administration, all of the toxic DHBTs at approximately their LD$_{50}$ dose evoke a very characteristic neurobehavioral response that develops rapidly after drug injection. This response includes episodes of hyperactivity (rapid running, rolling, jumping) and shivering. By contrast, this characteristic neurobehavioral response develops only 10–30 min after icv administration of 2,5-bis-S-CyS-DA. This compound is the most easily oxidized of all cysteinyl conjugates of DA (Table 1) to give DHBT-6 (major product) and DHBT-2 (Shen and Dryhurst 1996). It has been speculated, therefore, that 2,5-bis-S-CyS-DA might be oxidized in vivo to toxic DHBT-6 and DHBT-2 (Table 2) in order to explain the delayed neurobehavioral response (Shen and Dryhurst 1996).

5-S-Glu-DA is rapidly metabolized in rat brain in vivo to give, initially, 5-S-CyS-DA, which is then more slowly metabolized to 5-S-(N-acetylcysteinyl)dopamine (Shen et al. 1996). The initial reaction in this metabolic pathway involves the γ-GT-mediated cleavage of the γ-glutamyl bond of 5-S-Glu-DA to give 5-S-cysteinylglyci-

TABLE 1. Voltammetric peak oxidation
potentials (E_p) at pH 7.4 for cysteinyl
and glutathionyl conjugates of DA and
various dihydrobenzothiazines [a]

Compound[b]	E_p (mV vs. SCE)
DA	125
5-S-CyS-DA	69
2-S-CyS-DA	79
2,5-bis-S-CyS-DA	57
2,5,6-tris-S-CyS-DA	≈100
5-S-Glu-DA	134[c]
2,5-bis-S-Glu-DA	116[c]
2,5,6-tris-S-Glu-DA	182[c]
DHBT-1	68
DHBT-2	66
DHBT-4	≈100
DHBT-5	77
DHBT-6	50
DHBT-7	112
DHBT-8	122
DHBT-9	128
DHBT-10	155
BT-1	590

[a] E_p values measured in pH 7.4 phosphate
buffer at a pyrolytic graphite electrode using
a sweep rate of 10 mV s^{-1} unless otherwise
noted. Data from Zhang and Dryhurst (1994,
1995a), and Shen et al. (1997).
[b] Structures are shown in figures.
[c] Sweep rate:100 mV s^{-1}.

nyldopamine, which is then hydrolyzed by cysteinylglycine dipeptidase to 5-S-CyS-DA (Fig. 7). The N-acetylation of 5-S-CyS-DA is then catalyzed by cysteine conjugate N-acetyltransferase. 5-S-(N-Acetylcysteinyl)dopamine and its 2-S- and 2,5-bis-S-congeners are also toxic (lethal) when administered icv into mouse brain (Table 2) (Shen et al. 1996).

Although 5-S-CyS-DA levels, relative to DA, increase in the Parkinsonian SN, the absolute concentrations of this conjugate are not significantly different to those measured in control patients (Fornstedt et al. 1989). One interpretation of this observation is that 5-S-CyS-DA is further oxidized in the Parkinsonian SN, so that the levels of the conjugate measured reflect only a small fraction of that formed in the cytoplasm of pigmented dopaminergic cell bodies. Because 5-S-CyS-DA appears to be non-toxic when injected into mouse brain but is easily oxidized to DHBT-1, which is toxic (Zhang and Dryhurst 1994; Shen and Dryhurst 1996), some preliminary (unpublished) studies have been carried out to assess the neurotoxicity of the latter compound in rat brain. The results of these studies have been somewhat inconclusive. In general, it appears that, 7 days following unilateral injections of DHBT-1 into the SN, the levels

TABLE 2. LD$_{50}$ values in mice for
cysteinyldopamines and dihydrobenzothiazines
formed upon oxidation of DA in the
presence of CySH[a]

Compound	LD$_{50}$ (μg)[b]
2,5-bis-S-CyS-DA	37
DHBT-1	14
DHBT-2	70
DHBT-5	5
DHBT-6	17
DHBT-8	88
DHBT-9	18

TABLE 2. *Continued.*

Compound	LD_{50} $(\mu g)^b$
DHBT-10	1.5
5-S-(N-acetylcysteinyl)DA	14
2-S-(N-acetylcysteinyl)DA	25
2,5-bis-S-(N-acetylcysteinyl)DA	42

[a]Data from Zhang and Dryhurst (1994), Shen and Dryhurst (1996), and Shen et al. (1996, 1997).

[b]Compounds, dissolved in 5 μL of isotonic saline, were injected into the left lateral ventricle of male albino Sprague–Dawley mice weighing approximately 30 g.

of 5-HT are reduced in the ipsilateral striatum, although the levels of DA and NE do not appear to be significantly altered. However, the γ-GT-mediated translocation of CySH into pigmented dopaminergic SN cells should result in the *intraneuronal* (cytoplasmic) formation of cysteinyldopamines, DHBTs, and their ultimate oxidation products.

In view of this assumption and the fact that a mitochondrial complex I deficiency appears to contribute to the degeneration of pigmented dopaminergic cell bodies in the Parkinsonian SN (Mizuno et al. 1989; Schapira et al. 1990a,b), the influence of DHBT-1 on mitochondrial respiration has been studied (Li and Dryhurst 1997). Using an oxygen electrode system to measure oxygen consumption, it has been demonstrated that DHBT-1 inhibits ADP-stimulated oxidation of malate and pyruvate (state

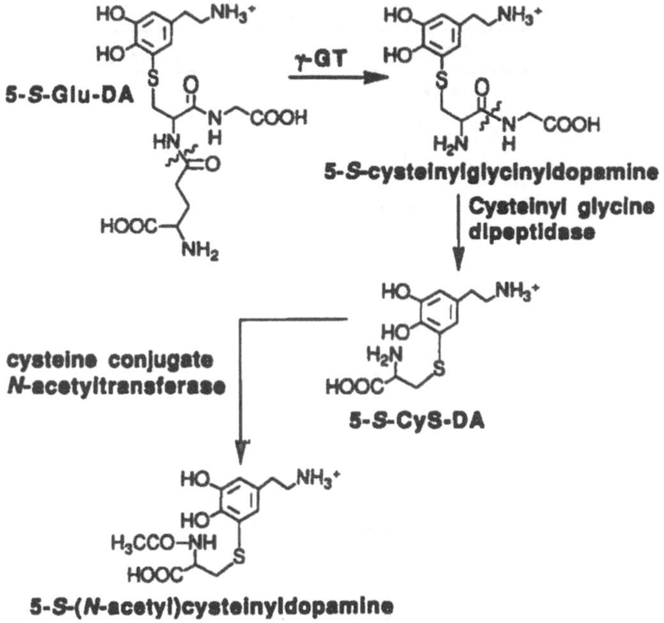

FIG. 7. Metabolism of 5-*S*-Glu-DA in rat brain.

3 or complex I respiration) when incubated for 5 min at 27°C with rat brain mitochondria with an IC_{50} of approximately 0.8 mM. Under identical conditions the well-known dopaminergic neurotoxin MPP^+ inhibited complex I respiration with an IC_{50} of approximately 0.3 mM. Incubation of DHBT-1 with freeze-thawed rat brain mitochondria (i.e., mitochondrial membranes), in the presence and the absence of KCN and/or NADH, results in a time-dependent, irreversible decrease of NADH-CoQ_1 reductase (complex I) activity. Unlike MPP^+ (Cleeter et al. 1992), DHBT-1 evokes the most profound irreversible inhibition of complex I when the mitochondrial respiratory chain is maintained in its more physiologically relevant nonreduced state (i.e., in the absence of KCN and added NADH) (Li and Dryhurst 1997). The irreversible inhibition of complex I caused by DHBT-1 under the latter conditions can be attenuated but not blocked by GSH, ascorbic acid, SOD, or catalase.

Together, these results suggest that DHBT-1 can cross the mitochondrial membrane and evoke damage to complex I by mechanisms that are only partly related to oxygen radicals. Preliminary experiments also indicate that an enzyme system associated with rat brain mitochondrial membranes catalyzes the oxidation of DHBT-1. Since the inhibition of complex I by DHBT-1 is time-dependent and little of this compound remains when maximum inhibition is observed, it is possible that an intermediate and/or product of DHBT-1 contributes to the irreversible inhibition of NADH-CoQ_1 reductase.

8. A NEW HYPOTHESIS FOR THE PATHOGENESIS OF PARKINSON'S DISEASE

Based upon currently available information, it appears that there are at least three distinct factors associated with the pathogenesis of PD. These are: (i) factors that predispose an individual to the development of PD, (ii) mechanisms that might trigger the pathological processes, and (iii) the fundamental mechanisms that underlie mitochondrial complex I damage and oxidative stress.

8.1 Factors That Might Predispose an Individual to PD

Because PD has a predilection for pigmented catecholaminergic neurons, it may reasonably be concluded that neuromelanin—or, more probably, intraneuronal (cytoplasmic) conditions that promote formation of this polymer—are a key factor in the predisposition of dopaminergic SN and noradrenergic LC cells to degeneration in PD. However, in view of the report that certain very heavily pigmented cell bodies in certain regions of the SN are somewhat resistant to degeneration in PD (Kastner et al. 1992), it may also be concluded that anatomic location is also a factor of relevance to the vulnerability of dopaminergic neurons to degeneration. Conditions that promote formation of indolic neuromelanin include a weak intraneuronal antioxidant system, including low cytoplasmic levels of GSH and CySH. Only under such conditions is it possible for significant autooxidation of DA to alkali-insoluble indolic neuromelanin to occur. The more heavily neuromelanin-pigmented cells in certain regions of the SN are preferentially vulnerable to degeneration in PD (Hirsch et al. 1988; Kastner et al. 1992). Thus, it may be concluded that those cells that sustain the highest basal levels of DA autooxidation in these regions are those most vulnerable to degeneration.

Genetic factors, perhaps expressed as an inability to effectively detoxify and eliminate certain environmental toxicants and xenobiotics (Steventon et al. 1989; Waring et al. 1989), also appear to play a role in predisposing an individual to PD (Golbe et al. 1990; Johnson et al. 1990). Finally, the absence of an effective BBB in both the SN and LC, but in very few other areas of the brain (Ross et al. 1995), might provide a third factor that predisposes pigmented catecholaminergic cell bodies in these structures to degeneration in PD.

8.2 Mechanism That Might Trigger Pathological Processes in the Substantia Nigra and Locus Ceruleus

A plausible mechanism that might trigger the pathological processes that cause the selective degeneration of SN and LC cells in PD might be inferred from the link between chronic exposure to high levels of agricultural, industrial, and perhaps other environmental chemicals or xenobiotics and an increased incidence of PD (Barbeau 1983; Barbeau et al. 1985; Tanner et al. 1989). Thus, chronic exposure of an individual genetically equipped with defective detoxification enzyme systems to certain environmental/xenobiotic chemicals might permit these substances to be transported by the circulatory system into the brain and selectively enter the brain parenchyma in the SN and LC because of the ineffective BBB in these structures (Ross et al. 1995). It is conceivable that entry of certain environmental toxicants into the SN and LC in this way might serve as one mechanism to trigger the neuropathological processes that underlie PD. Our preliminary brain microdialysis experiments discussed in connection with Fig. 6 suggest that a consequence of such a chemical insult might be the elevated release of GSH from nigral glial cells and upregulation of γ-GT with resultant increases of extracellular CySH levels. Subsequent translocation of CySH/CySSCy into axons, terminals, and neuronal cell bodies in the SN would be a normal protective response designed to increase cytoplasmic levels of GSH, a powerful antioxidant and scavenger of electrophilic toxicants (Meister and Anderson 1983; Sagara et al. 1993, 1996). However, the cytoplasm of dopaminergic cell bodies in the SN (and noradrenergic cell bodies in the LC) is unusual in that it contains an endogenous electrophile in the form of DA-o-quinone formed by autooxidation of DA as the initial step in a complex pathway leading to neuromelanin pigment (Fig. 1).

8.3 Neuropathological Mechanisms That Might Underlie PD

The results of in vitro chemical studies suggest that an elevated translocation of CySH into the cytoplasm of pigmented cell bodies in the SN should divert the normal neuromelanin pathway by scavenging DA-o-quinone to give 5-S-CyS-DA (major metabolite) and other cysteinyl conjugates of DA (Fig. 2) (Zhang and Dryhurst 1994; Shen and Dryhurst 1996; Shen et al. 1997). The observation that nigral γ-GT activity is significantly elevated even at advanced (terminal) stages of PD (Sian et al. 1994b) might imply that such a translocation of CySH into pigmented SN cells takes place over long periods of time. The massive, anatomically selective loss of GSH without a corresponding increase of GSSG (Riederer et al. 1989; Sofic et al. 1992; Sian et al. 1994a) [a very early event in the pathogenesis of PD (Dexter et al. 1993)], and the increased 5-S-CyS-DA/DA concentration ratio (Fornstedt et al. 1989) that occur in the Parkinsonian SN are all compatible with release of GSH from glia, upregulation of γ-GT, and hydrolysis of GSH to CySH that is then irreversibly lost by reaction with DA-o-quinone upon translocation into pigmented dopaminergic cell bodies. Furthermore, such a CySH-mediated diversion of the neuromelanin pathway to give 5-S-CyS-DA and other cysteinyldopamines might provide an explanation for the apparent deposition of less dark brown/black indolic melanin polymer in SN cell bodies in the Parkinsonian brain than in the nondiseased brain (Mann and Yates 1983; Kastner et al. 1992).

The 5-S-CyS-DA/DA ratio measured in Parkinsonian SN tissue (Fornstedt et al. 1989) and the 5-S-CyS-DA/HVA ratio measured in the CSF of PD patients (Cheng et al. 1996) are both elevated compared to the ratios measured in controls. However, the absolute concentrations of 5-S-CyS-DA measured in the SN and CSF of PD patients are not significantly different from control levels. These observations might

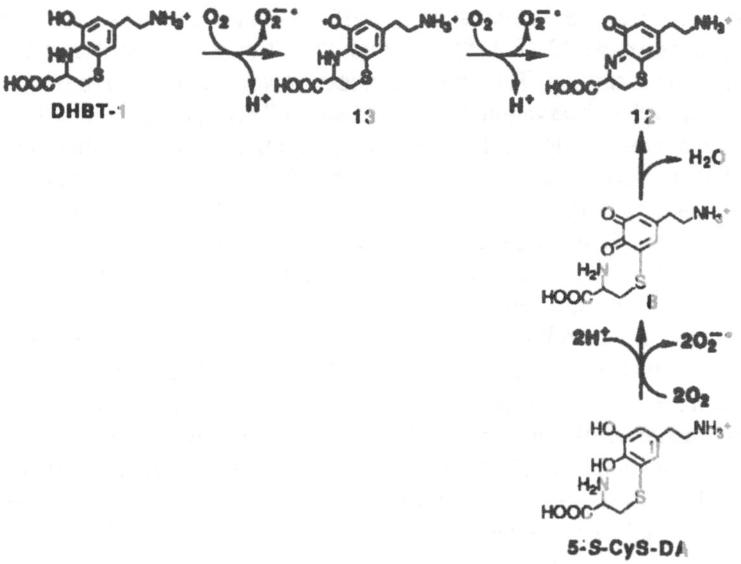

FIG. 8. Oxidations of DHBT-1 and 5-*S*-CyS-DA to electrophilic *o*-quinone imine **12**.

indicate that the 5-*S*-CyS-DA levels measured in PD represent only a remnant of the conjugate that is actually formed owing to its further metabolism. However, 5-*S*-CyS-DA is not a substrate for MAO (Fornstedt et al. 1989). Nevertheless, 5-*S*-CyS-DA and several other cysteinyl conjugates of DA have significantly lower oxidation potentials than DA (Table 1). Thus, under conditions where DA is autoxidized in the cytoplasm of neuromelanin-pigmented SN cells, 5-*S*-CyS-DA and other cysteinyldopamines would be expected to be even more easily autoxidized. Such facile autooxidation reactions would be expected to generate $O_2^{-\bullet}$ and thence other ROS byproducts that might contribute to oxidative stress, one of the pathological processes that occur in the Parkinsonian SN (Dexter et al. 1989, 1994a; Jenner et al. 1992; Sanches-Ramos and Ames 1994).

However, of potentially far greater importance is the facile oxidation of 5-*S*-CyS-DA to DHBT-1 and many other toxic cysteinyldopamines and DHBTs (Figs. 3 and 4; Table 2). DHBT-1 is able to cross the mitochondrial membrane and evoke a time-dependent irreversible inhibition of NADH-CoQ$_1$ reductase (complex I) (Li and Dry-hurst 1997). In these experiments, incubation of 500 μM DHBT-1 with freeze-thawed rat brain mitochondria (200 μg protein) at 30°C in 20 mM, pH 8.0 phosphate buffer in the absence of KCN and added NADH resulted in a statistically significant and irreversible inhibition of NADH-CoQ$_1$ reductase when the latter enzyme activity was assayed after 60 min. Higher concentrations of DHBT-1 evoked more profound inhibition of complex I. The mechanism(s) by which DHBT-1 evokes irreversible inhibition of complex I is not yet understood. However, rat brain mitochondrial membrane preparations catalyze the oxidation of DHBT-1 (Li and Dryhurst 1997), presumably to o-quinone imine 12 (Fig. 8), although the ultimate metabolites of this reaction remain to be identified. The time-dependent irreversible inhibition of complex I by DHBT-1 and the observation that antioxidants (GSH, ascorbic acid, SOD, catalase) attenuate but do not block this inhibition suggest that at least two mechanisms might contribute to NADH-CoQ₁ reductase damage. One might involve irreversible (covalent) binding of DHBT-1, its oxidative intermediate 12, or unknown products to the complex I site. Mitochondrial membranes also catalyze oxidation of 5-S-CyS-DA (unpublished results). Furthermore, low concentrations of 5-S-CyS-DA (0.5–5 mM) also cause a time-dependent, irreversible inhibition of complex I.

Together, these preliminary results suggest that 12, formed by mitochondrial-catalyzed oxidation of both DHBT-1 and 5-S-CyS-DA (Fig. 8), or unknown products derived from this o-quinone imine might be the species that inhibits complex I. A second mechanism that might contribute to the irreversible inhibition of NADH-CoQ₁ reductase evoked by DHBT-1 (and 5-S-CyS-DA) involves the generation of oxygen radicals from the mitochondrial respiratory chain, an effect evoked by other complex I inhibitors such as rotenone and MPP⁺ (Turrens and Bovaris 1980; Hasegawa et al. 1990; Cleeter et al. 1992).

9. SUMMARY

In this review, a new hypothesis is advanced for the pathogenesis of PD. It is suggested that the key step in initiating the neuropathological processes in this disorder is a elevated release of GSH from nigral glial cells and upregulation of γ-GT with resultant elevated translocation of CySH into neuromelanin-pigmented dopaminergic neurons in the SN pars compacta. It is further suggested that the entry of certain environmental/xenobiotic chemicals into the SN is a trigger that initiates the translocation of CySH into pigmented SN cells. A genetically inherited (Golbe et al. 1990; Johnson et al. 1990) inability to detoxify and eliminate certain environmental/xenobiotic chemicals (Steventon et al. 1989; Waring et al. 1989), the absence of an effective BBB in the SN (Ross et al. 1995), and chronic exposure to such chemicals might together permit such substances to enter the brain selectively in the SN. Other factors might also be able to trigger the elevated translocation of CySH into nigral SN cells. These might include severe and repetitive brain trauma such as that experienced by boxers.

Such an elevated translocation of CySH into nigral neurons probably represents a normal defensive response to a chemical or other brain insult as the first step in increasing cytoplasmic levels of GSH (Meister and Anderson 1983). However, dopaminergic cell bodies in the SN are unusual in that they are pigmented with neuromelanin formed by the autooxidation of cytoplasmic DA to DA-o-quinone that subsequently polymerizes (Fig. 1). Thus, it is proposed that the influx of CySH into pigmented dopaminergic cells diverts the normal neuromelanin pathway by scavenging DA-o-quinone to give 5-S-CyS-DA, which can be further easily oxidized to other cysteinyldopamines and DHBTs (Figs. 2–4). Such a translocation of CySH into pigmented dopaminergic cells is supported by the upregulation of γ-GT (Sian et al. 1994b), irreversible loss of nigral GSH without a corresponding increase of GSSG (Riederer et al. 1989; Jenner et al. 1992; Sofic et al. 1992; Dexter et al. 1994b; Sian et al. 1994a), and the increased 5-S-CyS-DA/DA concentration ratio (Fornstedt et al. 1989), all of which occur in the Parkinsonian SN. Furthermore, such a CySH-mediated metabolic diversion could account for the deposition of less indolic neuromelanin pigment in the cytoplasm of dying dopaminergic cells in the Parkinsonian SN (Mann and Yates 1983; Kastner et al. 1992).

At least two early products of the hypothesized CySH-mediated diversion of the neuromelanin pathway, 5-S-CyS-DA and DHBT-1, are able to evoke irreversible inhibition to mitochondrial complex I. This is significant because a deficiency in mitochondrial complex I respiration appears to be an important factor in SN cell death in PD (Schapira et al. 1990a,b). All other factors being equal, an elevated translocation of CySH into SN cells would be expected to generate higher yields of mitochondrial toxins (5-S-CyS-DA and DHBT-1) in those neurons that sustain the highest basal levels of DA autooxidation. This, therefore, might explain the higher vulnerability of the more heavily neuromelanin-pigmented SN cells in certain anatomic regions of the brain to degeneration in PD (Hirsch et al. 1988; Kastner et al. 1992).

The hypothesis advanced in this review might also provide a basis for understanding the processes that cause oxidative stress that contributes to the degeneration of SN neurons in PD (Dexter et al. 1989, 1994a; Jenner et al. 1992; Sanches-Ramos and Ames 1994). Thus, the CySH-mediated diversion of the neuromelanin pathway would be expected to lead to at least two metabolites, 5-S-CyS-DA and DHBT-1, that are irreversible inhibitors of mitochondrial complex I. Other complex I inhibitors, in addition to DHBT-1 (Li and Dryhurst 1997), are known to mediate the generation of ROS from the mitochondrial respiratory chain (Turrens and Bovaris 1980; Hasegawa et al. 1990; Cleeter et al. 1992).

A second consequence of the hypothesized release of glial GSH and upregulation of γ-GT would be elevated extracellular levels of glutamate (Glu) in addition to CySH. A significant elevation of Glu would be expected to evoke excitotoxic damage to nigral neurons mediated by its interaction with NMDA receptors (Olney and Ho 1970; Olney, Ho, and Rhee 1971; Choi 1988; Horowski et al. 1994), in part as a consequence of intraneuronal $O_2^{-\bullet}$ generation (Coyle and Puttfarcken 1993; Lafon-Cazal et al. 1993).

It may be of particular relevance that studies with cortical and hippocampal neuronal cultures have demonstrated that uptake of DA or other biogenic amine neurotransmitters followed by their intraneuronal oxidation by an unknown enzyme, possibly prostaglandin synthase (Hastings and Zigmond 1992; Zigmond and Hastings 1992; Hastings 1995), is a key step in mediating Glu oxidative neurotoxicity (Maher and Davis 1996). Intraneuronal oxidation of the biogenic amines has been proposed as the mechanism that generates cytotoxic levels of ROS in this process. Dopaminergic cell bodies in the SN contain relatively high levels of cytoplasmic DA and hence Glu oxidative excitotoxicity and ROS generation might be expected to be particularly severe in these neurons. Decreased mitochondrial complex I activity in dopaminergic cells, perhaps mediated by endotoxins formed as a result of the CySH-mediated diversion of the neuromelanin pathway, would be expected to cause reduced ATP production. Such a reduction in energy stores in SN cells might then lead to a reduced trans-membrane potential that relieves the voltage-sensitive block of NMDA receptors by Mg^{2+}. Under these conditions, the excitotoxic effects caused by both normal and elevated extracellular levels of Glu would be expected to become even more pronounced (Novielli et al. 1988; Beal et al. 1993). Thus, intraneuronal autoxidation of cysteinyldopamines and DHBTs, inhibition of mitochondrial complex I by 5-S-Cys-DA, DHBT-1, and perhaps other related metabolites, and Glu-mediated excitotoxicity potentiated by intraneuronal DA (which all generate ROS) would all be expected to result from the elevated release of GSH and upregulation of γ-GT in the SN, and therefore might account for the selective oxidative damage that occurs in this structure in PD. Furthermore, oxidative stress resulting from these processes might also be a trigger that initiates apoptosis in dopaminergic SN cells (Ratan, Murphy, and Baraban 1994a) in PD (Anglade et al. 1997; Charles et al. 1997). Interestingly, in culture, depletion of intraneuronal GSH appears to be a key factor in oxygen-radical-mediated apoptosis in cortical neurons (Ratan et al. 1994a,b). The hypothesis proposed in this review for the pathogenesis of PD suggests that a chemical or other insult to the SN cannot elevate intraneuronal levels of GSH in the cytoplasm of pigmented dopaminergic cells, because its biosynthetic precursor or the tripeptide itself is scavenged by DA-o-quinone. ·

Dopaminergic cell bodies in the SN constitute approximately 5 percent of the total nigral cell count. Thus, the relatively large decrease of complex I activity measured in the Parkinsonian SN (Schapira et al. 1990a,b, 1992) probably implies that glial complex I is also defective. If, indeed, an elevated translocation of CySH into dopaminergic cells is responsible for formation of mitochondrial endotoxins such as 5-S-CyS-DA and DHBT-1, then such toxins would be expected to leak out of dying or dead neurons and perhaps affect surrounding glia (Schapira et al. 1992). The very large decrease of GSH levels in the Parkinsonian SN might, in part, be related to irreversible inhibition of glial complex I, reduced ATP production, and resulting decreased biosynthesis of GSH.

Recently, Montine, Farris, and Graham (1995) have reported that the products of oxidation of DA and other catechols can covalently crosslink neurofilaments in vitro and that this might be the mechanism underlying Lewy-body formation in the Parkinsonian brain. It is also conceivable that the electrophilic o-quinones formed by the facile oxidation of cysteinyl dopamines and/or o-quinone imines formed by oxidation of DHBTs might also be capable of crosslinking neurofilaments and Lewy-body formation.

Very little is known about changes that occur in the LC in the Parkinsonian brain. However, it is conceivable that noradrenergic cell bodies in the LC are also vulnerable to degeneration in PD for exactly the same reasons as are dopaminergic cell bodies in the SN. Thus, LC cell bodies are pigmented with neuromelanin as a result of autoxidation of cytoplasmic NE, and there is apparently no effective BBB in this structure (Ross, Romrell, and Kaye 1995). Ongoing studies in this laboratory indicate that 5-S-cysteinylnorepinephrine and the DHBT formed by oxidative intramolecular cyclizations of this conjugate are mitochondrial complex I toxins in vitro.

PD is a neurodegenerative brain disorder that affects not only nigrostriatal dopaminergic neurons and noradrenergic neurons that project from the LC, but also serotonergic, cholinergic, and other pathways. Leakage of mitochondrial toxins from dying or dead pigmented dopaminergic and noradrenergic cell bodies might, therefore, contribute to the degeneration of proximate neurons. Additionally, elevated extracellular Glu and release of DA and NE from dying dopaminergic and noradrenergic cells might together mediate oxidative excitotoxicity that affects proximate neurons.

ACKNOWLEDGMENTS

The work carried out in the author's laboratory and described in this review was supported by the U.S. National Institutes of Health, Grant NS29886. Additional support was provided by the Vice President for Research and the Research Council at the University of Oklahoma. The author would also like to express his appreciation to Drs. Fa Zhang, Xue-Ming Shen, Hong Li, Jilin Han, Bing Xia, and Monika Z. Wrona, and to Dr. Fu-Chou Cheng and his colleagues at the Veterans General Hospital in Taichung, Taiwan, for their contributions to studies into the possible pathogenesis of PD.

REFERENCES

Agrup, G., Falck, B., Kennedy, B.-M., Rorsman, H., Rosengren, A.-M., and Rosengren, E. 1975. Formation of cysteinyldopa from glutathionyldopa in melanoma. *Acta Dermato-Venereologica (Stockholm)* 55:1–3.

Ambani, L. M., Van Woert, M. H., and Murphy, S. 1975. Brain peroxides and catalase in Parkinson's disease. *Archives of Neurology* 32:114–8.

Anglade, P., Vyas, S., Agid, Y., and Hirsch, E. C. 1997. Apoptosis in dopaminergic neurons of the human substantia nigra during normal aging with reference to Parkinson's disease. *Movement Disorders* 12(Suppl. I): P012.

Arendash, G. W., Olanow, C. W., and Sengstock, G. J. 1993. Intranigral iron infusion in rats: a progressive model for excess nigral iron levels in Parkinson's disease? In *Iron in central nervous system disorders*, ed. P. Riederer, and M. B. H. Youdim, 87–101. New York: Springer-Verlag.

Bannai, S. 1984. Transport of cysteine and cystine in mammalian cells. *Biochimica et Biophysica Acta* 779:289–306.

Barbeau, A. 1983. Etiology of Parkinson's disease: A research strategy. *Canadian Journal of Neurological Sciences* 11:24–8.

Barbeau, A. 1984. Manganese and extrapyramidal disorders. *Neurotoxicology* 5:13–36.

Barbeau, A., Roy, M., Paris, S., Cloutier, T., Plasse, L., and Poirier, J. 1985. Ecogenetics of Parkinson's disease: 4-hydroxylation of debrisoquine. *Lancet* 1213–6.

Beal, M. F., Hyman, B. T., and Koroshetz, W. 1993. Do defects in mitochondrial energy metabolism underlie the pathology of neurodegenerative diseases? *Trends in Neuroscience* 16:125–31.

Ben-Shachar, D., and Youdim, M. B. H. 1991. Intranigral iron injection induces behavioral and biochemical "Parkinsonism" in rats. *Journal of Neurochemistry* 57:2133–5.

Ben-Shachar, D., Riederer, P., and Youdim, M. B. H. 1991. Iron–melanin interaction and lipid peroxidation: implications for Parkinson's disease *Journal of Neurochemistry* 57:1609–14.

Bernheimer, H., Birkmayer, W., Hornykiewicz, O., Jellinger, K., and Seitelberger, F. 1973. Brain dopamine and the syndromes of Parkinson and Huntington. Clinical, morphological and neurochemical correlations. *Journal of the Neurological Sciences* 20:415–55.

Birkmayer, W., and Riederer, P. 1985. *Die Parkinson-Krankheit: Biochemie, Klinik, Therapie,* 2nd ed., 60–101. New York: Springer-Verlag.

Calabrese, V., and Fariello, R. G. 1988. Regional distribution of malonaldehyde in mouse brain. *Biochemical Pharmacology* 37:2287–8.

Carlsson, A. 1991. Role of brain monoamines and other neurotransmitters in normal and pathological aging. In *Basic Clinical and Therapeutic Aspects of Alzheimer's and Parkinson's Diseases*, Vol. 1, eds. T. Nagatsu, A. Fisher and M. Yoshida, 13–18, New York: Plenum Press.

Carstam, R., Brinck, C., Hindemith-Augustsson, A., Rorsman, H., and Rosengren, E. 1991. The neuromelanin of the human substantia nigra. *Biochimica et Biophysica Acta* 1097:152–60.

Cash, R., Raisman, R., Ploska, A., and Agid, Y. 1985. High and low affinity [3H]imipramine binding site in control and Parkinsonian brains. *American Journal of Pharmacology* 117:71–80.

Chandra, S. V., and Shukla, G. S. 1981. Concentration of striatal catecholamines in rats given manganese chloride through drinking water. *Journal of Neurochemistry* 36:683–7.

Charles, P. D., Kerr, L. D., Whetsell, W. O., Davis, T. L., Sakon, K., Scharar, S., and Robertson, D. 1997. Evidence for apoptopic cell death in Parkinson's disease. *Movement Disorders* 12 (Suppl. I):P014.

Cheng, F.-C., Kuo, J.-S., Chia, L.-C., and Dryhurst, G. 1996. Elevated 5-S-cysteinyldopamine/homovanillic acid ratio and reduced homovanillic acid in cerebrospinal fluid: possible markers for and potential insights into the pathoetiology of Parkinson's disease. *Journal of Neural Transmission* 103:433–46.

Choi, D. W. 1988. Glutamate neurotoxicity and diseases of the nervous system. *Neuron* 1:623–34.

Chui, H. C., Mortimer, J. A., Slager, U., Zarow, C., Bondareff, W., and Webster, D. D. 1986. Pathologic correlates of dementia in Parkinson's disease. *Archives of Neurology* 43:991–5.

Cleeter, M. W. J., Cooper, J. M., and Schapira, A. H. V. 1992. Irreversible inhibition of mitochondrial complex I by 1-methyl-4-phenylpyridinium: evidence for free radical involvement. *Journal of Neurochemistry* 58:786–9.

Coyle, J. T., and Puttfarcken, P. 1993. Oxidative stress, glutamate and neurodegenerative disorders. *Science* 262:689–95.

D'Amato, R. J., Lipman, Z. P., and Snyder, S. H. 1986. Selectivity of the parkinsonian neurotoxin MPTP: toxic metabolite MPP+ binds to neuromelanin. *Science* 231:987–9.

D'Amato, R. J., Zweig, R. M., Whitehouse, P. J., Wenk, G. L., Singer, H. S., Mayeux, R., Price, D. L., and Snyder, S. H. 1987a. Aminergic systems in Alzheimer's disease and Parkinson's disease. *Annals of Neurology* 22:229–36.

D'Amato, R. J., Alexander, G. M., Schwartzman, R. J., Kitt, C. A., Price, O. L., and Snyder, S. H. 1987b. Evidence for neuromelanin involvement in MPTP-induced neurotoxicity. *Nature* 327:324–6.

Dexter, D. T., Carter, C. J., Wells, F. R., Javoy-Agid, F., Lees, A. J., Jenner, P., and Marsden, C. D. 1989. Basal lipid peroxidation in substantia nigra is increased in Parkinson's disease. *Journal of Neurochemistry* 52:381–9.

Dexter, D. T., Carayon, A., Vidailhet, M., Ruberg, M., Agid, F., Agid, Y., Lees, A., Wells, F. R., and Marsden, C. D. 1990. Decreased ferritin levels in brain in Parkinson's disease. *Journal of Neurochemistry* 55:16–20.

Dexter, D. T., Sian, J., Jenner, P., and Marsden, C. D. 1993. Implications of alterations in trace element levels in brain in Parkinson's disease and other neurological disorders affecting the basal ganglia. *Advances in Neurology* 60:273–81.

Dexter, D. T., Holley, A. E., Flitter, W. D., Slater, T. F., Wells, F. R., Daniel, S. E., Lees, A. J., Jenner, P., and Marsden, C. D. 1994a. Increased levels of lipid hydroperoxides in the Parkinsonian substantia nigra: an HPLC and ESR study. *Movement Disorders* 9:92–7.

Dexter, D. T., Sian, J., Rose, S., Hindmarsh, G., Mann, V. M., Cooper, J. M., Wells, F. R., Daniel, S. E., Lees, A. J., Schapira, A. H. V., Jenner, P., and Marsden, C. D. 1994b. Indices of oxidative stress and mitochondrial function in individuals with incidental Lewy body disease. *Annals of Neurology* 35:38–44.

Dryhurst, G., Kadish, K. M., Scheller, F., and Renneberg, R. 1982. *Biological electrochemistry, I:* 116–79. New York: Academic Press.

Fariello, R. G., Calabrese, V., and Nappi, G. 1988a. Oxidative stress and energy transduction defects as causes of selective neuronal degeneration. In *Neurodegenerative disorders: the role played by endotoxins and xenobiotics*, ed. G. Nappi, 81–92. New York: Raven Press.

Fariello, R. G., Ghilardi, O., Peschechera, A., Ramucci, M. T., and Angelucci, T. 1988b. Regional distribution of ubiquinones and tocopherols in the mouse brain: lowest content of ubiquinols in the substantia nigra. *Neuropharmacology* 27:1077–80.

Fearnley, J. M., and Lees, A. J. 1991. Aging and Parkinson's disease: substantia nigra regional selectivity. *Brain* 114:2283–301.

Fearnley, J., and Lees, A. 1994. Pathology of Parkinson's disease. In *Neurodegenerative diseases*, ed. D. B. Calne, 545–554, Philadelphia: W.B. Saunders.

Fehling, C., Hansson, C., Poulsen, J., Rorsman, H., and Rosengren, E. 1981. Formation of glutathionyldopa in albino rats after DOPA injection. *Acta Dermato-Venereologica (Stockholm)* 61:339–42.

Ferraro, T. N., Golden, G. T., DeMattei, M., Hare, T. A., and Fariello, R. G. 1986. Effect of 1-methyl-4-phenyl-1,2,3,6-tetrahydropyridine (MPTP) on levels of glutathione in the extrapyramidal system of the mouse. *Neuropharmacology* 25:1071–4.

Forno, L. S. 1982. Pathology of Parkinson's disease. In *Movement disorders*, ed. C. D. Marsden, and S. Fahn, 2:25–40. London: Butterworth Scientific.

Forno, L. S. 1986. The Lewy body in Parkinson's disease. *Advances in Neurology* 45:35–42.

Fornstedt, B., and Carlsson, A. 1989. A marked rise in 5-*S*-cysteinyl-dopamine levels in guinea pig striatum following reserpine treatment. *Journal of Neural Transmission* 76:155–61.

Fornstedt, B., Rosengren, E., and Carlsson, A. 1986. Occurrence and distribution of 5-*S*-cysteinyl derivatives of dopamine, DOPA, and DOPAC in the brains of eight mammalian species. *Neuropharmacology* 25:451–4.

Fornstedt, B., Brun, A., Rosengren, E., and Carlsson, A. 1989. The apparent autoxidation rate of catechols in dopamine-rich regions of human brains increases with the degree of depigmentation of substantia nigra. *Journal of Neural Transmission: Parkinson's Disease and Dementia Section* 1:279–95.

Fornstedt, B., Pilebad, E., and Carlsson, A. 1990a. In vivo autoxidation of dopamine in guinea pig striatum increases with age. *Journal of Neurochemistry* 55:655–9.

Fornstedt, B., Bergh, I., Rosengren, E., and Carlsson, A. 1990b. An improved HPLC–electrochemical detection method for measuring brain levels of 5-*S*-cysteinyldopamine, 5-*S*-cysteinyl-3,4-dihydroxy-phenylalanine, and 5-*S*-cysteinyl-3,4-dihydroxyphenylacetic acid. *Journal of Neurochemistry* 54:578–86.

Gaspar, P., Duyckaerts, C., Alvarez, C., Javoy-Agid, F., and Berger, B. 1991. Alteration of dopaminergic and noradrenergic innervations in motor cortex in Parkinson's disease. *Annals of Neurology* 30:365–74.

Gerlach, M., and Riederer, P. 1993. The pathophysiological basis of Parkinson's disease. In *Inhibitors of Monoamine Oxidase B: Pharmacology and Clinical use in Neurodegenerative Diseases*, ed. I. Szelenyi, 25–50, Basel: Birkhauser, Verlag.

Gerlach, M., and Riederer, P. 1996. Animal models of Parkinson's disease: an empirical comparison with the phenomenology of the disease in man. *Journal of Neural Transmission* 103:987–1041.

Gerlach, M., Riederer, P., Przuntek, H., and Youdim, M. B. H. 1991. MPTP mechanisms of neurotoxicity and their implications for Parkinson's disease. *European Journal of Pharmacology* 208:273–86.

Gerlach, M., Jellinger, K., and Riederer, P. 1994. The possible role of noradrenergic deficits in selected signs of Parkinson's disease. In *Noradrenergic Mechanisms in Parkinson's Disease*, eds. M. Briley and M. Marien, 59–71. Boca Raton: CRC Press.

Gibb, W. R. G. 1989. The diagnostic relevance of Lewy bodies and other inclusions in Parkinson's disease. In *Early diagnosis and preventive therapy in Parkinson's disease*, ed. H. Przuntek, and P. Riederer, 171–180. New York: Springer-Verlag.

Gibb, W. R. G., and Lees, A. J. 1988. The relevance of the Lewy body to the pathogenesis of idiopathic Parkinson's disease. *Journal of Neurology, Neurosurgery, and Psychiatry* 51:757.

Gibb, W. R. G., Fearnley, J. M., and Lees, A. J. 1990. The anatomy and pigmentation of the human substantia nigra in relation to selective neuronal vulnerability. *Advances in Neurology* 53:31–4.

Golbe, L. I., Iorio, G., Bonavita, V., Miller, D. C., and Duvoisin, R. C. 1990. A large kindred with autosomal dominant Parkinson's disease. *Annals of Neurology* 27:276–82.

Graham, D. G. 1978. Oxidation pathways for catecholamines in the genesis of neuromelanin and cytotoxic quinones. *Molecular Pharmacology* 14:633–43.

Graham, D. G., Tiffany, S. M., Bell, W. R., and Gutnecht, W. F. 1978. Autoxidation versus covalent binding of quinones as the mechanism of toxicity of dopamine, 6-hydroxydopamine and related compounds towards C1300 neuroblastoma cells in vitro. *Molecular Pharmacology* 14:644–53.

Halliwell, B., and Gutteridge, J. M. C. 1987. *Free radicals in biology and medicine*. Oxford: Clarendon Press.

Harris, Z. L., Takahashi, Y., Miyajima, H., Serizawa, M., MacGillivray, R. T. A., and Gitlin, J. D. 1995. Aceroplasminemia: molecular characterization of this disorder of iron metabolism. *Proceedings of the National Academy of Sciences of the U.S.A.* 92:2539–43.

Hasegawa, E., Takeshige, K., Oishi, T., Murai, Y., and Minikami, S. 1990. 1-Methyl-4-phenylpyridinium (MPP+) induces NADH-dependent superoxide formation, and enhances NADH-dependent peroxidation in bovine heart submitochondrial particles. *Biochemical and Biophysical Research Communications* 170:1049–55.

Hastings, T. G. 1995. Enzymatic oxidation of dopamine: the role of prostaglandin H synthase. *Journal of Neurochemistry* 64:919–24.

Hastings, T. G., and Zigmond, M. J. 1992. Prostaglandin synthase-catalyzed oxidation of dopamine. *Society for Neuroscience Abstracts* 18:1444.

Hedera, P., and Whitehouse, P. J. 1994. Neurotransmitters in neurodegeneration. In *Neurodegenerative diseases*, ed. D. B. Calne, 97–117. Philadelphia: W.B. Saunders.

Hirsch, E. C., Graybiel, A. M., and Agid, Y. 1988. Melanized dopaminergic neurons are differentially susceptible to degeneration in Parkinson's disease. *Nature* 334:345–8.

Hirsch, E. C., Brandel, J.-P., Galle, P., Javoy-Agid, F., and Agid, Y. 1991. Iron and aluminum increase in the substantia nigra of patients with Parkinson's disease: an x-ray microanalysis. *Journal of Neurochemistry* 56:446–51.

Hornykiewicz, O. 1988. Neurochemical pathology and etiology of Parkinson's disease: basic facts and hypothetical possibilities. *Mount Sinai Journal of Medicine* 55:11–20.

Hornykiewicz, O. 1989. Aging and neurotoxins as caustive factors in idiopathic Parkinson's disease— a critical analysis of neurochemical evidence. *Progress in Neuro-Psychopharmacology, Biology, and Psychiatry* 13:319–28.

Hornykiewicz, O., and Kish, S. J. 1986. Biochemical pathophysiology of Parkinson's disease. *Advances in Neurology* 45:19–34.

Horowski, R., Wachtel, H., Turski, L., and Löschmann, P.-A. 1994. Glutamate excitotoxicity as a possible pathogenetic mechanism in chronic neurodegeneration. In *Neurodegenerative diseases*, ed. D. B. Calne, 163–75. Philadelphia: W.B. Saunders.

Ikeda, H., Markey, C. J., and Markey, S. P. 1992. Search for neurotoxins structurally related to 1-methyl-4-phenylpyridine (MPP+) in the pathogenesis of Parkinson's disease. *Brain Research* 575:285–98.

Jellinger, K. 1988. Pathology of Parkinson's syndrome. In *Handbook of Experimental Pharmacology*, ed. D. B. Calne, 88:47–112. Berlin: Springer-Verlag.

Jellinger, K. 1990. New developments in the pathology of Parkinson's disease. *Advances in Neurology* 53:1–16.

Jellinger, K. A. 1991. Pathology of Parkinson's disease. Changes other than the nigrostriatal pathway. *Molecular and Chemical Neuropathology* 14:153–97.

Jellinger, K., Kienzl, E., Rumpelmair, G., Riederer, P., Stachelberger, H., Ben-Schachar, D., and Youdim, M. B. H. 1992. Iron–melanin complex in substantia nigra of Parkinsonian brains: an x-ray microanalysis. *Journal of Neurochemistry* 59:1168–71.

Jenner, P., Dexter, D. T., Sian, J. Schapira, A. H. V., and Marsden, D. C. 1992. Oxidative stress as a cause of nigral cell death in Parkinson's disease and incidental Lewy body disease. *Annals of Neurology* 32 (Suppl.) S82–7.

Johnson, W. G., Hodge, S. E., and Duvoisin, R. 1990. Twin studies and the genetics of Parkinson's disease—a reappraisal. *Movement Disorders* 5:187–94.

Kastner, A., Hirsch, E. C., Lejeune, O., Javoy-Agid, F., Rascol, O., and Agid, Y. 1992. Is the vulnerability of neurons in the substantia nigra of patients with Parkinson's disease related to their neuromelanin content? *Journal of Neurochemistry* 59:1080–9.

Kinemuchi, H., Fowler, C. J., and Tipton, K. F. 1987. The neurotoxicity of 1-methyl-4-phenyl-1,2,3,6-tetrahydropyridine (MPTP) and its relevance to Parkinson's disease. *Neurochemistry International* 11:359–73.

Kish, S. J., Morito, C. H., and Hornykiewicz, O. 1985. Glutathione peroxidase activity in Parkinson's disease brain. *Neuroscience Letters* 58:343–6.

Kish, S. J., Rajput, A., Gilbert, J., Rozdilsky, B., Chang, L.-C., Shannak, K., and Hornykiewicz, O. 1986. GABA is elevated in striatal but not extrastriatal regions in Parkinson's disease: Correlations with striatal dopamine loss. *Annals of Neurology* 20:26–31.

Lafon-Cazal, M., Pietri, S., Calcusi, M., and Bockaert, J. 1993. NMDA-dependent superoxide production and neurotoxicity. *Nature* 364:535–7.

Li, H., and Dryhurst, G. 1997. Irreversible inhibition of mitochondrial complex I by 7-(2-aminoethyl)-3,4-dihydro-5-hydroxy-2H-1,4-benzothiazine-3-carboxylic acid (DHBT-1): a putative nigral endotoxin of relevance to Parkinson's disease. *Journal of Neurochemistry* 69:1530–41.

Liccione, J. J., and Maines, M. D. 1988. Selective vulnerability of glutathione metabolism and cellular defense mechanisms in rat striatum to manganese. *Journal of Pharmacology and Experimental Therapeutics* 247:156–61.

Liu, R.-M., Hu, H., Robinson, T. W., and Forman, H. J. 1996. Increased γ-glutamylcysteine synthetase and γ-glutamyl transpeptidase activities enhance resistance of rat lung epithelial L2 cells to quinone toxicity. *American Journal of Respiratory Cell and Molecular Biology* 14:192–7.

Lloyd, K. G., and Hornykiewicz, O. 1973. L-Glutamic acid decarboxylase in Parkinson's disease: effect of L-DOPA therapy. *Nature* 243:521–3.

Lowe, J. 1994. Lewy bodies. In *Neurodegenerative diseases*, ed. D. B. Calne, 51–69. Philadelphia: W.B. Saunders.

Maher, P., and Davis, J. B. 1996. The role of monoamine metabolism in oxidative glutamate toxicity. *Journal of Neuroscience* 16:6394–401.

Mann, D. M. A., and Yates, P. O. 1983. Possible role of neuromelanin in the pathogenesis of Parkinson's disease. *Mechanisms of Ageing and Development* 21:193–203.

Marstein, S., Jellum, E., Nesbakken, R., and Perry, T. L. 1981. Biochemical investigation of biopsied brain tissue and autopsied organs from patients with pyroglutamic acidemia (5-oxoprolinemia). *Clinica and Chimica Acta* 111:219–28.

Marttila, R. J., Lorentz, H., and Rinne, U. K. 1988. Oxygen toxicity protecting enzymes in Parkinson's disease. Increase of superoxide-dismutase-like activity in the substantia nigra and basal nucleus. *Journal of Neurolical Science* 86:321–331.

McGeer, P. L., McGeer, E. G., and Suzuki, J. S. 1976. Aging and extrapyramidal function. *Archives of Neurology* 34:33–5.

Meister, A. 1988. Glutathione metabolism and its selective modification. *Journal of Biological Chemistry* 263:17205–8.

Meister, A., and Anderson, M. E. 1983. Glutathione. *Annual Review of Biochemistry* 52:711–60.

Mithöfer, K., Sandy, M. S., Smith, M. T., and Di Monte, D. 1992. Mitochondrial poisons cause depletion of reduced glutathione in isolated hepatocytes. *Archives of Biochemistry and Biophysics* 295:132–6.

Mizuno, Y., Ohta, S., Tanaka, M., Takamiya, S., Suzuki, K., Sata, T., Oya, H., Ozawa, T., and Kagawa, Y. 1989. Deficiencies in complex I subunits of the respiratory chain in Parkinson's disease. *Biochemical and Biophysical Research Communications* 163:1450–5.

Montine, T. J., Farris, D. B., and Graham, D. G. 1995. Covalent crosslinking of neurofilament proteins by oxidized catechols as a potential mechanism of Lewy body formation. *Journal of Neuropathology and Experimental Neurology* 54:311–9.

Nicklas, W. J., Vyas, I., and Heikkila, R. E. 1985. Inhibition of NADH-linked oxidation in brain mitochondria by 1-methyl-4-phenylpyridine, a metabolite of the neurotoxin 1-methyl-4-phenyl-1,2,3,6-tetrahydropyridine. *Life Sciences* 36:2503–8.

Novielli, A., Reilly, J. A., Lysko, P. G., and Henneberry, R. C. 1988. Glutamate becomes neurotoxic when intracellular energy levels are reduced. *Brain Research* 451:205–12.

Odh, G., Carstam, R., Paulson, J., Wittbjer, A., Rosengren, E., and Rorsman, H. 1994. Neuromelanin of the human substantia nigra: a mixed-type melanin. *Journal of Neurochemistry* 62:2030–6.

Olanow, C. W. 1990. Oxidation reactions in Parkinson's disease. *Neurology* 40:32–7.

Olney, J. W., and Ho, O. L. 1970. Brain damage to infant mice following oral intake of glutamate, aspartate or cysteine. *Nature* 227:609–10.

Olney, J. W., Ho, O. L., and Rhee, V. 1971. Cytotoxic effects of acidic and sulfur-containing amino acids on the infant mouse central nervous system. *Experimental Brain Research* 14:61–76.

Parenti, M., Flauto, C., Parati, E., Vescovi, A., and Groppetti, A. 1986. Manganese neurotoxicity: effects of L-DOPA and pargyline treatments. *Brain Research* 367:8–13.

Parenti, M., Rusconi, L., Cappabianca, V., Parati, E. A., and Groppetti, A. 1988. Role of dopamine in manganese neurotoxicity. *Brain Research* 473:236–40.

Parkinson, J. D. 1817. *The shaking palsy.* London: Sherwood, Neely and Jones.

Perry, T. L., Godin, D. A., and Hansen, S. 1982. Parkinson's disease: a disorder due to nigral glutathione deficiency? *Neuroscience Letters* 33:305–10.

Perry, T. L., Javoy-Agid, F., Agid, Y., and Fibiger, H. C. 1983. Striatal GABAergic neuronal activity is not reduced in Parkinson's disease. *Journal of Neurochemistry* 40:1120–3.

Perry, T. L., Yong, V. W., Jones, K., and Wright, J. M. 1986. Manipulation of glutathione content fails to alter dopaminergic nigrostriatal neurotoxicity of 1-methyl-4-phenyl-1,2,3,6-tetrahydropyridine (MPTP) in the mouse. *Neuroscience Letters* 70:261–5.

Philbert, M. A., Beiswanger, C. M., Waters, D. K., Reuhl, K. R., and Lowndes, H. E. 1990. Cellular and regional distribution of reduced glutathione in the nervous system of the rat: histochemical localization by mercury orange and o-phthalaldehyde-induced histofluorescence. *Toxicology and Applied Pharmacology* 107:215–27.

Pilas, B., Sarna, T., Kalyanarman, B., and Swatz, H. M. 1988. The effect of melanin on iron-associated decomposition of hydrogen peroxide. *Free Radicals in Biology and Medicine* 4:285–93.

Pryor, W. 1986. Oxy-radicals and related species: their formation, lifetimes, and reactions. *Annual Review of Physiology* 48:657–67.

Raps, S. P., Lai, J. C. K., Hertz, L., and Cooper, A. J. L. 1989. Glutathione is present in high concentrations in cultured astrocytes but not in cultured neurons. *Brain Research* 493:398–401.

Ratan, R. R., Murphy, T. H., and Baraban, J. M. 1994a. Oxidative stress induces apoptosis in embryonic cortical neurons. *Journal of Neurochemistry* 62:376–9.

Ratan, R. R., Murphy, T. H., and Baraban, J. M. 1994b. Macromolecular synthesis inhibitors prevent oxidative stress-induced apoptosis in embryonic cortical neurons by shunting cysteine from protein synthesis to gluathione. *Journal of Neuroscience* 14:4385–92.

Riederer, P., Sofic, E., Rausch, W.-D., Schmidt, B., Reynolds, G. P., Jellinger, K., and Youdim, M. B. H. 1989. Transition metals, ferritin, glutathione, and ascorbic acid in Parkinsonian brains. *Journal of Neurochemistry* 52:515–20.

Rodgers, A. D., and Curzon, G. 1975. Melanin formation by human brain in vitro. *Journal of Neurochemistry* 24:1123–9.

Rosengren, E., Linder-Eliasson, E., and Carlsson, A. 1985. Detection of 5-S-cysteinyldopamine in human brain. *Journal of Neural Transmission* 63:247–53.

Ross, H. R., Romrell, L. J., and Kaye, G. I. 1995. *Histology. A text and atlas.* 11:282. Baltimore: Williams and Wilkins.

Sagara, J.-I., Miura, K., and Bannai, S. 1993. Maintenance of neuronal glutathione by glial cells. *Journal of Neurochemistry* 61:1672–6.

Sagara, J.-I., Makino, N., and Bannai, S. 1996. Glutathione efflux from cultured astrocytes. *Journal of Neurochemistry* 66:1876–81.

Saggu, H., Cooksey, J., Dexter, D., Wells, F. R., Lees, A. J., Jenner, P., and Marsden, C. D. 1989. A selective increase in particulate superoxide dismutase activity in Parkinsonian substantia nigra. *Journal of Neurochemistry* 53:692–7.

Salazar, M., Sokoloski, T. D., and Patil, P. N. 1978. Binding of dopaminergic drugs by the neuromelanin of the substantia nigra, synthetic melanins, and melanin grannules. *Federation Proceedings* 37:2403–7.

Sanches-Ramos, A., and Ames, B. N. 1994. A marker of oxyradical-mediated DNA damage (8-hydroxy-2-deoxyguanosine) is increased in the nigrostriatum of Parkinson's disease brain. *Neurodegeneration* 3:197–204.

Sayre, L. M. 1989. Biochemical mechanism of action of the dopaminergic neurotoxin 1-methyl-4-phenyl-1,2,3,6-tetrahydropyridine (MPTP). *Toxicology Letters* 48:121–49.

Scatton, B., Rouquier, F., Javoy-Agid, F., and Agid, Y. 1982. Dopamine deficiency in cerebral cortex in Parkinson's disease. *Neurology* 32:1039–40.

Scatton, B., Javoy-Agid, F., Rouquier, L., Dubois, B., and Agid, Y. 1983. Reduction of cortical dopamine, noradrenaline, serotonin and their metabolites in Parkinson's disease. *Brain Research* 275:321–8.

Schapira, A. H. V., Cooper, J. M., Dexter, D., Clark, J. B., Jenner, P., and Marsden, C. D. 1990a. Mitochondrial complex I deficiency in Parkinson's disease. *Journal of Neurochemi* 54:823–7.

Schapira, A. H. V., Mann, D. M., Cooper, J. M., Dexter, D., Daniel, S. E., Jenner, P., Clark, J. B., and Marsden, C. D. 1990b. Anatomic and disease specificity of NADH-CoQ₁ reductase (complex I) deficiency in Parkinson's disease. *Journal of Neurochemistry* 55:2142–5.

Schapira, A. H. V., Mann, V. M., Cooper, J. M., Krige, D., Jenner, P. J., and Marsden, C. D. 1992. Mitochondrial function and Parkinson's disease. *Annals of Neurology* 32:S116–24.

Scheulen, M., Wollenbert, P., Kappus, H., and Remmer, H. 1975. Irreversible binding of DOPA and dopamine metabolites to protein by rat liver microsomes. *Biochemical and Biophysical Research Communications* 66:1396–400.

Schoenberg, B. S. 1986. Descriptive epidemiology of Parkinson's diseases: distribution and hypothesis formulation. *Advances in Neurology* 45:277–83.

Sengstock, G. J., Olanow, C. W., Dunn, A. J., and Arandesh, G. W. 1992. Iron induces degeneration of nigrostriatal neurons. *Brain Research Bulletin* 28:645–9.

Sengstock, G. J., Olanow, C. W., Menzies, R. A., Dunn, A. J., and Arendash, G. W. 1993. Infusion of iron into the rat substantia nigra: nigral pathology and dose-dependent loss of striatal dopaminergic markers. *Journal of Neuroscience Research* 35:76–82.

Sengstock, G. J., Olanow, C. W., Dunn, A. J., Barone, S., and Arendash, G. W. 1994. Progressive changes in striatal dopaminergic markers, nigral volume, and rotational behavior following iron infusion into rat substantia nigra. *Experimental Neurology* 130:82–94.

Shen, X.-M., and Dryhurst, G. 1996. Further insights into the influence of L-cysteine on the oxidation chemistry of dopamine: reaction pathways of potential relevance to Parkinson's disease. *Chemical Research Toxicology* 9:751–63.

Shen, X.-M., Xia, B., Wrona, M. Z., and Dryhurst, G. 1996. Synthesis, redox properties, in vivo formation and neurobehavioral effects of N-acetylcysteinyl conjugates of dopamine. Possible metabolites of relevance to Parkinson's disease. *Chemical Research in Toxicology* 9:1117–26.

Shen, X.-M., Zhang, F., and Dryhurst, G. 1997. Oxidation of dopamine in the presence of cysteine: characterization of new toxic products. *Chemical Research in Toxicology* 10:147–55.

Sian, J., Dexter, D., Lees, A., Daniel, S., Agid, Y., Javoy-Agid, F., Jenner, P., and Marsden, C. D. 1994a. Alterations in glutathione levels in Parkinson's disease and other neurodegenerative disorders of the basal ganglia. *Annals of Neurology* 36:348–55.

Sian, J., Dexter, D. T., Lees, A. J., Daniel, S., Jenner, P., and Marsden, C. D. 1994b. Glutathione-related enzymes in brain in Parkinson's disease. *Annals of Neurology* 36:356–61.

Simonian, N. A., and Coyle, J. T. 1996. Oxidative stress in neurodegenerative diseases. *Annual Review of Pharmacology and Toxicology* 36:83–106.

Skellerud, K., Marstein, S., Schrader, H., Brundlet, P. J., and Jellum, E. 1980. The cerebral lesions in a patient with generalized glutathione deficiency and pyroglutamic aciduria(5-oxoprolinuria). *Acta Neuropathologica* 52:235–8.

Slivka, A., Mytilineou, C., and Cohen, G. 1987. Histochemical evaluation of glutathione in brain. *Brain Research* 409:275–84.

Sofic, E., Riederer, P., Heisen, H., Bechmann, H., Reynolds, G. P., Habenstreit, G., and Youdim, M. B. H. 1988. Increased iron(III) and total iron content in postmortem substantia nigra in Parkinsonian brains. *Journal of Neural Transmission* 74:199–205.

Sofic, E., Lange, K. W., Jellinger, K., and Riederer, P. 1992. Reduced and oxidized glutathione in the substantia nigra of patients with Parkinson's disease. *Neuroscience Letters* 142:128–30.

Steventon, G. B., Heafield, M. T. E., Waring, R. H., and Williams, A. C. 1989. Xenobiotic metabolism in Parkinson's disease. *Neurology* 39:883–7.

Tagliavini, F., Pilleri, G., Bouras, C., and Constantinidis, J. 1984. The basal nucleus of Meynert in idiopathic Parkinson's disease. *Acta Neurologica Scandinavica* 69:20–8.

Tanner, C. M., Chen, B., Wang, W., Peng, M., Liu, M., Liang, X., Kao, L.-C., Gilley, D. W., Goetz, C. G., and Schoenberg, B. S. 1989. Environmental factors and Parkinson's disease. *Neurology* 39:660–4.

Tipton, K., and Singer, T. P. 1993. Advances in our understanding of the mechanisms of the neurotoxicity of MPTP and related compounds. *Journal of Neurochemistry* 61:1191–206.

Torack, R. M., and Morris, J. C. 1988. The association of ventral tegmental area histopathology with adult dementia. *Archives of Neurology* 45:497–501.

Tse, D. C. S., McCreery, R. L., and Adams, R. N. 1976. Potential oxidative pathways of brain catecholamines. *Journal of Medical Chemistry* 19:37–40.

Turrens, J. F., and Bovaris, A. 1980. Generation of superoxide anion by the NADH dehydrogenase of bovine heart mitochondria. *Biochemical Journal* 191:421–7.

Uhl, G. R., Hedreen, J. C., and Price, D. L. 1985. Parkinson's disease: loss of neurons from the ventral tegmental area contralateral to therapeutic surgical lesions. *Neurology* 35:1215–18.

Van Woert, M. H., and Ambani, L. M. 1974. Biochemistry of neuromelanin. *Advances of Neurology* 5:215–23.

Wakamatsu, K., Ito, S., and Nagatsu, T. 1991. Cysteinyldopamine is not incorporated into neuromelanin. *Neuroscience Letters* 131:57–60.

Waring, R. H., Sturman, S. G., Smith, M. C. G., Steventon, G. B., Heafield, M. T. E., and Williams, A. C. 1989. S-Methylation in motoneuron disease and Parkinson's disease. *Lancet* 356–7.

Wesemann, W., Blaschke, S., Clement, H.-W., Grote, C. H. R., Weiner, N., Kolasiewicz, W., and Sontag, K. H. 1993. Iron and neurotoxic intoxication: comparative in vitro and in vivo studies. In *Iron in central nervous system disorders.*, ed. P. Riederer, and M. B. H. Youdim, 79–86. New York: Springer-Verlag.

Wesemann, W., Blaschke, S., Solbach, M., Grote, C., Clement, H.-W., and Riederer, P. 1994. Intranigral injected iron progressively reduces striatal dopamine metabolism. *Journal of Neural Transmission (P-D Section)* 8:209–14.

Whitehouse, P. J., Hedreen, J. C., White, C. L., and Price, D. L. 1983. Basal forebrain in the dementia of Parkinson's disease. *Annals of Neurology* 13:243–8.

Wick, M. M. 1978. Dopamine: a novel antitumor agent active against B-16 melanoma in vivo. *Journal of Investigative Dermatology* 71:163–4.

Yong, V. W., Perry, T. L., and Krisman, A. A. 1985. Depletion of glutathione in brainstem of mice caused by

N-methyl-4-phenyl-1,2,3,6-tetrahydropyridine is prevented by antioxidant pretreatment. *Neuroscience Letters* 63:56–60.

Youdim, M. B. H., Ben-Shachar, D., and Riederer, P. 1989. Is Parkinson's disease a progressive siderosis of substantia nigra resulting in iron and melanin induced neurodegenerations. *Acta Neurologica Scandinavica* 126:47–55.

Young, T. E., and Babbitt, B. W. 1983. Electrochemical study of the oxidation of α-methyldopamine, α-methylnoradrenaline and dopamine. *Journal of Organic Chemistry* 48:562–6.

Zhang, F., and Dryhurst, G. 1993. Oxidation chemistry of dopamine: possible insights into the age-dependent loss of dopaminergic nigrostriatal neurons. *Bioorganic Chemistry* 21:392–410.

Zhang, F., and Dryhurst, G. 1994. Effects of L-cysteine on the oxidation chemistry of dopamine: reaction pathways of potential relevance to idiopathic Parkinson's disease. *Journal of Medical Chemistry* 37:1084–98.

Zhang, F., and Dryhurst, G. 1995a. Influence of glutathione on the oxidation chemistry of the catecholaminergic neurotransmitter dopamine. *Journal of Electroanalytical Chemistry* 398:117–28.

Zhang, F., and Dryhurst, G. 1995b. Reactions of cysteine and cysteine derivatives with dopamine-o-quinone and further insights into the oxidation chemistry of 5-S-cysteinyldopamine: potential relevance to idiopathic Parkinson's disease. *Bioorganic Chemistry* 23:193–216.

Zigmond, M. J., and Hastings, T. G. 1992. A method for measuring dopamine-protein conjugates as an index of protein oxidation. *Society for Neuroscience Abstracts* 18:1443.

Glutathione in the Nervous System
Edited by Christopher A. Shaw
Copyright © 1998 Taylor & Francis

16

Free-Radical Toxicity in Amyotrophic Lateral Sclerosis

Merit E. Cudkowicz and Robert H. Brown

Day Neuromuscular Research Center and Neurology Service, Massachusetts General Hospital, Charleston, Massachusetts 02109

Richard A. Smith

The Center for Neurologic Study, La Jolla, California 92037

1. INTRODUCTION

Amyotrophic lateral sclerosis (ALS) is a degenerative disease of motor neurons. The median age of onset is 55, and the median survival is 3 to 5 years; the disease is lethal. The majority of ALS cases are sporadic (SALS). However, 10–15% of cases are inherited as an autosomal dominant trait (FALS) (Mulder et al. 1986). The cause of SALS is unknown. In approximately 20% of cases of FALS, the primary defects are mutations in the gene encoding cytosolic, copper–zinc superoxide dismutase (SOD1) (Rosen et al. 1993). The detection of mutations in SOD1 in patients with FALS suggests that motor-neuron death in this disease is a consequence of perturbations of free-radical homeostasis and resulting oxidative toxicity to motor neurons. A corollary implication is that, because it resembles FALS both clinically and pathologically (Munsat 1984), SALS may also be a free-radical disease. There is now in vitro and in vivo evidence of increased free-radical production in ALS. These insights may be critical in guiding the development of treatment strategies for ALS.

The course of ALS is progressive and more or less linear, although there is variation among patients. For example, patients with mutations in codon 4 of the SOD1 gene (alanine for valine) show a uniformly aggressive course, with mean survival of about one year. For most patients with ALS (both familial and sporadic) the factors determining the rate of disease progression are not understood. It is conceivable that one such factor modifying the course of the disease is an individual's repertoire of oxidative defenses.

2. FAMILIAL AMYOTROPHIC LATERAL SCLEROSIS

Currently more than fifty different mutations in SOD1 have been reported in FALS. SOD1 is a metalloenzyme of about 153 amino acids that is expressed in all eukaryotic cells. The primary function of SOD1 is to convert superoxide anion to hydrogen peroxide (H_2O_2) (Halliwell et al. 1995; Huber et al. 1977). Hydrogen peroxide is then detoxified by cytosolic glutathione peroxidase (GSHPx) or peroxisomal catalase to form water. Superoxide is potentially toxic by itself. Moreover, through formation of H_2O_2, superoxide anion can lead to the production of highly reactive hydroxyl radicals (OH•). These are generated via interaction with reduced transition metals such as Fe^{2+} or Cu^{1+} (Fenton chemistry) or via interaction with Fe^{3+} (Haber–Weiss reaction). Superoxide anion may also combine nonenzymatically with nitric oxide to generate peroxynitrite (ONOO–). This reaction is extremely rapid and diffusion-limited (Beckman et al. 1990). In turn, ONOO– may produce reactive species. If protonated, for example, ONOO– may produce OH• (Kooy et al. 1994).

The mechanisms whereby the FALS-associated mutations in SOD1 trigger ALS are not well understood (Robberecht et al. 1994). Although the mutations are commonly associated with reduced SOD1 activity (Robberecht et al. 1994; Bowling et al. 1993, 1995), resulting, in part, from a reduction in the stability of the mutant protein (Borchelt et al. 1994), several observations strongly suggest that the critical defect in the mutant SOD1 molecule is *gain of a toxic function*. These include (1) an autosomal dominant inheritance pattern in FALS, (2) that some of the mutant SOD1 molecules (e.g., A90D) have normal or even supranormal (G37R) dismutase activity (Borchelt et al. 1994), (3) that some of the mutant molecules can fully rescue yeast from the phenotype of loss of wild-type SOD1 function (Nishida, Gralla, and Valentine 1994), and (4) that mice expressing mutant SOD1 molecules die in early "adulthood" (3 to 4 months of age) from a motor-neuron degeneration that reproduces many aspects of human ALS (Gurney et al. 1994, 1996).

The mechanisms of cytotoxicity of the mutant SOD1 molecule are unclear. With rare exceptions the mutations occur outside of the active site of the enzyme and appear to alter the folding and stability of the protein (Deng et al. 1993; Lyons et al. 1996). The half-life of the mutant protein is diminished (Borchelt et al. 1994), and the mutant protein has altered metal binding (Nishida et al. 1994; Carri et al. 1994). The mutations may allow the active-site copper to become more accessible to (1) hydrogen peroxide, leading to increased hydroxyl radical formation, or (2) peroxynitrite, leading to nitration of critical tyrosine residues in proteins essential to neuronal viability (e.g., heavy subunit of neurofilament or tyrosine kinase receptors) (Beckman et al. 1990, 1993; Ischiropoulos et al. 1992).

3. IN VITRO EVIDENCE OF FREE-RADICAL TOXICITY IN ALS

In support of the peroxidation hypothesis, two groups have recently shown that four different mutant SOD1 proteins generate hydroxyl radicals more readily than the wild-type molecule (Wiedau-Pazos et al. 1996; Yim et al. 1996). Wiedau-Pazos and colleagues found that two of the mutant SOD1 enzymes associated with FALS generated increased hydroxyl radicals, as measured by the formation of hydroxy adducts of a model substrate (spin trap 5,5'-dimethyl-1-pyrroline N-oxide—DMPO). Chelation of the copper ion (with diethyldithiocarbamate and penicillamine) reduces formation of these adducts. Recently, Roos and colleagues delivered and expressed human wild-type or mutant SOD1 genes into primary neurons and PC12 cells using recombinant adenoviruses. Expression of the mutant SOD1 protein induced apoptotic cell death. The cells were rescued from cell death by copper chelators, SOD mimics, enhancers of catalase, and agents known to have antiapoptotic effect. Microfluorimetry confirmed increased generation of radical species in cells infected with the mutant SOD1 cDNA (Roos et al. 1997).

Further support for the role of free-radical toxicity in ALS comes from studies of cerebrospinal fluid (CSF) in ALS. The CSF in ALS patients is toxic for neurons in culture (Couratier et al. 1993). This toxicity is blocked by antioxidant drugs, including vitamin E and the xanthine oxidase inhibitor allopurinol (Terro et al. 1996).

4. IN VIVO EVIDENCE OF FREE-RADICAL TOXICITY IN ALS

Reported in vivo evidence of oxidative stress in ALS includes alterations in (1) protective antioxidant enzymes and (2) markers of oxidative damage to protein, DNA, and lipids. Alterations in antioxidant enzymes have been found in patients with SALS (Przedborski et al. 1996; Ince et al. 1994). Glutathione peroxidase (GSHPx) is the major protective enzyme against hydrogen peroxide toxicity. A reduction in GSHPx activity may cause cytotoxicity by increasing levels of hydrogen peroxide (Michiels et al. 1994). Przedborski and colleagues measured SOD1, catalase, and GSHPx activities in post mortem brain from nine SALS and nine control subjects. Glutathione peroxidase activity was significantly reduced, by approximately 40%, in the precentral gyrus but not the cerebellar cortex of subjects with SALS (Przedborski et al. 1996). Superoxide dismutase and catalase activity were unchanged in both brain regions. Glutathione peroxidase activity in the precentral gyrus correlated positively with disease duration, suggesting that reduction in GSHPx activity might influence the rate of disease progression (Przedborski et al. 1996). However, other investigators have reported that GSHPx activity is normal or even elevated in the spinal cord of ALS patients (Ince et al. 1994; Fujita et al. 1996). Ince and colleagues measured GSHPx activity in spinal-cord tissue from 10 SALS subjects and 16 controls. Glutathione peroxidase activity was increased by 26% in the SALS group (Ince et al. 1994). Fujita and colleagues measured SOD and GSHPx activities in the spinal cords of five SALS subjects and nine normal controls. They found no significant difference in either SOD1 or GSHPx activities (Fujita et al. 1996).

Glutathione binding sites in the dorsal and ventral horns of the cervical spinal cord have been reported to be increased in patients with SALS. Spinal-cord sections from five SALS and five control subjects were examined for glutathione binding sites. Glutathione binding was significantly increased (by 16%) in both the dorsal and ventral horns of cervical spinal cords from SALS patients. Saturation binding studies indicated that the increase in glutathione binding was due to an increase in glutathione receptor numbers and not affinity (Lanius et al. 1993).

We recently determined that glutathione levels in cerebrospinal fluid from 12 subjects with ALS were not different from levels found in control subjects (Cudkowicz and Brown, unpublished). Ince et al. (1994) studied selenium and iron, both of which have important roles in free-radical metabolism. They measured the concentration of selenium and iron in lumbar spinal cord from 38 SALS and 22 controls and found a 42% elevation in selenium levels and a 39% elevation in iron levels.

Przedborski and colleagues measured red-blood-cell catalase, SOD1, and GSHPx activity in 31 SALS, 18 FALS, and 24 control subjects. Catalase activity was normal in red blood cells from sporadic and familial ALS patients. Red-blood-cell GSHPx activity was decreased only in SALS patients treated with insulin-like growth factor. SOD1 activity in red blood cells was reduced in the FALS patients with mutations of Cu/Zn-SOD1 but was normal in SALS patients and FALS patients without identified Cu/Zn-SOD1 mutations (Przedborski et al. 1996). Mitchell and colleagues detected a significant, progressive decline in whole-blood GSHPx activity that paralleled the disease progression in 19 ALS patients studied for 24 weeks (Mitchell et al. 1993).

Free radicals are highly reactive and typically short-lived. It is difficult to measure their levels directly. Accordingly, several biochemical parameters are used to gauge indirectly the extent of oxidative damage to various cellular constituents; these include markers of oxidative damage to DNA, proteins, and lipids. Oxidative damage to DNA results in chemical changes in constituent bases. One that can be readily measured is 8-hydroxy-2-deoxyguanosine (OH^8dG) (Ames 1987). Protein oxidation can be quantitated by measuring protein carbonyl groups in plasma and tissue (Floyd and Carney 1992; Halliwell and Aruoma 1995). 3-Nitrotyrosine is a marker for protein oxidation mediated by peroxynitrite (Schulz et al. 1995). Malondialehyde is a biochemical marker of lipid peroxidation (Gutteridge and Halliwell 1990). The enzyme heme oxygenase-1 (HO-1) is induced by oxidative stress (Dwyer, Nishimura, and Lu 1995); elevated HO-1 levels are thus often interpreted as evidence of oxidative toxicity.

Recently, it has been demonstrated that protein carbonyl groups are significantly increased in the frontal cortex (Bowling et al. 1993), motor cortex (Ferrante et al.

1997), and spinal cord (Shaw et al. 1995) from SALS patients as compared to controls and patients with FALS. Levels of OH^8dG were similarly increased in the SALS motor cortex but were unchanged in FALS patients (Ferrante et al. 1997). Immunohistochemical studies showed increased neuronal staining for HO-1, malondialdehyde-modified protein, and OH^8dG in both SALS and FALS spinal cord (Ferrante et al. 1997; Beal et al. 1996, 1997). Several groups have found increased tyrosine nitration in spinal-cord tissue of both SALS and FALS patients with SOD1 mutations (Beal et al. 1996; Abe et al. 1995; Chou, Wang, and Komgi 1996) as well as in transgenic mice overexpressing the G93A SOD1 mutation (Beal et al. 1997). Beal and colleagues found significant increases in concentrations of malondialdehye in cerebral cortex of transgenic ALS mice. They also detected increased immunostaining for 3-nitrotyrosine, heme oxygenase-1, and malondialdehyde-modified protein throughout the spinal cord of the mice (Beal et al. 1996).

Transgenic ALS mice that express a gene encoding mutant SOD1 protein develop motor-neuron degeneration similar to the human disease (Gurney et al. 1994). Onset of symptoms is seen at a mean age of 95 days with a mean survival of 134 days. Administration of vitamin E to the ALS mice produced a significant (13-day) delay in time to disease onset without an effect on survival (Gurney et al. 1994). Riluzole and gabapentin, two antiglutamatergic agents, prolonged survival without an effect on onset (Gurney et al. 1996).

5. THERAPY

Antioxidant therapies were among the first to be advocated for the treatment of ALS. One of these, vitamin E, has a venerable history dating to the 1930s, when deficiency syndromes were demonstrated and the drug was chemically synthesized. At least eight compounds with antioxidant properties, each with isomers, are collectively referred to as vitamin E (Rock et al. 1996). The most potent of these is α-tocopherol. The principal source of vitamin E in western diets is food oils. Absorption of the vitamin after oral administration is inefficient, but is improved when consumed with fat. After uptake by the liver, vitamin E reenters the circulation in association with lipoproteins.

Most of the storage of vitamin E occurs in adipose tissue. Due to its lipophilic nature, vitamin E exerts its principal neuroprotective effects at cellular membranes, where free radicals generate a tocopheroxy radical. This can be regenerated in the presence of reduced glutathione or possibly vitamin C. Based on these and similar considerations, physicians have long advocated the use of vitamin E for the treatment of neurodegenerative disorders. In the 1940s, Wechsler reported that vitamin E could arrest the course of ALS, but the failure of the drug to exert a favorable effect in the instance of Lou Gehrig put this result in better perspective (Wechsler 1940). There was a resurgence of interest in this subject in the 1960s when pancreatic enzyme and vitamin E replacement were advocated as a treatment for ALS (Quick and Greer 1967). Follow-up studies at the NIH failed to confirm these reports, but only a few of patients were studied and a control group was not included (Dorman et al. 1969).

Recently, a role for the use of vitamin E in the treatment of Alzheimer's disease has been established in controlled studies (Sano et al. 1997). Dietary supplementation with 2000 IU per day lengthened the time to institutionalization and slowed the deterioration of activities of daily living. Cognitive changes were not spared.

Along with vitamin C, glutathione contributes to the reducing environment of the cell. Its constituent amino acids are taken up by cells, which subsequently synthesize glutathione in a two-step reaction involving ATP. A feedback mechanism limits the synthesis of glutathione, which exerts its antioxidant effect by generating reduced forms of vitamins C and E and mitigating the oxidative damage brought about by the production of hydrogen peroxide and lipid peroxidases (Meister 1992). For this reason, glutathione represents a potential treatment target, but due to the poor intracellular uptake of GSH, it is necessary to utilize other drugs as vehicles to deliver cysteine. *N*-acetyl cysteine (NAC), which has been shown to exhibit neuroprotective effects, is the accepted therapy for the treatment of acetaminophen toxicity. Acetaminophen hepatotoxicity is mediated by oxidant stress (Rumack et al. 1981).

Recently, orally administered NAC has been shown to slow the progression of nerve and muscle deterioration associated with the murine Wobbler mutant (Henderson et al. 1996). Treated and control animals were sacrificed after nine weeks of treatment with 1% NAC, delivered per os. The axon diameters of the facial nerve, which predictably diminish in Wobbler mutants, were maintained in NAC-treated animals. Using choline acetyltransferase as a marker of motor neurons, a similar sparing effect was noted when cervical cell counts were compared between treated and untreated animals. Further, treatment with NAC enhanced the levels of GSHPX in cervical spinal cord. Glutathione peroxidase, which appears to be regulated by neurotrophins, may exert some of its neuroprotective effect by moderating programmed cell death (Buttke and Sandstrom 1994). Noteworthy is the finding that exogenous administration of NAC can have a similar effect on GSHPX in normal mice.

A placebo-controlled treatment trial of ALS with NAC has recently been completed (Louwerse et al. 1995). One hundred eleven patients were administered placebo or 50 mg/kg of daily NAC by subcutaneous injection over a 12-month period. The primary endpoints were survival and rate of progression, as determined by manual muscle testing, forced vital capacity, and the activities of daily living. Prolonged survival was noted in the treatment group, but this was not significant, and no other significant differences in other outcome measures were observed.

Although the etiology for sporadic ALS is not known, circumstantial evidence supports the notion that excitotoxicity is a factor. The levels of glutamate and aspartate have been reported to be elevated in the spinal fluid of ALS patients, and recent evidence suggests that glutamate recycling mechanisms may be impaired (Rothstein 1995). In a model system employing explanted rat spinal cord, glutamate toxicity has been demonstrated to selectively destroy motor neurons (Rothstein and Kuncl 1995). This was prevented by the administration of compounds that moderate the release or binding of glutamate or downregulate the glutamate receptor. Antioxidants such as *N*-acetylcysteine were also protective.

Glutamate-based therapies have been studied in ALS patients. Gabapentin, an antiepileptic medication that dampens the release of glutamate, was tested in 152 patients (Miller et al. 1996). A trend towards decreased motor deterioration was seen, but this was not significant. A follow-up trial is now in progress. Riluzole, another drug that inhibits glutamate release, slows the progression of weakness in transgenic mice that carry the mutant SOD1 gene associated with familial ALS (Gurney et al. 1996). Although riluzole did not slow motor deterioration in ALS patients in a multinational controlled trial, the drug increased the probability of survival. The effect was modest, adding about three months of life expectancy (Lacomblez et al. 1996).

If indeed there is oxidative injury in CNS tissues in ALS, one might speculate that an antioxidant protein like SOD1 would be beneficial, even if levels of endogenous brain tissue and CSF SOD1 activity are not significantly reduced. Preliminary studies in both animal and man have demonstrated the feasibility of administering SOD1 intrathecally or intraventricularly (Cudkowicz et al. 1997; Smith et al. 1995). This has been done by bolus or continuous infusion, using an implanted pump. Monkeys administered 0.5 mg of bovine SOD (bSOD) intracranially daily for a month did not exhibit clinical signs of toxicity (Balis, Saifer, and Smith, unpublished). Serial spinal-fluid white counts varied from 1 to 7 cells/mm^3, a normal range for animals with indwelling ventricular catheters. Similar results have been seen in sheep, suggesting that animal studies involving intrathecal or intraventricular administration of proteins are predictive of results in humans. The pharmacokinetic results in animals also provide predictive information regarding the doses that can be administered to humans. At steady state, CSF SOD1 values in the monkey who received 2.5 mg/day (Fig. 1) were similar to those in a human who received 30 mg/day. Unfortunately, there was no response in the three patients treated with intrathecal SOD or the single patient treated intraventricularly. Although installation of proteins into the spinal fluid does bypass the blood–brain barrier, it does not guarantee that the administered

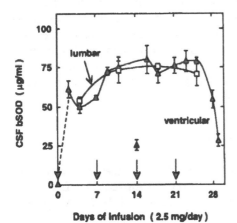

FIG. 1. bSOD levels in monkey CSF during and after intraventricular infusion.

protein will be taken up intracellularly. However, SOD1 might still exert a favorable effect on membranes even if it remained localized to the extracellular space.

Although they have not been tested in ALS, a number of innovative strategies are under consideration for the treatment of hypoxic insults to the nervous system and other disorders where excitotoxic or free-radical mechanisms appear to be causally linked to the degenerative process. Free-radical spin traps represent one such novel strategy (Schulz et al. 1995). These compounds form stable adducts after reacting with free radicals. They have been shown to attenuate damage caused by experimental ischemia and neurotoxic agents that inhibit energy production. Another novel strategy is to employ antisense DNA to downregulate glutamate receptors or some other target. The elegance of this approach is that the treatment is highly selective and it can be directed specifically to the CNS. Finally, it should be noted that therapies that may not appear to be working through their effect on oxidative stress may actually be doing so, at least indirectly. For example, bcl-2, a protooncogene that inhibits apoptosis, protects immortalized neural cells exposed to agents that oxidize cell membranes or induce free-radical formation (Zhong et al. 1993). Further research is needed to exploit these and other findings for therapeutic purposes.

6. FUTURE DIRECTIONS

Molecular biology has had a profound effect on our understanding of ALS. One hopes that insights gained from the study of ALS will enhance our understanding of motor-neuron and other neuron selective degenerative diseases. It is self-evident that treatment strategies will be easier to develop when the causes of ALS are fully identified. Although this may represent the ideal for crafting a treatment strategy, there may be reason to believe that more general treatment strategies, such as enhancing cellular oxidative defenses, will be beneficial. While many of these notions can now be tested in animal models, large-scale human trials are required to demonstrate a treatment effect. To surmount this obstacle it is critical to develop better clinical trial methodologies. Identification of a biologic marker of disease activity would provide a needed endpoint. Such markers should enhance our ability to monitor therapeutic effects, as in the case of AIDS, where the measurement of viral burden has proved to be so important. Based on the rate of progress, it is certain that these and other advances will revolutionize therapy for ALS and kindred diseases, which have thus far been poorly understood and poorly treated.

ACKNOWLEDGMENTS

M.E.C. is the recipient of NIH grant K08NS01896, the American Academy of Neurology Clinical Investigator Award, and grants from the Amyotrophic Lateral Sclerosis Association and the Muscular Dystrophy Association. R.H.B. is supported by grants from the Amyotrophic Lateral Sclerosis Association, the Muscular Dystrophy Association, the Pierre L. Bourgknect ALS Research Foundation, the Myrtle May

MacLellan ALS Research Foundation, and the C.B. Day Investment Co., and by NIA grant 1P01Ag12992-01 and NIH grants 1P01NS31248-01 and RINS34913. Research at the Center for Neurologic Study is generously sponsored by the National Institutes of Health (1 R41 NS36481-01), the Thagard Foundation, Ron Tuttle, Ph.D., Vysis Corporation, and Susan and Charles Donnelly, to whom this manuscript is dedicated.

REFERENCES

Abe, K., Pan, L., Watanabe, M., Kato, T., and Itoyama, Y. 1995. Induction of nitrotyrosine-like immunoreactivity in the lower motor neuron amyotrophic lateral sclerosis. *Neuroscience Letters* 199: 152–4.

Ames, B. N. 1987. Endogenous oxidative DNA damage, aging and cancer. *Free Radical Research Communications* 71:121–8.

Beal, M. F., Shinobu, L., Schulz, J., Matthew, R., Thomas, C., Kowall, N., Gurney, M., and Ferrante, R. 1997. Increased 3-nitrotyrosine and oxidative damage in mice with a human Cu,Zn superoxide dismutase. *Neurology* 48(3):A149–50.

Beal, M. F., Ferrante, R., Matthews, R., Kowall, N., and Brown, R. 1996. 3-Nitrotyrosine in ALS. *Society for Neuroscience Abstracts* 22:1943.

Beckman, J., Beckman, T., Chen, J., Marshall, P. A., Freeman, B. A. 1990. Apparent hydroxyl radical production by peroxynitrite: Implications for endothelial injury from nitric oxide and superoxide. *Proceedings of the National Academy of Sciences of the U.S.A.* 87:1620–4.

Beckman, J. S., Carson, M., Smith, C. D., and Kuppenol, W. H. 1993. ALS, SOD, and peroxynitrite. *Nature* 364:584.

Borchelt, D. R., Lee, M. K., Slunt, H. S., Guarnieri, M., Xu, Z. S., Wong, P. C., Brown, R. H., Jr., Price, D. L., Sisodia, S. S., and Cleveland, D. W. 1994. Superoxide dismutase 1 with mutations linked to familial amyotrophic lateral sclerosis possesses significant activity. *Proceedings of the National Academy of Sciences of the U.S.A.* 91:8292–6.

Bowling, A. C., Schulz, J. B., Brown, R. H., Jr., and Beal, M. F. 1993. Superoxide dismutase activity, oxidative damage and mitochondrial energy metabolism in familial and sporadic amyotrophic lateral sclerosis. *Journal of Neurochemistry* 61:2322–9.

Bowling, A. C., Barkowski, E. E., McKenna-Yasek, D., Sapp, P., Horvitz, H. R., Beal, M. F., and Brown, R. H., Jr. 1995. Superoxide dismutase concentration and activity in familial amyotrophic lateral sclerosis. *Journal of Neurochemistry* 64:2366–9.

Buttke, T. M., and Sandstrom, P. A. 1994. Oxidative stress as a mediator of apoptosis. *Immunology Today* 15:7–10.

Carri, M., Battistoni, A., Polizio, F., Desideri, A., and Rotilio, G. 1994. Impaired copper binding by H46R mutant of human Cu,Zn superoxide dismutase, involved in amyotrophic lateral sclerosis. *FEBS letters* 356:314–16.

Cudkowicz, M. E., Warren, L., Francis, J. W., Lloyd, K. J., Friedlander, R. M., Borges, L. F., Kassem, N., Munsat, T. L., and Brown, R. H. 1997. Intrathecal administration of recombinant human superoxide dismutase 1 (rhSOD1) in amyotrophic lateral sclerosis (ALS): A preliminary safety and pharmacokinetic study. *Neurology* 49:213–22.

Chou, S., Wang, H., and Komgi, K. 1996. Co-localization of NOS and SOD1 in neurofilament accumulation within motor neurons of amyotrophic lateral sclerosis: An immunohistochemical study. *Journal of Chemical Neuroanatomy* 10(3–4):249–58.

Couratier, P., Hugon, J., Sindou, P., Vallat, J., and Dumas, M. 1993. Cell culture evidence for neuronal degeneration in amyotrophic lateral sclerosis being linked to glutamate AMPA/kainate receptors. *Lancet* 341(8840):265–8.

Deng, H.-X., Hentati, A., Tainer, J. A., Iqbal, Z., Cayabyab, A., Hung, W.-Y., Getzoff, E. D., Hu, P., Herzfeldt, B., Roos, R. P., Warner, C., Deng, G., Soriano, E., Smyth, C., Parge, H. E., Ahmed, A., Roses, A. D., Hallewell, R. A., Pericak-Vance, M. A., and Siddique, T. 1993. Amyotrophic lateral sclerosis and structural defects in Cu,Zn superoxide dismutase. *Science* 261:1047–51.

Dorman, J. D., Engel, W. K., and Fried, D. M. 1969. Therapeutic trial in amyotrophic lateral sclerosis: Lack of benefit with pancreatic extract and DL-alpha Tocopherol in 12 patients. *JAMA* 209:257–8.

Dwyer, B., Nishimura, R., and Lu, S. 1995. Differential expression of heme oxygenase-1 in cultured cortical neurons and astrocytes determined by the aid of a new heme oxygenase antibody. Response to oxidative stress. *Molecular Brain Research* 30:37–47.

Ferrante, R., Browne, S., Shinobu, L., Bowling, A., Baik, M., MacGarvey, U., Kowall, N., Brown, R., and Beal, M. 1997. Evidence of increased oxidative damage in both sporadic and familial ALS. *Journal of Neurochemistry.* In press.

Floyd, R. A., and Carney, J. M. 1992. Free radical damage to protein and DNA: Mechanisms involved and relevant observations on brain undergoing oxidative stress. *Annals of Neurology* 32:S22–7.

Fujita, K., Yamauchi, K., Shibayama, K., Ando, M., Honda, M., and Nagata, Y. 1996. Decreased cytochrome c oxidase activity but unchanged superoxide dismutase and glutathione peroxidase activities in the spinal cords of patients with amyotrophic lateral sclerosis. *Journal of Neuroscience Research* 45:276–81.

Gurney, M. E., Pu, H., Chiu, A. Y., Dal Canto, M. C., Polchow, C. Y., Alexander, D. D., Caliendo, J., Hentati, A., Kwon, Y. W., Deng, H.-X., Chen, W., Zhai, P., Sufit, R. L., and Siddique, T. 1994. Motor neuron degeneration in mice that express a human Cu,Zn superoxide dismutase mutation. *Science* 264:1772–5.

Gurney, M., Cutting, F., Zhai, P., Doble, A., Taylor, C., Andrus, P., and Hall, E. 1996. Benefit of vitamin E, riluzole and gabapentin in a transgenic model of familial ALS. *Annals of Neurology* 39(2):147–57.

Gutteridge, J., and Halliwell, B. 1990. The measurement and mechanism of lipid peroxidation in biological systems. *Trends in Biochemical Sciences* 15:129–35.

Halliwell, B. 1992. Reactive oxygen species and the central nervous system. *Journal of Neurochemistry* 59:1609–23.

Halliwell, B., Aeschbach, R., Loliger, J., Aruoma, O. I. 1995. The characterization of antioxidants. *Food and Chemical Toxicology* 33(7):601–17.

Henderson, J. T., Javaheri, M., Kopko, S., and Roder, J. C. 1996. Reduction of lower motor neuron degeneration in wobbler mice by N-acetyl-L-cysteine. *Journal of Neuroscience* 16(23):7574–82.

Huber, W., Menander-Huber, K. B., Saifer, M. G., and Dang, P. H. 1977. Studies on the clinical and laboratory pharmacology of drug formulations of bovine Cu-Zn superoxide dismutases (orgotein). In *perspectives in inflammation*, ed. D. A. Willoughby et al., pp. 527–40, Lancaster: MTP Press.

Ince, P., Shaw, P., Candy, J., Mantle, D., Tandon, L., Ehmann, W., and Markesbery, W. 1994. Iron, selenium and glutathione peroxidase activity are elevated in sporadic motor neuron disease. *Neuroscience Letters* 182:87–90.

Ischiropoulos, H., Zhu, L., Chen, J., Tsai, M., Martin, J., Smith, C., and Beckman, J. 1992. Peroxynitrite-mediated tyrosine nitration catalyzed by superoxide dismutase. *Archives of Biochemistry and Biophysics* 16:149–56.

Kooy, N., Royall, J., Ischiropoulos, H., and Beckman, J. 1994. Peroxynitrite mediated oxidation of dihydrorhodamine 123. *Free Radicals in Biology and Medicine* 16:149–56.

Lacomblez, L., Bensimon, G., Leigh, P. N., Guillet, P., and Meininger, V. 1996. Dose ranging study of riluzole in amyotrophic lateral sclerosis. *The Lancet* 347(9013):1425–31.

Lanius, R., Krieger, C., Wagey, R., and Shaw, C. 1993. Increased 35S glutathione binding sites in spinal cords from patients with amyotrophic lateral sclerosis. *Neuroscience Letters* 163(1):89–92.

Louwerse, E. S., Weverling, G. J., Bossuyt, P. M. M., Meyjes, F. E. P., and Vianney de Jong, J. M. B. 1995. Randomized, double-blind, controlled trial of acetylcysteine in amyotrophic lateral sclerosis. *Archives of Neurology* 52:559–64.

Lyons, T., Liu, H., Goto, J., Nersissian, A., Roe, J., Graden, J., Café, C., Ellerby, L., Bredesen, D., Gralla, E., and Valentine, J. 1996. ALS-mutant copper–zinc superoxide dismutase proteins with defective zinc binding sites and altered redox behavior. *Proceedings of the National Academy of Sciences of the U.S.A.* 93(22):12240–4.

Miller, R. G., Moore, D., Young, L. A., Armon, C., Barohn, R. J., Bromberg, M. B., Bryan, W. W., Gelinas, D. F., Mendoza, M. C., Neville, H. E., Parry, G. J., Petajan, J. H., Ravits, J. M., Ringel, S. P., Ross, M. A., and the WALS Study Group. 1996. Placebo-controlled trial of gabapentin in patients with amyotrophic lateral sclerosis. *Neurology* 47:1383–8.

Mitchell, J., Gatt, J., Phillips, T., Houghton, E., Rostron, G., and Wignall, C. 1993. Cu/Zn superoxide dismutase, free radicals, and motoneuron disease. *Lancet* 342:1051–2.

Mulder, D. W., Kurland, L. T., Offord, K. P., and Beard, C. M. 1986. Familial adult motor neuron disease: Amyotrophic lateral sclerosis. *Neurology* 36:511–17.

Munsat, T. L. 1984. Adult motor neuron diseases. *Merritt's Textbook of Neurology.* 7th ed., ed. L. P. Rowland, pp. 548–52. Philadelphia: Lea and Febiger.

Nishida, C., Gralla, E., and Valentine, J. 1994. Characterization of three yeast copper–zinc superoxide dismutase mutants analogous to those coded for in familial amyotrophic lateral sclerosis. *Proceedings of the National Academy of Sciences of the U.S.A.* 91:9906–10.

Przedborski, S., Donaldon, D., Jakowee, M., Kish, S., Guttman, M., Kosoklija, G., and Hays, A. 1996. Brain superoxide dismutase, catalase and glutathione peroxidase activities in amyotrophic lateral sclerosis. *Annals of Neurology* 39:158–65.

Quick, D. T., and Greer, M. 1967. Pancreatic dysfunction in patients with amyotrophic lateral sclerosis. *Neurology* 17:112–6.

Robberecht, W., Sapp, P., Viaene, M. K., Rosen, D., McKenna-Yasek, D., Haines, J., Horvitz, R., Theys, P., and Brown, R. H., Jr. 1994. Cu/Zn superoxide dismutase activity in familial and sporadic amyotrophic lateral sclerosis. *Journal of Neurochemistry* 62:384–7.

Rock, C. L., Jacob, R. A., and Bowen, P. E. 1996. Update on the biological characteristics of the antioxidant micronutrients: Vitamin C, vitamin E, and the carotenoids. *Journal of the American Dietetic Association* 96:693–700.

Roos, R., Lee, J., Bindokas, V., Jordan, J., Miller, R., Ma, L., Weihl, C., Habib, A., and Ghadge, G. 1997. Gene delivery by replication-deficient recombinant adenoviruses (AdVs) in the Study of Cu,Zn superoxide dismutase type 1 (SOD)-linked familial amyotrophic lateral sclerois (FALS). *Neurology* 48(3):A150.

Rosen, D. R., Siddique, T., Patterson, D., Figlewicz, D. A., Sapp, P., Hentati, A., Donaldson, D., Goto, J., O'Regan, J. P., Deng, H.-X., Rahmani, Z., Krizus, A., McKenna-Yasek, D., Cayabyab, A., Gaston, S. M., Berger, R., Tanzi, R. E., Halperin, J. J., Herzfeldt, B., Van den Bergh, R., Hung, W.-Y., Bird, T., Deng, G., Mulder, D. W., Smyth, C., Laing, N. G., Soriano, E., Pericak-Vance, M., Haines, J., Rouleau, G. A., Gusella, G. S., Horvitz, H. R., and Brown, R. H., Jr. 1993. Mutations in Cu/Zn superoxide dismutase are associated with familial amyotrophic lateral sclerosis. *Nature* 362:59–62.

Rothstein, J. D. 1995. Excitotoxic mechanisms in the pathogenesis of amyotrophic lateral sclerosis. *Advances in Neurology* 68:7–20.

Rothstein, J. D., and Kuncl, R. W. 1995. Neuroprotective strategies in a model of chronic glutamate-mediated motor neuron toxicity. *Journal of Neurochemistry* 65(2):643–51.

Rumack, B. H., Peterson, R. C., Koch, G. G., and Amara, I. A. 1981. Acetaminophen overdose. *Archives of Internal Medicine* 141:380–5.

Sano, M., Ernesto, C., Thomas, R. G., Klauber, M. R., Schaffer, K., Grundman, M., Woodbury, P., Growdon, J., Cotman, C. W., Pfeiffer, E., Schneider, L. S., and Thal, L. J. 1997. A controlled trial of selegiline, alpha-tocopherol, or both as treatment for alzhcimcr's disease. *The New England Journal of Medicine* 336(17):1216–22.

Schulz, J. B., and Beal, M. F., 1995. Neuroprotective effects of free radical scavengers and energy repletion in animal models of neurodegenerative disease. *Annals of the New York Academy of Sciences* 765:100–110.

Schulz, J. B., Henshaw, D. R., Siwek, D., Jenkins, B. G., Ferrante, R. J., Cipolloni, P. B., Kowall, N. W., Rosen, B. R., and Beal, M. F. 1995. Involvement of free radicals in excitotoxicity in vivo. *Journal of Neurochemistry* 64(5):2239–47.

Schulz, J. B., Matthews, R. T., Jenkins, B. G., Brar, P., and Beal, M. F. 1995. Improved therapeutic window for treatment of histotoxic hypoxia with a free radical spin trap. *Journal of Cerebral Blood Flow and Metabolism* 15(6):948–52.

Shaw, P., Ince, P., Falkous, G., and Mantle, D. 1995. Oxidative damage to protein in sporadic motor neuron disease spinal cord. *Annals of Neurology* 38:691–5.

Smith, R., Balis, F., Ott, K., Elsberry, D., Sherman, M., and Saifer, M. 1995. Pharmacokinetics and tolerability of vertricularly administered superoxide dismutase in monkeys and preliminary clinical observations in familial ALS. *Journal of the Neurological Sciences* 129:13–18.

Terro, F., Lesort, M., Viader, F., Ludolph, A, and Hugon, J. 1996. Antioxidant drugs block in vitro the neurotoxicity of CSF from patients with amyotrophic lateral sclerosis. *NeuroReport* 7:1970–2.

Wechsler, I. S. 1940. Recovery in amyotrophic lateral sclerosis. *Neurology* 114:948–950.

Wiedau-Pazos, M., Goto, J., Rabizadeh, S., Gralla, E., Roe, J., Valentine, J., and Bredesen, D. 1996. Altered reactivity of superoxide dismutase in familial amyotrophic lateral sclerosis. *Science* 271:515–18.

Yim, M., Kang, J., Yim, H., Kwak, H., Chock, P., and Stadtman, E. 1996. A gain-of-function mutation of an amyotrophic lateral sclerosis–associated Cu,Zn-superoxide dismutase mutant: An enhancement of free radical formation due to a decrease in K_m for hydrogen peroxide. *Proceedings of the National Academy of Sciences of the U.S.A.* 93:5709–14.

Zhong, L. T., Sarafian, T., Kane, D. J., Charles, A. C., Mah, S. P., Edwards, R. H., and Bredesen, D. E. 1993. bcl-2 inhibits death of central neural cells induced by multiple agents. *Proceedings of the National Academy of Science* 90:4533–7.

Glutathione in the Nervous System
Edited by Christopher A. Shaw
Copyright © 1998 Taylor & Francis

17

Oxidative Stress and Neurological Diseases: Is Glutathione Depletion a Common Factor?

Jaswinder S. Bains and Christopher A. Shaw

Departments of Ophthalmology and Physiology and Neuroscience Program, The University of British Columbia, Vancouver, British Columbia, Canada V6T 1Z3

ABBREVIATIONS

AD Alzheimer's disease
ALS amyotrophic lateral sclerosis (Lou Gehrig's disease)
BSO buthionine sulfoximine
CSF cerebrospinal fluid

GSH reduced glutathione
GSSG oxidized glutathione
IP_3 1,4,5-inositol triphosphate
MPP^+ 1-methyl-4-phenylpyridinium
MPTP 1-methyl-4-phenyl-1,2,3,6-tetrahydropyridine
PD Parkinson's disease
PKC protein kinase C
ROS reactive oxygen species
SOD superoxide dismutase

1. INTRODUCTION

A common pathological hallmark of various neurodegenerative disorders such as amyotrophic lateral sclerosis (ALS), Alzheimer's disease (AD), and Parkinson's disease (PD) is the loss of particular subsets of neurons. Neurodegeneration of these neural subsets may be a consequence of various forms of neural death, for example, necrosis and apoptosis. In turn, such forms of death may be the result of a variety of a cellular insults, such as excitotoxicity or oxidative stress.

Necrosis is caused by catastrophic toxic or traumatic events accompanied by inflammation and rapid collapse of internal homeostasis (Bonfoco et al. 1995). A common example of necrosis is instantaneous cell death following chemical fixation of tissues (Buja, Eigenbrodt, and Eigenbrost 1993). Severe ischemia (60 min of hypoxia) results in infarction and necrosis in young rat brain (Dragunow et al. 1994). In contrast, apoptosis is a programmed cell death characterized by cell shrinkage, membrane blebbing, and genomic fragmentation (Ellis, Yuan, and Horvitz 1991; Nagata 1997). Apoptosis is frequently observed in neural development. For example, as many as 80 percent of the original pool of neurons are lost in some areas of the nervous system during development (see Zaman and Ratan, chapter 6, this volume). Excitotoxicity results from the overactivation of cells by excitatory amino acids (EAAs) such as glutamate and cysteine or their agonists (Olney 1978; Choi 1988; Whetsell 1996). Excitotoxic cell death is often attributed to extreme stimulation of NMDA receptors leading to increased calcium (Ca^{+2}) flux. Examples include degeneration of neural retina in neonatal mice following systemic administration of either glutamate or aspartate (Lucas and Newhouse 1957) and the observation that high concentrations of glutamate and other excitatory amino acids selectively kill neurons by their depolarizing action (Olney 1978). In the absence of EAA stimulation, disturbances of intracellular Ca^{+2} homeostasis may also result in the alterations of cell function, cell blebbing, lysis, and ultimately cell death (Orrenius et al. 1992). Although each of these processes has distinct causes, stimuli, and pathological outcomes, emerging evidence from a number of studies suggests that oxidative stress could be involved in a final common pathway for a number of such processes in a wide variety of acute and chronic neurological diseases as well as in normal aging. The role of oxidative stress in neurological diseases has been discussed in a number of previous articles (Evans 1993; Olanow 1992; Halliwell 1992) and chapters in this volume (e.g., see Dryhurst, chapter 15).

In the following section, we will describe the role of oxidative stress in general in the induction of different forms of neuronal degeneration in various neurological disorders. In the context of oxidative stress and these disorders, we will then discuss the specific role of the free-radical scavenger glutathione and suggest how GSH depletion may be a common early event underlying some neurodegenerative diseases. In addition, we will discuss how GSH's other roles as redox modulator of ionotropic receptor function and putative neurotransmitter may also contribute to the origin of various neurological disorders.

2. REACTIVE OXYGEN SPECIES, ANTIOXIDANT DEFENSES, AND OXIDATIVE STRESS-MEDIATED CELL DEATH

2.1 Reactive Oxygen Species and Antioxidant Defenses

Although oxygen is essential for life, it produces highly toxic radicals, collectively termed reactive oxygen species (ROS), during cellular respiration. A free radical is defined as any ion or molecule capable of independent existence, and that contains one or more unpaired electrons. Examples are the superoxide radical ($O_2^{-\bullet}$), the hydroxyl radical (•OH), and the nitric oxide radical (NO•). Natural defense mechanisms of the organism protect it against ROS. Typically, equilibrium exists between generation of ROS and antioxidant defenses that maintain a homeostatic control of the cell's oxidative state. When this balance is altered to favor ROS, oxidative stress occurs (Sies 1991), resulting in a buildup of oxidatively modified molecules that can cause cellular dysfunction. For a number of reasons, neurons may be particularly susceptible (Evans 1993) and cell death can result (Simonian and Coyle 1996).

In addition to oxidative metabolism, environmental stimuli such as radiation or the presence of transition metals (M) can also induce oxidative stress. For the latter, hydrogen peroxide reacts with iron or copper to give rise to the very toxic •OH radical:

$$M^{n+} + H_2O_2 \longrightarrow M^{(n+1)+} + \bullet OH + {}^-OH$$

Cellular defenses against the oxidants include enzymatic and nonenzymatic antioxidant mechanisms that protect the cell from oxidative injury. The enzymatic defenses include superoxide dismutase (SOD), which converts the superoxide radical ($O_2^{-\bullet}$) to form hydrogen peroxide (H_2O_2). SOD exists in three forms: an extracellular and an intracellular copper/zinc (Cu/Zn) SOD, a mitochondrial SOD, and a manganese (Mn) SOD. At low concentrations, H_2O_2 is converted to water and molecular oxygen by glutathione peroxidase (GSH-Px) which uses reduced glutathione (GSH) as a proton donor; in the process, GSH is oxidized to the dimer glutathione disulfide (GSSG). GSSG is catalyzed by glutathione reductase and reduced back to GSH (Fig. 1). High concentrations of H_2O_2 are removed by catalase. Nonenzymatic antioxidant defenses

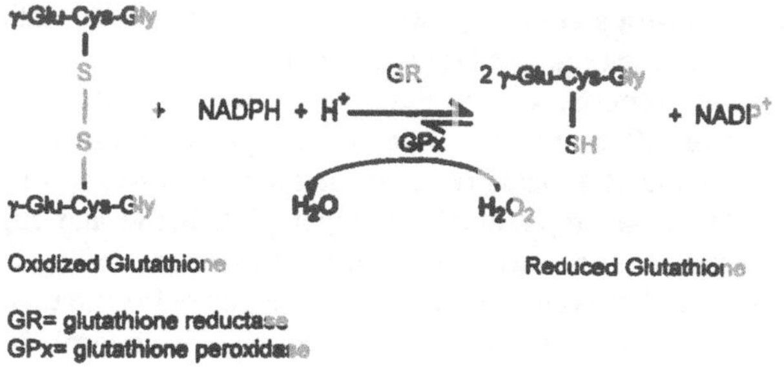

FIG. 1. Oxidation–reduction pathway: relationship of oxidized (GSSG) to reduced (GSH) glutathione.

include ascorbic acid (vitamin C, a hydrophilic antioxidant) and α-tocopherol (one of the vitamin E family, a hydrophobic antioxidant concentrated in membranes). These antioxidants act synergistically with GSH (Chen and Tappel 1994; Meister 1994).

Glutathione (GSH), a tripeptide composed of L-glutamate, L-cysteine, and glycine, is considered to be the most prevalent and important intracellular nonprotein thiol/ sulfhydryl compound in mammalian cells (Meister 1988; Wu, Murphy, and Chiueh 1994). A key role for GSH is as free-radical scavenger against the hydroxyl radical. This role is crucial, as there are no known enzymatic defenses against this species of radical. The ability of GSH to nonenzymatically scavenge both singlet oxygen and •OH (Coyle and Puttfarcken 1993) provides a first line of defense against such free radicals.

Reduced GSH strongly modulates the redox state (ratio of oxidizing to reducing equivalents) of the cell, a role that is critical for cell survival (Chance, Sies, and Boveris 1979). In addition to these antioxidative actions, GSH is involved in DNA synthesis and repair, protein synthesis, amino acid transport, enhancement of immune functions, and enzyme activation (Lomaestro and Malone 1995). More recently, a role for GSH in neural activity as a neuromodulator (Ogita et al. 1995) or a neurotransmitter (Shaw, Pasqualotto, and Curry 1996) has been reported.

2.2 Oxidative Stress: Necrosis and Apoptosis

As discussed above, necrosis is a passive form of "accidental" cell death that follows physical damage. It usually involves groups of cells and has the deleterious end result of causing tissue inflammation. Necrosis is characterized by a selective loss of membrane permeability, resulting in swelling of organelles and rupture of the plasma membrane (Simonian and Coyle 1996). Necrotic cell death can be induced by oxidative stress. Free radical action on lipids can disrupt membrane ion gradients, leading to rupture of the plasma membrane. For example, neurons from the hypothalamic cell line GT1–7 undergo necrosis following exposure to the GSH-depleting agent buthionine sulfoximine (BSO) (Kane et al. 1993).

In contrast to necrosis, apoptosis is an active form of programmed cell death involving individual cells that are surrounded by healthy neighbors. Apoptosis requires ATP and new protein synthesis (Earnshaw 1995). Increasing evidence indicates that oxidative stress can also induce apoptosis (see Zaman and Ratan, chapter 6, this volume). In some cases, cells undergo apoptosis when exposed to H_2O_2 (Hockenbery et al. 1993). Cortical neurons grown in media that are low in cysteine, which depletes GSH, also undergo apoptosis (Ratan, Murphy, and Baraban 1994). Oxidative DNA damage is also considered to be a trigger for apoptosis (Ward 1977; Myers 1980): Oxidized DNA activates a nuclear enzyme, poly(ADP ribose) polymerase (PARP), which is involved in DNA repair (Schraufstatter et al. 1986). Proteolysis of PARP precludes its recruitment to sites of DNA damage and induces apoptosis (Nicholson et al. 1995). Antioxidants can prevent apoptotic cell death. Other studies have shown that the antioxidant N-acetyl cysteine prevents apoptosis induced by tumor necrosis factor α in neuroblastoma cells (Talley et al. 1995), and B-cell leukemia/lymphoma 2 (Bcl-2) overexpression protects B cells (lymphocytes formed in the bone marrow of mammals) from undergoing apoptosis following interleukin-3 deprivation (Hockenbery et al. 1993).

2.3 Excitotoxicity and Oxidative Stress

Excitotoxicity refers to neuronal cell death caused by excessive activation of excitatory amino acid receptors by glutamate or its agonists. Three subtypes of glutamatergic ionotropic receptors exist and are named for their pharmacological agonists: N-methyl-D-aspartate (NMDA), and the non-NMDA agonists quisqualate or α-amino-3-hydroxy-5-methyl-4-isoxazolepropionic acid (AMPA) and kainic acid (KA) (Watkins and Olverman 1987). A form of slow/weak excitotoxicity has been proposed to account for neurodegenerative disorders that evolve gradually over a period of years (Albin and Greenamyre 1992; Beal 1992a). Slow excitotoxicity can also occur as result of a defect in energy metabolism. An inability to maintain ATP levels may lead to partial neuronal depolarization, relief of the voltage-dependent Mg^{+2} block of NMDA receptors, and persistent receptor activation by ambient glutamate levels (Novelli et al. 1988; Zeevalk and Nicklas 1990). Although sodium influx through glutamate-gated ion channels may mediate the acute form of neuronal degeneration, calcium appears to be critical for delayed cell death induced by NMDA and non-NMDA receptors (Choi 1995). The occurrence of both apoptotic and necrotic cell death has been reported following excitotoxic striatal lesions and in Huntington's disease striatum (Portera-Cailliau et al. 1995). A fourth class of glutamate receptors, the metabotropic receptors, exert their effects by means of G-protein-initiated biochemical events rather than through ion channels. Although the metabotropic glutamate receptors (mGluR) do not directly mediate the neurotoxic effects of glutamate, activation of subtypes of these receptors may potentiate or attenuate neurotoxicity (Schoepp and Conn 1993).

A number of recent studies suggest that oxidative stress may act as a mediator of excitotoxic cell death following the stimulation of ionotropic glutamate receptors

(Dykens, Stern, and Trenker 1987; Beal 1992b; Coyle and Puttfarcken 1993; Beal 1995). When considering such studies, it is important to consider that three lines of evidence are critical for distinguishing between oxidative stress as an epiphenomenon versus a causal event in neural death: in the latter, (i) oxygen radicals should be generated during the period of irreversible damage, (ii) evidence of oxidative damage should be present, and (iii) free-radical scavengers or inhibitors of processes generating oxygen radicals should prevent neuronal degeneration. All three criteria are satisfied for both NMDA and non-NMDA receptor agonists, although the multiple sources of oxygen radicals and variations in intrinsic mechanisms of protection may account for apparent contradictory results in different experimental settings. Activation of NMDA and KA receptors increase free-radical damage to lipids; this damage can be prevented by antioxidants (Dykens, Stern, and Trenker 1987; Miyamoto et al. 1989; Monyer, Hartley, and Choi 1990; Miyamoto and Coyle 1990; Puttfarcken, Getz, and Coyle 1993). These findings suggest that glutamate receptor-linked processes can activate intracellular pathways that produce free radicals, uncouple mitochondrial electron transport, depress the neuronal defense system against free-radical damage, or a combination of all of these. Furthermore, free radicals themselves can increase the release and decrease the reuptake of glutamate, leading to the increase of glutamate in the synaptic cleft (Volterra et al. 1994). Thus, a self-perpetuating cycle begins in which activation of glutamate receptors increases free radicals, which may then lead to further receptor activation (Simonian and Coyle 1996).

3. ROLE OF OXIDATIVE STRESS IN VARIOUS NEUROLOGICAL DISEASES

Alterations in oxidative damage or changes in antioxidant status in nervous tissue have been implicated in the pathogenesis of neurodegenerative disorders such as amyotrophic lateral sclerosis (ALS) (Rosen et al. 1993), Parkinson's disease (PD) (Adams et al. 1991; Gotz et al. 1994; Jenner and Olanow 1996), and Alzheimer's disease (AD) (Ceballos et al. 1990; Benzi and Moretti 1995a; Mattson 1995). Although we will not discuss other disorders in this chapter, similar mechanisms have been suggested to underlie various other neurological disorders, such as epilepsy, ischemia (Charriaut-Marlangue et al. 1996), and Down's syndrome (Busciglio and Yankner 1995).

3.1 Amyotrophic Lateral Sclerosis

Amyotrophic lateral sclerosis (motor neuron disease, ALS), or Lou Gehrig's disease, is a motor neuron degenerative disease occurring mainly in the fifth or sixth decade of life. The age of onset for this and other neurological diseases raises the probability of an age-dependent contributory factor(s). It affects approximately 5–10 per 100,000 people (see review by de Belleroche, Orrell, and Virgo 1996). The disease usually begins asymmetrically and distally in one limb, most commonly the leg, and then appears to spread within the neuraxis to involve contiguous groups of motor neurons.

TABLE 1. *Factors leading to oxidative stress in various neurological disease*

Amyotrophic lateral sclerosis
Loss of motor neurons (upper and lower) in brain stem, motor cortex, and spinal cord
    (see review by de Belleroche, Orrell, and Virgo 1996)
SOD-1 gene mutations (Rosen et al. 1993)
Glutmate toxicity (Coyle and Puttfarcken 1993)

Parkinson's disease
Loss of dopaminergic neurons (Forno et al. 1981)
GSH content decreases in substantia nigra (Perry, Godin, and Hansen 1982; Riederer et al. 1989)
Mitochondrial complex I defect (Parker, Boyson, and Parks 1989; Schapira et al. 1990a,b)

Alzheimer's disease
Loss of cholinergic and pyramidal neurons (Joseph and Roth 1992; Weinstock 1995)
Lipid peroxidation (Gotz et al. 1992; Hajimohammadreza and Brammer 1990)
Advanced glycation end products (AGE) (Munch et al. 1997; Yan et al. 1995, 1996)
Intracellular NFT formation in entorhinal, hippocampus, and neocortex (Mattson, Rychlik, and Engle 1990)
Extracellular deposition of β-amyloid (Joachim and Selkoe 1992; Mattson et al. 1993)
Mitochondrial complex III defect (COX) (Parker, Filley, and Parks 1990)
Mitochondrial NADH dehydrogenase subunit 2 defect (Lin et al. 1992)
Aluminum deposition (Birchall and Chappell 1988; Candy et al. 1986, Zatta, Zanoni, and Favarato 1992)
Apolipoprotein E genotype (Strittmatter et al. 1993; Sanan et al. 1994)
Activation of microglia (Chao, Hu, and Peterson 1995; Evans et al. 1992a,b)

The progression of the disease is rapid, and most patients die within 3–5 years of onset (Williams and Windebank 1991). Pathologically, ALS is characterized by progressive loss of motor neurons in the spinal cord, brain stem, and motor cortex, which results in muscle atrophy and loss of motor functions.

ALS may occur either as the sporadic or as the familial form, the majority being the former. Approximately 5–10 percent of ALS cases are familial (FALS), and of these 20–25 percent are associated with dominantly inherited mutations in SOD-1, the gene that encodes human Cu/Zn-superoxide dismutase (Rosen et al. 1993). Most mutations occur in exons 1, 2, 4, and 5, and not in the active site formed by exon 3 (Deng et al. 1993). However, more recently discovered mutations have been located at the active site (e.g., D125H) or are important in copper binding (e.g., H48Q) (Enayat et al. 1995). The mechanism by which SOD point mutations lead to neuronal degeneration could be attributable to reduced levels of SOD-1 activity or to the gain of an unidentified adverse function (de Belleroche, Orrell, and Virgo 1996). Evidence for a gain of adverse function comes from transgenic studies in which a SOD-1 mutation associated with human ALS induces a phenotype like that of motor neuron disease (MND) but without a corresponding loss of enzyme activity (Gurney et al. 1994). The possibility that the high levels of SOD-1 activity in these animals trigger the neuronal degeneration is unlikely, however, as mice overexpressing normal SOD-1 to a similar extent do not develop MND. These results suggests that the mutant SOD-1 is somehow cytotoxic to the mice overexpressing it (Brown 1995).

The following possibilities have been suggested to explain the gain of function of SOD-1 in the pathogenesis of ALS:

1. The mutant enzyme may be so unstable that it participates in forming toxic cytoplasmic aggregates.

2. The mutant enzyme may have an altered substrate affinity that leads to high levels of toxic reaction products. For example, Beckman et al. (1993) have proposed that the mutant SOD-1 may have an enhanced affinity for peroxynitrite and can produce elevated levels of nitronium ions, which in turn catalyze nitration of tyrosine on proteins such as neurofilaments and tyrosine kinases.
3. If the stability of the mutant enzyme is reduced, there may be subnormal copper or zinc buffering by the enzyme and resulting neurotoxicity from one or both of these metals (Wong et al. 1995; Brown 1995). Normal SOD-1 binds copper and zinc very avidly. Mutations in SOD-1 may decrease binding of copper and zinc, leading to an increase in the intracellular concentration of these free metals, both of which can be neurotoxic (Sheinberg 1988; Koh and Choi 1988).

A few studies have examined brains of individuals with ALS for evidence of oxidative damage. One such study showed an elevation in protein carbonyl groups, a marker of oxidative damage, in frontal cortex of individuals with sporadic ALS (Bowling et al. 1993). Gurney et al. (1994) have developed a transgenic mouse model of FALS in which the mutated SOD (Cu/Zn SOD) is overexpressed. Recently, the same group has shown dopaminergic neuronal degeneration in the midbrain of transgenic mice, with mutant SOD (Kostic et al. 1997). These latter data may suggest commonalities of various forms of neurological damage linked by a common mechanism of oxidative stress.

3.2 Parkinson's Disease

Parkinson's disease (PD) is a slowly progressive neurodegenerative disorder, occurring in old age, mainly characterized by loss of dopaminergic neurons in the substantia nigra pars compacta and the frequent presence of Lewy bodies (Benzi and Moretti 1995b; see also Benzi and Moretti, chapter 11; Sian et al., chapter 14; Dryhurst, chapter 15, this volume). Lewy bodies are round cytoplasmic inclusions, composed of masses of neurofilaments (Lewy 1912; Duffy and Tennyson 1965; Goldman and Yen 1986), found principally in neurons of substantia nigra of patients with PD (Den Hartog Jager and Bethlem 1980; Forno 1987; see also Dryhurst, chapter 15, this volume). Although the pathogenesis of PD is multifactorial, two main hypotheses are currently debated: increased oxidative stress and altered mitochondrial activity. The first hypothesis suggests that the formation of ROS may lead to neuronal damage by shifting the cellular oxidation–reduction (redox) equilibrium toward oxidative stress (Olanow 1990). The second, and more recent, hypothesis concerns a deficiency of mitochondrial activity and the consequent failure of energy production and changes in neuronal metabolism (Di Monte 1991). As we will discuss below, these hypotheses are not mutually exclusive.

Oxidative stress can be an important primary or secondary cause of degeneration in pigmented neurons in substantia nigra in PD. Evidence from post mortem studies at the end of illness points to oxidative stress as a component of the pathological mechanisms leading to nigral cell death. Increased lipid peroxidation, impaired glutathione metabolism, and enhanced superoxide activity in PD substantia nigra have been reported (Dexter et al. 1994; Jenner et al. 1992). The proposal of oxidative stress also finds support from the finding that high concentrations of iron are present in substantia nigra. As described above (section 2), iron or other transition metals may act as a catalyst for oxidative production of free radicals (Olanow et al. 1989). However, the importance of oxidative stress in the pathogenesis of PD has been challenged on various grounds. First, clinical administration of deprenyl to patients with early idiopathic Parkinsonism has not been conclusive (Calne 1992). Further, Kish and colleagues (Kish et al. 1992b) did not find any relationship between the rate of synthesis of dopamine and the regional loss of dopaminergic nerve terminals, which should be indicative of dopamine depletion. According to a free radical hypothesis, the loss of neurons should be highest in brain areas where the rate of turnover of dopamine and conversion of dopamine to dopamine radicals is maximum (for a discussion of related points, see Dryhurst, chapter 15, this volume). However, no such relationship was found. Thus, oxidative stress alone may not be sufficient to explain the pathogenesis of idiopathic Parkinsonism in all conditions. However, as noted above, the recent results of Kostic et al. (1997) suggest commonalities related to oxidative stress across different neurological disorders. We will return to this last point in the final section of this chapter.

3.3 Alzheimer's Disease

Alzheimer's disease (AD) is the most common cause of age-related intellectual impairment. *Early onset* of AD occurs before 65 years; *late onset*, after that age. It is characterized by degeneration of neurons, especially pyramidal neurons in the hippocampus, entorhinal cortex, and neocortical areas, and by a loss of cholinergic neurons in the median forebrain. Two major hallmarks of AD are extracellular deposits of β-amyloid protein in plaques, and abnormal intracellular cytoskeletal filaments, the neurofibrillary tangles (NFT). The amyloid $\beta/A4$ protein is a 39–43-amino-acid peptide that forms the core of the senile plaques in AD brain and is thought to play a crucial role in AD pathogenesis [reviewed by Yankner (1996)]. The observation that $A\beta$ can be toxic to neurons in vitro (Yankner et al. 1989) has led to the hypothesis that neuronal degeneration in AD is due to $A\beta$ toxicity. Aggregates of $A\beta$ peptide can be toxic in neuronal cultures (Pike et al. 1993; Lorenzo and Yankner, 1994), favoring the hypothesis that $A\beta$ may be a primary cause of neuronal degeneration (Yankner et al. 1989; Roher et al. 1991). A variety of results indicate that oxidative stress may contribute to $A\beta$ toxicity. Increases in H_2O_2 were detected in cells following exposure to $A\beta$, and both vitamin E and catalase prevented cell death (Behl et al. 1992, 1994). Free radical peptides that can inactivate oxidation-sensitive enzymes and initiate lipid peroxidation have been detected in aqueous solutions of $A\beta$ (Butterfield et al. 1994; Hensley et al. 1994). These latter data indicate that amyloid itself can generate free radicals. Moreover, soluble $A\beta$ has been found to aggregate in vitro by the addition of metal-catalyzed oxidation systems (Dyrks et al. 1992), which can be inhibited by antioxidants.

Several potential sources of oxidative stress can be considered in the pathogenesis of Alzheimer's disease. First, the concentration of iron (a potent catalyst of oxyradical generation) is increased in NFT-bearing neurons (Good et al. 1992). Second, the increased concentration of iron would result in increased protein modifications, catalyzed by metal ions and by reducing sugars (Stadtman 1992; Wolff, Bascal, and Hunt 1989; Nagraj et al. 1991). Third, microglial cells are activated and increased in number in Alzheimer's disease, and represent a major source of free radicals (Colton and Gilbert 1987). Fourth, the increased concentration of aluminum in NFT-laden neurons has a stimulatory effect on iron-induced lipid peroxidation (Oteiza 1994). Aluminum salts can induce the enhanced release of iron (Gutteridge et al. 1985), which in turn induces lipid peroxidation (Ohyashiki, Karino, and Matsui 1993; Oteiza 1994). Membrane disturbances that are observed in degenerating neurons and neurites, possibly resulting from lipid peroxidation, are expected to lead to an influx of Ca^{+2} into the extracellular space. This increase in cytosolic Ca^{+2} can destabilize cytoskeleton and membranes, and activate Ca^{+2}-dependent degradative enzymes (e.g., phospholipases, proteases, and endonucleases) (Okabe et al. 1989; Orrenius et al. 1992). Evidence from a recent study suggests that β-amyloid peptide causes neurotoxic effect, both directly, by inducing oxidant stress, and indirectly, by activating microglia (Yan et al. 1996). The authors identified a receptor for advanced glycation end products (RAGE) that mediates effects of the β-amyloid peptide on neurons and microglia. They further showed that the increased expression of RAGE in AD is relevant to the pathogenesis of neuronal dysfunction and death.

The NFT is another major pathologic hallmark of AD. Polymerization of tau, the main component of intracellular NFT, has also been associated with oxidative stress. In vitro studies showed that oxidation of tau induces dimerization and polymerization of the protein into filaments. In addition, Yan et al. (1994) demonstrated that advanced glycation end products (AGEs) were colocalized with NFTs in AD. They also found that AGE-recombinant tau generates ROS when transfected into neuroblastoma cells.

Parker and colleagues (Parker, Filley, and Parks 1990) showed a reduction in cytochrome oxidase (complex IV of the electron transport chain in the mitochondrion) activity in platelets of AD patients. Several other studies have found significant decreases in cytochrome oxidase activity in different regions of AD brain (Kish et al. 1992a; Mutisya, Bowling, and Beal 1994; Parker et al. 1994). A decrease in mRNA of cytochrome oxidase subunits I and III was found in the midtemporal gyrus in AD (Chandrasekaran et al. 1994), and similarly cytochrome II expression is reduced in the AD hippocampus (Simonian and Hyman 1993). It is noteworthy that cytochrome oxidase subunit II mRNA and cytochrome oxidase activity are preferentially expressed in layer II and VI neurons of the entorhinal cortex, which are prone to NFT formation (Chandrasekaran et al. 1992).

Cell loss in AD could be due to apoptosis mediated by oxidative stress. Neurons in culture undergo apoptosis following withdrawal of trophic factors (Edwards, Buckmaster, and Tolkovsky 1991). Similarly, hippocampal neurons show characteristics of apoptotic cell death when exposed to β-amyloid (Loo et al. 1993). More recently, Smale and coworkers (Smale et al. 1995) have provided direct evidence of apoptosis occurring in AD. Their data suggested that apoptosis may be involved in both the primary neuronal cell loss and in the glial response that is a component of AD. A study of cell death in AD showed that the number of neurons with DNA fragmentation was about 50 times greater in AD than in control brains (Lassmann et al. 1995). Although these changes are believed to be indicative of necrosis (Steller 1995) rather than apoptosis, they could represent the early stages of apoptosis (Oberhammer et al. 1993).

Most recently, Sano et al. (1997) provided clinical evidence that selegiline or α-tocopherol slows the progression of Alzheimer's disease. These data strongly suggest a role for oxidative stress in Alzheimer's disease and suggest possible therapeutic approaches based on the control of oxidative stress.

4. GLUTATHIONE IN THE NERVOUS SYSTEM

Glutathione appears to serve multiple roles in the nervous system. The traditional one is that of free-radical scavenger. As cited above, free radicals can be generated by a variety of normal cellular processes. For neural cells these include oxidative phosphorylation, breakdown products of neurotransmitters such as dopamine and serotonin, overactivation of neurons by calcium or excitatory amino acids, and β-amyloid production. Oxidative stress arising from free-radical formation can affect the ratio of reduced to total GSH (the GSH status) of the cell as GSH is depleted to combat such radicals. As will be discussed in section 5, diminished GSH status has been linked with normal aging as well as neurodegenerative diseases (see Benzi and Moretti, chapter 11, this volume).

By processes still not fully understood, GSH provides neuroprotection against excitatory amino acid excitotoxicity as in stroke, ischemia, and epilepsy (see Cooper, chapter 5; Zaman and Ratan, chapter 6; Cuénod and Do, chapter 13, this volume). In addition, GSH appears to modulate ionotropic receptor action (see Ogita et al., chapter 7; Janáky et al., chapter 8, this volume). GSH may also have excitatory neurotransmitter/neuromodulator actions in some neural pathways (see Pasqualotto et al., chapter 9, this volume; Pasqualotto and Shaw 1996).

4.1 Glutathione, Oxidative Stress, and Mitochondrial Damage

Approximately 90 percent of total glutathione is localized in the cytosolic fraction, and the rest is compartmentalized within mitochondria (Reed 1990). The mitochondrial pool of GSH is likely to be involved in maintaining intramitochondrial protein thiols in the reduced state. These protein thiols are essential for a number of functions of these organelles, including selective membrane permeability to Ca^{+2}. Excessive production of H_2O_2 within mitochondria may lead to depletion of mitochondrial GSH, oxidation of protein thiols, and impairment of mitochondrial function, providing a clear example of the relationship between glutathione status, oxidative stress, and mitochondrial damage. This relationship may have implications for the degeneration of dopaminergic neurons, as substrates of the enzyme monoamine oxidase (MAO) may be sources of H_2O_2 within mitochondria and may cause a decrease in mitochondrial GSH (Sandri, Panfili, and Ernster 1990).

Schapira and coworkers (Schapira et al. 1990a,b) found a selective reduction of NADH-CoQ reductase activity, an enzyme specific to complex I. This loss was limited to substantia nigra and not found in any other region of the brain in PD. Nor was this abnormality found in multiple system atrophy, in which nigral degeneration also occurs (Fahn and Cohen 1992).

It is intriguing that the loss of GSH may cause mitochondrial damage (Heales et al. 1995). It is also likely that the converse situation occurs, namely, that impairment of mitochondrial function may lead to decrease in cytosolic GSH. GSH synthesis requires ATP, and thus a deficiency of energy supplied by mitochondria is likely to affect the cellular turnover of glutathione. Experimental evidence for this suggestion was provided by Mithofer et al. (1992) in an in vitro study where GSH and GSSG were measured in cells exposed to three mitochondrial poisons [potassium cyanide (KCN), antimycin A, and 1-methyl-4-phenylpyridinium (MPP^+)]. All of these compounds caused a decrease in GSH and a stoichiometric increase in GSSG and could be counterbalanced by addition of substrates for glycolytic production of ATP. The GSH loss induced by KCN, antimycin A, or MPP^+ was not due to oxidation to GSSG, but rather appeared to be correlated to intracellular levels of ATP (Di Monte, Chan, and Sandy 1992). Such relationships may be particularly crucial for neurons, and it is noteworthy that studies by Ottersen and coworkers (Hjelle et al., chapter 4, this volume) have provided evidence that mitochondria in substantia nigra in rat and cat show higher immunoreactivity for GSH. These data lend further support to the hypothesis that the oxidative stress and neuronal damage observed in the substantia nigra of patients with PD could be caused by a mitochondrial defect in GSH.

4.2 Glutathione Modulation in Amyotrophic Lateral Sclerosis

As cited above, oxidative stress may be an important factor in the pathogenesis of ALS (Przedborski et al. 1992; Gsell et al. 1995). Motor neurons are especially vulnerable to free radical attack and impaired defense mechanisms, which may lead to neuronal cell death (Richter, Park, and Ames 1988; Fujita et al. 1996). Changes in GSH status, by decreased synthesis or increased degradation, could lead to a failure to combat free-radical formation. Similarly, increased GSH degradation can increase the concentration of free glutamate. As discussed in section 2, increased glutamate may cause hyperactivity of neurons, resulting in excitotoxic death (Coyle and Puttfarcken 1993). Oxidative stress and excitotoxic mechanisms may both come into play in the events leading to motor-neuron death in ALS.

The very specific neuronal populations that are progressively lost in ALS and related disorders may possess unique characteristics that render them susceptible to insult and ultimately result in their demise. Such characteristics might be the types and ratios of cell surface receptors, including those for neurotransmitters. As neurotransmitter receptors are sensitive to the actions of neurotransmitters and neuromodulators (Hucho 1993), the possibility exists that a receptor-directed insult to neurons could account for some of the changes observed in ALS. Further, the nature of receptor modification may provide important clues to the stages leading to neuron death.

Our own data on several receptor populations and their regulatory enzymes may address this possibility. Glutathione receptor regulation studies showed that phorbol ester increases [^{35}S]GSH binding in human spinal cord (Lanius et al. 1994; Wagey, Krieger, and Shaw 1994). As phorbol ester is a relatively specific activator of protein kinase C (PKC), the data suggest that GSH receptors may be regulated by phosphorylation. In addition, GSH applied to cultured astrocytes increases 1,4,5-inositol triphosphate (IP$_3$), suggesting that GSH may stimulate specific receptors leading to changes in second-messenger activity (Guo, McIntosh, and Shaw 1992). IP$_3$ activation is part of a cascade leading to Ca^{+2} release from internal stores and is also associated with diacylglycerol production. We have also noted that the ionotropic NMDA-receptor population in spinal cord is regulated by PKC (see also Bannai 1984; Krieger et al. 1993, 1996). Measurements of NMDA receptor levels in spinal-cord sections from ALS victims have shown a significant 45-percent reduction in binding of [^{3}H]MK-801 (a noncompetitive NMDA receptor antagonist) compared to cords from control patients. This reduction could be totally reversed by activation of PKC (Krieger et al. 1993). In binding experiments on tissue from the same patient population, the GSH receptor number was up by approximately 16 percent (Fig. 2; see also Lanius et al. 1993).

The above data show that PKC regulates both NMDA and GSH receptor populations in human spinal cord, both of which are altered in the disease. PKC is itself greatly elevated in ALS spinal-cord tissue (Lanius et al. 1995). If so, why is one receptor population elevated while the other is decreased? Our very speculative working hypothesis is that the above changes are related via alterations in overall GSH status. A decline in GSH has two major consequences: loss of free radical scavenging and loss of GSH as neurotransmitter (see Pasqualotto et al., chapter 9, this volume). The first action leads to oxidative stress as discussed previously. The loss of GSH as neurotransmitter causes a decline in activity of target motor neurons (see Hjelle et al., Fig. 7, chapter 4, this volume). A combination of these events, along with the possibility of a diminished trophic factor release, culminates in motor-neuron death. Changes in NMDA and GSH receptor levels and in PKC reflect the compensatory events in the surviving cells. For the latter, increased activity of PKC on affected and survivor cells may generate more free radicals, further exacerbating the rate of neural degeneration (Robert 1996; Kuo et al. 1995).

In regard to our working model, some other recent observations may be relevant: Both reduced and oxidized GSH produce a concentration-dependent increase in free calcium in cultured rat brain neurons (Leslie et al. 1992). GSH-induced changes in free calcium were completely blocked by preexposure to an NMDA-receptor antagonist

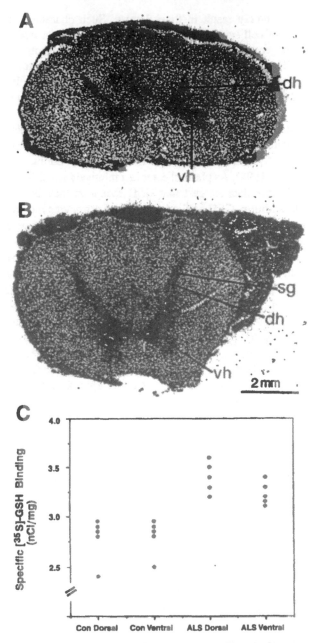

FIG. 2. [^{35}S]GSH binding in cervical spinal cord: (A) control; (B) ALS; (C) graph representing the densitometric values of [^{35}S]GSH binding. Calibration was done using measured film values of Amersham [^{14}C] microscale in actual polymer activity. Data from individual subjects are shown as filled circles. [^{35}S]GSH binding was increased significantly in both dorsal and ventral horns from ALS subjects, compared with controls. Bar represents 2 mm. dh = dorsal horn, vh = ventral horn, sg = substantia gelatinosa, Con = control, ALS = amyotrophic lateral sclerosis.

and extracellular magnesium ions. Furthermore, both reduced and oxidized GSH displaced bound [^{3}H]CGP 39653 from membrane fractions of hippocampal and cortical homogenates (Leslie et al. 1992; Ogita et al. 1995), indicating that GSH also binds to the NMDA receptor.

Under normal conditions most oxygen free radicals are generated in mitochondria as a by-product of electron transport and oxidative phosphorylation. Cells are protected by oxygen-free-radical scavengers: superoxide dismutase, reduced glutathione, and glutathione peroxidase. As oxidative stress has been implicated in ALS, the activity of antioxidant enzymes—particularly SOD, which scavenges superoxide, and catalase or GSH-Px, both of which detoxify harmful peroxide—may be affected during the process of tissue degeneration occurring in the CNS of ALS patients. Data on enzymatic activities in ALS are limited. Przedborski et al. (1996) showed that mean Cu/Zn-SOD activity was reduced in familial ALS patients with SOD mutations; however, no change was detected in ALS cases, either sporadic or familial, without SOD mutations. They further reported that catalase and GSH-Px activities were normal in ALS patients. Similar results have also been reported by Bowling and colleagues (Bowling et al. 1993). GSH-Px activity was reduced only in sporadic ALS patients treated with insulinlike growth factor I (Przedborski et al. 1996). The activities of GSH-Px show a wide variation, and some studies reported no detectable GSH-Px in anterior horn tissue obtained from ALS patients (Shaw et al. 1995). Data from a recent report suggest that GSH-Px activity does not change; however, cytochrome oxidase activity is decreased in the spinal cord of patients with ALS (Fujita et al. 1996).

Overall, the studies cited above support claims of impaired oxidation/reduction mechanisms in ALS and suggest that free-radical pathogenesis acting on receptor regulatory processes may be part of the cascade leading to motor-neuron degeneration in this disorder (Lanius et al. 1993).

4.3 Glutathione and Parkinson's Disease

The glutathione pathway includes a key reaction against the accumulation of oxygen radicals, in the reduction of H_2O_2 to H_2O, catalyzed by the enzyme GSH-Px. In this reaction GSH acts as the electron donor and is oxidized to GSSG. Under normal conditions, GSSG is rapidly reduced back to GSH by glutathione reductase at the expense of NADPH, which keeps the ratio of GSH to GSSG almost constant (Fig. 1). A complete absence of reduced GSH has been reported in the substantia nigra in PD (Perry, Godin, and Hansen 1982; Perry and Yong 1986; Perry, Hansen, and Jones 1988). The authors attributed the dramatic depletion of reduced GSH to the action of some environmental toxin that required the glutathione for its own metabolism. However, these findings have been criticized on methodological grounds (Slivka, Spina, and Cohen 1987). More recently, Jenner et al. (1992) have found a 40-percent reduction in reduced glutathione in substantia nigra of PD patients.

Turnover of dopamine by monoamine oxidase (MAO) and accumulation of hydrogen peroxide via the oxidation–reduction (Fig. 1) pathway is believed to be involved in degeneration of dopaminergic nigrostriatal neurons (Olanow 1990). The relationship between increased dopamine and H_2O_2 production on one hand, and oxidation of GSH to GSSG on the other has been provided by Spina and coworkers (Spina and Cohen 1989). If H_2O_2 formation via MAO, or if oxidative stress in general, is involved in neurodegeneration in PD, changes in brain glutathione may be expected to occur. The reduction of GSH is correlated with the severity of Parkinsonism (Riederer et al. 1989). However, the GSSG content does not increase in substantia nigra in tandem with GSH depletion, which indicates the irreversible consumption of the reduced form, an alternative route of GSH depletion in the cascade of neuronal loss (see Sian et al., chapter 14, this volume).

The protoxin (inactive toxin that is metabolized to produce active toxic products) 1-methyl-4-phenyl-1,2,3,4-tetrapyridinium (MPTP) produces selective destruction of the dopaminergic nigrostriatal pathway, inducing Parkinson's-like symptoms in humans and primates (Ballard, Tetrud, and Langston 1985; Langston 1987). The specificity for degeneration of the dopamine system lies in the affinity of its toxic products, 1-methyl-4-phenylpyridinium (MPP^+), to be taken up into dopamine neurons and terminals. MPP^+ inhibits complex I in mitochondria, reduces ATP synthesis, and also generates free electrons and increased amounts of the superoxide anion radical, leading to lipid peroxidation. It has been shown that toxic effects of MPTP (and MPP^+) are nullified in transgenic mice with high Cu/Zn-SOD activity, indicating that MPP^+ may exert its effects through oxyradicals (Przedborski et al. 1992).

Several studies have shown deficiency of GSH in the substantia nigra of patients with PD (Di Monte, Chan, and Sandy 1992; for reviews, see Sian et al., chapter 14; Dryhurst, chapter 15, this volume). Because of the scavenging activity of GSH against accumulation of oxygen radicals, its decrease in the brains of Parkinsonian patients has been interpreted as a sign of oxidative stress.

Dryhurst (chapter 15, this volume) has proposed three distinct factors that may be associated with the pathogenesis of PD. These include (i) factors that predispose an individual to the development of PD (e.g., low GSH and cysteine, genetic factors, and an ineffective blood–brain barrier), (ii) factors that may trigger the pathological processes (e.g., environmental toxin), and (iii) the mechanisms underlying mitochondrial-complex damage and oxidative stress (e.g., elevated translocation of cysteine into pigmented bodies, selective GSH loss). He concludes that the key step in neuropathological processes in PD may be the elevated release of GSH from nigral glial cells and upregulation of γ-glutamyltranspeptidase, resulting in enhanced translocation of cysteine into neuromelanin-pigmented dopaminergic neurons in substantia nigra. This increased translocation of cysteine may be responsible for higher vulnerability of the more heavily neuromelanin-pigmented SN cells in PD.

4.4 Glutathione and Alzheimer's Disease

As cited above, different neural populations appear to be highly susceptible to oxidative stress and free-radical damage due to low levels of GSH. In AD, cellular toxicity of β-amyloid is related directly to free-radical damage, which can be prevented by a free-radical scavenger, vitamin E (Behl et al. 1992). However, it has been reported that GSH increases with age in AD (Makar et al. 1995), which could be the compensatory response due to enhanced oxidative stress in aged AD brain. Another study showed that oxidative stress and GSH depletion induced by diamide can enhance the susceptibility of AD patients to the hepatotoxic potential of a drug, velnacrine maleate, used for the treatment of AD (Casey et al. 1995).

Evidence for a specific role for GSH in the etiology of AD includes direct demonstration of alterations in GSH status or regulatory enzyme action (for a review, see Benzi and Moretti 1995b) as well as some intriguing indirect correlational data. These latter data include observations that AD is correlated with environmental factors that may act as stressors [for example, aluminum in drinking water (McLachlan et al. 1996) and head injury as in trauma and ischemia (Willmore and Triggs 1991; Evans 1993)] and has a clear age dependence, increasing in prevalence with aging. Interestingly, iron ions (which are released by some forms of injury) induce hydroxyl-radical formation (Willmore and Triggs 1991). Aluminum can induce lipid peroxidation by stimulating ROS (Evans et al. 1992a,b; Fraga et al. 1990) and by accelerating iron-induced lipid peroxidation (Gutteridge 1985; Ohyashiki, Karino, and Matsui 1993). Further, the increase in prevalence of AD with advancing age is paralleled by aspects of GSH biochemistry, which appear to show a strong age dependence—GSH status and/or regulatory-enzyme action declining in the brains of aged individuals (Benzi and Moretti 1995a,b; Benzi et al. 1989; Ceballos-Picot et al. 1996; see also Benzi and Moretti, chapter 11, this volume). As noted above, Makar et al. (1995) found an increase in GSH with aging in AD brains, but not with normal aging. Different patterns of changes in GSH during normal aging and AD aging indicate different roles of GSH status in the two conditions.

Historically, the notion that GSH status may be involved in AD had a contentious experimental beginning, as many of the earliest studies attempting to link GSH status to AD were contradictory at various levels. For example, the total brain levels of GSH appeared to be unaffected in AD (e.g., see Balazs and Leon, 1994 and Perry et al. 1987), whereas GSH peroxidase and reductase were elevated in different brain regions (Lovell et al. 1995). Serum GSH-Px activity, however, could be either decreased (Jeandel et al. 1989; Perrin et al. 1990) or increased in AD (Ceballos-Picot et al. 1996). Some of these discrepancies may have arisen from differing techniques: For GSH levels, the measurements of total GSH rather than the crucial reduced GSH/GSSG ratio may hide real effects. Further, variations in the post mortem interval for tissue preparation may also contribute frequently to experimental discrepancies.

It is notable, however, that GSH-Px in erythrocytes/serum from AD patients appears altered. These data might suggest a genomic alteration as a factor in the propensity for altered GSH synthesis/action. Alternatively, they may suggest that far greater effects are to be found in brain, where changes in GSH status may reflect abnormalities in GSH transporters at the blood–brain barrier (Kannan et al. 1990; see Kannan et al., chapter 3, this volume) or in GSH synthesis or transport within the brain itself; detection at the periphery may depend on other experimental or interpatient variables (e.g., see Lovell et al. 1995).

Overall, although there is evidence for oxidative stress (perhaps arising from an altered GSH status) in AD, consistent data have been difficult to achieve. In turn, even if the data were more consistent, conclusive support for a causal rather than a coincident role has not been reported. However, as noted by Benzi and Moretti (1995b), even if not causal, oxidative stress, perhaps mediated by alterations in GSH status, might contribute to cascading neurological dysfunction and the progressive nature of the disease.

5. A WORKING MODEL OF GSH DEPLETION AND NEUROLOGICAL DISEASE

Recently, Meister and colleagues (Meister 1994; Boesgaard et al. 1994) have reported that the compound L-buthionine(S, R)sulfoximine (BSO), a transition-state inhibitor of γ-glutamylcysteine synthetase, blocks the synthesis of GSH (Jain et al. 1991; Broquist 1992). These workers have reported that in newborn rats, BSO administration (orally or by intraperitoneally injection, 2–4 mmol/kg) leads to significant decreases in reduced GSH. The consequence of GSH depletion for the development of various structures, including the brain, have been detailed in a series of publications (Jain et al. 1991; Martensson and Meister 1991). In brain, GSH depletion in young animals leads to neuronal pathology characterized by an enlargement and degeneration of cortical cell mitochondria. Similar effects were noted in adult animals, but the latter data were more cursory.

The notion that oxidative damage leads to neurological diseases and the proposal for antioxidant treatment have been described by Halliwell (1992) and Olanow (1992). Both these authors have defined a role for oxidative stress leading to neuronal destruction. By controlling the excess of free radicals using dietary intervention within a well-defined range of antioxidant actions, further oxidative damage might be prevented. In the following, we extend the above hypotheses of oxidative damage and provide a GSH-depletion model for neurological diseases.

The previous sections have drawn attention to the effects of oxidative stress and the potential role for neuronal damage in various neurological disorders. Glutathione, as a critical component of antioxidant defense, has been linked directly to oxidative stress, and evidence that alterations in GSH status may play a role in neurological diseases is growing. We note, however, that oxidative stress may be only one way that changes in GSH status could affect neural function and survival. In addition to its critical role as an antioxidant and/or free-radical scavenger, GSH may act as a redox modulator of ionotropic receptors or as a neuroprotectant against the effects of

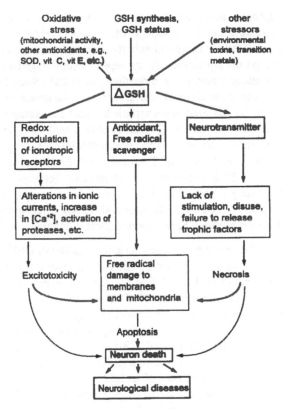

FIG. 3. A working model of GSH depletion and neurologic disease illustrating multiple roles of GSH.

glutamatergic excitotoxicity, and may also be a unique neurotransmitter. These roles are summarized in Fig. 3.

We propose as a working hypothesis that impairments of GSH status occurring at particular neural populations sets in motion a cascade of biochemical abnormalities culminating in various forms of neuronal cell death. The model has the following elements:

1. Factors that determine the site of degeneration are: selective alterations in GSH status due to abnormalities in synthesis (precursor pools or synthetic enzyme levels), GSH degradation (e.g., production of excessive glutamate or cysteine), GSH recovery from GSSG via GSH reductase, and alterations in local transport of GSH.

2. The severity of the effect (specific neuronal vulnerability) may depend on: the presence of stressors (e.g., transition metals and/or aluminum); age; the interaction with, and potential for, compensation by other antioxidant defenses; the other roles of GSH utilized by selective neural populations (e.g., action against free radicals produced by neurotransmitter breakdown, alterations in the redox modulation of some

ionotropic receptor populations giving rise to changes in membrane currents, decreases in GSH neuroprotection against EAAs, loss of GSH neurotransmitter function with subsequent alterations in neural activity and/or trophic-factor release).

3. Depending on the above factors, the type of neural degeneration may be either necrotic or apoptotic and may strike more than one neural subset. In regard to this last point, we note that in some circumstances, various of the neurological disorders, e.g., ALS, AD, PD, may be clustered in the same individual; for example, Guamanian cases are a combination of ALS and PD (Figlewicz et al. 1994). Dementia in patients with motor-neuron disease can sometimes be associated with coexisting conditions, such as Alzheimer's disease (von Poppe and Tennstedt 1963; Frecker et al. 1990), spongiform encephalopathy, and Pick's disease (von Poppe and Tennstedt 1963; Minauf and Jellinger 1969). Many Lewy-body diseases are also accompanied by dementia of Alzheimer type (AD) (Kalra 1996; Samuel et al. 1996). Overlapping of various other neurological diseases has also been reported (Feany, Mattiace, and Dickson 1996).

The model may be tested by creating an animal model of GSH depletion in which various stressors and animal age are considered [in contrast, a recent study that reported no direct effect of GSH depletion on neural degeneration in an animal model of PD failed to consider these points (Toffa et al. 1997)]. For different neurological diseases, different experimental conditions can be set. GSH depletion can be achieved by three approaches: first, GSH can be directly complexed to an electrophilic agent such as diethylmaleate (DEM); second, GSH synthesis can be inhibited by a specific inhibitor, such as buthionine sulfoximine (BSO); and third, oxidative stress can be induced, particularly in the presence of an inhibitor of GSH reductase such as N,N-bis(2-chloroethyl)-N-nitrosourea (Deneke and Fanburg 1989). Targeting specific cells for GSH depletion may lead to the development of particular neurodegenerative outcomes. For example, we speculate that depleting GSH from dopaminergic neurons may produce a model for Parkinson's disease.

Replacement of GSH may ameliorate PD symptoms in animals and humans, and we note that evidence for this in human PD patients has recently been demonstrated (Sechi et al. 1996). In relation to this last point, and to our model in which alterations in GSH's various other roles may all contribute to neuronal dysfunction, the results in Sechi et al. (1996), demonstrating improvement of PD symptoms following GSH replacement, are just as consistent with GSH administration serving as a replacement neurotransmitter as they are with its providing protection from free radical damage.

6. THERAPEUTIC APPROACHES TO NEUROLOGICAL DISEASES

As cited above, Sechi and colleagues suggest that GSH administration to PD patients may serve to slow the progress of the disease (Sechi et al. 1996). Such data suggest that similar positive benefits may occur in response to early treatment of AD and ALS. In relation to AD, a recent study (Sano et al. 1997) has found favorable effects of vitamin E on cognitive functions of the AD patients. It should be noted, however, that since all of these disorders have done considerable neural damage before detection, the best outcome from GSH treatment at this late stage would be to halt further disease progression.

Clearly, it would be advantageous if prophylactic treatment could be initiated before significant neural damage occurs. The determination and measurement of altered GSH status may allow for the best possible long-term outcome and should be a clinical goal. Possible sites for the examination of GSH status may be in serum or CSF of persons at risk (e.g., children of those expressing familial forms of disease).

Another factor that should be considered is the development of more efficacious and potent GSH analogs and agonists: the former to provide long-term antioxidant/redox action, the latter as replacements for neurotransmitters. We note that attempts to provide GSH analogs as antioxidants have already been reported. A Japanese pharmaceutical company has developed an analog, for GSH and shown that it protects against cerebral ischemia in rats (Yamamoto et al. 1993). This new compound, YM737 [(*N*-r-L-glutamyl-L-cysteinyl)glycine 1-isopropyl ester sulfate monohydrate] can be more readily transported than GSH. With the development of this analog for GSH, the idea for GSH therapy in various neurological diseases seems more promising.

Early GSH-status therapy may prevent the development of such disorders or halt them at a sufficiently rudimentary stage to prevent the unleashing of degenerative biochemical cascades. Such treatment might be accomplished by treatment with GSH (Sechi et al. 1996), other GSH analogs such as GSH monoester, or the above YM 737, or via gene therapy designed to enhance the normal production of GSH.

7. SUMMARY

There is increasing evidence that oxidative stress may play an important role in the pathogenesis of ALS, PD, and AD. Selective neuronal loss in these neurodegenerative disorders may reflect the highly localized effects of ROS within different cells and points to a confluence of factors that may render certain neuronal populations uniquely vulnerable.

Alteration of GSH status could be a common factor in various diseases. For example, a decrease in GSH in dopaminergic neurons may be a contributory factor for PD pathogenesis. Age is also a common risk factor for many if not all neurodegenerative diseases. Further, overlapping of various diseases in some cases is evident from the literature, for example, Guamanian cases, Lewy-body diseases, and dementia with PD and AD. Thus, a common mechanism of neurodegeneration may exist. We propose that GSH may be such a common pathway leading to the demise of selective populations of neurons. A GSH-depletion model may serve to test this hypothesis.

ACKNOWLEDGMENTS

We wish to thank Ms. J. C. McEachern and Dr. B. A. Pasqualotto for their comments on the manuscript, and Dr. Ken Curry for his continuing advice. This work was supported by an operating grant from Natural Science and Engineering Research

Council (NSERC) of Canada and the Amyotrophic Lateral Sclerosis Association (CAS).

REFERENCES

Adams, J. D., Jr., Klaidman, L. K., Odunze, I. N., Shen, H. C., and Miller, C. A. 1991. Alzheimer's and Parkinson's disease. Brain levels of glutathione, glutathione disulfide, and vitamin E. *Molecular Chemistry and Neuropathology* 14:213–25.

Albin, R. L., and Greenamyre, J. T. 1992. Alternative excitotoxic hypotheses. *Neurology* 42:733–8.

Balazs, L., and Leon, M. 1994. Evidence of an oxidative challenge in the Alzheimer's brain. *Neuochemical Research* 19:1131–7.

Ballard, P. A., Tetrud, J. W., and Langston, J. W. 1985. Permanent human Parkinsonism due to 1-methyl-4-phenyl-1,2,3,6-tetrahydropyridine (MPTP): seven cases. *Neurology* 35:949–56.

Bannai, S. 1984. Transport of cysteine and cystine in mammalian cells. *Biochemica et Biophysica Acta* 779:289–306.

Beal, M. F. 1992a. Does impairment of energy metabolism result in excitotoxic neuronal death in neurodegenerative illnesses? *Annals of Neurology* 31:119–30.

Beal, M. F. 1992b. Mechanisms of excitotoxicity in neurologic diseases. *FASEB Journal* 6:3338–44.

Beal, M. F. 1995. Aging, energy, and oxidative stress in neurodegenerative diseases. *Annals of Neurology* 38:357–66.

Beckman, J. S., Carson, M., Smith, C. D., and Koppenol, W. H. 1993. ALS, SOD and peroxynitrite. *Nature* 364:584.

Behl, C., Davis, J., Cole, G. M., and Schubert, D. 1992. Vitamin E protects nerve cells from amyloid β protein toxicity. *Biochemical and Biophysical Research Communications* 186:944–52.

Behl, C., Davies, J. B., Lesley, R., and Schubert, D. 1994. Hydrogen peroxide mediates amyloid β protein toxicity. *Cell* 77:817–27.

Benzi, G., and Moretti, A. 1995a. Are reactive oxygen species involved in Alzheimer's disease? *Neurobiology of Aging* 16:661–74.

Benzi, G., and Moretti, A. 1995b. Age- and peroxidative stress–related modifications of the cerebral enzymatic activities linked to mitochondria and the glutathione system. *Free Radicals in Biology and Medicine* 19:77–101.

Benzi, G., Pastoris, O., Marzatico, F., and Villa, R. F. 1989. Age-related effect induced by oxidative stress on the cerebral glutathione system. *Neurochemical Research* 14:473–81.

Birchall, J. D., and Chappell, J. S. 1988. Aluminium, chemical physiology, and Alzheimer's disease. *Lancet* II:1008–10.

Boesgaard, S., Aldershvile, J., Poulsen, H. E., Loft, S., Anderson, M., and Meister, A. 1994. Nitrate tolerance *in vivo* is not associated with depletion of arterial or venous thiol levels. *Circulation Research* 74:115–20.

Bonfoco, E., Krainc, D., Ankarcona, M., Nicotera, P., and Lipton, S. 1995. Apoptosis and necrosis: two distinct events induced, respectively, by mild and intense insults with N-methyl-D-aspartate or nitric oxide/superoxide in cortical cell cultures. *Proceedings of the National Academy of Sciences of the U.S.A.* 92:7162–6.

Bowling, A. C., Schulz, J. B., Brown, R. H., and Beal, M. F. 1993. Superoxide dismutase 1 with mutations linked to familial and sporadic amyotrophic lateral sclerosis. *Journal of Neurochemistry* 61:2322–5.

Broquist, H. P. 1992. Buthionine sulfoximine, an experimental tool to induce glutathione deficiency and ascorbate in their role as antioxidants. *Nutrition Reviews* 50:110–11.

Brown, R. H., Jr. 1995. Amyotrophic lateral sclerosis: recent insights from genetics and transgenic mice. *Cell* 80:687–92.

Buja, L. M., Eigenbrodt, L., and Eigenbrost, E. H. 1993. Apoptosis and necrosis. Basic types and mechanisms of cell death. *Archives of Pathology and Laboratory Medicine* 117:1208–14.

Busciglio, J., and Yankner, B. A. 1995. Apoptosis and increased generation of reactive oxygen species in Down's syndrome neurons in vitro. *Nature* 378:776–9.

Butterfield, D., Hensley, K., Harris, M., Mattson M., and Carney, J. 1994. β-Amyloid peptide free radical fragments initiate synaptosomal lipoperoxidation in a sequence-specific fashion: implications to Alzheimer's disease. *Biochemical and Biophysical Research Communications* 200:710–15.

Calne, D. B. 1992. The free radical hypothesis in idiopathic Parkinsonism: evidence against it. *Annals of Neurology* 32:799–803.

Candy, J. M., Klinowski, J., Perry, R. H., Perry, E. K., Fairbairn, A., Oakley, A. E., Carpenter, T. A., Atack, J. R., Blessed, G., and Edwardson, J. A. 1986. Aluminosilicates and senile plaque formation in Alzheimer's disease. *Lancet* I:354–6.

Casey, S. A., Brewster, D., Viau, C., and Acosta, D. 1995. Effect of glutathione depletion and oxidative stress in in vitro cytotoxicity of velnacrine maleate. *Toxicology Letters* 76:257–65.

Ceballos, I., Agid, F., Delacourte, A., Defossez, A., Nicole, A., and Sinet, P. M. 1990. Parkinson's disease and Alzheimer's disease: neurodegenerative disorders due to brain antioxidant system deficiencies? In *Advances in experimental biology and medicine: antioxidants in therapy and preventive medicine*, ed. I. Emerit, L. Packer, and C. Auclair. 493–8. New York: Plenum.

Ceballos-Picot, I., Merad-Boudia, M., Nicole, A., Thevenin, M., Hellier, G., Legrain, S., and Berr, C. 1996. Peripheral antioxidant enzyme activities and selenium in elderly subjects and in dementia of Alzheimer's type—place of the extracellular glutathione peroxidase. *Free Radicals in Biology and Medicine* 20:579–87.

Chance, B., Sies, H., and Boveris, A. 1979. Hydroperoxide metabolism in mammalian organs. *Physiological Reviews* 59:527–603.

Chandrasekaran, K., Stoll, J., Brady, D. R., and Rapoport, S. I. 1992. Localization of cytochrome oxidase (COX) activity and COX mRNA in the hippocampus and entorhinal cortex of the monkey brain: correlation with specific neuronal pathway. *Brain Research* 579:333–6.

Chandrasekaran, K., Giordano, T., Brady, D. R., Stoll, J., Martin, L. J., and Rapoport, S. I. 1994. Impairment in mitochondrial cytochrome oxidase gene expression in Alzheimer's disease. *Molecular Brain Research* 24:336–40.

Chao, C. C., Hu, S., and Peterson, P. K. 1995. Glia, cytokines, and neurotoxicity. *Critical Reviews in Neurobiology* 9:189–205.

Charriaut-Marlangue, C., Aggoun-Zouaoui, D., Represa, A., and Ben-Ari, Y. 1996. Apoptotic features of selective neuronal death in ischemia, epilepsy and gp 120 toxicity. *Trends in Neuroscience* 19:109–14.

Chen, H., and Tappel, A. L. 1994. Protection by vitamin E, selenium, trolox C, ascorbic acid palmitate, acetylcysteine, coenzyme Q, beta-carotene, canthaxanthin, and (+)-catechin against oxidation damage to liver slices measured by oxidized heme proteins. *Free Radicals in Biology and Medicine* 16:437–44.

Choi, D. W. 1988. Glutamate neurotoxicity and diseases of the nervous system. *Neuron* 1:623–34.

Choi, D. W. 1995. Calcium: still center-stage in hypoxic–ischemic neuronal death. *Trends in Neuroscience* 18:58–60.

Colton, C. A., and Gilbert, D. L. 1987. Production of superoxide anions by a CNS macrophage, the microglia. *FEBS Letters* 223:284–8.

Coyle, J. T., and Puttfarcken, P. 1993. Oxidative stress, glutamate, and neurodegenerative disorders. *Science* 262:689–95.

de Belleroche, J., Orrell, R. W., and Virgo, L. 1996. Amyotrophic lateral sclerosis: recent advances in understanding disease mechanism. *Journal of Neuropathology and Experimental Neurology* 55:747–57.

Deneke, S. M., and Fanburg, E. L. 1989. Regulation of cellular glutathione. *American Journal of Physiology* 257 (Lung Cell Molecular Physiology 1):L163–73.

Deng, H., Henatati, A., Tainer, J., Iqbal, Z., Cayabyab, A., Hung, W.-Y., Getzoff, E. D., Hu, P., Herzfeldt, B., Roos, R. P., Warner, C., Deng, G., Soriano, E., Smyth, C., Parge, H. E., Ahmed, A., Roses, A. D., Hallewell, R. A., Pericak-Vance, M. A., and Siddique, T. 1993. Amyotrophic lateral sclerosis and structural defects in Cu,Zn superoxide dismutase. *Science* 261:1047–51.

Den Hartog Jager, W. A., and Bethlem, J. 1980. The distribution of Lewy bodies in the central and autonomic nervous systems in idiopathic paralysis agitans. *Journal of Neurological and Neurosurgical Psychiatry* 23:283–90.

Dexter, D. T., Sian, J., Rose, S., Hindmarsh, J. G., Mann, V. M., Cooper, J. M., Wells, F. R., Daniel, S. E., Lees, A. J., Shapira, A. H. V., Jenner, P., and Marsden, C. D. 1994. Indices of oxidative stress and mitochondrial function in individuals with incidental Lewy body disease. *Annals of Neurology* 35:38–44.

Di Monte, D. A. 1991. Mitochondrial DNA and Parkinson's disease. *Neurology* 41(Suppl. 2):38–42.

Di Monte, D. A., Chan, P., and Sandy, M. S. 1992. Glutathione in Parkinson's disease: a link between oxidative stress and mitochodrial damage? *Annals of Neurology* S111–15.

Dragunow, M., Belharz, E., Sirimane, E., Lawlor, P., Williams, C., Bravo, R., and Gluckman, P. 1994. Immediate–early gene protein expression in neurons undergoing delayed death but not necrosis, following hypoxia–ischemic injury to the young rat brain. *Molecular Brain Research* 25:19–33.

Duffy, P. E., and Tennyson, V. M. 1965. Phase and electron microscopic observations of Lewy bodies and melanin granules in the substantia nigra and locus ceruleus in Parkinson's disease. *Journal of Neuropathology and Experimental Neurology* 24:398–414.

Dykens, J. A., Stern, A., and Trenker, E. 1987. Mechanisms of kainate toxicity to cerebellar neurons in vitro is analogous to reperfusion tissue injury. *Journal of Neurochemistry* 49:1222–8.

Dyrks, T., Dyrks, E., Hartman, T., Masters, C., and Beyreuther, K. 1992. Amyloidogenicity of beta A4 and β A4-bearing amyloid protein precursor fragments by metal-catalyzed oxidation. *Journal of Biological Chemistry* 267:18210–17.

Earnshaw, W. C. 1995. Apoptosis: lessons from in vitro systems. *Trends in Cell Biology* 5:217–20.

Edwards, S. N., Buckmaster, A. E., and Tolkovsky, A. M. 1991. The death programme in cultured sympathetic neurons can be suppressed at the posttranslational level by nerve growth factor, cyclic AMP, and depolarization. *Journal of Neurochemistry* 57:2140–3.

Ellis, R. E., Yuan, J., and Horvitz, H. R. 1991. Mechanisms and functions of cell death. *Annual Review of Cell Biology* 7:663–98.

Enayat, Z. E., Orrell, R. W., Claus, A., Ludolph, A., Powell, P. N., Leigh, P. N., and de Bellorche, J. 1995. Two novel mutations in the gene for copper zinc superoxide dismutase in UK families with amyotrophic lateral sclerosis. *Human Molecular Genetics* 4:1239–40.

Evans, P. H. 1993. Free radicals in brain metabolism and pathology. *British Medical Bulletin* 49:577–87.

Evans, P. H., Peterhans, E., Burge, T., and Klinowski, J. 1992a. Aluminosilicate-induced free radical generation by murine brain glial cell in vitro: potential significance in the aetiopathogenesis of Alzheimer's dementia. *Dementia* 3:1–6.

Evans, P. H., Yano, E., Klinowski, J., and Peterhans, E. 1992b. Oxidative damage in Alzheimer's dementia, and the potential etiopathgenic role of aluminosilicates, microglia and micronutrient interactions. *EXS* 62:178–89.

Fahn, S., and Cohen, G. 1992. The oxidant stress hypothesis in Parkinson's disease: evidence supporting it. *Annals of Neurology* 32:804–12.

Feany, M. B., Mattiace, L. A., and Dickson, D. W. 1996. Neuropathologic overlap of supranuclear palsy, Pick's disease and corticobasal degeneration. *Journal of Neuropathology and Experimental Neurology* 55:53–67.

Figlewicz, D. A., Garruto, R. M., Krizus, A., Yanagihara, R., and Rouleau, G. A. 1994. The Cu/Zn superoxide dismutase gene in ALS and Parkinsonism–dementia of Guam. *Neuroreport* 5:557–60.

Forno, L. S. 1981. Pathology of Parkinson's disease. In *Movement disorders, neurology* 2, ed. C. D. Marsden, and S. Fahn. 21–40. London: Butterworth Scientific.

Forno, L. S. 1987. Lewy bodies in Parkinson's disease. *Advances in Neurology* 45:35–43.

Fraga, C. G., Oteiza, P. I., Golub, M. S., Gershwin, M. E., and Keen, C. L. 1990. Effect of aluminum on brain lipid peroxidation. *Toxicology Letters* 51:213–19.

Frecker, M. F., Fraser, F. C., Anderman, E., and Pryse-Phillips, W. E. M. 1990. Association between Alzheimer's disease and amyotrophic lateral sclerosis? *Canadian Journal of Neurological Science* 17:12–14.

Fujita, K., Yamaguchi, M., Shibayama, K., Ando, M., Honda, M., and Nagata, Y. 1996. Decreased cytochrome c oxidase activity but unchanged superoxide dismutase and glutathione peroxidase activities in spinal cord of patients with amyotrophic lateral sclerosis. *Journal of Neuroscience Research* 45:276–81.

Goldman, J. E., and Yen, S.-H. 1986. Cytoskeletal protein abnormalities in neurodegenerative diseases. *Annals of Neurology* 19:209–23.

Good, P. F., Perl, D. P., Bierer, L. M., and Schmeidler, J. 1992. Selective accumulation of aluminum and iron in the neurofibrillar tangles of Alzheimer's disease: a laser microprobe (LAMMA) study. *Ann. Neurol.* 31:286–92.

Gotz, M. E., Freyberger, A., Hauer, E. Burger, R., Sofic, E., Gsell, W., Hecker, S., Jellinger, K., Hebenstreit, G., Frolich, L., Beckman, H., and Riederer, P. 1992. Susceptibility of brains from patients with Alzheimer's disease to oxygen-stimulated lipid peroxidation and differential scanning calorimetry. *Dementia* 3:213–22.

Gotz, M. E., Dirr, A., Gsell, W., Burger, R., Janetzky, B., Freyberger, A., Reichmann, H., Rausch, W.-D., and Riederer, P. 1994. Influence of N-methyl-4-phenyl-1,2,3,6-tetrahydropyridine, lipoic acid and L-deprenyl on the interplay between cellular redox systems. *Journal of Neural Transmission (Supplement)* 43:145–62.

Gsell, W., Conard, R., Hickethier, M., Sofic, E., Fralich, L., Wichart, I., Jellinger, K., Moll, G., Ransmayer, G., Beckman, H., and Riederer, P. 1995. Decreased catalase activity but unchanged superoxide dismutase activity in brains of patients with dementia of Alzheimer type. *Journal of Neurochemistry* 64: 1216–23.

Guo, N., McIntosh, C., and Shaw, C. A. 1992. Glutathione: new candidate neuropeptide in the central nervous system. *Neuroscience* 51:835–42.

Gurney, M. E., Pu, H., Chiu, A. Y., Dal Canto, M. C., Polchow, C. Y., Alexander, D. D., Caliendo, J., Hentati, A., Kwon, Y. W., Deng, H.-X., Chen, W., Zhai, P., Sufit, R. L., and Siddique, T. 1994. Motor neuron degeneration that express a human Cu,Zn superoxide dismutase mutation. *Science* 264: 1772–5.

Gutteridge, J. M. C., Quinlan, G. J., Clark, I., and Halliwell, B. 1985. Aluminium salts accelerate peroxidation of membrane lipids stimulated by iron salts. *Biochimica et Biophysica Acta* 835:441–7.

Hajimohammadreza, I., and Brammer, M. 1990. Brain membrane fluidity and lipid peroxidation in Alzheimer's disease. *Neuroscience Letters* 112:333–7.

Halliwell, B. 1992. Reactive oxygen species and central nervous system. *Journal of Neurochemistry* 59:1609–23.

Heales, S. J. R., Davies, S. E. C., Bates, T. E., and Clark, J. B. 1995. Depletion of brain glutathione is accompanied by impaired mitochondrial function and decreased N-acetyl aspartate concentration. *Neurochemical Research* 20:31–8.

Hensley, K., Carney, J. M., Mattson, M. P., Aksenova, M., Harris, M., Wu, J. F., Floyd, R. A., and Butterfield, D. A. 1994. A model for beta-amyloid aggregation and neurotoxicity based on free radical generation by the peptide: relevance to Alzheimer disease. *Proceedings of the National Academy of Sciences of the U.S.A.* 91:3270–4.

Hockenbery, D. M., Oltvai, Z. N., Yin, X.-M., Millian, C. L., and Korsmeyer, S. J. 1993. Bcl-2 functions in an antioxidant pathway to prevent apoptosis. *Cell* 75:241–51.

Hucho, F. 1993. Transmitter receptors—general principles and nomenclature. In *Neurotransmitter receptors*, ed. F. Huch. 3–59. New York: Elsevier Science.

Jain, A., Martensson, J., Stole, E., Auld, P. A. M., and Meister, A. 1991. Glutathione deficiency leads to mitochondrial damage in brain. *Proceedings of the National Academy of Sciences of the U.S.A.* 88:1913–17.

Jeandel, C., Nicolas, M. B., Dubois, F., Nabet-Belleville, F., Penin, F., and Cuny, G. 1989. Lipid peroxidation and free radical scavengers in Alzheimer's disease. *Gerontology* 35:275–82.

Jenner, P., and Olanow, C. W. 1996. Oxidative stress and the pathogenesis of Parkinson's disease. *Neurology* 47(Suppl.):S161–70.

Jenner, P., Dexter, D. T., Sian, J., Schapira, A. H. V., and Marsden, C. D. 1992. Oxidative stress as a cause of nigral cell death in Parkinson's disease and incidental Lewy body disease. *Annals of Neurology* 32:S82–7.

Joachim, C. L., and Selkoe, D. J. 1992. The seminal role of β-amyloid in the pathogenesis of Alzheimer's disease. *Alzheimer Disease and Associated Disorders* 6:7–34.

Joseph, J. A., and Roth, G. S. 1992. Cholinergic systems in aging: the role of oxidative stress. *Clinical Neuropathology (Supplement)* 15:508A–9A.

Kalra, S. 1996. Lewy body disease and dementia. *Archives of Internal Medicine* 156:487–93.

Kane, D. J., Sarafin, T. A., Anton, R., Hahn H., Gralla, E. B., Valentine, J. S., Ord, T., and Bredesen, D. E. 1993. Bcl-2 inhibition of neural death: decreased generation of reactive oxygen species. *Science* 262:1274–7.

Kannan, R., Kuhlenkamp, J. F., Jeandidier, E., Trinh, H., Ookhtens, M., and Kaplowitz, N. 1990. Evidence for carrier-mediated transport of glutathione across the blood–brain barrier in the rat. *Journal of Clinical Investigation* 85:2009–13.

Kish, S. J., Bergeron, C., Rajput, A., Deck, H. N., and Hornykiewicz, O. 1992a. Brain cytochrome oxidase in Alzheimer's disease. *Journal of Neurochemistry* 59:776–9.

Kish, S. J., Shannak, K., Rajput, A., Dozic, S., Mastrogiacomo, F., Chang, L.-J., Wilson, J. M., DiStefano, L. M., and Nobrega, J. N. 1992b. Aging produces a specific pattern of striatal dopamine loss: implications for the etiology of idiopathic Parkinson's disease. *Journal of Neurochemistry* 58:642–8.

Koh, J., and Choi, D. 1988. Zinc alters excitatory amino acid neurotoxicity on central neurons. *Journal of Neuroscience* 8:2164–71.

Kostic, V., Gurney, M. E., Deng, H.-X., Siddique, T., Epstein, C. J., and Przedborski, S. 1997. Midbrain dopaminergic degeneration in a transgenic mouse model of familial amyotrophic lateral sclerosis. *Annals of Neurology* 41:497–504.

Krieger, C., Wagey, R., Lanius, R. A., and Shaw, C. A. 1993. Activation of PKC reverses apparent NMDA receptor reduction in ALS. *Neuroreport* 4:931–4.

Krieger, C., Lanius, R. A., Pelech, S. L., and Shaw, C. A. 1996. Amyotrophic lateral sclerosis: the involvement of intracellular Ca^{+2} and protein kinase C. *Trends in Pharmacological Science* 17:114–20.

Kuo, M. L., Lee, K. C., Lin, J. K., and Huang, T. S. 1995. Pronounced activation protein kinase C, ornithine decarboxylase and c-jun proto-omncogene by paraquat-generated active oxygen species WI-38 human lung cells. *Biochimica and Biophysica Acta* 1268:229–36.

Langston, J. W. 1987. MPTP: the promise of a new neurotoxin. In *Movement disorders 2*, ed. C. D. Marsden, and S. Fahn, 73–90, London: Buttersworth.

Lanius, R. A., Krieger, C., Wagey, R., and Shaw, C. A. 1993. Increased [^{35}S]glutathione binding sites in spinal cords from patients with amyotrophic lateral sclerosis. *Neuroscience Letters* 163:89–92.

Lanius, R. A., Shaw, C. A., Wagey, R., and Krieger, C. 1994. Characterization, distribution, and protein kinase C–mediated regulation of [^{35}S]glutathione binding sites in mouse and human spinal cord. *Journal of Neurochemistry* 63:155–60.

Lanius, R. A., Paddon, H. B., Mezei, M., Wagey, R., Krieger, C., Pelech, S. L., and Shaw, C. A. 1995. Increased PKC activity in patients with amyotrophic lateral sclerosis. *Society for Neuroscience Abstracts* 21:492.

Lassmann, H., Bancher, C., Breitschopf, H., Weigel, J., Bobinski, M., Jellinger, K., and Wisniewski, H. M. 1995. Cell death in Alzheimer's disease evaluated by DNA fragmentation in situ. *Acta Neuropathologica* 89:35–41.

Leslie, S. W., Brown, L. M., Trent, R. D., Lee, Y.-H., Morris, J.L., Jones, T.W., Randall, P. K., Lau, S. S., and Monks, T. J. 1992. Stimulation of N-methyl-D-aspartate receptor-mediated calcium entry into dissociated neurons by reduced and oxidized glutathione. *Molecular Pharmacology* 41:308–314.

Lewy, F. H. 1912. Paralysis agitans I: Pathologische anatomie. In *Handbuch der Neurologie*, ed. M. Lewandowsky, 920–33. New York: Springer.

Lin, F.-H, Lin, R., Wisnieski, H. M., Hwang, Y.-W., Grundke-Iqbal, F., Healy-Louie, G., and Iqbal, K. 1992. Detection of point mutations in codon 331 of mitochondrial NADH dehydrogenase subunit 2 in Alzheimer's brains. *Biochemical and Biophysical Research Communications* 182:238–46.

Lomaestro, B. M., and Malone, M. 1995. Glutathione in health and disease: pharmacotherapeutic issues. *Annals of Phamacotherapy* 29:1263–73.

Loo, D. T., Coplani, A., Pike, C. J., Whittemore, E. R. Walencewicz, A. J., and Cotman, C. W. 1993. Apoptosis is induced by β-amyloid in cultured central nervous system neurons. *Proceedings of the National Academy of Sciences of the U.S.A.* 99:7951–5.

Lorenzo, A., and Yankner, B. A. 1994. β-Amyloid neurotoxicity requires fibril formation and is inhibited by Congo red. *Proceedings of the National Academy of Sciences of the U.S.A.* 91:12243–7.

Lovell, M. A., Ehmann, W. D., Butler, S. M., and Markesbery, W. R. 1995. Elevated thiobarbituric acid–reactive substances and antioxidant enzyme activity in the brain in Alzheimer's disease. *Neurology* 45:1594–601.

Lucas, D. R., and Newhouse, J. P. 1957. The toxic effect of sodium L-glutamate on the inner layers of the retina. *AMA Archives of Ophthalmology* 58:193–210.

Makar, T. K., Cooper, A. J. L., Tofel-Grehl, B., Thaler, H. T., and Blass, J. P. 1995. Carnitine, carnitine acetyltransferase, and glutathione in Alzheimer brain. *Neurochemical Research* 20:705–11.

Martensson, J., and Meister, A. 1991. Glutathione deficiency decreases tissue ascorbate levels in newborn rats: ascorbate spares glutathione and protects. *Proceedings of the National Academy of Sciences of the U.S.A.* 88:4656–60.

Mattson, M. P. 1995. Free radicals and disruption of neuronal ion homeostasis in AD: a role for amyloid β-peptide? *Neurobiology of Aging* 16:679–82.

Mattson, M. D., Rychlik, B., and Engle, M. G. 1990. Possible involvement of calcium and inositol phospholipid signaling pathways in neurofibrillary degeneration. In *Alzheimer's disease: basic mechanisms, diagnosis and therapeutic strategies*, ed. K. Iqbal, D. R. C. McLachlan, B. Winblad, and H. M. Wisnieski, 191–8. New York: Wiley.

Mattson, M. P., Barger, S. W., Cheng, B., Lieberburg, I., Smith-Swintosky, V. L., and Rydel, R. E. 1993. β-Amyloid precursor protein metabolites and loss of neuronal Ca^{+2} homeostasis in Alzheimer's disease. *Trends in Neuroscience* 16:409–14.

McLachlan, D. R., Bergeron, C., Smith, J. E., Boomer, D., and Rifat, S. L. 1996. Risk for neuropathologically confirmed Alzheimer's disease and residual aluminum in municipal drinking water employing weighted residential histories. *Neurology* 46:401–5.

Meister, A. 1994. Glutathione–ascorbic acid antioxidant system in animals. *Journal of Biological Chemistry* 269:9397–400.

Meister, A. 1988. Glutathione metabolism and its selective modification. *Journal of Biological Chemistry* 263:17205–8.

Minauf, M., and Jellinger, K. 1969. Kombination von amyotrophischer Lateralsklerose mit Pickscher Krankheit. *Archiv fur Psychiatrie und Nervenkrankheiten* 212:279–88.

Mithofer, K., Sandy, M. S., Smith, M. T., and Di Monte, D. 1992. Mitochondrial poisons cause depletion of reduced glutathione in isolated hepatocytes. *Archives of Biochemistry and Biophysics* 295:132–6.

Miyamoto, M., and Coyle J. 1990. Idebenone attenuates neuronal degeneration induced by intrastriatal injection of excitotoxins. *Experimental Neurology* 108:38–45.

Miyamoto, M., Murphy, T. H., Schnaar, R. L., and Coyle, J. T. 1989. Antioxidants protect against glutamate-induced cytotoxicity in a neuronal cell line. *Journal of Pharmacology and Experimental Therapeutics* 250:1132–40.

Monyer, H., Hartley, D. M., and Choi, D. W. 1990. 21-Aminosteroids attenuate excitotoxic neuronal injury in cortical cell cultures. *Neuron* 5:121–6.

Munch, G., Thome, J., Foley, P., Shinzel, R., and Riederer, P. 1997. Advanced glycation end products in ageing and Alzheimer's disease. *Brain Research Reviews* 23:134–43.

Mutisya, E. M., Bowling, A. C., and Beal, M. F. 1994. Cortical cytochrome oxidase activity is reduced in Alzheimer's disease. *Journal of Neurochemistry* 63:2179–84.

Myers, L. S. 1980. Free radical damage to nucleic acids and their components: the direct absorption of energy. In *Free radicals in biology*, ed. W. Pryor, 4:95–114. New York: Academic Press.

Nagata, S. 1997. Apoptosis by death factor. *Cell* 88:355–65.

Nagraj, R. H., Sell, D. R., Prabhakaram, M., Ortwerth, B., and Monnier, V. M. 1991. High correlation between pentosidine protein crosslinks and pigmentation implicates ascorbate oxidation in human lens senescence and cataractogenesis. *Proceedings of the National Academy of Sciences of the U.S.A.* 88:10257–61.

Nicholson, D. W., Ali, A., Thornberry, N. A., Vaillancourt, J. P., Ding, C. K., Gallant, M., Gareau, Y., Griffin, P. R., Labelle, M., Lazebnik, Y. A., Munday, N. A., Raju, S. M., Smulson, M. E., Yamin, T.-T., Yu, V. L., and Miller, D. K. 1995. Identification and inhibition of the ICE/CED-3 protease necessary for mammalian apoptosis. *Nature* 376:37–43.

Novelli, A. Reilley, J. A., Lysko, P. G., and Henneberry, R.C. 1988. Glutamate becomes neurotoxic via the *N*-methyl-D-aspartate receptor when intracellular energy levels are reduced. *Brain Research* 451:205–12.

Oberhammer, F., Fritsch, G., Schmied, M., Pavelka, M., Printz, D., Purchio, T., Lassmann, H., and Schulte-Hermann, R. 1993. Condensation of the chromatin at the membrane of an apoptotic nucleus is not associated with activation of an endonuclease. *Journal of Cell Science* 104:317–26.

Ogita, K., Enomoto, R., Nakahara, F., Ishitsubo, N., and Yoneda, Y. 1995. A possible role of glutathione as an endogenous agonist at the *N*-methyl-D-aspartate recognition domain in rat brain. *Journal of Neurochemistry* 64:1088–96.

Ohyashiki, T., Karino, T., and Matsui, K. 1993. Stimulation of Fe^{+2}-induced lipid peroxidation in phosphatidylcholine liposomes by aluminium ions at physiological pH. *Biochemica et Biophysica Acta* 1170:182–8.

Okabe, E., Sugihara, M., Tanaka, K., Sasaki, H., and Ito, H. 1989. Calmodulin and free radicals interaction with steady-state calcium accumulation and passive calcium permeability of cardiac sarcoplasmic reticulum. *Journal of Pharmacology and Experimental Therapeutics* 250:286–92.

Olanow, C. W. 1990. Oxidation reactions in Parkinson's disease. *Neurology* 40(Suppl. 3):32–7.

Olanow, C. W. 1992. An introduction to the free radical hypothesis in Parkinson's disease. *Annals of Neurology* 32:S2–9.

Olanow, C. W., Holgate, R. C., Murtaugh, R., and Martinez, C. 1989. MR images in Parkinson's disease and aging. *Advances Neurology* 36:155–64.

Olney, J. W. 1978. Neurotoxicity of excitatory amino acids. In *Kainic acid as a tool in neurobiology*, ed. E. G. McGeer, J. W. Olney, and P. L. McGeer, 1–15. New York: Raven Press.

Orrenius, S., Burkitt, M. J., Kass, G. E. N., Dypbukt, J. M., and Nicotera, P. 1992. Calcium ions and oxidative cell injury. *Annals of Neurology* 32:S33–42.

Oteiza, P. I. 1994. A mechanism for the stimulatory effect of aluminum on iron-induced lipid peroxidation. *Archives of Biochemistry and Biophysics* 308:374–9.

Parker, W. D., Jr., Boyson, S. J., and Parks, J. K. 1989. Abnormalities of the electron transport chain in idiopathic Parkinson's disease. *Annals of Neurology* 26:719–23.

Parker, W. D., Jr., Filley, C. M., and Parks, J. K. 1990. Cytochrome oxidase deficiency in Alzheimer's disease. *Neurology* 40:1302–3.

Parker, W. D., Jr., Parks, J. K., Filley, C. M., and Kleinschmidt-DeMasters, B. K. 1994. Electron transport chain defects in Alzheimer's disease brain. *Neurology* 44:1090–6.

Pasqualotto, B. A., and Shaw, C. A. 1996. Regulation of ionotropic receptors by protein phosphorylation. *Biochemical Pharmacology* 51:1417–25.

Perrin, R., Briancon, S., Jeandel, C., Artur, Y., Minn, A., Penin, F., and Siest, G. 1990. Blood activity of Cu/Zn superoxide dismutase, glutathione peroxidase and catalase in Alzheimer's disease: a case-control study. *Gerontology* 36:306–13.

Perry, T. L., and Yong, V. W. 1986. Idiopathic Parkinson's disease, progressive supranuclear palsy and glutathione metabolism in the substantia nigra of patients. *Neuroscience Letters* 67:269–74.

Perry, T. L., Godin, D. V., and Hansen, S. 1982. Parkinson's disease: a disorder due to nigral glutathione deficiency. *Neuroscience Letters* 33:300–10.

Perry, T. L., Yong, V. W., Bergeron, C., Hansen, S., and Jones, K. 1987. Amino acids, glutathione, and glutathione transferase activity in the brains of patients with Alzheimer's disease. *Annals of Neurology* 21:331–6.

Perry, T. L., Hansen, S., and Jones, K. 1988. Brain amino acids and glutathione in progressive supranuclear palsy. *Neurology* 38:943–6.

Pike, C. J., Burdick, D., Walencewicz, A. J., Glabe, C. G., and Cotman, C. W. 1993. Neurodegeneration induced by β-amyloid peptides in vitro: the role of peptide assembly state. *Journal of Neuroscience* 13:1676–87.

Portera-Cailliau, C., Hedreen, J., Price, D., and Koliatsos, V. 1995. Evidence for apoptotic cell death in Huntington's disease and excitotoxic animal models. *Journal of Neuroscience* 15:3775–87.

Przedborski, S., Jackson-Lewis, V., Kostic, V., Carlson, E., Epstein, C. J., and Cadet, J. L. 1992. Superoxide dismutase, catalase and glutathione peroxidase activities in copper/zinc-superoxide dismutase transgenic mice. *Journal of Neurochemistry* 58:1760–7.

Przedborski, S., Donaldson, D. M., Murphy, P. L., Hirsch, O., Lange, D., Naini, A. B., McKenna-Yasek, D., and Brown, R. H., Jr. 1996. Blood superoxide dismutase, catalase and glutathione peroxidase activities in familial and sporadic amyotrophic lateral sclerosis. *Neurodegeneration* 5:57–64.

Puttfarcken, P. S., Getz, R. L., and Coyle, J. T. 1993. Kainic acid induced lipid peroxidation: protection with butylated hydroxytoluene and U 78517F in primary culture of cerebellar granule cells. *Brain Research* 624:223–32.

Ratan, R. R., Murphy, T. H., and Baraban, J. M. 1994. Oxidative stress induces apoptosis in embryonic cortical neurons. *Journal of Neurochemistry* 62:376–9.

Reed, D. J. 1990. Glutathione: toxicological implications. *Annual Review of Pharmacological Toxicology* 30:603–31.

Richter, C., Park, J.-W., and Ames, B. N. 1988. Normal oxidative damage to mitochondrial and nuclear DNA is extensive. *Proceedings of the National Academy of Sciences of the U.S.A.* 85:6465–7.

Riederer, P., Sofic, E., Rausch, W. D., Schmidt, B., Reynold, G. P., Jellinger, K., and Youdim, M. B. H. 1989. Transition metals: ferritin, glutathione, and ascorbic acid in Parkinsonian brains. *Journal of Neurochemistry* 52:515–20.

Robert, L. 1996. Aging of vascular wall and atherogenesis: role of the elastin–laminin receptor. *Atherosclerosis* 123:169–79.

Roher, A. E., Ball, M. J., Bhave, S. V., and Wakade, A. R. 1991. β-Amyloid from Alzheimer's disease brain inhibits sprouting and survival of sympathetic neurons. *Biochemical and Biophysical Research Communications* 174:572–9.

Rosen, D. R., Siddique, T., Patterson, D., Figlewicz, D. A., Sapp, P., Hentati, A., Donaldson, D., Goto, J., O'Regan, J. P., Deng, H.-X., Rahmani, Z., Krizus, A., McKenna-Yasek, D., Cayabyab, A., Gaston, S. M., Berger, R., Tanzi, R. E., Halperin, J. J., Herzfeldt, B., Van den Bergh, R., Hung, W.-Y., Bird, T., Deng, G., Mulder, D. W., Smyth, C., Laing, N. G., Soriano, E., Pericak-Vance, M. A., Haines, J., Rouleau, G. A., Gusella, J. S., Horvitz, H. R., and Brown, R. H., Jr. 1993. Mutations in Cu/Zn superoxide dismutase gene are associated with familial amyotrophic lateral sclerosis. *Nature* 362:59–62.

Samuel, W., Galasko, D., Masliah, E., and Hansen, L. A. 1996. Neocortical Lewy body variant of Alzheimer's disease. *Journal of Neuropathology and Experimental Neurology* 55:44–52.

Sanan, D. A., Weisgraber, K. H., Russell, S. J., Mahley, R. W., Huang, D. Saunder, A. M., Schemechal, D., Wisniesky, T., Fragione, B., Roses, A. D., and Strittmatter, W. J. 1994. Apolipoprotein E associates with β-amyloid peptide of Alzheimer's disease to form novel monofibril: isoform apo E4 associates more efficiently than apo E3. *Journal of Clinical Investigation* 94:860–9.

Sandri, G., Panfili, E., and Ernster, L. 1990. Hydrogen peroxide production by monoamine oxidase in isolated rat brain mitochondria, its effect on glutathione levels and Ca^{+2} efflux. *Biochimica et Biophysica Acta* 1035:300–5.

Sano, M., Ernesto, C., Thomas, R. G., Klauber, M. R., Schafer, K., Grundman, M., Woodbury, P., Growdon, J., Cotman, C. W., Pfeiffer, E., Schneider, L. S., and Thal, L. J. 1997. A controlled trail of selegiline, alpha-tocopherol, or both as treatment for Alzheimer's disease. *New England Journal of Medicine* 336:1216–22.

Schapira, A. H. V., Cooper, J. M., Dexter, D., Clark, J. B., Jenner, P., and Marsden, C. D. 1990a. Mitochondrial complex I deficiency in Parkinson's disease. *Journal of Neurochemistry* 54:823–7.

Schapira, A. H. V., Mann, V. M., Cooper, J. M., Dexter, D., Daniel, S. E., Jenner, P., Clark, J. B., and Marsden, C. D. 1990b. Anatomic and disease specificity of NADH CoQ I reductase (complex I) deficiency in Parkinson's disease. *Journal of Neurochemistry* 55:2142–5.

Schoepp, D. D., and Conn, P. J. 1993. Metabotropic glutamate receptors in brain function and pathology. *Trends in Pharmacological Sciences* 14:13–20.

Schraufstatter, I. U., Hyslop P. A., Hinshaw, D. B., Spragg, R. G., Sklar, L. A., and Cochrane, C. G. 1986. Hydrogen peroxide–induced injury of cells and its prevention by inhibitors of poly (ADP-ribose) polymerase. *Proceedings of the National Academy of Sciences of the U.S.A.* 832:4908–12.

Sechi G., Deledda M. G., Bua, G., Satta, W. M., Deiana, G. A., Pes, G. M., and Rosati, G. 1996. Reduced intravenous glutathione in the treatment of early Parkinson's disease. *Progress in Neuropsychopharmacological Biology and Psychiatry* 20:1159–70.

Shaw, C. A., Pasqualotto, B. A., and Cury, K. 1996. Glutathione-induced sodium currents in neocortex. *Neuroreport* 7:1149–52.

Shaw, I. C., Fitzmaurice, P. S., Mitchell, J. D., and Lynch, P. G. 1995. Studies on cellular free radical protection mechanisms in the anterior horn from patients with amyotrophic lateral sclerosis. *Neurodegeneration* 4:391–6.

Sheinberg, H. 1988. The neurotoxicity of copper. In *Metal neurotoxicity*, eds. S. Bondy and K. Prasad, 55–60. Boca Raton, FL: CRC Press.

Sies, H., ed. 1991. *Oxidative stress: oxidants and antioxidants*. New York: Academic Press.

Simonian, N. A., and Coyle, J. T. 1996. Oxidative stress in neurodegenerative diseases. *Annual Review of Pharmacology and Toxicology* 36:83–106.

Simonian, N. A., and Hyman, B. T. 1993. Functional alterations in Alzheimer's disease: diminution of cytochrome oxidase in hippocampal formation. *Journal of Neuropathology and Experimental Neurology* 52:580–5.

Slivka, A., Spina, M. B., and Cohen, G. 1987. Reduced and oxidized glutathione in human and monkey brain. *Neuroscience Letters* 74:112–18.

Smale, G., Nichols, N. R., Brady, D. R., Finch, C. E., and Horton, W. E., Jr. 1995. Evidence for apoptotic cell death in Alzheimer's disease. *Experimental Neurology* 133:225–30.

Spina, M. B., and Cohen, G. 1989. Dopamine turnover and glutathione oxidation: implications for Parkinson's disease. *Proceedings of the National Academy of Sciences of the U.S.A.* 86:1398–1400.

Stadtman, E. R. 1992. Protein oxidation and aging. *Science* 257:1220–4.

Steller, H. 1995. Mechanisms of genes and cellular suicide. *Science* 267:1445–9.

Strittmatter, W. J., Weisgraber, K. H., Huang, D. Y., Dong, L.-M., Salveson, G. S., Pericak-Vance, M., Schmechel, D., Saunders, A. M., Goldgaber, D., and Roses, A. D. 1993. Binding of human apolipoprotein E to synthetic β peptide: isoform-specific effects and implications for late-onset Alzheimer's disease. *Proceedings of the National Academy of Sciences of the U.S.A.* 90:8098–102.

Talley, A. K., Dewhurst, S., Perry, S. W., Dollard, S. C., Gummuluru, S., Fine, S. M., New, D., Epstein, L. G., Gendelman, H. E., and Gelbard, H. A. 1995. Tumor necrosis factor alpha–induced apoptosis in human neuronal cells: protection by antioxidant N-acetylcysteine and the genes bcl-2 and crmA. *Molecular and Cellular Biology* 15:2359–66.

Toffa, S., Kunikowska, G. M., Zeng, B.-Y., Jenner, P., and Marsden, C. D. 1997. Glutathione depletion in rat brain does not cause nigrostriatal pathway degeneration. *Journal of Neural Transmission* 104: 67–75.

Volterra A., Trotti, C., Tromba, C., Floridi, S., and Racagni, G. 1994. Glutamate uptake inhibition by oxygen free radicals in rat brain cortical astrocytes. *Journal of Neurochemistry* 14:2924–32.

von Poppe, W., and Tennstedt, A. 1963. Klinische und pathologisch-anatomische Untersuchungen über Kombinations formen praseniler Hirnatrophien (Pick, Alzheimer) mit spinalen atrophisierenden Prozessen. *Psychiatria et Neurologia* 145:322–34.

Wagey, R., Krieger, C., and Shaw, C. A. 1994. Abnormal phosphorylation activity in the pathogenesis of ALS. *Society for Neuroscience Abstracts* 20:1648.

Ward, J. 1977. Molecular mechanisms of radiation-induced damage to nucleic acids. *Advances in Radiation Biology* 5:181.

Watkins, J. C., and Olverman, H. J. 1987. Agonists and antagonists for excitatory amino acids. *Trends in Neuroscience* 10:265–72.

Weinstock, M. 1995. The pharmacotherapy of Alzheimer's disease based on the cholinergic hypothesis: an update. *Neurodegeneration* 4:349–356.

Whetsell, W. O., Jr. 1996. Current concepts of excitotoxicity. *Journal of Neuropathology and Experimental Neurology* 55:1–13.

Williams, D. B., and Windebank, A. J. 1991. Motor neuron disease (amyotrophic lateral sclerosis). *Mayo Clinic Proceedings* 66:54–82.

Willmore, L. J., and Triggs, W. J. 1991. Iron-induced lipid peroxidation and brain injury responses. *International Journal of Developmental Neuroscience* 9:175–80.

Wolff, S. P., Bascal, Z., and Hunt, J. V. 1989. In *The millard reaction in ageing, diabetes and nutrition*, ed. K. Yagi, 259–275. Alan R. Liss.

Wong, P., Pardo, C., Borchelt, D., Lee, M., Copeland, N., Jenkins, N. A., Sisodia, S. S., Cleveland, D. W., and Price, D. L. 1995. An adverse property of a familial ALS-linked SOD 1 mutation causes motor neuron disease. *Neuron* 14:1105–16.

Wu, R.-M., Murphy, D. L., and Chiueh, C. C. 1994. Protection of nigral neurons against MPP$^+$-induced oxidative injury by deprenyl (selegiline), U-78517F, and DMSO. *New Trends in Clinical Neuropharmacology* 8:187–8.

Yamamoto, M., Sakamoto, N., Iwai, A., Yatsugi, S., Hidaka, K., Noguchi, K., and Yuasa, T. 1993. Protective actions of YM737, a new glutathione analog, against cerebral ischemia in rats. *Research Communications in Chemical Pathology and Pharmacology* 81:221–32.

Yan, S.-D., Chen, X., Schmidt, A.-M., Brett, J., Godman, G., Zou, Y.-S., Scott, C. W., Caputo, C., Frappier, T., Smith, M. A., Perry, G., Yen, S.-H., and Stern, D. 1994. Glycated tau protein in Alzheimer's disease: a mechanism for induction of oxidative stress. *Proceedings of the National Academy of Sciences of the U.S.A.* 91:7787–91.

Yan, S.-D, Yan, S. F., Chen, X., Fu, J., Chen, M., Kuppusamy, P., Smith, M., Perry, G., Godman, G. C., Nawroth, P., Zweier, J. L., and Stern, D. 1995. Non-enzymatically glycated tau in Alzheimer's disease induces neuronal oxidant stress resulting in cytokine gene expression and release of amyloid β-peptide. *Nature Medicine* 693–9.

Yan, S.-D., Chen, X., Fu, J., Chen, M., Zhu, H., Roher, A., Slattery, T., Zhao, L., Nagashima, M., Morser, J., Mighell, A., Nawroth, P., Stern, D., and Schmidt, A. N. 1996. RAGE and amyloid β peptide neurotoxicity in Alzheimer's disease. *Nature* 382:685–91.

Yankner, B. A. 1996. Mechanisms of neuronal degeneration in Alzheimer's disease. *Neuron* 16:921–32.

Yankner, B. A., Dawes, L. R., Fisher, S., Villa-Komaroff, L., Oster-Granite, M. L., and Neve, R. L. 1989. Neurotoxicity of a fragment of the amyloid precursor associated with Alzheimer's disease. *Science* 245:417–20.

Zatta, P., Zanoni, S., and Favarato, M. 1992. The identification of aluminum (III) in the core of mature plaques from Alzheimer's disease. *Neurobiology of Aging* 13(Suppl. 1):S117.

Zeevalk, G. D., and Nicklas, W. J. 1990. Mechanisms underlying initiation of excitotoxicity associated with metabolic inhibition. *Journal of Pharmacology and Experimental Therapeutics* 257:870–8.

Index

9 780367 447885